Sustainability Sciences in Asia and Africa

Agroecosystem Dynamics and Sustainable Practices

Series Editors

Sustainability as a scientific endeavor requires broad understanding of the interconnections in our global environment. It is therefore positioned at the crossroads of research in natural sciences, technology and human sciences. This book series supports the global efforts by providing timely coverage of the progress, opportunities, and challenges towards achieving sustainable development goals (SDGs) in Asia and Africa. It fosters the research in transdisciplinary academic fields spanning across agriculture and global food security, health and diseases, access to clean water and affordable energy, climate resilience and effects of global climate change on biodiversity, water and land. This series is planned to narrate the stories of scientific development and progression towards SDGs from the regions of Asia and Africa. The overall aim of the series is to provide a comprehensive publication platform and act as a knowledge engine in the growth of sustainability sciences with a special focus on developing nations. The series constitute of four sub-series each focusing on a specific theme closely linked to a set of SDGs. Each sub-series is leaders and experts in agricultural sciences, environmental sciences, biomedicine and medical sciences respectively. Books in this series are useful for environmentalists, biotechnologists, plant scientists, medical practitioners, conservation biologists, policy makers, NGOs and scientific bodies working in the areas of sustainability sciences in the developing regions.

This book series publishes research associated with the prevailing opportunities and challenges faced by agroecosystems in Asia and Africa. It also explores the future scope of agroecological techniques and offers guidance on sustainable management of these systems. The series discusses important areas such as sustainable agricultural production, soil health, water resource management, biodiversity conservation, animal health and productivity, insect management strategies, genome editing and climate change mitigation. Further, it includes the challenges faced by African agriculture in arid and semi-arid regions, particularly in response to climate change. It highlights issues such as water shortages, soil degradation, desertification, and the need for sustainable farming techniques.

Sustainable agroecosystem maintenance entails strategies that improve overall environmental health and economic viability while limiting negative environmental impacts. The series takes a multidisciplinary approach to investigate creative ways for boosting sustainable agriculture, increasing food security, and improving lives for people in global south. This book series acts as a global publishing platform to collate and disseminate knowledge on fostering resilient and productive agroecosystems, thus enhancing accessibility, and fostering informed decision-making.

Heba I. Mohamed
Editor

Elicitors for Sustainable Crop Production

Overcoming Abiotic Stress Challenges in Plants

Springer

ARC

Editor
Heba I. Mohamed
Biological and Geological Sciences
Faculty of Education
Ain Shams University
Heliopolis, Cairo, Egypt

ISSN 2730-6771 ISSN 2730-678X (electronic)
Sustainability Sciences in Asia and Africa
ISSN 3059-247X ISSN 3059-2488 (electronic)
Agroecosystem Dynamics and Sustainable Practices
ISBN 978-981-95-3532-3 ISBN 978-981-95-3533-0 (eBook)
https://doi.org/10.1007/978-981-95-3533-0

This Springer imprint is published by the registered company Springer Nature Singapore Pte Ltd.
The registered company address is: 152 Beach Road, #21-01/04 Gateway East, Singapore 189721, Singapore

Preface

Agriculture and food security continue to face serious challenges due to crop losses induced by abiotic stress. Abiotic stresses such as drought, salinity, extreme temperatures, and nutrient deficiencies disrupt plant physiological and biochemical processes, ultimately reducing yield and quality. With climate change intensifying the frequency and severity of these stressors, the impact on global food systems is becoming increasingly critical. Addressing these challenges requires the development of resilient crop varieties, sustainable agronomic practices, and innovative biotechnological solutions to ensure food security for a growing global population.

Climate change poses a threat to agriculture, and its effects on the sustainability and financial success of agricultural production systems are significant. The impact of abiotic stresses on crops is expected to grow due to climate change. Abiotic stress refers to the negative impact of non-living environmental factors on plants. Common types of abiotic stress include drought, salinity, extreme temperatures (heat or cold), heavy metals, and radiation. These stresses can severely affect plant growth, development, and productivity by disrupting physiological and biochemical processes. For instance, drought stress reduces water availability, leading to decreased photosynthesis, wilting, and even plant death. Salinity stress causes ionic imbalances and osmotic stress, affecting nutrient uptake and leading to stunted growth. High or low temperatures can damage cell membranes, proteins, and enzymes, impairing overall metabolism.

Elicitors can be classified into biotic (e.g., chitosan, microbial extracts), abiotic (e.g., salicylic acid, jasmonic acid, hydrogen peroxide), and chemical or synthetic compounds (e.g., nanoparticles, phytohormones). These molecules act as signaling agents that stimulate antioxidant enzyme systems (like SOD, CAT, and APX), promote the accumulation of osmolytes (such as proline and glycine betaine), and upregulate genes associated with stress tolerance. For instance, chitosan, a natural polysaccharide derived from chitin, has been widely reported to enhance drought and salt tolerance by improving stomatal regulation, increasing chlorophyll content, and reducing reactive oxygen species (ROS) damage. Similarly, salicylic acid and jasmonic acid are known to modulate hormone signaling pathways and trigger protective responses under salinity and heat stress conditions. Microbial elicitors, such as those produced by plant growth-promoting rhizobacteria (PGPR), can induce systemic tolerance by producing phytohormones and enhancing nutrient uptake, especially under nutrient-deficient or saline soils. Recent advances also highlight the use of nanoparticle-based elicitors, which not only improve nutrient delivery but also mimic hormone-like effects that support plant growth under stress. For example, silicon and zinc oxide nanoparticles have been shown to enhance heat and drought tolerance by stabilizing cellular membranes and boosting antioxidant capacity.

Overall, elicitors represent a sustainable and eco-friendly strategy to enhance plant resilience to abiotic stress. Their integration into agricultural systems can reduce reliance on synthetic agrochemicals, improve crop yield and quality, and contribute to global food security amid growing climatic challenges.

This book explores the biosynthesis of biostimulants (elicitors) and their role in reducing abiotic stress and increasing plant growth and productivity. It provides a summary of recent research in biostimulants, molecular and physiological interactions, and field monitoring of abiotic stress in plants. This will be the first comprehensive book to provide a summary of the latest research on elicitors assisting in abiotic stress management in crops, with a focus on sustainability in agriculture. The current volume is interdisciplinary and will be very useful for students, teachers, researchers, plant physiologists, environmental scientists, and agrochemical companies. It will also be of immense value to graduate students in agriculture, biotechnology, and environmental university programs. Professionals or practitioners in research institutes involved in plant science, material science, and crop production research will also benefit from this book.

Cairo, Egypt Heba I. Mohamed

Contents

Editors and Contributors

About the Editor

Heba I. Mohamed is Professor of Plant Physiology, at the Faculty of Education, Biological and Geological Sciences Department, Ain Shams University. She completed her MSc and PhD in Plant Physiology, at the Faculty of Education, Ain Shams University. Her research interests include biotic and abiotic stresses, plant biochemistry, the use of eco-friendly compounds to alleviate plant stress, plant secondary metabolites, and genetic differences between different genotypes. She has edited 6 books in Springer Nature published. She has published 40 book chapters, 10 review articles, and more than 120 research articles in international peer-reviewed journals. She is a reviewer of international journals. She also is an editor of the *Phyton* journal. She has obtained a certificate of recognition in honor of achievement in an international publication that supports Ain Shams University World Ranking of 25/6/2012, 2017, 2018, 2019–2025. She obtained the Country Encouragement Prize in Cairo in 2018. In addition, according to Stanford University's worldwide database rating in 2021, 2022, 2023, and 2024, she was ranked in the top 2% of the most influential scientists in the world by Elsevier and Stanford University.

Contributors

Mahmoud S. Abdelmoneim Botany and Microbiology Department, Faculty of Science, Assiut University, Assiut, Egypt

Alessandra Adessi Department of Agriculture, Food Environment & Forestry (DAGRI), University of Florence, Florence, Italy

Haseeb Ahmad College of Agriculture, Guangxi University, Nanning, China

Sana AlHajj Ahmad Faculty of Agricultural and Food Sciences, American University of Beirut, Beirut, Lebanon

Farah K. Ahmed Nanobiotechnology Department, Faculty of Postgraduate Studies for Nanotechnology, Cairo University, Giza, Egypt

Lourd Abd Al Ahad Faculty of Agricultural and Food Sciences, American University of Beirut, Beirut, Lebanon

Muhammad Tanveer Altaf Department of Field Crops, Faculty of Agriculture, Recep Tayyip Erdoğan University, Rize/Pazar, Turkey

Department of Plant Production and Technologies, Faculty of Agricultural Sciences and Technologies, Sivas University of Science and Technology, Sivas, Turkey

Saleh M. AlTurki Department of Arid Land Agriculture, College of Agricultural and Food Science, King Faisal University, Al-Ahsa, Saudi Arabia

A. K. M. Aminul Islam Department of Genetics and Plant Breeding, Faculty of Agriculture, Bangabandhu Sheikh Mujibur Rahman Agricultural University, Gazipur, Bangladesh

Moh Sajid Ansari Department of Botany, Faculty of Life Sciences, Aligarh Muslim University, Aligarh, Uttar Pradesh, India

Faheem Shehzad Baloch Department of Biotechnology, Faculty of Science, Mersin University, Yenişehir, Mersin, Turkey

Department of Plant Resources and Environment, Jeju National University, Jeju, South Korea

Celaleddin Barutçular Department of Field Crops, Faculty of Agriculture, Çukurova University, Adana, Turkey

Manal Bencheikh Laboratory of Phoeniciculture Research "Phoenix", Faculty of Nature and Life Sciences, University of Kasdi Merbah, Ouargla, Algeria

Zakaria Boual Laboratory of Preservation of Ecosystems in Arid and Semi-Arid Areas, Faculty of Nature and Life Sciences, University of Kasdi Merbah-Ouargla, Ouargla, Algeria

Fatma Zohra Chakou Laboratory of Preservation of Ecosystems in Arid and Semi-Arid Areas, Faculty of Nature and Life Sciences, University of Kasdi Merbah-Ouargla, Ouargla, Algeria

Soumen Chakraborty Department of Agronomy, Institute of Agricultural Sciences, Banaras Hindu University, Varanasi, Uttar Pradesh, India

Matilde Ciani Department of Agriculture, Food Environment & Forestry (DAGRI), University of Florence, Florence, Italy

Sourav Das Division of Soil Science and Agricultural Chemistry, Indian Agricultural Research Institute, New Delhi, India

Mona F. A. Dawood Botany and Microbiology Department, Faculty of Science, Assiut University, Assiut, Egypt

Mattaparthi Lakshmi Durga Division of Agricultural Engineering, ICAR-Indian Institute of Agricultural Research, New Delhi, India

Israa El Ghandour Faculty of Agricultural and Food Sciences, American University of Beirut, Beirut, Lebanon

Mohamed M. El-Mogy Department of Arid Land Agriculture, College of Agricultural and Food Science, King Faisal University, Al-Ahsa, Saudi Arabia

Rania Francis Department of Agriculture and Food Engineering, School of Engineering, Holy Spirit University of Kaslik, Jounieh, Lebanon

Prakash I. Gangashetty International Crops Research Institute for the Semi-Arid Tropics (ICRISAT), Patancheru, Hyderabad, Telangana, India

Nihal Gören-Saglam Istanbul University, Faculty of Science, Department of Biology, Division of Botany, Vezneciler, Istanbul, Turkey

Karima Guehaz Laboratory of Preservation of Ecosystems in Arid and Semi-Arid Areas, Faculty of Nature and Life Sciences, University of Kasdi Merbah-Ouargla, Ouargla, Algeria

Sabine Hilal Faculty of Agricultural and Food Sciences, American University of Beirut, Beirut, Lebanon

Jaweria Iqbal Department of Plant Breeding and Genetics, Faculty of Agricultural Sciences and Technology, Bahauddin Zakariya University, Multan, Pakistan

Amna Jamil Department of Horticultural Sciences, The Islamia University of Bahawalpur, Bhawalpur, Pakistan

Muhammad Faheem Jan Guangxi Key Laboratory of Agro-Environment and Agro-Products Safety, College of Agriculture, Guangxi University, Nanning, China

Suresh Kadaru Medium Maturity Rice Breeding Program, International Rice Research Institute (IRRI), Los Banos, Laguna, Philippines

Abrar A. Khan Department of Botany, Faculty of Life Sciences, Aligarh Muslim University, Aligarh, Uttar Pradesh, India

Adnan Khan Department of Botany, Faculty of Life Sciences, Aligarh Muslim University, Aligarh, Uttar Pradesh, India

Ehtisham Hassan Khan MOA Key Laboratory of Crop Ecophysiology and Farming System in the Middle Reaches of the Yangtze River, College of Plant Science and Technology, Huazhong Agricultural University, Wuhan, China

Sabyasachi Koley Department of Soil Science and Agricultural Chemistry, Institute of Agricultural Sciences, Banaras Hindu University, Varanasi, Uttar Pradesh, India

Vandana Kumari Department of Soil Science, Advance Centre on Sericulture, Dr. Kalam Agricultural College (Bihar Agricultural University), Kishanganj, Bihar, India

Ming Li College of Agriculture, Northeast Agricultural University, Harbin, China

Waqas Liaqat Department of Field Crops, Faculty of Agriculture, Recep Tayyip Erdoğan University, Rize/Pazar, Turkey

Changzhuang Liu College of Agriculture, Northeast Agricultural University, Harbin, China

Debdut Manna Division of Biochemistry, Indian Agricultural Research Institute, New Delhi, India

Asia Maqbool College of Horticulture, Hebei Agricultural University, Baoding, Hebei, China

Heba I. Mohamed Biological and Geological Science Department, Faculty of Education, Ain Shams University, Cairo, Egypt

Partha Mondal Department of Agricultural Biochemistry, Bidhan Chandra Krishi Viswavidyalaya, Mohanpur, West Bengal, India

Aicha Mouane Department of Biology, Faculty of Natural and Life Sciences, University of El Oued, El Oued, Algeria

Sayon Mukherjee Department of Soil Science and Agricultural Chemistry, Institute of Agricultural Sciences, Banaras Hindu University, Varanasi, Uttar Pradesh, India

Debabrata Nath International Rice Research Institute (IRRI)-South Asia Regional Centre (ISARC), Varanasi, Uttar Pradesh, India

Gareth J. Norton School of Biological Science, University of Aberdeen, Aberdeen, Scotland, UK

Dibyajyoti Panda Department of Soil Science and Agricultural Chemistry, Institute of Agricultural Sciences, Banaras Hindu University, Varanasi, Uttar Pradesh, India

Abhik Patra Krishi Vigyan Kendra, Narkatiaganj, West Champaran, Dr. Rajendra Prasad Central Agricultural University, Pusa, Bihar, India

Gorantla Prathap Reddy Department of Agronomy, Institute of Agricultural Sciences, Banaras Hindu University, Varanasi, Uttar Pradesh, India

Keyanna Radi Faculty of Agricultural and Food Sciences, American University of Beirut, Beirut, Lebanon

Nahed M. Rashed Department of Arid Land Agriculture, College of Agricultural and Food Science, King Faisal University, Al-Ahsa, Saudi Arabia

Madhurya Ray Department of Plant Physiology, Institute of Agricultural Sciences, Banaras Hindu University, Varanasi, Uttar Pradesh, India

Tony Kevork Sajyan Faculty of Agricultural and Food Sciences, American University of Beirut, Beirut, Lebanon

Department of Agriculture and Food Engineering, School of Engineering, Holy Spirit University of Kaslik, Jounieh, Lebanon

Sumi Sarkar Department of Genetics and Plant Breeding, Faculty of Agriculture, Gazipur Agricultural University, Gazipur, Bangladesh

Arop Sarker Department of Molecular Biotechnology, School of Bioscience, University of Skövde, Högskolevägen, Skövde, Sweden

Muhammad Sheraz Fodder and Forrage Program, Crop Sciences Institute, NARC, Islamabad, Pakistan

Wang Shuai College of Agriculture, Northeast Agricultural University, Harbin, China

Alia Telli Laboratory of Preservation of Ecosystems in Arid and Semi-Arid Areas, Faculty of Nature and Life Sciences, University of Kasdi Merbah-Ouargla, Ouargla, Algeria

Aslihan Temel Istanbul University, Faculty of Science, Department of Molecular Biology and Genetics, Vezneciler, Istanbul, Turkey

Unaiza Department of Field Crops, Faculty of Agriculture, Institute of Natural and Applied Sciences, Çukurova University, Adana, Turkey

Randa A. Zarban Department of Plant Biotechnology, College of Agricultural and Food Science, King Faisal University, Al-Ahsa, Saudi Arabia

Miaoli Zhang College of Agriculture, Northeast Agricultural University, Harbin, China

Chapter 1
Plant Defense Mechanisms Against Abiotic Stresses

Sumi Sarkar, Prakash I. Gangashetty, Arop Sarker, Suresh Kadaru, Gareth J. Norton, and A. K. M. Aminul Islam

Abstract Plants face numerous abiotic stresses like drought, heat, salinity, nutrient deficiencies, etc., which can negatively influence their growth and yield. To address the above challenges, plants utilize a range of defense strategies that fall into morphological, physiological, biochemical, and molecular categories. Key adaptations involve physiological responses such as osmotic regulation to preserve cellular integrity under stress, the synthesis of protective metabolites, and enhanced root structure development to improve water and nutrient absorption. Biochemically, the synthesis of protective compounds such as antioxidants, osmoprotectants, and stress-associated hormones plays a key role in reducing damage caused by these stressors and regulating the plant's response. Furthermore, molecular responses, including the initiation of stress-associated genes and signaling pathways, are essential in coordinating these defenses. Additionally, secondary metabolite synthesis contributes to defensive responses against oxidative stress, boosting overall plant resilience. Understanding the complex interplay among these mechanisms can

S. Sarkar · A. K. M. Aminul Islam (✉)
Department of Genetics and Plant Breeding, Faculty of Agriculture, Gazipur Agricultural University, Gazipur, Bangladesh
e-mail: sumisarkar@bsmrau.edu.bd; aminulgpb@bsmrau.edu.bd

P. I. Gangashetty
International Crops Research Institute for the Semi-Arid Tropics (ICRISAT), Hyderabad, India
e-mail: Prakash.Gangashetty@icrisat.org

A. Sarker
Department of Molecular Biotechnology, School of Bioscience, University of Skövde, Högskolevägen, Skövde, Sweden

S. Kadaru
Medium Maturity Rice Breeding Program, International Rice Research Institute (IRRI), Los Banos, Laguna, Philippines
e-mail: s.kadaru@iir.org

G. J. Norton
School of Biological Science, University of Aberdeen, Aberdeen, UK
e-mail: g.norton@abdn.ac.uk

H. I. Mohamed (ed.), *Elicitors for Sustainable Crop Production*, Sustainability Sciences in Asia and Africa, https://doi.org/10.1007/978-981-95-3533-0_1

provide insights into how plants maintain balance and increase resilience, which is vital for adapting to changing environments. Investigating these defense mechanisms is crucial for the development of crop varieties that are capable of surviving against biotic stress, mainly the challenges of climate change effects and declining natural resources. This chapter brings together recent research efforts focused on elucidating tolerance against abiotic stress. Insights gained from this chapter will provide valuable knowledge in addressing the challenges posed by the abiotic stresses and improving the sustainability of agriculture to face the effects of climate change.

Keywords Stresses · Signaling · Hormones · Antioxidants · Osmolytes · Genes · Defense mechanism

1.1 Introduction

The productivity of agriculture faces threats from various stress factors that are linked to climate change as the demand for crop products continues rising (Dresselhaus and Hückelhoven 2018). In most countries, precipitation has been seen to increase over the past century, particularly during the winter season, but the summer season has become drier, and the days have doubled with the temperature above 30°C (Malhi et al. 2021). These weather conditions often benefit the pathogen and have a negative impact on plant productivity and fertility. The plants must manage complex interactions with numerous environmental factors and adapt to these changing environments, reducing the available resources for biomass and seed production, and thus yield. Exposure to abiotic stress disrupts plant metabolism, leading to physiological costs, reduced fitness, and decreased productivity. Additionally, combining or altering different abiotic stress factors makes plants more susceptible to other stresses (Zandalinas et al. 2022). Abiotic stress significantly impacts growth, and severe stresses like salinity and drought can result in complete yield losses (Kul et al. 2020). Resistance to abiotic stress has been documented in many crops, utilizing different mechanisms, for example, through defense priming, achieved via chemical stimuli and genetic manipulation.

A critical aspect of abiotic stress and plant resistance is the detection of stress to allow a rapid and efficient response. Upon recognition of stress conditions, plants activate their basic protection mechanisms, leading to an intricate signaling cascade that differs depending on stress category (Fig. 1.1). Plants activate certain ion channels and kinase pathways when they are exposed to abiotic stress and accumulate reactive oxygen species (ROS) and various hormones such as ethylene (ET), salicylic acid (SA), jasmonic acid (JA), and abscisic acid (ABA), and alter their genetic mechanism to enhance defense and tolerance (Park and Shin 2022). Current research focuses on recognizing variable plant responses to a particular stress, although simultaneous stresses present a more complex scenario. Plants operate diverse signaling pathways based on stress, and their response to multiple stresses cannot be

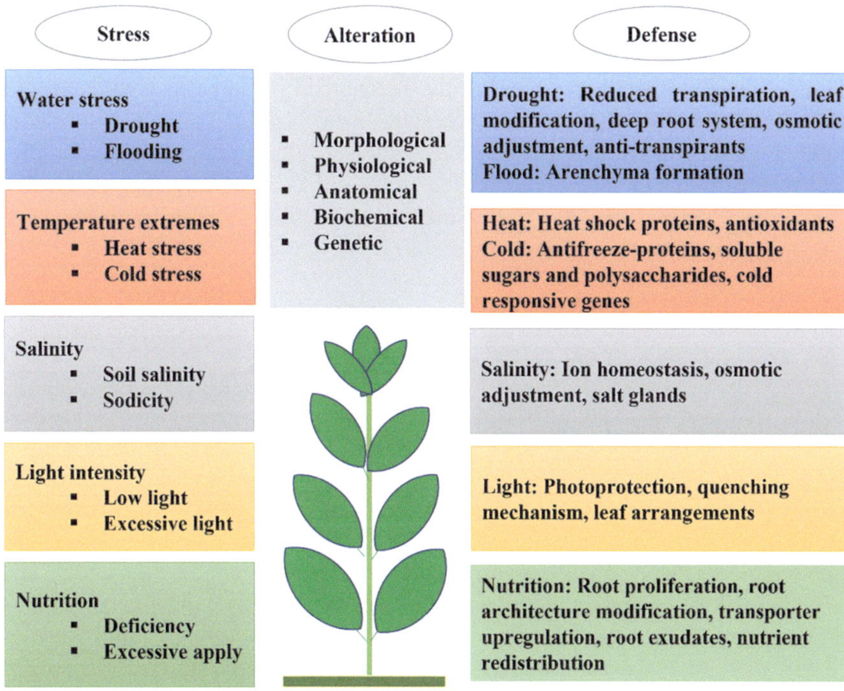

Fig. 1.1 Overview of different abiotic stresses, common alterations, and defense responses of the plant. (Kumari et al. 2022a, b; Mathivanan 2021)

anticipated based on single-stress responses. Studies have attempted to replicate natural conditions, but field conditions are uncontrolled, and therefore, multiple stresses can impact the defense mechanism of plants, whereas the primary defense response may be prompted when plants face a single stress rather than multiple stresses. Sensitivity also varies based on field conditions, plant development stage, stress intensity, and plant species. The interactions between defense mechanisms rely on the precise stress combinations and their simultaneity, with responses being additive, synergistic, or antagonistic, influencing susceptibility to other stresses (REF).

A combination of stressors may have negative additive effects, where greater damage is caused by the second stress or precedes antagonistic responses. However, exposure to several stresses may also result in cross-tolerance, where plants tolerant to one stress become more tolerant to the subsequent stress (Ramegowda et al. 2020). These conditions necessitate substantial investments in the research and development of plant stress tolerance. Therefore, this chapter also emphasizes combined tolerance against abiotic stresses as a module of prompt resistance for plant defense. Understanding plant responses against numerous stresses and applying this information in contemporary plant breeding programs is crucial for sustaining and improving yields.

1.2 Abiotic Stress

Plants' growth and yield are significantly threatened by numerous abiotic stresses from environmental changes like drought, heat, salinity, and heavy metals (Table 1.1). To survive and grow in these stresses, plants adopt several complex mechanisms to detect and counteract the stress. Effective abiotic stress responses depend on the plant's ability to sense signals, triggering signaling pathways, and activating resistance genes (Imran et al. 2021).

1.3 Plant Defense Strategies Against Stresses

1.3.1 Escape, Avoidance, and Adaptation

Plants have developed a range of complex immune response pathways that permit them, as sessile organisms, to survive under abiotic stresses. Plant abiotic stress interactions stimulate broad-spectrum defense responses both at the cellular and molecular levels to confirm that plant cells are protected from the challenges they face (Kooyers 2015) (Fig. 1.2). Plants can adapt to abiotic stress by rapidly maturing and reproducing before the stress begins. Consequently, flowering time is a crucial trait linked to stress escape in annual herbaceous plants (Chen et al. 2023). Flowering earlier in a shortened growing season due to stress likely leads to better fitness, either through increased seed production or larger seed size. Many plant species exhibit well-characterized adaptive stress responses. For example, drought conditions led to strong selection for earlier flowering in two populations of *Brassica napus* (Kayacetin 2021). However, flowering time following stress may not always be adaptive. Few studies have explicitly tested the adaptive responses to drought, and more empirical and theoretical research is needed to understand when such acclimations might be beneficial.

Escaping stress can evolve not only through inherent differences in developmental timing but also through flexible responses to environmental cues. Plants set their flowering based on cues that include temperature, vernalization, photoperiod, and water availability (Tun et al. 2021). Adaptive stress escape often relies on temperature and photoperiod cues, while water availability, considered a secondary cue, might be crucial for accurately assessing the stress conditions. Various morphological and physiological traits are responsible for the avoidance mechanism in plants under abiotic stress (Branch et al. 2024). Several traits are associated with improved transpiration efficiency under different abiotic stresses. For instance, drought-avoiding plants from arid regions often have better water-use efficiency (WUE) due to enhanced photosynthetic capacity and reduced transpiration. Transpiration efficiency, which reflects performance, can be measured at the leaf level using both instantaneous WUE and a long-term indicator of transpiration efficiency. Plants may also slow down or halt growth in response to early signs of stress (Bhattacharya

Table 1.1 Causes and major effects of abiotic stresses on plants

Abiotic stress	Causes	Major effects	References
Drought	Depreciation in rainfall or lack of available irrigation water, consequent dry spell	Stomatal closure, reduced CO_2 intake, limited carboxylation, increased photorespiration, reduced tissue water potential, oxidative damage to chloroplast and plasma membrane, obstruction in ATP synthesis, reduced nutrient uptake	Wahab et al. (2022) and Ault (2020)
Heat	Most crops have an ideal temperature range. Temperatures above this range, whether in the air or the soil, during the day or at night, may cause heat stress to the plant and a sufficiently long situation may permanently modify its capability to function or grow normally	Reduction in growth, photosynthesis rate, chlorophyll content, stomatal conductance, water-use efficiency, reduced mineral uptake, reduction in carbohydrate metabolism, poor seed development, sterility of the flower, increased oxidative stress	Łysiak and Szot (2023) and Ali et al. (2020)
Cold	Temperature falls below the optimum range causing damage to crop plants	Inhibits seed germination, cell membrane damage and ion leakage, cell dehydration because of ice crystals, protein structure, and stability disruption	Raza et al. (2023) and Etesami et al. (2023)
Salinity	Inhibits K^+ uptake, an essential element, but increases the accumulation of cl^- and Na^+, leading to toxicity. Excess Na^+ uptake competes with other nutrient elements, particularly K^+. High salinity reduces water uptake	Ionic homeostasis is a reason for K^+ deficiency that leads to a lower K^+/Na^+ ratio in plants under salinity stress, decreased shoot and root growth, increased ROS accumulation	Kibria and Hoque (2019), Zhao et al. (2021a, b) and Venkatappa et al. (2021)
Waterlogging	Surface layer of soil saturated with water, water table rising, overirrigation, excessive precipitation	Reduced root respiration, root damage, restricted root growth, limited water and nutrient uptake	Amer (2021) and Kaur et al. (2020)
Floods	Heavy rainfall due to tropical weather disturbances, deforestation, inappropriate agricultural practices, insufficient plans for drainage channels and constructions, deficient maintenance of drainage facilities, construction of settlements in flood plains	Flooding reduces oxygen in soil, reducing normal metabolism and potentially leading to anoxic soils. Growth and development of crops increase disease infections and nitrogen losses	Venkatappa et al. (2021) and Haraldsen (2022)

(continued)

Table 1.1 (continued)

Abiotic stress	Causes	Major effects	References
Heavy metals/ metalloids	Industrial waste, pesticides, soil pedogenesis, hydrothermal activities, volcanic eruptions, vehicle pollution	Chlorosis, necrosis, nutrient decrease, protein denaturation, cell detoxification, enzymatic activity disruption, decreased root biomass, germination rate inhibition	Alsafran et al. (2022), Vasilachi et al. (2023) and Pande et al. (2022)
Radiation	High-energy ultraviolet radiation, nuclear energy, industrial devices used for scientific research and measurement	Chromosomal aberrations, DNA damage, growth reduction, reduction in reproductive rate. The incidence of abnormalities in the developmental process or decreased viability of the offspring. Reduced seed germination, mortality, including both life span reduction and acute lethality, direct burn damage to exposed tissue	Singh et al. (2020)
Nutrient deficiency/ toxicity	Unsatisfactory environment, nutritional starvation, wrong or excess application of manure, fertilizers, pesticides, imbalanced irrigation, soil pH below or over the range, soil temperature	Leaf yellowing and browning, stunted growth and development, reduced flower and fruit setting	Rácz and Radócz (2019) and DeTurk (1941)

2022). Additionally, adaptations for stress avoidance might include a higher root-to-shoot growth ratio before stress conditions arise. Our understanding of the genetic mechanisms and fitness trade-offs related to stress resistance has advanced significantly over the past decade. Notably, quantitative trait loci (QTLs) are identified in various plant species related to the abiotic stress escape and avoidance strategies, shedding light on the roles of genetic mechanisms and gene-by-environment interactions in stress resistance evolution (Chen et al. 2024). Greenhouse and field research was also initiated to determine the fitness consequences of various stress escape and avoidance mechanisms. These findings, primarily derived from natural populations, highlight the value of wild populations in uncovering novel stress resistance mechanisms for both ecologists and agronomists. However, there are still gaps in our knowledge that need addressing. Specifically, there is an absence of an inclusive understanding of how environmental signals are integrated into physiological responses, and the majority of genes or epigenetic factors responsible for variation in stress responses have yet to be pinpointed. This challenge is relevant for genetic engineers and breeders, who must also consider the potential trade-offs of environment-dependent effects.

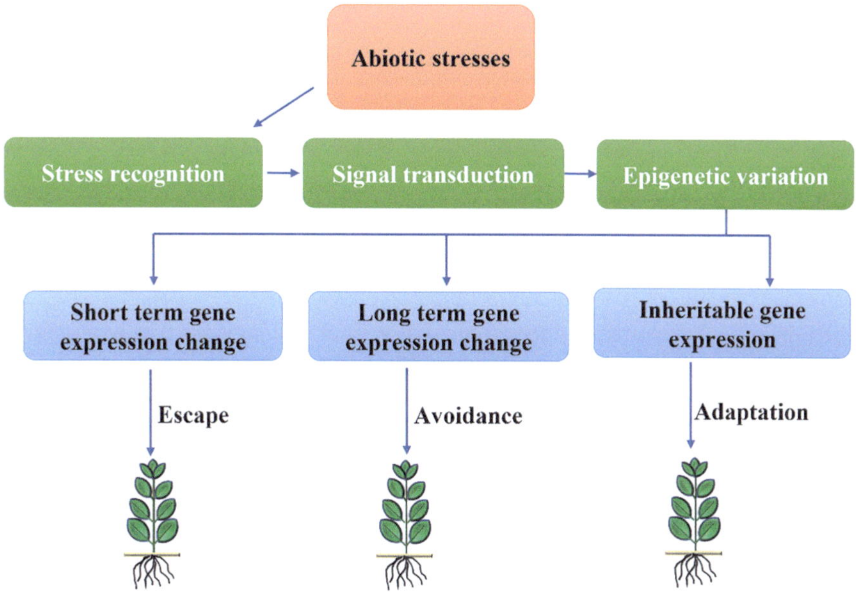

Fig. 1.2 Three major defense strategies of plants under abiotic stresses. (Ashapkin et al. 2020)

Although the issue of assessing losses across various environments is not new to traditional, trait-based, or molecular-assisted breeding programs, it remains crucial due to the gene-by-environment interactions associated with stress-related traits. Future research needs to explore how seasonal context and stress intensity impact the evolution of stress strategies and plant fitness in natural populations. There has been less focus on transient or intermittent stress events, and understanding plant responses and adaptations to these conditions could offer valuable insights. It remains a priority to know the role of adaptive constraints in the development process of stress tolerance. Natural populations provide valuable opportunities to investigate genetic constraints between different stress resistance strategies and determine when these constraints are adaptive. Variation in genetic correlations among different populations offers a natural setting to study the adaptive significance of these combinations (Kristensen et al. 2020). Identifying these constraints will expose the indirect benefit of selecting for specific traits, such as early flowering or heading, and could reveal new avenues for breeding stress-resistant crops that also maintain high yields. It is hard to expect that an ideal crop would simultaneously employ both stress escape and avoidance strategies. Instead, an optimal variety should allocate resources to maximize yield under favorable conditions while being adaptable to switch to a stress strategy as needed during early stress onset. The most effective stress strategy will vary based on plant species and regional stress conditions (Rivero et al. 2022). Therefore, efforts should focus on developing or identifying varieties that can adapt to initial stress conditions without compromising yield in optimal environments. These insights highlight the importance of understanding the benefits, costs, and gene-by-environment interactions for advancing crop development and predicting how populations will adapt to changing and unpredictable environments.

1.3.2 *Tolerance Mechanism*

Tolerance mechanism is another effective plant defense strategy that works properly under abiotic stress. This mechanism plays an effective role in the production of biomass and keeps physiological functions active by enduring extra stress. Tolerance mechanism generally emphasizes internal activities and maintaining cellular activities in place of other actions, which are also responsible for decreasing the exposure to stress (Qiao et al. 2024). Biochemical, molecular, and physiological activities increase the ability of the plants to work against abiotic stresses (Saharan et al. 2022).

Osmotic adjustment plays an important role under drought as well as salinity stress, which is known as osmoregulation, where two categories of osmoregulators are engaged, including inorganic osmoregulators, such as K^+, Na^+, and Ca^{2+}, and organic osmoregulators, for example, soluble sugars (SSs), betaine, proline, and polyols (Zhang et al. 2023). Oxygen radicals and the derivatives of oxygen radicals known as reactive oxygen species (ROS) are produced by the plants, which help in protecting them from abiotic stress under the changing environment by responding to the transmission of cellular signaling (Hasanuzzaman et al. 2020). Moreover, accumulation of excessive free radicals (superoxide anion, $O_2^{\cdot-}$; alkoxy radical, $RO^{\cdot}$; hydroxyl radical, OH; and hydroperoxyl radical, $HO_2^{\cdot}$) and nonradical molecules (hydrogen peroxide, H_2O_2 and singlet oxygen, $^1O_2$) directly induces the oxidative stress in plant species, which reduces the yield of the crops (Mittler et al. 2011; Ghoran et al. 2024). Different types of enzymatic antioxidants, such as catalase (CAT), peroxide (POD), and superoxide dismutase (SOD), play a role in the detoxification of reactive oxygen species (Jomova et al. 2024). Depending on the active site, SOD is divided into three isoenzymes according to the cofactors Zn/CU/Fe/Mn, which are known as Zn/CU-SOD, Fe-SOD, and Mn-SOD (Mishra et al. 2023). The working mechanism of SOD shows $O_2^{\bullet-}$ to O_2 and H_2O_2 (Mittler et al. 2011). Organelles such as nuclei, mitochondria, chloroplasts, cytoplasm, peroxisomes, and apoplast contain SOD like other oxygen-metabolizing cells (Singh et al. 2019). Due to high reactivity, not only the enzymatic antioxidant but also nonenzymatic antioxidants such as glutathione and ascorbate act as detoxifiers (Hasanuzzaman et al. 2019).

1.4 Plants' Recovery Mechanisms from Stresses

Plants have the ability to alter their recovery mechanisms under different abiotic stresses. These mechanisms include morphological, anatomical, physiological, and biochemical alterations, which, combined, help a plant cope with a stress condition (Fig. 1.3)..

1.4.1 Developmental (Morphological and Anatomical) Mechanisms

Plants adopt various adaptive mechanisms in response to stress, such as morphological and developmental changes, in addition to physiological and biochemical adjustments. The basic morphological and anatomical defense mechanisms that occur commonly under abiotic stress are variations in the anatomy of leaves, like the thickness of the cell wall and the exposed chloroplast surface area, shortening the

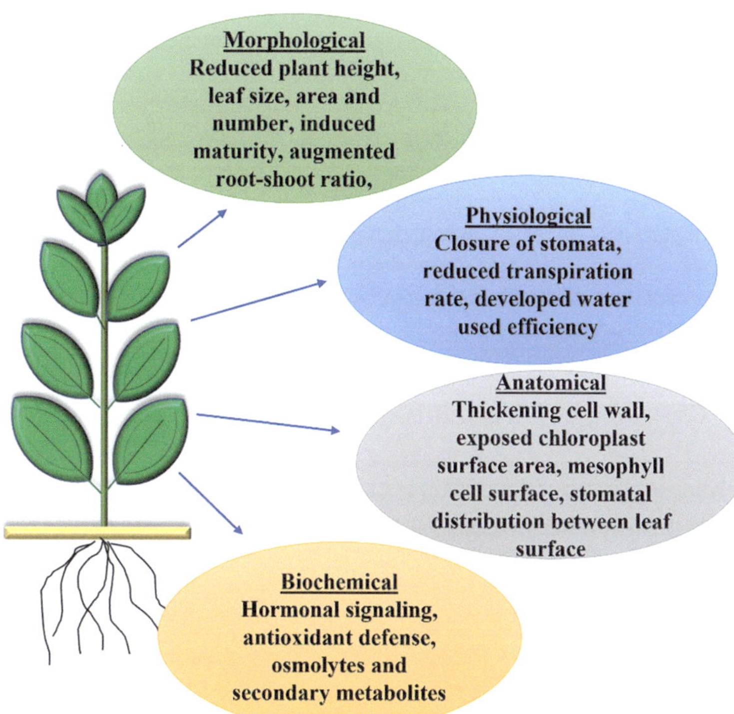

Fig. 1.3 Recovery mechanisms of plants at different levels. (Kashyap et al. 2021)

maturity period, and leaf rolling to reduce effective leaf area, thus reducing transpiration (Liu and Yang 2020).

Under different abiotic stresses, plants undergo morphological and anatomical changes for their recovery. For instance, shoot and root modification takes place under drought stress (Arif et al. 2020). The plants may develop deep root systems, increase root hair density to access groundwater, and improve nutrient absorption. Some plants develop additional roots from non-root tissues such as stems, branches, or leaves to stabilize and access resources. Alterations in leaf structure are also known as morphological adaptations under different abiotic stress (Chaudhry and Sidhu 2022). Plants may produce smaller leaves or reduced leaf surface area to minimize water loss. Leaf shape modifications, such as needle-like or succulent leaves, can reduce water loss and improve heat tolerance. Some plants adjust the orientation of leaves to minimize direct sunlight exposure and produce a thicker waxy cuticle on leaf surfaces to reduce water loss through transpiration. Leaves and stems may produce hair-like structures under abiotic stress, called trichomes, that help reflect sunlight to reduce water loss as well as heat absorption (Jasleen and Kariyat 2020).

Plants can also produce a dense cover of fine hairs on leaf surfaces to lower the temperature of leaves (Oliveira et al. 2020). Stem and shoot modifications take place under different abiotic stresses. In arid conditions, some plants have fleshy stems that store water. Under severe stress, some plants may reduce their overall growth and allocate resources to surviving rather than growing (Kumari et al. 2022a, b). Plants may alter leaf arrangement to optimize light capture and reduce shading of lower leaves, which can improve overall photosynthesis efficiency. Adjusting canopy density can help regulate internal temperature and reduce water loss (Klupczyńska and Pawłowski 2021). Variations in seed size and structures, like protective seed coats, can enhance survival under extreme conditions. These developmental adaptations are often complemented by physiological and biochemical responses, for instance, the accumulation of osmoprotectants or changes in enzyme activity, to provide a comprehensive strategy for coping with abiotic stress.

1.4.2 Physiological Mechanisms

Plants, when exposed to different kinds of stress, protect themselves through physiological defense mechanisms. These physiological changes are more prominent when plants face combined abiotic stress. However, the changes may either lead to tolerance or susceptibility based on the category and harshness of stress. The major significant defensive technique for plants under abiotic stress is stomatal closure. Stomatal closure plays a defensive role to protect plants from numerous stresses (Bharath et al. 2021). By closing their stomata, plants can retain water, reduce water loss, and limit pathogen entry. This mechanism is particularly important under abiotic stresses like drought. Under drought conditions, water losses due to transpiration are reduced by stomatal closure, enabling plants to preserve moisture. This response is facilitated by signaling molecules like abscisic acid (ABA) that trigger the stomatal closure to protect against excess water loss (Hsu et al. 2021). A major physiological defense technique of plants is the adjustment of osmotic pressure through osmolyte accumulation under abiotic stress. Plants synthesize and accumulate osmoprotectants like glycine betaine, soluble sugars, and proline (Zulfiqar et al. 2019). The osmoprotectants help to adjust the turgor pressure of the cell, stabilize protein molecules and cell membranes, and finally, protect cells from dehydration. Increased levels of soluble sugars, sucrose, and fructose also play a crucial role in the adjustment of osmotic pressure and the protection of cellular structures.

Increased antioxidant activity is a vital recovery strategy of plants during abiotic stress, as the stresses lead to the production of different reactive species (ROS) (Fig. 1.4). ROS serve as signaling molecules that are important for plant growth and development, but ROS can increase to an extreme level under abiotic stress, which is responsible for cellular damage. Effective ROS signaling can be ensured by plants through the local synthesis of ROS to initiate cellular responses and to scavenge ROS to inhibit it from extending to damaging levels (Singh et al. 2022). The plant generally responds to increased ROS by increasing the production of antioxidants, for example, glutathione, ascorbic acid, and various enzyme antioxidants (catalase, superoxide dismutase, peroxidase, etc.) to alleviate oxidative damage (Sharma et al. 2022).

Signal transduction and hormonal responses play a crucial physiological role. The abscisic acid is a key hormone in abiotic stress response, regulating stomatal closure, activating genes responsive to stress, and modulating other physiological responses for conservation or saving of water and enhancing tolerance to stress (Khan Zeba D 2020). Different enzymes also involved in osmotic adjustment and repair processes are upregulated to help the plant manage stress. Proline, an amino acid, accumulates in cells to protect against dehydration and maintain cellular turgor (Mohammadi Alagoz et al. 2023). Under abiotic stress, ABA signaling can trigger ROS production in plant cells (Postiglione and Muday 2020), which occurs via numerous pathways, such as stimulation of NADPH oxidases, which generate superoxide radicals. ROS serve as secondary messengers in ABA signaling pathways. They amplify the signal of ABA and initiate downstream responses such as activating specific transcription elements that enhance the expression of genes related to stress (Sachdev et al. 2021). ROS signaling mediated by ABA is crucial for the closure of stomata, which helps reduce water loss. This response is part of the instant response of plants to stress, such as drought, for conserving water and maintaining cellular turgor. Increased ROS levels activate antioxidant systems in plants. Genes that code for antioxidant enzymes such as catalase, peroxidase, superoxidase, and superoxide dismutase are upregulated to combat oxidative stress (Kerchev and Breusegem 2022). Together, ROS and ABA promote gene expression responsible for tolerance to stress, including those that encode protective proteins, osmolytes, and enzymes that detoxify harmful substances (Fig. 1.4).

The plant initiates premature leaf senescence (aging) in severe abiotic stress and abscission (leaf drop) to reduce water loss and conserve resources (Iqbal et al. 2020). The mechanisms jointly assist plants to manage stress and improve their

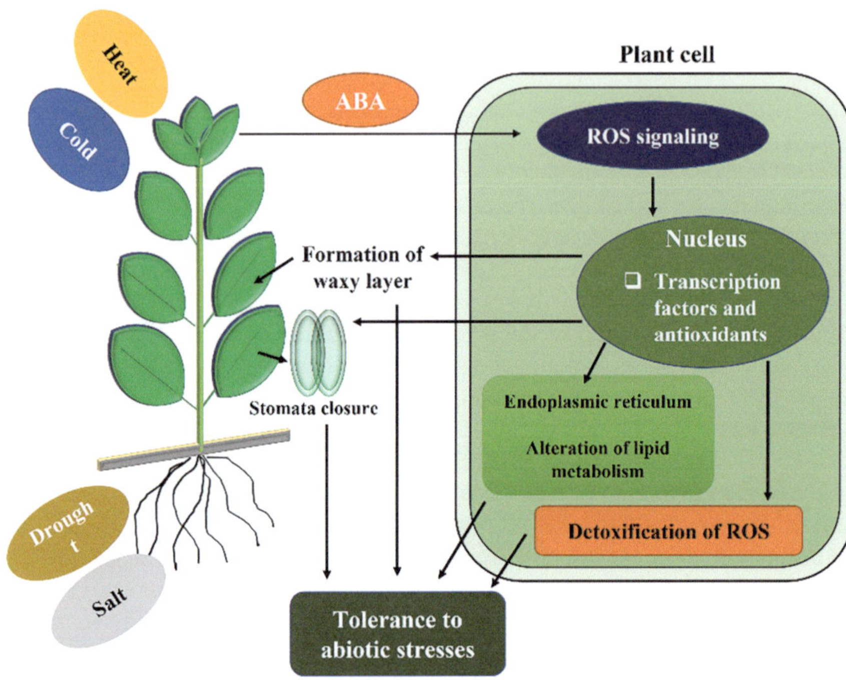

Fig. 1.4 Physiological basis of ABA activation for antioxidant defense. (Li et al. 2022)

chances of survival and reproduction in stressed environments. Different plant species may employ a combination of these strategies, depending on their evolutionary adaptations and the severity of the stress.

1.4.3 Biochemical Mechanisms

Plants go through biochemical changes and defense mechanisms for their self-protection under various stresses. The reaction to any stress by plants is a complex process regulated by a well-arranged signaling network involving various biochemical components. For instance, under drought stress, the process begins when signaling molecules (for example, ABA) are perceived by the definite receptors in the guard cells. The primary receptors are pyrabactin-resistant-1 (PYR-1) and PYR-1-like (PYL) proteins in the case of ABA (Zhai et al. 2023). Upon ABA binding, the receptor interacts with and activates the protein kinase OST1 (OpenStomata1). OST1 is an important player in the abscisic acid signaling pathway and enhances the activity of various downstream targets. OST1 activates NADPH oxidase (RBOHD/F), leading to a rise in ROS levels in the guard cells. The reactive oxygen species are important secondary messengers in the signaling cascade (Postiglione and Muday 2020). The increased ROS levels contribute to a rise in nitric oxide (NO) and calcium ions (Ca^{2+}) within guard cells. These secondary messengers play vital roles in the regulation of ion channels (Malik et al. 2020). Activation of Ca-dependent protein kinase (CDPKs) is activated by increased levels of Ca^{+}, which in turn activate slow anion channels (SLAC1) and S-type anion channels (SLAH3) (Hussain et al. 2020). These channels facilitate the outflow of anions from the guard cells. OST1 also modulates and enhances the efflux of potassium ions (K^{+}) from the guard cells in the potassium (K^{+}) channels (Kashtoh and Baek 2021). The consequence of loss of K+ and anions results in the decrease of osmotic pressure within the guard cells, leading to loss of turgor pressure and stomatal closure. However, the activity of OST1 may not always increase in response to these elicitors. High CO_2 and humidity can also influence stomatal responses, potentially interacting with the ABA signaling pathway and affecting guard cell behavior. OST1 can directly modulate ion channels to regulate stomatal movement independently of ROS/NO/Ca^{2+} pathways. Stomatal closure can also occur via OST1-independent pathways involving Plant Elicitor Peptides (Peps) and elevated CO_2, and through mitogen-activated protein kinase (MAPK) pathways (Melotto et al. 2024). Additional research is necessary to understand the full signaling pathways and how OST1 interacts with them to mediate stomatal responses. Understanding these mechanisms in detail could provide valuable insights for improving plant stress resilience and optimizing water use in crops.

1.4.4 Defense Mechanism Under Combined Stress

Some combinations of stresses can be advantageous for plants compared to the effects of each stress applied individually. For example, drought stress often reduces the stomatal closure but, on the other hand, enhances the plant's tolerance to ozone (O_3) stress (Hoshika et al. 2020). The reduction in stomatal conductance may play a vital role in counteracting the combined effects of drought and O_3 tolerance (Grulke and Heath 2020). Additionally, the ROS levels increased either through O_3 or drought alone but were reduced when both stresses were present simultaneously, bringing ROS concentrations closer to control levels (Pellegrini et al. 2021). This reduction in ROS was related to the increased levels of reduced glutathione and ascorbate, which are antioxidants that help manage oxidative stress (Pisoschi et al. 2021).

A second example is where two stress interactions are between salinity and heat stresses. The combined salinity and heat stresses can significantly protect tomato plants from the harmful effects of salinity stress alone (Sousa et al. 2022). This protective effect is associated with the increased accumulation of trehalose and glycine betaine (Giordano et al. 2021). These compounds help maintain a lower Na^{+} ratio, which improves water status in the cell and photosynthesis compared to salinity stress alone. There are several functions of glycine betaine, such as protecting photosystem II (PSII) from photo- and heat-induced damage and inhibiting salt-induced potassium (K^+) loss (Huang et al. 2020). Trehalose also contributes to maintaining photosynthetic capacity and PSII efficiency (Ziwei et al. 2022). Both trehalose and glycine betaine are important for protecting cells from oxidative stress. In fact, under the collective stress of salinity and heat, the accumulation of peroxide (H_2O_2) and protein oxidation was reduced, indicating that the plant's protection against the combination of stresses is partly due to reduced oxidative stress (Lopez-Delacalle et al. 2021).

1.4.5 Genetic Defense Mechanisms

The changes that happen inside a plant under stress conditions are potentially due to changes at the gene level. Several stress-tolerant genes have been identified that confer tolerance against abiotic stresses (Table 1.2). For abiotic stress, genes such as USP2, HSP70, SgNCED1, BADH, PVNCED1, HVA1, ALO, and LeNCED1 have been identified (Ahmad et al. 2021). In rice, the *OsSDG721* gene is found to encode a TRITHORAX-like protein, which positively regulates tolerance to salinity (Zhao et al. 2022). The gene OsmiR535 is responsible against tolerance to drought in rice (Tripathy et al. 2021). A heat shock gene, *TaHsfA2e-5D,* was isolated from chromosome 5D of wheat and characterized with a target to explore its functions in drought and heat stress response (Guzmán et al. 2022). From the gene expression profiling, *TaHsfA2e-5D* was found to be expressed in numerous tissues of wheat, most notably in root tissues during the reproductive phase. The expression of gene *TaHsfA2e-5D* was highly upregulated at the seedling stage in wheat by heat, drought, salt, cold, and multiple phytohormones (Tripathy et al. 2021). MYB transcription factors play a critical role during the reaction in plants under stress conditions. MYB-CC transcription factor contains a conserved MYB DNA-binding domain along with a coiled-coil (CC) domain. In maize, ZmMYB-CC10 reduced oxidative damage, thus providing drought tolerance (Xie et al. 2023).

1.5 Signaling as a Defense Response Against Multiple Stresses

1.5.1 Reactive Oxygen Species

When plants detect stress, they quickly generate reactive oxygen species (ROS), which acts as key signaling molecules within the cells (Das and Dutta 2024). Plants carefully control ROS production to prevent cellular damage. Although ROS are often viewed as harmful under stress, lower levels of ROS can facilitate stress responses effectively (Mittler et al. 2022). ROS play a role in cell-to-cell contact through intensifying signals by the transmembrane protein "Respiratory Burst

Table 1.2 Genes and their defensive role against abiotic stress. (Wang et al. 2021a, b, c)

Crop	Gene	Stress tolerance	References
Rice	*OsSRL1, OsSRL2, OsERA1, OsmiR535*	Drought	Wang et al. (2021a, b, c)
	OsDST	Drought and salinity	Zhang et al. (2022)
	OsRR22, OsHDA710, OsNAC04, OsSDG721	Salinity	Mi et al. (2023) and Zhao et al. (2022)
Wheat	*TaG6PDH, Ta6PGDH*	Cold	Rasheed et al. (2022)
	TaCLPB, TaHsfA2	Heat	Tian et al. (2021) and Erdayani et al. (2020)
	TaHSP23.9	Heat and salt	Guo et al. (2020)
	TaDREB2, TaERF3	Drought	Wang et al. (2020)
	TaHsfA2e-5D	Drought and heat	Trono and Pecchioni (2022)
	TabZIP15	Salinity	Bi et al. (2022)
Maize	*ZmICE1*	Cold	Bi et al. (2022)
	ZmHsf01, ZmCDPK7	Heat	Jiang et al. (2022) and Zhao et al. (2021a, b)
	ZmERF21, ZmMYB, ZmbHLH124	Drought	Zhang et al. (2020a, b), Wang et al. (2022) and Sun et al. (2022)
	ZmbZIP111	Salinity	Wei et al. (2021)
Tomato	*SlBBX17*	Heat	Wang et al. (2022)
	SlHSP17.7, VaPYL9, LeCOR413PM2	Cold	Bo et al. (2022), Zhang et al. (2020a, b) and Nai et al. (2022)
	SlNPR1	Drought	Zhang et al. (2021)
	SlHKT1;2, SlHyPRP1	Salinity	Krishna et al. (2022)

Oxidase Homologue D (RBOHD)" and serve as the secondary messengers through changing the structures of proteins and stimulating genes related to protection (Czarnocka and Karpiński 2018).

The ROS plays an important role in enabling cross-tolerance between various stress response pathways. ROS production is detected by transcription factors sensitive to ROS, which then activate stress response genes (Waszczak et al. 2018). ROS promote tolerance by triggering stress response elements such as mitogen-activated protein kinase (MAPKs), antioxidant enzymes, dehydrins, transcription elements, and proteins associated with low temperatures, heat shock, and pathogen defense (Matsuo et al. 2015). Stress tolerance can also be enhanced through chemical priming that modifies ROS signaling. For instance, applying brassinosteroids in cucumber plants boosts H_2O_2 levels, thereby priming them for enhanced tolerance against abiotic stress (Gechev et al. 2005). Similarly, priming with H_2O_2 for salt tolerance leads to moderately increased levels of S-nitrosylated and oxidized proteins in citrus plants, which remain stable under stress conditions, unlike in untreated plants that exhibit greater stress sensitivity (Cui et al. 2011).

To counteract the buildup of ROS, plants developed a refined scavenging system having both nonenzymatic and enzymatic mechanisms (Fig. 1.6). Nonenzymatic antioxidant metabolites include flavonoids, carotenoids, betalains, and vitamin E (Groß et al. 2013). Key enzymes in the ROS scavenging system include catalases (CATs), superoxidase dismutases (SODs), and numerous peroxidases (PODs) (Khan 2016). Superoxidase dismutases (SODs) convert $O_2^{\bullet-}$ into H_2O_2, which CATs and PODs then reduce to water (Tavanti et al. 2021). The ascorbate-glutathione (ASA-GSH) cycle, the enzymes dehydroascorbate reductase (DHAR), ascorbate peroxidase (APX), glutathione reductase (GR), and monodehydroascorbate reductase (MDHAR) also play an important role (Mishra and Sharma 2019). Additional detoxifying enzymes include ferritins and glutathione S-transferase (GST). Engineering these detoxifying enzymes could boost multi-stress tolerance in plants (Sano 2017). For example, overexpressing APXs in transgenic rice, mustard, and peanut plants has improved drought resistance, salinity resistance, and resistance to high light (He et al. 2018). Similarly, GSTs, which also possess glutathione peroxidase (GPX) activity, revealed increased tolerance to cold, heat, heavy metals, paraquat, and ultraviolet (UV) light (Gill and Tuteja 2010).

1.5.2 *Mitogen-Activated Protein Kinase*

Upon detecting stress, the flow of mitogen-activated protein kinase (MAPK) is stimulated and serves a vital role in regulatory pathways of stress response (Zhang and Zhang 2022). These mitogens activate protein kinase, which is highly conserved across eukaryotes and is critical for signal transduction in various processes of the cell under different abiotic stresses (Taj et al. 2010). MAPKs like MPK3, MPK4, and MPK6 respond to numerous abiotic stresses, playing a critical role in operating cross-tolerance between stress reactions (Kmiecik et al. 2016). MPK3 and MPK6 are particularly important for fully primed defense responses, suggesting they are crucial for mediating tolerance against subsequent abiotic stresses (Wang et al. 2023). Overexpression of the gene OsMPK5 and abscisic acid (ABA)-induced kinase activity contribute to increased drought, salinity, and cold tolerance in rice. MPK1 and MPK2 are stimulated in plants in response to ultraviolet-B radiation, wounding, and pathogens, enhancing the defense responses. MAPK signaling cooperates with ROS and ABA signaling pathways too, leading to the improvement of plant defense and initiating cross-acclimation to both stresses (Kumar et al. 2019).

1.5.3 *Hormonal Response*

Specific hormones regulate various types of stress in plants, enabling them to mount defense responses tailored to specific environmental conditions. Abscisic acid (ABA) is counted as a primary hormone implicated in the perception and response to different abiotic stresses (Dar et al. 2017). Increased concentrations of ABA control the abiotic stress response network. When plants first encounter abiotic stress, they maintain water potential through ABA-induced stomatal closure. ABA hormone, along with jasmonic acid (JA), salicylic acid (SA), and ethylene (ET), significantly contributes to the defense response to abiotic stress (Ben Rejeb et al. 2014).

Jasmonic acid with related compounds (jasmonates (JAs)) is especially significant for mediating plant defenses and responses to stress and has been extensively studied. Jasmonates (JAs), including JA and its derivatives, have diverse controlling activities in the growth and development of plants, such as elongation of the axis during embryogenesis, development of flowers, senescence of leaves, formation of roots, and opening of the stomata (Ahmad et al. 2016). Beyond these roles, JAs are critical for the enhancement of plant tolerance to stress through specific signaling pathways in response to numerous stress environments (Wang et al. 2021a, b, c). The pathways are activated by molecular patterns such as herbivore-related molecular patterns (HAMPs), microbe-related molecular patterns (MAMPs), and damage-associated molecular patterns (DAMPs) that originate after attacking organisms, damage to cells, and abiotic stresses (Fig. 1.5) (Abdoolnabi and Fathipour 2021).

The receptor-active conjugated complex of jasmonic acid-isoleucine (JA-Ile) is central to JA signaling pathways. The environmental impulses are conveyed by the surface of the cell receptors, leading to the synthesis of JA-Ile from lipids of the plastid (Wasternack and Kombrink 2010). This, in turn, initiates interactions with downstream transcription factors, promoting plant growth and development and response toward specific protections in the plant. The jasmonic acid in the signaling pathway, JA-Ile, facilitates the interaction between jasmonate zinc-finger (JAZ) proteins and the coronatine insensitive 1 (COI1) protein within the Skp1p-cullin-F-box protein (SCF) complex (Sengupta et al. 2016; Choudhary et al. 2023). This interaction causes the disintegration of JAZ proteins (JAZx, JAZy, JAZz) and activates JAZ-interacting transcription factors (TFs: TFa, TFb, TFc) (Wang et al. 2021a, b, c). These factors regulate gene expression involved in responding to abiotic stresses, thereby enhancing definite protective mechanisms.

1.5.4 Polyamine Response

Polyamines are essential organic compounds that help plants respond to drought, extreme temperature, cold, salt stress, and other abiotic stressors. Important polyamines include spermidine, putrescine, and spermine, which are involved in numerous physiological processes that aid in stress adaptation (Yadav et al. 2024). Polyamines stabilize cellular structures and help maintain cell turgor during stress by acting as osmoprotectants (Sengupta et al. 2016). They enhance the production of osmolytes that alleviate dehydration effects. These compounds exhibit antioxidant properties, effectively scavenging ROS that accumulate during stressful conditions. By minimizing oxidative damage, polyamines protect key cell components such as membranes, proteins, and DNA (Choudhary et al. 2023). Polyamines can

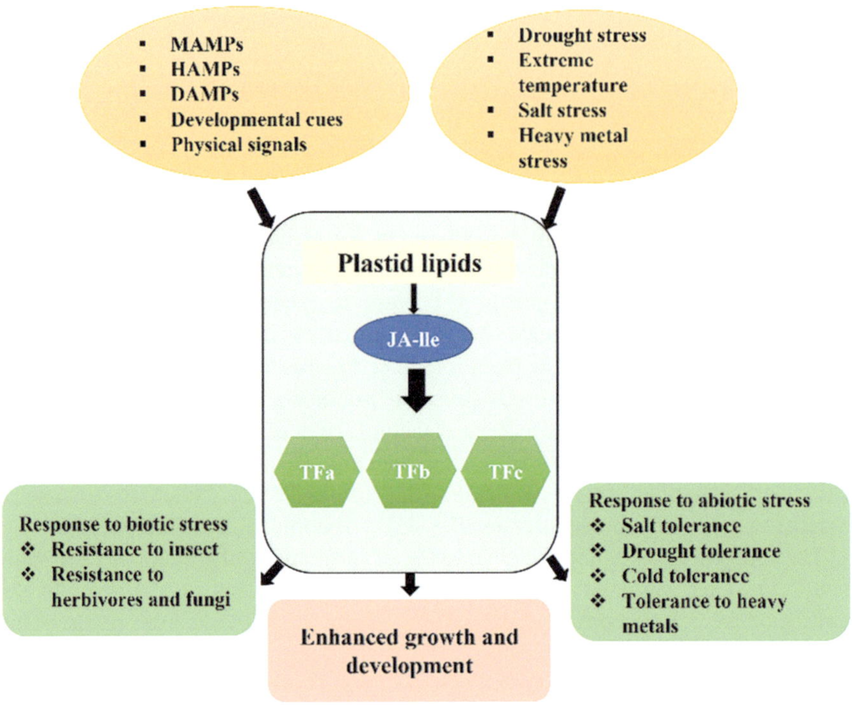

Fig. 1.5 Endogenous and exogenous jasmonic acid regulation mechanism in response to abiotic stresses. (Ben Rejeb et al. 2014; Wang et al. 2021a, b, c)

affect gene expression associated with stress responses of plants. They often serve as the signaling molecules, which regulate the gene expression related to stress tolerance, with those encoding protective enzymes and proteins (Lenis et al. 2017).

Under stress, polyamines promote the regulation of cell growth and cell division of plants and play a role in synthesizing proteins and nucleic acids, which are critical for the function and repair of the cell. Polyamines help to maintain the integrity of the cell membrane and fluidity under stress conditions and are essential for proper cellular function and communication. These compounds can trigger various stress-responsive pathways, including the production of stress hormones like ABA, which is critical for responses to drought and salinity (Kuznetsov and Shevyakova 2007). Polyamines also interact with other signaling molecules in plants, such as NO and H_2O_2, to bolster stress tolerance mechanisms (Pál et al. 2015). Overall, polyamines are crucial for conferring plants the ability to withstand abiotic stress by enhancing their resilience and adaptability. Ongoing research aims to uncover the specific pathways and mechanisms through which polyamines exert their protective effects, offering valuable insights for improving crop stress tolerance through breeding and biotechnological methods.

1.5.5 Heat Shock Proteins (HSPs)

The exact mechanism by which ROS are recognized by heat shock factor (HSF) proteins is not fully understood. It is hypothesized that HSFs respond to H_2O_2, as HSF trimers are stabilized by H_2O_2 through reversible oxidation of the residues of cysteine by forming Cys-Cys bonds (Spormann et al. 2021). Research has shown that the activity of HSFA4A and HSFA8, two HSFs, is influenced by the REDOX state of the cytoplasm, suggesting that these factors might act as sensors for cytoplasmic hydrogen peroxide during stress (Yurina 2023).

HSFs across different plant species can activate the gene expression that encodes antioxidant enzymes (SOD, APX, catalase) or increase glutathione levels (Fortunato et al. 2023). For example, HSFA4A can upregulate other transcription factors responsible for the protection against oxidative stress (Driedonks et al. 2015). HSFA2 has been shown to regulate ascorbate peroxidase 1 (APX1) expression, with HSFA2 overexpression leading to increased transcription of APX1, while the defective mutant of the HSFA2 gene shows reduced synthesis of APX1 (Jacob et al. 2017). This aligns with findings that Arabidopsis lines overexpressing APX1 exhibit improved tolerance to oxidative and thermal stress. Therefore, the expression of ROS-induced HSF contributes to increased resistance of plants against abiotic stress and reduces hydrogen peroxide levels through ROS detoxification, triggering a feedback mechanism that regulates HSF expression (Wang et al. 2016).

The interaction between HSF regulation and REDOX control may enhance cross-acclimatization to various stress conditions. HSF activity is regulated at the transcriptional and posttranscriptional levels. Alternative splicing can produce different isoforms of HSFs during transcriptional regulation (Rosenkranz et al. 2022). For example, in tomato plants, HSFA2 is found in two isoforms: HSFA2-II, which governs the early stress response, and HSFA2-I, which is involved in acclimatization and thermotolerance (Mesihovic et al. 2022). The activity of HSF proteins and their localization and metabolism are influenced by these modifications. Numerous MAPKs mediate ROS signals and phosphorylate plant transcription factors (TFs). For instance, phosphorylation of HSFA2 by MPK6 kinase in Arabidopsis facilitates its nuclear translocation and transcription activation under heat stress conditions (Evrard et al. 2013). The attachment of small ubiquitin-like modifiers (SUMO) affects the activity of eukaryotic proteins and their localization and stability (Nabil Elrouby 2015). SUMOYlated HSFs such as HSFA2, HSFB2B, and HSFC1 have been identified in Arabidopsis under heat stress (Guo et al. 2016). Sumoylation of HSFA2, in particular, was found to reduce its ability to activate heat shock protein (HSP) transcription.

Thus, HSFs are part of a multilayered regulatory complex that orchestrates plants' responses to environmental stress. The extensive range of the HSF gene and its varied functions likely reflect evolutionary adaptations that allowed plants to thrive on land. Over time, this gene family has diversified significantly, enabling plants to develop resistance toward a wide range of stresses and integrate these HSFs into different regulatory pathways (REF).

1.5.6 Osmolytes

The accumulation of osmotin occurs in plants under both stressed and nonstressed conditions. When there is no stress, osmotin functions as a housekeeping protein, supporting basic cellular metabolism (Bashir et al. 2020). This multidomain protein has various domains that each perform specific functions, either independently or in collaboration with other domains. A key domain of osmotin is similar to protein kinases, which are essential in eukaryotic cells both for signaling and regulatory processes, including enzyme activation, macromolecule assembly, protein localization, and degradation (Ullah et al. 2018).

Osmotin acts as an osmoprotectant in abiotic-stressed conditions (Fig. 1.6). Under salt stress, osmotin was first found to be induced by low water potential (Hao et al. 2021). Mostly stimulated osmotin and osmotin-like proteins (OLPs) were detected under different abiotic stresses, including drought. Osmotin protects enzymes and enhances the function of chaperone proteins, aiding in the stabilization and proper folding of proteins under stressful conditions like drought, extreme temperatures, and salinity (Rasheed et al. 2024). The above issues are important since abiotic stress can cause denaturation of proteins. The osmotin shields cells from the harmful effects of stress by facilitating correct protein folding and preventing aggregation, enhancing plant resilience in the changing environments (Zulfiqar et al. 2019).

Additionally, osmotin helps to adjust the osmotic potential of plant cells by accumulating or compartmentalizing solutes through metabolic and structural changes

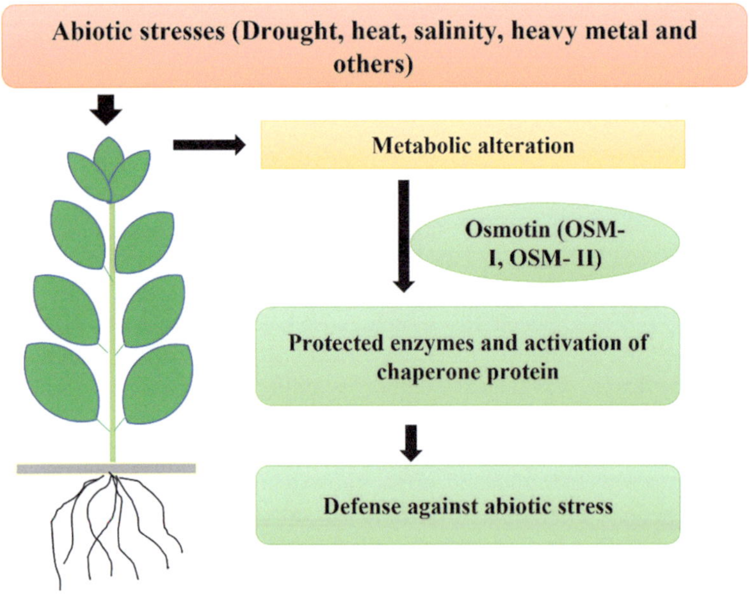

Fig. 1.6 Defense response of osmotin during abiotic stresses. (Bashir et al. 2020)

during osmotic adjustments (Mansour and Salama 2019). During abiotic stress conditions, osmotin plays a vital role in plants' immune systems. Osmotin also activates defense mechanisms by functioning as a transcriptional regulator under stress conditions, influencing the gene expression related to numerous stress-responsive enzymes. On top of that, osmotin acts as a signaling molecule through intracellular receptors. It also protects native proteins under abiotic stresses, aids in repairing denatured proteins (Dharshini et al. 2020), and helps to stabilize cellular structures and functions.

1.5.7 Secondary Metabolites

To manage abiotic stresses, such as drought, extreme temperatures, and salinity, secondary metabolites like phenolics, glycosides, alkaloids, terpenoids, and nitrogen compounds are essential. (Jan et al. 2021). Phenolic compounds are essential for helping plants tolerate abiotic stress. Phenolics enhance oxidative stress resistance by scavenging ROS and protect plants from UV light, thus protecting cellular structures (Hajam et al. 2023). These compounds absorb UV radiation, safeguarding plants from its harmful effects and improving survival in bright conditions (Singh et al. 2020). During drought, phenolics help plants maintain osmotic potential, allowing plants to keep turgor pressure. They are involved in signaling pathways that activate genes related to stress response, enhancing overall resilience. By contributing to lignin production, phenolics increase cell wall rigidity, boosting structural integrity under stress.

Another secondary compound, an alkaloid, may function as a signaling molecule and is associated with improved stress tolerance during stress responses. Many alkaloids show antioxidant effects, helping to neutralize ROS generated during stress, thus safeguarding cellular components from damage (Hasanuzzaman et al. 2020). Alkaloids activate specific stress-responsive genes and help maintain osmotic balance within cells, aiding water retention during drought and enhancing hydration. In saline conditions, alkaloids assist in regulating ion transport, mitigating the toxic effects of excess sodium, and promoting plant health (Ishtiyaq et al. 2021). They can influence metabolic pathways that boost energy production and stress adaptation, helping plants cope with adverse conditions.

Terpenoids, another essential secondary metabolite, help minimize water loss by affecting stomatal closure, thereby enhancing drought tolerance. They are vital for stabilizing cellular membranes and protecting plants from damage caused by extreme temperatures (Yadav et al. 2021). Many terpenoids exhibit antioxidant properties to reduce oxidative stress during challenging conditions. Terpenoids activate specific defense pathways to modulate responses of plants under abiotic stress. Terpenoids also influence the growth and development of plants under stressful environments by promoting survival and adaptation. All of these secondary metabolites are crucial for improving the tolerance of plants under various abiotic stresses through protective, signaling, and different physiological mechanisms.

1.6 Molecular Defense Response Against Stress

Gene expression changes the stress detection process by the reprogramming of molecular machinery regulated by transcription factors (He et al. 2020). This reprogramming is crucial for establishing an effective defense mechanism in plants, as many genes play multifunctional roles and can confer tolerance to multiple stresses (Hrmova and Hussain 2021). These defense-related genes are regulated by definite phytohormones, for example, abscisic acid, salicylic acid, jasmonic acid, and ethylene. For instance, salicylic acid (SA) pretreatment enhances the resilience of eggplant seedlings against oxidative damage from cold stress by boosting the activity of antioxidant enzymes and the expression of genes (Altaf et al. 2023). Additionally, acetyl-SA application increased catalase activity and reduced H_2O_2 levels in potatoes under extreme heat stress. Salicylic acid also mitigates damage of oxidative stress through scavenging of excess reactive oxygen species, thereby protecting the membranes of plant cells, increasing chlorophyll content, and lowering malondialdehyde and H_2O_2 levels in tomatoes exposed to heat stress. Jasmonic acid (JA) reduces oxidative damage by decreasing expression of respiratory burst oxidase (RBO) and P-type ATPase transporter genes (Jahan et al. 2019). Additionally, jasmonic acid controls the activity of the antioxidant system and contributes to enhanced plant stress tolerance.

Interestingly, many pathogenesis-related (PR) genes are also activated in response to stress conditions, contributing to resistance against diseases. The pathogenesis-proteins play an important role in the defense system of plants against pathogens, and expression of the gene is significantly upregulated during pathogen attacks (Ahmad et al. 2022). For instance, overexpression of specific transcription factors in plants subjected to cold stress and infection activates cold-responsive PR-genes that protect plants from both stresses (Manasa et al. 2022). Additionally, induction of certain transcription factors under abiotic stress leads to an increase in PR-protein levels. Abiotic stress triggers the expression of some transcription factors (TFs) such as C-repeat binding factors, No apical meristem (NAC), ATAF, CUP-SHAPED COTYLEDON, and MYB, which are mediated by abscisic acid (ABA). Remarkably, these transcription factors have been widely studied for their role in primary signaling pathways under abiotic stress that are essential for tolerance under stress (Singh and Chandra 2021). CBF, along with several PR proteins, is induced under cold stress conditions. The DREB transcription factor can mediate cross-tolerance between biotic and abiotic stresses, thereby improving the resistance of Arabidopsis to cold (Liu et al. 2024). Recent research has suggested that the WHIRLY1 protein and REDOX-Responsive Transcription Factor1 (RRTF1) can play roles in communicating between the nucleus and plastids (Karpinska et al. 2022). WHIRLY1 detects REDOX changes in plastids and transmits this information to the nucleus independently of NPR1, potentially aiding in acclimation and adaptation to new stresses (Sevilla et al. 2023). Similarly, REDOX-Responsive Transcription Factor1 induced by abiotic stresses may prime distant leaves to better defend against subsequent stresses (Ramesh et al. 2020).

1.6.1 *Genes to Overcome Abiotic Stress*

Abiotic stresses significantly impact crop yields and agricultural productivity. These stresses contribute to substantial reductions in farm outputs and economic losses. Researchers are increasingly focusing on the comprehension of plant metabolic and genetic responses to environmental changes, including during cell division and at the gene level. To enhance water-use efficiency in plant cells, it is crucial to develop strategies that optimize available water resources. Identification and isolation of genes responsible for stress tolerance using transcriptomic techniques could facilitate the development of transgenic plants. For example, overexpression of the drought-responsive gene CaDSR6 from *Capsicum annuum* in Arabidopsis has been shown to enhance tolerance to drought compared to wild-type plants (Islam et al. 2021). Additionally, genes involved in stress responses to NaCl, such as SNAC1, regulate signaling pathways related to protein phosphates, phosphatidylinositol-3-phosphate-5-kinase, and abscisic acid receptors, contributing to drought tolerance in wheat (Trono and Pecchioni 2022). Abiotic stress can adversely affect photosynthetic machinery, including photosystems and pigments, which leads to changes in photosynthetic gene expression. For instance, the violaxanthin deepoxidase gene (VDE), which regulates temperature response, has been shown to enhance nonphotochemical quenching, the Fv/Fm ratio, and the action of the xanthophyll cycle, thereby reducing photoinhibition in photo system I (PSI) and photo system II (PSII) under cold stress (Wang et al. 2021a, b, c).

1.7 Conclusion and Prospects

Understanding plant defense mechanisms is paramount for sustainable agriculture, as it enables the development of crop varieties resilient to environmental changes and capable of withstanding them, while minimizing the need for chemical inputs. By unraveling the molecular basis of the responses against stress, researchers can detect targets for genetic engineering and genome editing technologies, leading to stress-tolerant crop breeding with improved yield and quality. Future research in plant stress responses should focus on elucidating the complex interactions and regulatory networks governing stress signaling pathways and exploring novel approaches for boosting plant resilience to multiple stresses. Integration of omics technologies, systems biology approaches, and computational modeling will facilitate the analysis of comprehensive stress-responsive gene expression, protein dynamics, and metabolic pathways, providing insights into the mechanisms related to stress tolerance and adaptation. Using genome editing technology such as CRISPR-Cas9, which has been widely used for various applications, including mutation detection, gene modification, and functional gene analysis (Biswas et al. 2021), will allow for gene pyramiding, knockout of genes, delivery of proteins, and gene expression regulation, as well as epigenomic modification in various organisms.

References

Abdoolnabi B, Fathipour Y (2021) Induced resistance and defense primings. In: Omkar (ed) Molecular approaches for sustainable insect pest management. Springer, Singapore, pp 73–139. https://doi.org/10.1007/978-981-16-3591-5_3

Ahmad P, Rasool S, Gul A, Sheikh SA, Akram NA, Ashraf M, Kazi AM, Gucel S (2016) Jasmonates: multifunctional roles in stress tolerance. Front Plant Sci 7. https://doi.org/10.3389/fpls.2016.00813

Ahmad M, Ali Q, Hafeez MM, Malik A (2021) Improvement for biotic and abiotic stress tolerance in crop plants. Biol Clin Sci Res J 1:2021. https://doi.org/10.54112/bcsrj.v2021i1.50

Ahmad A, Wang R, Mubeen S, Akram W, Hu D, Yasin NA, Khan M, Wu T (2022) Comparative transcriptomics reveals defense acquisition in Brassica rapa by synchronizing brassinosteroids metabolism with PR1 expression. Eur J Plant Pathol 162(4):869–884. https://doi.org/10.1007/s10658-021-02443-0

Ali S, Rizwan M, Saleem Arif M, Ahmad R, Hasanuzzaman M, Ali B, Hussain a (2020) Approaches in enhancing Thermotolerance in plants: an updated review. J Plant Growth Regul 39. https://doi.org/10.1007/s00344-019-09994-x

Alsafran M, Hamzah Saleem M, Aljabri H, Rizwan M, Usman K (2022) Principles and applicability of integrated remediation strategies for heavy metal removal/recovery from contaminated environments. J Plant Growth Regul 42. https://doi.org/10.1007/s00344-022-10803-1

Altaf MA, Shahid R, Kumar R, Altaf MM, Kumar A, Khan LU, Saqib M, Nawaz MA, Saddiq B, Bahadur S, Tiwari RK (2023) Phytohormones mediated modulation of abiotic stress tolerance and potential crosstalk in horticultural crops. J Plant Growth Regul 42(8):4724–4750. https://doi.org/10.1007/s00344-022-10812-0

Amer R (2021) Spatial relationship between irrigation water salinity, waterlogging, and cropland degradation in the arid and semi-arid environments. Remote Sens 13(1047):2021. https://doi.org/10.3390/rs13061047

Arif Y, Singh P, Siddiqui H, Bajguz A, Hayat S (2020) Salinity induced physiological and biochemical changes in plants: an omic approach towards salt stress tolerance. Plant Physiol Biochem 156:64–77. https://doi.org/10.1016/j.plaphy.2020.08.042

Ashapkin VV, Kutueva LI, Aleksandrushkina NI, Vanyushin BF (2020) Epigenetic mechanisms of plant adaptation to biotic and abiotic stresses. Int J Mol Sci 21(20):2020. https://doi.org/10.3390/ijms21207457

Ault TR (2020) On the essentials of drought in a changing climate. Science 368(6488):256–260. https://doi.org/10.1126/science.aaz5492

Bashir MA, Silvestri C, Ahmad T, Hafiz IA, Abbasi NA, Manzoor A, Cristofori V, Rugini E (2020) Osmotin: a cationic protein leads to improve biotic and abiotic stress tolerance in plants. Plants 9(8):2020. https://doi.org/10.3390/plants9080992

Ben Rejeb I, Pastor V, Mauch-Mani B (2014) Plant responses to simultaneous biotic and abiotic stress: molecular mechanisms. Plants 3(4):458–475. https://doi.org/10.3390/plants3040458

Bharath P, Gahir S, Raghavendra AS (2021) Abscisic acid-induced stomatal closure: an important component of plant defense against abiotic and biotic stress. Front Plant Sci 12. https://doi.org/10.3389/fpls.2021.615114

Bhattacharya A (2022) Effect of low-temperature stress on germination, growth, and phenology of plants: a review. In: Physiological processes in plants under low temperature stress. Springer, Singapore, pp 1–106. https://doi.org/10.1007/978-981-16-9037-2_1

Bi H, Miao J, He J, Chen Q, Qian J, Li H, Xu Y (2022) Characterization of the wheat heat shock factor TaHsfA2e-5D conferring heat and drought tolerance in Arabidopsis. Int J Mol Sci 23(5):2022. https://doi.org/10.3390/ijms23052784

Biswas D, Saha SC, Dey A (2021) CRISPR-Cas genome-editing tool in plant abiotic stress tolerance. PlantGene 26:100286. https://doi.org/10.1016/j.plgene.2021.100286

Bo C, Cai R, Fang X, Wu H, Ma Z, Yuan H, Cheng B, Fan J, Ma Q (2022) Transcription factor ZmWRKY20 interacts with ZmWRKY115 to repress expression of ZmbZIP111 for salt tolerance in maize. Plant J 111(6):1660–1675. https://doi.org/10.1111/tpj.15914

Branch HA, Moxley DR, Angert AL (2024) Regional differences in leaf evolution facilitate photosynthesis following severe drought. New Phytol 243(6):24572469. https://doi.org/10.1111/nph.19963

Chaudhry S, Sidhu GPS (2022) Climate change regulated abiotic stress mechanisms in plants: a comprehensive review. Plant Cell Rep 41(1):1–31. https://doi.org/10.1007/s00299-021-02759-5

Chen M, Zhang TL, Hu CG, Zhang JZ (2023) The role of drought and temperature stress in the regulation of flowering time in annuals and perennials. Agronomy 13(12):2023. https://doi.org/10.3390/agronomy13123034

Chen Y, Yue XL, Feng JY, Gong X, Zhang WJ, Zuo JF, Zhang YM (2024) Identification of QTNs, QTN-by-environment interactions, and their candidate genes for salt tolerance related traits in soybean. BMC Plant Biol 24(1):316. https://doi.org/10.1186/s12870-024-05021-8

Choudhary S, Wani KI, Naeem M, Khan MMA, Aftab T (2023) Cellular responses, osmotic adjustments, and role of Osmolytes in providing salt stress resilience in higher plants: polyamines and nitric oxide crosstalk. J Plant Growth Regul 42(2):539–553. https://doi.org/10.1007/s00344-022-10584-7

Cui JX, Zhou YH, Ding JG, Xia XJ, Shi KAI, Chen SC, Asami T, Chen Z, Yu JQ (2011) Role of nitric oxide in hydrogen peroxide-dependent induction of abiotic stress tolerance by brassinosteroids in cucumber. Plant Cell Environ 34(2):347–358. https://doi.org/10.1111/j.1365-3040.2010.02248.x

Czarnocka W, Karpiński S (2018) Friend or foe? Reactive oxygen species production, scavenging and signaling in plant response to environmental stresses. Free Radic Biol Med 122:4–20. https://doi.org/10.1016/j.freeradbiomed.2018.01.011

Dar NA, Amin I, Wani W, Wani SA, Shikari AB, Wani SH, Masoodi KZ (2017) Abscisic acid: a key regulator of abiotic stress tolerance in plants. Plant Gene 11:106–111. https://doi.org/10.1016/j.plgene.2017.07.003

Das PK, Dutta D (2024) Application of CRISPR technology in plant salt tolerance and membrane proteins. In: Basics of CRISPR/Cas mediated Plant genome editing. CRC Press, p 24

DeTurk EE (1941) Plant nutrient deficiency symptoms. Physiological basis. Ind Eng Chem 33(5):648–653. https://doi.org/10.1021/ie50377a022

Dharshini S, Hoang NV, Mahadevaiah C, Padmanabhan TS, Alagarasan G, Suresha GS, Kumar R, Meena MR, Ram B, Appunu C (2020) Root transcriptome analysis of Saccharum spontaneum uncovers key genes and pathways in response to low-temperature stress. Environ Exp Bot 171:103935. https://doi.org/10.1016/j.envexpbot.2019.103935

Dresselhaus T, Hückelhoven R (2018) Biotic and abiotic stress responses in crop plants, Nov. 19, 2018, MDPI Agriculture. https://doi.org/10.3390/agronomy8110267

Driedonks N, Xu J, Peters JL, Park S, Rieu I (2015) multi-level interactions between heat shock factors, heat shock proteins, and the Redox system regulate acclimation to heat. Front Plant Sci 6. https://doi.org/10.3389/fpls.2015.00999

Elrouby N (2015) Analysis of small ubiquitin-like modifier (SUMO) targets reflects the essential nature of protein SUMOylation and provides insight to elucidate the role of SUMO in plant development. Plant Physiol 169(2):1006–1017. https://doi.org/10.1104/pp.15.01014

Erdayani E, Nagarajan R, Grant NP, Gill KS (2020) Genome-wide analysis of the HSP101/CLPB gene family for heat tolerance in hexaploid wheat. Sci Rep 10(1):3948. https://doi.org/10.1038/s41598-020-60673-4

Etesami H, Jeong BR, Glick BR (2023) Potential use of bacillus spp. as an effective biostimulant against abiotic stresses in crops–a review. Curr Res Biotechnol 5:100128. https://doi.org/10.1016/j.crbiot.2023.100128

Evrard A, Kumar M, Lecourieux D, Lucks J, Koskull-Döring P, Hirt H (2013) Regulation of the heat stress response in Arabidopsis by MPK6-targeted phosphorylation of the heat stress factor HsfA2. PeerJ 1:e59. https://doi.org/10.7717/peerj.59

Fortunato S, Lasorella C, Dipierro N, Vita F, de Pinto MC (2023) Redox signaling in plant heat stress response. Antioxidants 12(3):2023. https://doi.org/10.3390/antiox12030605
Gechev T, Gadjev I, Dukiandjiev S, Minkov I (2005) Reactive oxygen species as signaling molecules controlling stress adaptation in plants. In: Handbook of photosynthesis, 2nd edn. CRC, Boca Raton, pp 209–218
Ghoran SH, Saed-Moucheshi A, Mozafari AA, Pessarakli M (2024) Association of oxygen free radicles with photosynthetic process. Handbook Photo. CRC Press, New York, pp 142–153. https://doi.org/10.1201/b22922
Gill SS, Tuteja N (2010) Reactive oxygen species and antioxidant machinery in abiotic stress tolerance in crop plants. Plant Physiol Biochem 48(12):909–930. https://doi.org/10.1016/j.plaphy.2010.08.016
Giordano M, Petropoulos SA, Rouphael Y (2021) Response and defence mechanisms of vegetable crops against drought, heat and salinity stress. Agriculture 11(5):2021. https://doi.org/10.3390/agriculture11050463
Groß F, Durner J, Gaupels F (2013) Nitric oxide, antioxidants and prooxidants in plant defense responses. Front Plant Sci 4. https://doi.org/10.3389/fpls.2013.00419
Grulke NE, Heath RL (2020) Ozone effects on plants in natural ecosystems. Plant Biol 22(S1):12–37. https://doi.org/10.1111/plb.12971
Guo M, Liu JH, Ma X, Luo DX, Gong ZH, Lu MH (2016) The Plant heat stress transcription factors (HSFs): structure, regulation, and function in response to abiotic stresses. Front Plant Sci 7:2016. https://doi.org/10.3389/fpls.2016.00114
Guo X, Yuan SN, Zhang HN, Zhang YY, Zhang YJ, Wang GY, Li YQ, Li G (2020) Heat-response patterns of the heat shock transcription factor family in advanced development stages of wheat (*Triticum aestivum* L.) and thermotolerance-regulation by TaHsfA2–10. BMC Plant Biol 20(1):364. https://doi.org/10.1186/s12870-020-02555-5
Guzmán MG, Cellini F, Fotopoulos V, Balestrini R, Arbona V (2022) New approaches to improve crop tolerance to biotic and abiotic stresses. Physiol Plant 174(1):e13547. https://doi.org/10.1111/ppl.13547
Hajam YA, Lone R, Kumar R (2023) Role of plant Phenolics against reactive oxygen species (ROS) induced oxidative stress and biochemical alterations. In: Lone R, Khan S, Mohammed Al-Sadi A (eds) Plant Phenolics in abiotic stress management. Springer, Singapore. https://doi.org/10.1007/978-981-19-6426-8_7
Hao S, Wang Y, Yan Y, Liu Y, Wang J, Chen S (2021) A review on plant responses to salt stress and their mechanisms of salt resistance. Horticulturae 7:132. https://doi.org/10.3390/horticulturae7060132
Haraldsen TK (2022) Flood damage on agricultural land and methods for restoration of agricultural soils after catastrophic floods in cold areas. In: Ferreira TM, Shi H (eds) Flood risk in a climate change context. IntechOpen., ch. 5, Rijeka. https://doi.org/10.5772/intechopen.109111
Hasanuzzaman M, Bhuyan MHMB, Anee TI, Parvin K, Nahar K, Mahmud JA, Fujita M (2019) Regulation of ascorbate-glutathione pathway in mitigating oxidative damage in plants under abiotic stress. Antioxidants (Basel) 8(9):384. https://doi.org/10.3390/antiox8090384
Hasanuzzaman M, Bhuyan MB, Zulfiqar F, Raza A, Mohsin SM, Mahmud JA, Fujita M, Fotopoulos V (2020) Reactive oxygen species and antioxidant defense in plants under abiotic stress: revisiting the crucial role of a universal defense regulator. Antioxidants 9(8):2020. https://doi.org/10.3390/antiox9080681
He M, He CQ, Ding NZ (2018) abiotic stresses: general defenses of land plants and chances for engineering multistress tolerance. Front Plant Sci 9. https://doi.org/10.3389/fpls.2018.01771
He F, Ru X, Wen T (2020) NRF2, a transcription factor for stress response and beyond. Int J Mol Sci 21(13):2020. https://doi.org/10.3390/ijms21134777
Hoshika Y, Fares S, Pellegrini E, Conte A, Paoletti E (2020) Water use strategy affects avoidance of ozone stress by stomatal closure in Mediterranean trees—a modelling analysis. Plant Cell Environ 43(3):611–623. https://doi.org/10.1111/pce.13700

Hrmova M, Hussain SS (2021) Plant transcription factors involved in drought and associated stresses. Int J Mol Sci 22(11):2021. https://doi.org/10.3390/ijms22115662
Hsu PK, Dubeaux G, Takahashi Y, Schroeder JI (2021) Signaling mechanisms in abscisic acid-mediated stomatal closure. Plant J 105(2):307–321. https://doi.org/10.1111/tpj.15067
Huang S, Zuo T, Ni W (2020) Important roles of glycinebetaine in stabilizing the structure and function of the photosystem II complex under abiotic stresses. Planta 251(2):36. https://doi.org/10.1007/s00425-019-03330-z
Hussain J, Bibi G, Shafiq S (2020) Calcium dependent protein kinases in plants. In: Protein kinases and stress signaling in plants. Wiley. ch. 8, pp 192–225. https://doi.org/10.1002/9781119541578.ch8
Imran QM, Falak N, Hussain A, Mun BG, Yun BW (2021) Abiotic stress in plants; stress perception to molecular response and role of biotechnological tools in stress resistance. Agronomy 11(8):2021. https://doi.org/10.3390/agronomy11081579
Iqbal A, Fahad S, Iqbal M, Alamzeb M, Ahmad A, Anwar S, Khan AA, Amanullah, Arif M, Inamullah, Shaheenshah (2020) Special adaptive features of plant species in response to drought," in Salt and drought stress tolerance in plants: signaling networks and adaptive mechanisms, M Hasanuzzaman Mirza and Tanveer (Eds.), Cham: Springer, pp. 77–118. doi:https://doi.org/10.1007/978-3-030-40277-8_4
Ishtiyaq S, Kumar H, Varun M, Ogunkunle CO, Paul MS (2021) Role of secondary metabolites in salt and heavy metal stress mitigation by halophytic plants: an overview. In: Hasanuzzaman M, Prasad MNV (eds) Handbook of bioremediation. Academic, pp 307–327. https://doi.org/10.1016/B978-0-12-819382-2.00020-X
Islam W, Sanaullah T, Khalid N, Aqeel M, Noreen S, Irshad MK, Ali N (2021) Plants, environmental constraints, and climate change. In: Climate change and plants. CRC Press, p 22
Jacob P, Hirt H, Bendahmane A (2017) The heat-shock protein/chaperone network and multiple stress resistance. Plant Biotechnol J 15(4):405–414. https://doi.org/10.1111/pbi.12659
Jahan MS, Wang Y, Shu S, Zhong M, Chen Z, Wu J, Sun J, Guo S (2019) Exogenous salicylic acid increases the heat tolerance in tomato (*Solanum lycopersicum* L) by enhancing photosynthesis efficiency and improving antioxidant defense system through scavenging of reactive oxygen species. Sci Hortic 247:421–429. https://doi.org/10.1016/j.scienta.2018.12.047
Jan R, Asaf S, Numan M, Lubna KKM (2021) Plant secondary metabolite biosynthesis and transcriptional regulation in response to biotic and abiotic stress conditions. Agronomy 11(5):2021. https://doi.org/10.3390/agronomy11050968
Jasleen RK, Kariyat (2020) Role of Trichomes in plant stress biology. In: Núñez-Farfán Juan PL, Valverde (eds) Evolutionary ecology of plant-herbivore interaction. Springer, Cham, pp 15–35. https://doi.org/10.1007/978-3-030-46012-9_2
Jiang H, Shi Y, Liu J, Li Z, Fu D, Wu S, Li M, Yang Z, Shi Y, Lai J, Yang X, Gong Z, Hua J, Yang S (2022) Natural polymorphism of ZmICE1 contributes to amino acid metabolism that impacts cold tolerance in maize. Nat Plants 8(10):1176–1190. https://doi.org/10.1038/s41477-022-01254-3
Jomova K, Alomar SY, Alwasel SH, Nepovimova E, Kuca K, Valko M (2024) Several lines of antioxidant defense against oxidative stress: antioxidant enzymes, nanomaterials with multiple enzyme-mimicking activities, and low-molecular-weight antioxidants. Arch Toxicol 98:1323–1367. https://doi.org/10.1007/s00204-024-03696-4
Karpinska B, Razak N, James EK, Morris JA, Verrall SR, Hedley PE, Hancock RD, Foyer CH (2022) WHIRLY1 functions in the nucleus to regulate barley leaf development and associated metabolite profiles. Biochem J 479(5):641–659. https://doi.org/10.1042/BCJ20210810
Kashtoh H, Baek KH (2021) Structural and functional insights into the role of guard cell ion channels in abiotic stress-induced stomatal closure. Plants 10(12):2021. https://doi.org/10.3390/plants10122774
Kashyap VH, Kohli I, Singh A, Bhattacharya A, Singh PK, Varma A, Joshi NC (2021) Physiological, biochemical, and morphological approaches to mitigate the effects of abiotic stress in plants. In: Rai AC, Rai A, Rai KK, Rai VP, Kumar A (eds) Stress tolerance in

horticultural crops. Woodhead Publishing, pp 193–212. https://doi.org/10.1016/B978-0-12-822849-4.00011-5
Kaur G, Singh G, Motavalli P, Nelson K, Orlowski J, Golden B (2020) Impacts and management strategies for crop production in waterlogged or flooded soils: a review. Agron J 112:2020. https://doi.org/10.1002/agj2.20093
Kayacetin F (2021) Effect of environment on seed and oil yield of Turkish Brassica rapa L. genotypes. Fresenius Environ Bull 30(1):60–69
Kerchev PI, Breusegem FV (2022) Improving oxidative stress resilience in plants. The Plant 109:359–372. https://doi.org/10.1111/tpj.15493
Khan MI (2016) Plant Betalains: safety, antioxidant activity, clinical efficacy, and bioavailability. Compr Rev Food Sci Food Saf 15(2):316–330. https://doi.org/10.1111/1541-4337.12185
Khan Zeba D S (2020) Role of heat shock proteins (HSPs) and heat stress tolerance in crop plants. In: Roychowdhury Rajib SS, Choudhury HM (eds) Sustainable agriculture in the era of climate change. Springer International Publishing, Cham, pp 211–234. https://doi.org/10.1007/978-3-030-45669-6_9
Kibria M, Hoque M (2019) A review on Plant responses to soil salinity and amelioration strategies. Open J Soil Sci 09:219–231. https://doi.org/10.4236/ojss.2019.911013
Klupczyńska EA, Pawłowski TA (2021) Regulation of seed dormancy and germination mechanisms in a changing environment. Int J Mol Sci 22(3):2021. https://doi.org/10.3390/ijms22031357
Kmiecik P, Leonardelli M, Teige M (2016) Novel connections in plant organellar signaling link different stress responses and signaling pathways. J Exp Bot 67(13):3793–3807. https://doi.org/10.1093/jxb/erw136
Kooyers NJ (2015) The evolution of drought escape and avoidance in natural herbaceous populations. Plant Sci 234:155–162. https://doi.org/10.1016/j.plantsci.2015.02.012
Krishna R, Ansari WA, Soumia PS, Yadav A, Jaiswal DK, Kumar S, Singh AK, Singh M, Verma JP (2022) Biotechnological interventions in tomato (*Solanum lycopersicum*) for drought stress tolerance: achievements and future prospects. Biotech 11(4):2022. https://doi.org/10.3390/biotech11040048
Kristensen TN, Ketola T, Kronholm I (2020) Adaptation to environmental stress at different timescales. Ann N Y Acad Sci 1476(1):5–22. https://doi.org/10.1111/nyas.13974
Kul R, Ekinci M, Turan M, Ors S, Yildirim E (2020) How abiotic stress conditions affects plant roots. In: Yildirim E, Turan M, Ekinci M (eds) Plant roots. IntechOpen, Rijeka, p ch. 4. https://doi.org/10.5772/intechopen.95286
Kumar M, Kesawat MS, Ali A, Lee SC, Gill SS, Kim HU (2019) Integration of Abscisic acid signaling with other signaling pathways in Plant stress responses and development. Plants 8(12):2019. https://doi.org/10.3390/plants8120592
Kumari VV, Banerjee P, Verma VC, Sukumaran S, Chandran MAS, Gopinath KA, Venkatesh G, Yadav SK, Singh VK, Awasthi NK (2022a) Plant nutrition: an effective way to alleviate abiotic stress in agricultural crops. Int J Mol Sci 23(15):2022. https://doi.org/10.3390/ijms23158519
Kumari A, Sharma B, Singh BN, Hidangmayum A, Jatav HS, Chandra K, Singhal RK, Sathyanarayana E, Patra A, Mohapatra KK (2022b) Physiological mechanisms and adaptation strategies of plants under nutrient deficiency and toxicity conditions. In: Aftab T, Roychoudhury A (eds) Plant perspectives to global climate changes. Academic, p 173194. https://doi.org/10.1016/B978-0-323-85665-2.00012-1
Kuznetsov VV, Shevyakova NI (2007) Polyamines and stress tolerance of plants. Plant Stress 1(1):50–71
Lenis YY, Elmetwally MA, Maldonado-Estrada JG, Bazer FW (2017) Physiological importance of polyamines. Zygote 25(3):244–255. https://doi.org/10.1017/S0967199417000120
Li S, Liu S, Zhang Q, Cui M, Zhao M, Li N, Wang S, Wu R, Zhang L, Cao Y, Wang L (2022) The interaction of ABA and ROS in plant growth and stress resistances. Front Plant Sci 13. https://doi.org/10.3389/fpls.2022.1050132
Liu S, Yang R (2020) Regulations of reactive oxygen species in plants abiotic stress: an integrated overview. In: Tripathi DK, Singh VP, Chauhan DK, Sharma S, Prasad SM, Dubey NK,

Ramawat N (eds) Plant life under changing environment. Academic Press, pp 323–353. https://doi.org/10.1016/B978-0-12-818204-8.00017-5
Liu F, Xi M, Liu T, Wu X, Ju L, Wang D (2024) The central role of transcription factors in ridging biotic and abiotic stress responses for plants' resilience. New Crops 1:100005. https://doi.org/10.1016/j.ncrops.2023.11.003
Lopez-Delacalle M, Silva CJ, Mestre TC, Martinez V, Blanco-Ulate B, Rivero RM (2021) Synchronization of proline, ascorbate and oxidative stress pathways under the combination of salinity and heat in tomato plants. Environ Exp Bot 183:104351. https://doi.org/10.1016/j.envexpbot.2020.104351
Łysiak GP, Szot I (2023) The use of temperature-based indices for estimation of fruit production conditions and risks in temperate climates. Agriculture (Basel) 13(5):960. https://doi.org/10.3390/agriculture13050960
Malhi GS, Kaur M, Kaushik P (2021) Impact of climate change on agriculture and its mitigation strategies: a review. Sustainability 13(3):2021. https://doi.org/10.3390/su13031318
Malik Z, Afzal S, Danish M, Abbasi GH, Bukhari SAH, Khan MI, Dawood M, Kamran M, Soliman MH, Rizwan M, Alhaithloulf HAS, Ali S (2020) Role of nitric oxide and calcium signaling in abiotic stress tolerance in plants. In: Protective chemical agents in the amelioration of plant abiotic stress. Wiley., ch. 28, pp 563–581. https://doi.org/10.1002/9781119552154.ch28
Manasa L, Panigrahy SM, Panigrahi KCS, Rout GR (2022) Overview of cold stress regulation in plants. Bot Rev 88(3):359–387. https://doi.org/10.1007/s12229-021-09267-x
Mansour MMF, Salama KHA (2019) Cellular mechanisms of plant salt tolerance. In: Giri Bhoopander A, Varma (eds) Microorganisms in saline environments: strategies and functions. Springer, Cham, pp 169–210. https://doi.org/10.1007/978-3-030-18975-4_8
Mathivanan S (2021) Abiotic stress-induced molecular and physiological changes and adaptive mechanisms in plants. In: Fahad S, Saud S, Chen Y, Wu C, Wang D (eds) Abiotic stress in plants. IntechOpen., ch. 15, Rijeka. https://doi.org/10.5772/intechopen.93367
Matsuo M, Johnson JM, Hieno A, Tokizawa M, Nomoto M, Tada Y, Godfrey R, Obokata J, Sherameti I, Yamamoto YY, Böhmer FD, Oelmüller R (2015) High Redox responsive transcription Factor1 levels result in accumulation of reactive oxygen species in *Arabidopsis thaliana* shoots and Roots. Mol Plant 8(8):1253–1273. https://doi.org/10.1016/j.molp.2015.03.011
Melotto M, Fochs B, Jaramillo Z, Rodrigues O (2024) Fighting for survival at the stomatal gate. Annu Rev Plant Biol 75(75):551–577. https://doi.org/10.1146/annurev-arplant-070623-091552
Mesihovic A, Ullrich S, Rosenkranz RR, Gebhardt P, Bublak D, Eich H, Weber D, Berberich T, Scharf KD, Schleiff E, Fragkostefanakis S (2022) HsfA7 coordinates the transition from mild to strong heat stress response by controlling the activity of the master regulator HsfA1a in tomato. Cell Rep 38(2):202 2. https://doi.org/10.1016/j.celrep.2021.110224
Mi X, Yu J, Zhang J, Zhong N (2023) Research progress of CRISPR/CAS9 application in rice cultivation. Highl Sci Eng Technol 74:297–301. https://doi.org/10.54097/4z9f5b17
Mishra P, Sharma P (2019) Superoxide Dismutases (SODs) and their role in regulating abiotic stress induced oxidative stress in plants. In: Reactive oxygen, nitrogen and sulfur species in plants. Wiley., ch. 3, pp 53–88. https://doi.org/10.1002/9781119468677.ch3
Mishra N, Jiang C, Chen L, Paul A, Chatterjee A, Shen G (2023) Achieving abiotic stress tolerance in plants through antioxidative defense mechanisms. Front Plant Sci 14:1110622. https://doi.org/10.3389/fpls.2023.1110622
Mittler R, Vanderauwera S, Suzuki N, Miller G, Tognetti VB, Vandepoele K, Gollery M, Shulaev V, Breusegem FV (2011) ROS signaling: the new wave? Trends Plant Sci 16:300–309. https://doi.org/10.1016/j.tplants.2011.03.007
Mittler R, Zandalinas SI, Fichman Y, Van Breusegem F (2022) Reactive oxygen species signaling in plant stress responses. Nat Rev Mol Cell Biol 23(10):663–679. https://doi.org/10.1038/s41580-022-00499-2
Mohammadi Alagoz S, Asgari Lajayer B, Ghorbanpour M (2023) Proline and soluble carbohydrates biosynthesis and their roles in plants under abiotic stresses. In: Ghorbanpour M, Adnan

Shahid M (eds) Plant stress Mitigators. Academic, pp 169–185. https://doi.org/10.1016/B978-0-323-89871-3.00027-6

Nai G, Liang G, Ma W, Lu S, Li Y, Gou H, Guo L, Chen B, Juan M (2022) Overexpression VaPYL9 improves cold tolerance in tomato by regulating key genes in hormone signaling and antioxidant enzyme. BMC Plant Biol 22(1):344. https://doi.org/10.1186/s12870-022-03704-8

Oliveira H, Correia, Pereira AR, Araújo, Mateus N, Freitas V, Oliveira J, Fernandes I (2020) Exploring the applications of the Photoprotective properties of Anthocyanins in biological systems. Int J Mol Sci 21(20):2020. https://doi.org/10.3390/ijms21207464

Pál M, Szalai G, Janda T (2015) Speculation: polyamines are important in abiotic stress signaling. Plant Sci 237:16–23. https://doi.org/10.1016/j.plantsci.2015.05.003

Pande A, Mun BG, Methela NJ, Rahim W, Lee DS, Lee GM, Hong JK, Hussain A, Loake G, Yun BW (2022) Heavy metal toxicity in plants and the potential NO-releasing novel techniques as the impending mitigation alternatives. Front Plant Sci 13. https://doi.org/10.3389/fpls.2022.1019647

Park CJ, Shin R (2022) Calcium channels and transporters: roles in response to biotic and abiotic stresses. Front Plant Sci 13. https://doi.org/10.3389/fpls.2022.964059

Pellegrini E, Cotrozzi L, Neri L, Baraldi R, Carrari E, Nali C, Lorenzini G, Paoletti E, Hoshika Y (2021) Stress markers and physiochemical responses of the Mediterranean shrub Phillyrea angustifolia under current and future drought and ozone scenarios. Environ Res 201:111615. https://doi.org/10.1016/j.envres.2021.111615

Pisoschi AM, Pop A, Iordache F, Stanca L, Predoi G, Serban AI (2021) Oxidative stress mitigation by antioxidants – An overview on their chemistry and influences on health status. Eur J Med Chem 209:112891. https://doi.org/10.1016/j.ejmech.2020.112891

Postiglione AE, Muday G (2020) The role of ROS homeostasis in ABA-induced guard cell signaling. Front Plant Sci 11:968. https://doi.org/10.3389/fpls.2020.00968

Qiao M, Hong C, Jiao Y, Hou S, Gao H (2024) Impacts of drought on photosynthesis on majr food crops and the related mechanisms of plant responses to drought. Plants 13:1808. https://doi.org/10.3390/plants13131808

Rácz D, Radócz L (2019) Nutrient deficiency and effects of various nutrition technologies on crop health. Acta Agraria Debreceniensis 2:109–113. https://doi.org/10.34101/actaagrar/2/3687

Ramegowda V, Da Costa MVJ, Harihar S, Karaba NN, Sreeman SM (2020) Abiotic and biotic stress interactions in plants: a cross-tolerance perspective. In: Hossain MA, Liu F, Burritt DJ, Fujita M, Huang B (eds) Priming-mediated stress and cross-stress tolerance in crop plants. Academic, pp 267–302. https://doi.org/10.1016/B978-0-12817892-8.00017-9

Ramesh SV, Kumar RR, Praveen S (2020) Plant transcriptional regulation in modulating cross-tolerance to stress. In: Hossain MA, Liu F, Burritt DJ, Fujita M, Huang B (eds) Priming-mediated stress and cross-stress tolerance in crop plants. Academic, pp 231–245. https://doi.org/10.1016/B978-0-12-817892-8.00015-5

Rasheed A, Li H, Nawaz M, Mahmood A, Hassan MU, Shah AN, Hussain F, Azmat S, Gillani SFA, Majeed Y, Qari SH (2022) Molecular tools, potential frontiers for enhancing salinity tolerance in rice: a critical review and future prospective. Front Plant Sci 13:2022. https://doi.org/10.3389/fpls.2022.966749

Rasheed Y, Khalid F, Ashraf H, Asif K, Maqsood MF, Naz N, Shahbaz M, Zulfiqar U, Ali Q, Rana S (2024) Enhancing Plant stress resilience with Osmolytes and nanoparticles. J Soil Sci Plant Nutr:1–36. https://doi.org/10.1007/s42729-024-01821-x

Raza A, Charagh S, Najafi Kakavand S, Abbas S, Shoaib Y, Anwar S, Sharifi S, Lu G, Siddique KHM (2023) Role of phytohormones in regulating cold stress tolerance: physiological and molecular approaches for developing cold-smart crop plants. Plant Stress 8:100152. https://doi.org/10.1016/j.stress.2023.100152

Rivero RM, Mittler R, Blumwald E, Zandalinas SI (2022) Developing climate-resilient crops: improving plant tolerance to stress combination. Plant J 109(2):373–389. https://doi.org/10.1111/tpj.15483

Rosenkranz RRE, Ullrich S, Löchli K, Simm S, Fragkostefanakis S (2022) Relevance and regulation of alternative splicing in Plant heat stress response: current understanding and future directions. Front Plant Sci 13:2022. https://doi.org/10.3389/fpls.2022.911277

Sachdev S, Ansari SA, Ansari MI, Fujita M, Hasanuzzaman M (2021) Abiotic stress and reactive oxygen species: generation, signaling, and defense mechanisms. Antioxidants 10:277. https://doi.org/10.3390/antiox10020277

Saharan BS, Brar B, Duhan JS, Kumar R, Marwaha S, Rajput VD, Minkina T (2022) Moleccular and physiological mechanism to mitigate abiotic stress conditions in plants. Life 12:1634. https://doi.org/10.3390/life12101634

Sano S (2017) Molecular and functional characterization of Monodehydro-ascorbate and Dehydroascorbate reductases. In: Anwar HM, Munné-Bosch (eds) Ascorbic acid in plant growth, development and stress tolerance. Springer, Cham, pp 129–156. https://doi.org/10.1007/978-3-319-74057-7_5

Sengupta A, Chakraborty M, Saha J, Gupta B, Gupta K (2016) Polyamines: Osmoprotectants in Plant abiotic stress adaptation. In: Iqbal N, Nazar RA, Khan N (eds) Osmolytes and plants acclimation to changing environment: emerging omics technologies. Springer, New Delhi. https://doi.org/10.1007/978-81-322-2616-1_7

Sevilla F, Martí MC, Brasi-Velasco SD, Jiménez A (2023) Redox regulation, thioredoxins, and glutaredoxins in retrograde signalling and gene transcription. J Exp Bot 74(19):5955–5969. https://doi.org/10.1093/jxb/erad270

Sharma SK, Singh D, Pandey H, Jatav RB, Singh V, Pandey D (2022) An overview of roles of enzymatic and nonenzymatic antioxidants in Plant. In: Tariq KRA, Hakeem (eds) Antioxidant defense in plants: molecular basis of regulation. Springer Nature, Singapore, pp 1–13. https://doi.org/10.1007/978-981-16-7981-0_1

Singh K, Chandra A (2021) DREBs-potential transcription factors involve in combating abiotic stress tolerance in plants. Biologia (Bratisl) 76(10):3043–3055. https://doi.org/10.1007/s11756-021-00840-8

Singh A, Kumar A, Yadav S, Singh IK (2019) Reactive oxygen species-mediated signaling during abiotic stress. Plant Gene 18:100173. https://doi.org/10.1016/j.plgene.2019.100173

Singh S, Fatima A, Tiwari S, Prasad SM (2020) Plant responses to radiation stress and its adaptive mechanisms. In: Tripathi DK, Singh VP, Chauhan DK, Sharma S, Prasad SM, Dubey NK, Ramawat N (eds) Plant life under changing environment. Academic, pp 105–122. https://doi.org/10.1016/B978-0-12-818204-8.00006-0

Singh A, Mehta S, Yadav S, Nagar G, Ghosh R, Roy A, Chakraborty A, Singh IK (2022) How to cope with the challenges of environmental stresses in the era of global climate change: an update on ROS stave off in plants. Int J Mol Sci 23:1995. https://doi.org/10.3390/ijms23041995

Sousa B, Rodrigues F, Soares C, Martins M, Azenha M, Neto TL, Santos C, Cunha A, Fidalgo F (2022) Impact of combined heat and salt stresses on tomato plants-insights into nutrient uptake and Redox homeostasis. Antioxidants 11(3). https://doi.org/10.3390/antiox11030478

Spormann S, Soares C, Teixeira J, Fidalgo F (2021) Polyamines as key regulatory players in plants under metal stress–a way for an enhanced tolerance. Ann Appl Biol 178(2):209–226. https://doi.org/10.1111/aab.12660

Sun S, Yao X, Liu X, Qiao Z, Liu Y, Li X, Jiang X (2022) Brassinolide can improve drought tolerance of maize seedlings under drought stress: by inducing the photosynthetic performance, antioxidant capacity and ZmMYB gene expression of maize seedlings. J Soil Sci Plant Nutr 22(2):2092–2104. https://doi.org/10.1007/s42729-022-00796-x

Taj G, Agarwal P, Grant M, Kumar A (2010) MAPK machinery in plants: recognition and response to different stresses through multiple signal transduction pathways. Plant Signal Behav 5(11):1370–1378. https://doi.org/10.4161/psb.5.11.13020

Tavanti TR, de Melo AAR, Moreira LDK, Sanchez DEJ, Silva RS, Silva RM, Reis ARD (2021) Micronutrient fertilization enhances ROS scavenging system for alleviation of abiotic stresses in plants. Plant Physiol Biochem 160:386–396. https://doi.org/10.1016/j.plaphy.2021.01.040

Tian Y, Peng K, Bao Y, Zhang D, Meng J, Wang D, Wang X, Cang J (2021) Glucose-6-phosphate dehydrogenase and 6-phosphogluconate dehydrogenase genes of winter wheat enhance the cold tolerance of transgenic Arabidopsis. Plant Physiol Biochem 161:86–97. https://doi.org/10.1016/j.plaphy.2021.02.005

Tripathy SK, Majhi PK, Ahamad A (2021) Achieving drought tolerance in Rice by targeted genome editing. Biotica Research Today 3(11):1065–1067

Trono D, Pecchioni N (2022) Candidate genes associated with abiotic stress response in plants as tools to engineer tolerance to drought, salinity and extreme temperatures in wheat: An overview. Plants 11(23):2022. https://doi.org/10.3390/plants11233358

Tun W, Yoon J, Jeon JS, An G (2021) Influence of climate change on flowering time. J Plant Biol 64(3):193–203. https://doi.org/10.1007/s12374-021-09300-x

Ullah A, Hussain A, Shaban M, Khan AH, Alariqi M, Gul S, Jun Z, Lin S, Li J, Jin S, Munis MFH (2018) Osmotin: a plant defense tool against biotic and abiotic stresses. Plant Physiol Biochem 123:149–159. https://doi.org/10.1016/j.plaphy.2017.12.012

Vasilachi IC, Stoleru V, Gavrilescu M (2023) Analysis of heavy metal impacts on cereal crop growth and development in contaminated soils. Agriculture 13(10):2023. https://doi.org/10.3390/agriculture13101983

Venkatappa M, Sasaki N, Han P, Abe I (2021) Impacts of droughts and floods on croplands and crop production in Southeast Asia–an application of Google earth engine. Sci Total Environ 795:148829. https://doi.org/10.1016/j.scitotenv.2021.148829

Wahab A, Abdi G, Saleem MH, Ali B, Ullah S, Shah W, Mumtaz S, Yasin G, Muresan CC, Marc RA (2022) Plants' physio-biochemical and Phyto-hormonal responses to alleviate the adverse effects of drought stress: a comprehensive review. Plants 11(13):2022. https://doi.org/10.3390/plants11131620

Wang X, Huang W, Yang Z, Liu J, Huang B (2016) Transcriptional regulation of heat shock proteins and ascorbate peroxidase by CtHsfA2b from African bermudagrass conferring heat tolerance in Arabidopsis. Sci Rep 6(1):28021. https://doi.org/10.1038/srep28021

Wang J, Gao X, Dong J, Tian X, Wang J, Palta J, Xu S, Fang Y, Wang Z (2020) Over-expression of the heat-responsive wheat gene TaHSP23.9 in transgenic Arabidopsis conferred tolerance to heat and salt stress. Front Plant Sci 11:2020. https://doi.org/10.3389/fpls.2020.00243

Wang Y, Mostafa S, Zeng W, Jin B (2021a) Function and mechanism of Jasmonic acid in plant responses to abiotic and biotic stresses. Int J Mol Sci 22(16):2021. https://doi.org/10.3390/ijms22168568

Wang X, Niu Y, Zheng Y (2021b) Multiple functions of MYB transcription factors in abiotic stress responses. Int J Mol Sci 22:6125. https://doi.org/10.3390/ijms22116125

Wang X, Ren P, Ji L, Zhu B, Xie G (2021c) OsVDE, a xanthophyll cycle key enzyme, mediates abscisic acid biosynthesis and negatively regulates salinity tolerance in rice. Planta 255(1):6. https://doi.org/10.1007/s00425-021-03802-1

Wang Z, Zhao X, Ren Z, Elwafa SFA, Pu X, Zhu Y, Dou D, Su H, Cheng H, Liu Z, Chen Y, Wang E, Shao R, Ku L (2022) ZmERF21 directly regulates hormone signaling and stress-responsive gene expression to influence drought tolerance in maize seedlings. Plant Cell Environ 45(2):312–328. https://doi.org/10.1111/pce.14243

Wang D, Wei L, Liu T, Zhao J, Tsuda K, Wang Y (2023) Suppression of ETI by PTI priming to balance plant growth and defense through an MPK3/MPK6-WRKYs-PP2Cs module. Mol Plant 16(5):903–918. https://doi.org/10.1016/j.molp.2023.04.004

Wasternack C, Kombrink E (2010) Jasmonates: structural requirements for lipid-derived signals active in Plant stress responses and development. ACS Chem Biol 5(1):63–77. https://doi.org/10.1021/cb900269u

Waszczak C, Carmody M, Kangasjärvi J (2018) Reactive oxygen species in plant signaling. Annu Rev Plant Biol 69:209–236. https://doi.org/10.1146/annurev-arplant-042817-040322

Wei S, Xia R, Chen C, Shang X, Ge F, Wei H, Chen H, Wu Y, Xie Q (2021) ZmbHLH124 identified in maize recombinant inbred lines contributes to drought tolerance in crops. Plant Biotechnol J 19(10):2069–2081. https://doi.org/10.1111/pbi.13637

Xie H, Zhang P, Jiang C (2023) Combined transcriptomic and metabolomic analyses of high temperature stress response of quinoa seedlings. BMC Plant Biol 23:292. https://doi.org/10.1186/s12870-023-04310-y

Yadav B, Jogawat A, Rahman MS, Narayan OP (2021) Secondary metabolites in the drought stress tolerance of crop plants: a review. Gene Rep 23:101040. https://doi.org/10.1016/j.genrep.2021.101040

Yadav P, Nehra A, Kalwan G, Bhardwaj D, Yasheshwar VR, Agarwala N, Tuteja N, Gill R, Ansari MW, Gill SS (2024) Harnessing Jasmonate, salicylate, and microbe synergy for abiotic stress resilience in crop plants. J Plant Growth Regul 44(1):40–61. https://doi.org/10.1007/s00344-023-11218-2

Yurina NP (2023) Heat shock proteins in Plant protection from oxidative stress. Mol Biol 57(6):951–964. https://doi.org/10.1134/S0026893323060201

Zandalinas SI, Balfagón D, Gómez-Cadenas A, Mittler R (2022) Plant responses to climate change: metabolic changes under combined abiotic stresses. J Exp Bot 73:3339–3354. https://doi.org/10.1093/jxb/erac073

Zhai X, Lan D, Jia M, Gan Z, Chen C, Zeng X, Chen J, Kai W (2023) Characterization and expression analysis of key abscisic acid signal transduction genes during kiwifruit development. Sci Hortic 309:111672. https://doi.org/10.1016/j.scienta.2022.111672

Zhang M, Zhang S (2022) Mitogen-activated protein kinase cascades in plant signaling. J Integr Plant Biol 64(2):301–341. https://doi.org/10.1111/jipb.13215

Zhang H, Li G, Hu D, Zhang Y, Zhang Y, Shao H, Zhao L, Yang R, Guo X (2020a) Functional characterization of maize heat shock transcription factor gene ZmHsf01 in thermotolerance. Peer J 8:e8926. https://doi.org/10.7717/peerj.8926

Zhang N, Zhao H, Shi J, Wu Y, Jiang J (2020b) Functional characterization of class I SlHSP17.7 gene responsible for tomato cold-stress tolerance. Plant Sci 298:110568. https://doi.org/10.1016/j.plantsci.2020.110568

Zhang L, Guo X, Zhang Z, Wang A, Zhu J (2021) Cold-regulated gene LeCOR413PM2 confers cold stress tolerance in tomato plants. Gene 764:145097. https://doi.org/10.1016/j.gene.2020.145097

Zhang G, Li G, Xiang Y, Zhang A (2022) The transcription factor ZmMYB-CC10 improves drought tolerance by activating ZmAPX4 expression in maize. Biochem Biophys Res Commun 604:1–7. https://doi.org/10.1016/j.bbrc.2022.02.051

Zhang Y, Xu J, Li R, Ge Y, Li Y, Li R (2023) Plants' response to abiotic stress: mechanisms and strategies. Int J Mol Sci 24:10915. https://doi.org/10.3390/ijms241310915

Zhao Y, Du H, Wang Y, Wang H, Yang S, Li C, Chen N, Yang H, Zhang Y, Zhu Y, Yang L, Hu X (2021a) The calcium-dependent protein kinase ZmCDPK7 functions in heat-stress tolerance in maize. J Integr Plant Biol 63(3):510–527. https://doi.org/10.1111/jipb.13056

Zhao S, Zhang Q, Liu M, Zhou H, Ma C, Wang P (2021b) Regulation of Plant responses to salt stress. Int J Mol Sci 22(9):2021. https://doi.org/10.3390/ijms22094609

Zhao H, Liu X, Wang J, Qian Q, Zhang G (2022) The coordinated regulation mechanism of rice plant architecture and its tolerance to stress. Front Plant Sci 13:1087378. https://doi.org/10.3389/fpls.2022.1087378

Ziwei YG, Sun ZM, Luo Y (2022) Exogenous trehalose differently improves photosynthetic carbon assimilation capacities in maize and wheat under heat stress. J Plant Interact 17(1):361–370. https://doi.org/10.1080/17429145.2022.2041119

Zulfiqar F, Akram NA, Ashraf M (2019) Osmoprotection in plants under abiotic stresses: new insights into a classical phenomenon. Planta 251(1):3. https://doi.org/10.1007/s00425-019-03293-1

Chapter 2
Abiotic Stress Elicitors: Mechanisms and Responses to Drought, Salinity, and Temperature

Muhammad Tanveer Altaf, Waqas Liaqat, Jaweria Iqbal, Muhammad Faheem Jan, Muhammad Sheraz, Amna Jamil, Asia Maqbool, Ehtisham Hassan Khan, Celaleddin Barutçular, and Faheem Shehzad Baloch

Abstract Environmental stresses like drought, salinity, and extreme temperatures present major obstacles to worldwide agriculture, negatively impacting plant growth, development, and yield. In reaction to these challenges, plants have evolved complex defense strategies, which are frequently activated and strengthened by external compounds referred to as elicitors. Abiotic stress elicitors, including chemical and physical stimuli, play a vital role in activating specific plant pathways that improve tolerance to adverse environmental conditions. This chapter explores the types of abiotic elicitors, both natural and synthetic, and their underlying mechanisms in enhancing plant responses to drought, salinity, and temperature stress. It also examines how elicitors induce protective mechanisms, such as the accumulation of osmoprotectants, activation of antioxidant defense systems, and regulation of stress-responsive genes. By delving into the molecular processes associated with elicitors, this chapter highlights the potential of these compounds to be applied in climate-smart agriculture to mitigate the negative impacts of abiotic stresses. The

M. T. Altaf (✉) · W. Liaqat (✉)
Department of Field Crops, Faculty of Agriculture, Recep Tayyip Erdoğan University, Rize/Pazar, Turkey
e-mail: muhammadtanveer.altaf@erdogan.edu.tr

J. Iqbal
Department of Plant Breeding and Genetics, Faculty of Agricultural Sciences and Technology, Bahauddin Zakariya University, Multan, Pakistan

M. F. Jan
Guangxi Key Laboratory of Agro-environment and Agro-products Safety, College of Agriculture, Guangxi University, Nanning, China

M. Sheraz
Fodder and Forrage Program, Crop Sciences Institute, Islamabad, Pakistan

A. Jamil
Department of Horticultural Sciences, The Islamia University of Bahawalpur, Bhawalpur, Pakistan

H. I. Mohamed (ed.), *Elicitors for Sustainable Crop Production*, Sustainability Sciences in Asia and Africa, https://doi.org/10.1007/978-981-95-3533-0_2

use of elicitors offers a sustainable and environmentally friendly approach to enhancing crop resilience, reducing the need for chemical inputs, and ensuring food security in a changing climate.

Keywords Climate change · Salicylic acid · Jasmonic acid · Physical elicitors · Stress tolerance

2.1 Introduction

Agricultural productivity encounters substantial challenges due to environmental stresses, negatively impacting plant development and yield. Among these, drought, salinity, and temperature extremes are particularly detrimental, leading to considerable yield losses (Liaqat et al. 2022). As climate change intensifies these stressors, comprehending plant responses to abiotic stresses has emerged as a pivotal area of research. This understanding is essential for ensuring food security and advancing agricultural sustainability (Liaqat et al. 2024a). Drought stress, characterized by inadequate water availability, disrupts fundamental plant processes such as photosynthesis, transpiration, and nutrient transport. This condition results in stomatal closure, diminished carbon dioxide uptake, and lowered photosynthesis, all of which can lead to decreased growth and yield loss (Liaqat et al. 2024b; Qiao et al. 2024). Salinity stress, caused by excessive salt concentrations in the soil, impairs water absorption and ionic balance within plants, resulting in ion toxicity, osmotic stress, and reduced growth. Increased salinity not only impedes a plant's capacity to take up water but also interferes with nutrient absorption, intensifying the overall stress impact (Balasubramaniam et al. 2023). Temperature stress, which includes both high heat and extreme cold conditions, impacts enzymatic activities, protein stability, and membrane integrity. Heat stress accelerates plant metabolism beyond optimal levels, while cold stress decelerates essential processes, affecting flowering, seed development, and overall plant productivity.

A. Maqbool
College of Horticulture, Hebei Agricultural University, Baoding, Hebei, China

E. H. Khan
MOA Key Laboratory of Crop Ecophysiology and Farming System in the Middle Reaches of the Yangtze River, College of Plant Science and Technology, Huazhong Agricultural University, Wuhan, China

C. Barutçular
Department of Field Crops, Faculty of Agriculture, Çukurova University, Adana, Turkey

F. S. Baloch
Department of Biotechnology, Faculty of Science, Mersin University, Mersin, Turkey

Department of Plant Resources and Environment, Jeju National University, Jeju, South Korea

To manage these stresses, plants have developed intricate defense mechanisms. They activate stress-responsive genes, produce stress-related proteins, accumulate compatible solutes, and bolster antioxidant defense systems (Chaffai et al. 2024). Additionally, plants rely on external signals or molecules, known as elicitors, to modulate and enhance their responses to abiotic stresses. Elicitors, which can be either naturally occurring or synthetic compounds, trigger plant defense mechanisms, thereby improving their ability to withstand harsh environmental circumstances (Meena et al. 2022).

Abiotic stress elicitors play a crucial role in mitigating the adverse effects of drought, salinity, and temperature on plants. These elicitors function by triggering specific signaling pathways that enhance the plant's physiological and biochemical responses to stress (Humbal and Pathak 2023). Certain abiotic elicitors, like salicylic acid, abscisic acid, and jasmonic acid, play a crucial role in regulating stress responses. Abscisic acid (ABA) is instrumental in managing stomatal closure during drought conditions, thereby conserving water and reducing water loss through transpiration (Cardoso et al. 2020). Salicylic acid (SA) and jasmonic acid (JA) regulate antioxidant activities and stress response gene expression (Hu et al. 2023; Wang et al. 2020). Abiotic elicitors can be categorized into chemical elicitors (such as plant growth regulators, antioxidants, and osmoprotectants) and physical elicitors (such as temperature treatments or controlled drought conditions). Notable chemical elicitors that are frequently studied include nitric oxide (NO), calcium ions (Ca^{2+}), and hydrogen peroxide (H_2O_2). These molecules play crucial roles in signal transduction pathways and can enhance the plant's defense mechanisms. They promote stress resistance by inducing the synthesis of stress-related proteins, secondary metabolites (SMs), and antioxidants (Malik et al. 2020; Liu et al. 2021).

Recently, the development of synthetic elicitors has garnered significant interest in agriculture as a promising strategy for enhancing crop resilience (Rani et al. 2023). Elicitors present a sustainable and environmentally friendly approach to bolstering crop resilience, thereby reducing dependency on pesticides and chemical fertilizers, which can adversely affect environmental and soil health. Furthermore, elucidating the mechanisms by which abiotic elicitors operate across diverse plant species can contribute to the development of stress-tolerant crops, especially those capable of thriving under the harsh environmental conditions intensified by climate change. This chapter explores various types of abiotic stress elicitors and their specific roles in plant responses to stress.

2.2 Stresses in Agriculture

Agriculture faces a myriad of stresses that threaten crop productivity and sustainability. These stresses can broadly be categorized into abiotic and biotic types. Abiotic stresses include factors like drought, salinity, temperature extremes, and nutrient deficiencies, which negatively affect plant growth and development (Hussain et al. 2025) (Fig. 2.1). These environmental challenges are becoming more

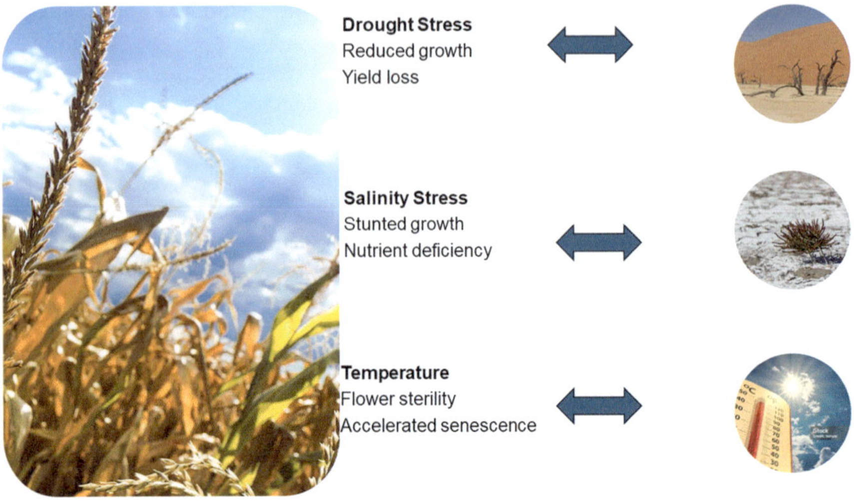

Fig. 2.1 Major environmental stresses and their consequences for crops

pronounced due to climate change, leading to unpredictable weather patterns and resource scarcity (Chaudhry and Sidhu 2022). On the other hand, biotic stresses involve living organisms such as pests, pathogens, and weeds that directly harm crops, causing significant yield losses (Surówka et al. 2020). Both types of stress require management strategies to minimize their impact and sustain agricultural productivity.

Abiotic stress elicitors, both natural and synthetic, offer a promising approach to enhance crop resilience by activating the plant's defense mechanisms. These compounds, when applied to crops, can precondition plants to better withstand environmental stresses by stimulating stress-responsive genes and protective pathways (Ahmad et al. 2019). Integrating abiotic stress elicitors into agricultural practices provides a proactive solution to lessen the damaging effects of salinity, drought, and temperature fluctuations, ultimately supporting sustainable crop production in the face of environmental challenges (Lone et al. 2023).

2.2.1 High Temperature

Temperature significantly influences plant distribution, growth, and development (Table 2.1). Plants generally thrive within an optimal temperature range of 25–30 °C, which is crucial for vigorous growth (Hatfield and Prueger 2015). Climate change is projected to increase average temperatures by 1.2–3.7 °C over the next 30–50 years, potentially causing heat stress that adversely affects seed set and development (Yamori et al. 2014). Variations in temperatures can disrupt intermolecular interactions and impair normal plant growth. Prolonged exposure to high temperatures can induce cellular disorders and alter metabolic activities, leading to

Table 2.1 Influence of elevated temperatures on different crop species

Crops	Treatment (temperature)	Stage	Impact	Reference
Sorghum	30/25 °C, D/N (day/night)	60 days after sowing (DAS) to maturity phase	Reduction in leaf size, early booting, heading, anthesis, and maturity, reduced number of grains/spikes, reduced grain size, and yield	Djanaguiraman et al. (2010)
Wheat	37/28 °C, D/N 20 days	Maturity and grain formation	Time to maturity and grain formation are reduced, leading to decreased yield	Rahman et al. (2009)
Rice	35/27 °C, D/N 14 days	Reproductive period	Reduction in ear development, destruction of cob extensibility, impairment of cellulose and hemicellulose production, diminished photosynthesis	Yin et al. (2010)
Indica Rice	38/30 °C, D/N	Maturity, propagative and harvesting period	Reduction in fruit weight and width, enhanced quantity of abnormal seeds per fruit	Cao et al. (2009)
Rice	40/30 °C, D/N	65 DAS to maturity period	Lowered antioxidant enzyme activity, photosystem II photochemistry, chlorophyll content, and yield	Mohammed and Tarpley (2010)
Tobacco	38/28 °C, D/N, 14 days	Flowering	Leaf conductance decreased, damage to the chloroplast, thylakoid, and plasma membranes, distorted mitochondrial matrix, membranes, and cristae	Tan et al. (2011)
Wheat	Above 33 °C, 10 days	Heading phase	Reduced fertility rates of pollen and spikelets	Hurkman et al. (2009)
Maize	33–40 °C	–	23% reduction with heat stress applied at early phase of grain filling	(Edreira et al. 2014)
Wheat	33–40 °C, 15 days	Silking and pre-anthesis	Affected plant and ear growth	Zhang et al. (2013)
Maize	33–40 °C	–	75% decrease in final kernel number/plant during silking, 52% decrease before anthesis	Rattalino-Edreira and Otegui (2013)
Okra	43 °C, 2 h	Initial growth period	Reduction in stomatal conductance, photosynthetic efficiency, and antioxidant enzymes activities	Gunawardhana and de Silva (2011)
Soybean	25–42.5 °C	Period of vegetative growth	Lowered CO_2 assimilation rate	Djanaguiraman et al. (2011)
Wheat	35 °C		26% reduction in peduncle water soluble carbohydrate content	Talukder et al. (2014)

reduced growth and development (Dos Santos et al. 2022). Heat stress notably impacts germination, the initial growth phase of plants, by decreasing germination percentage, growth, and viability due to cellular water loss. Studies have shown that elevated temperatures negatively affect the germination process in different plants, resulting in lowered kernel dry weight as well as altered nutritional composition (Kumar et al. 2023; Jumrani et al. 2024).

In *Origanum compactum*, high temperatures can completely inhibit seed germination (Laghmouchi et al. 2017). Research indicates that high temperatures can reduce growth rates in various plant species by decreasing relative growth rates. This decline is linked to increased production of reactive oxygen species (ROS) under high-temperature conditions (Devireddy et al. 2021). Heat stress has a major impact on plant physiology and metabolism, particularly when it interacts with other abiotic stresses like salinity and drought. Extreme temperature stress typically affects crops more severely during the reproductive stages than the vegetative stages, often leading to reduced yields due to pollen sterility (Liu et al. 2024). This stress causes the denaturation of cellular structures, protein instability, membrane breakdown, and metabolic disruptions, which in turn result in excessive production of ROS and intense oxidative stress in plants (Maity and Shrivastav 2024; Bita and Gerats 2013).

High temperatures activate cellular machinery in plants, triggering responses such as lipid signaling, Ca^{2+} influx, and cytoskeletal reorganization (Carmody et al. 2016). This response is linked to heat stress-mediated inhibition of Rubisco activase enzymes, which disrupts the C3 cycle of carbon fixation. Consequently, this disruption leads to the generation of excessive ROS and impairs plant repair processes (Apel and Hirt 2004). Additionally, heat stress reduces photosynthetic pigment levels and antioxidant enzyme activity (Lal et al. 2021). High temperatures impair membrane stability, denature proteins, and inactivate enzymes, leading to reduced plant productivity (Ifeduba et al. 2024). Specific crops, like wheat, show significant sensitivity during reproductive stages, with high temperatures causing reductions in growth attributes and yield components (Rajput et al. 2023). Maize hybrids exhibit varied responses to elevated temperatures, with some hybrids showing improved physiological performance under elevated CO2 conditions (Vanaja et al. 2024). Furthermore, high-temperature stress is projected to significantly reduce crop yields, threatening food security as global temperatures rise (Kamatchi et al. 2024). For instance, wheat varieties sown later in the season experienced severe yield reductions due to heat stress during critical growth phases (Rajput et al. 2023). Conversely, elevated CO_2 can enhance certain yield traits, potentially offsetting some negative impacts of high temperatures (Roy et al. 2024).

2.2.2 Salinity

Abiotic stressors, particularly salinity stress caused by excessive mineral salts like Na^+ and Cl^-, are contributing to a decline in crop yields. This issue is expected to worsen with climate change and the adoption of certain irrigation practices

(Khondoker et al. 2023). Salinity stress is prevalent in semiarid and arid regions, characterized by low rainfall, high rates of evaporation and transpiration, and limited water availability. The accumulation of salts disrupts plant metabolism, leading to necrosis and chlorosis. The severity of salinity stress varies depending on environmental factors, such as light intensity, soil conditions, and the plant species present (Tang et al. 2015; Balasubramaniam et al. 2023). Plants are classified as glycophytes or halophytes based on their ability to thrive in saline soils. Most crop plants, being glycophytes, cannot survive at NaCl concentrations of 100–200 mM, whereas halophytes exhibit enhanced resistance, tolerating NaCl levels of 300–500 mM (Acosta-Motos et al. 2017; Flowers and Colmer 2015).

Salinity, initially detected by plant roots, poses significant challenges to plant growth, with both immediate and long-term effects. In the short term, salinity reduces water availability, leading to osmotic stress, while prolonged exposure results in ion toxicity due to excessive salt accumulation. This disrupts key physiological processes such as germination, growth, photosynthesis, oxidative balance, water status, and overall productivity (Shahid et al. 2024; Parihar et al. 2015; Shekari et al. 2017).

Excessive salts in water can inhibit plant growth by reducing water uptake, damaging cells, and disrupting the transpiration stream, leading to ion toxicity (Parihar et al. 2015). Research has demonstrated reduced growth in different species (Hualpa-Ramirez et al. 2024). Furthermore, moderate-to-high salt concentrations have been shown to negatively impact shoot and root length in pistachio rootstocks, highlighting the importance of effective water management (Rahneshan et al. 2018). Severe salt stress reduces carbon assimilation and impairs growth by disrupting metabolic processes (Rahneshan et al. 2018). Salt stress also decreases photosynthetic rate and enzyme activity, which in turn affects protein synthesis (Qados 2011).

Salinity negatively impacts plants by causing an excessive buildup of Na^+ ions, which results in stomatal closure and the production of ROS in green tissues (Shabala and Munns 2017). This leads to oxidative damage, as seen in sorghum, and reduces chlorophyll content and photosynthesis due to elevated levels of H_2O_2, as observed in tomato plants (Nxele et al. 2017; Marsic et al. 2018). Chlorosis, a typical symptom of salinity stress, further hinders photosynthesis (Jha and Subramanian 2016). The ROS generated during salinity stress cause damage to vital macromolecules, such as membrane lipids, proteins, and nucleic acids (AbdElgawad et al. 2016). Additionally, this stress induces excessive accumulation of malondialdehyde (MDA) in plant tissues, contributing to oxidative stress by increasing Na^+, MDA, and H_2O_2 levels, as observed in maize (Ashraf et al. 2018). Salinity stress disrupts essential physiological processes, including photosynthesis, nutrient uptake, and hormone regulation, leading to reduced growth and yield (Chaudhary et al. 2024; Ahmed et al. 2024). Furthermore, specific crops, such as groundnuts, exhibit varying levels of salinity tolerance, with physiological metrics like relative water content (RWC) and electrolytic leakage serving as indicators of stress response (Table 2.2) (Joshi et al. 2024). Salinity can cause yield reductions of up to 70% in legumes and nearly 50% in wheat, emphasizing the need for effective management strategies (Salmanpour et al. 2024).

Table 2.2 Physiological and biochemical reactions of plants to salinity stress

Crop/plant	Stress intensity	Effects	References
Cucumber	Moderate (120 mM)	A reduction in water potential, enhanced abscisic acid levels, and increased electrolytic leakage have been noted	Kang et al. (2014)
Soybean	Low (25–100 mM)	Decreased cortical thickness, shoot and root biomass, plant stature, and leaf count	Dolatabadian et al. (2011)
Common/ garden thyme	Low (30–90 mM)	Reduction in plant dry matter, K^+ and Ca^{2+} content	Bistgani et al. (2019)
Salad bean	Low (50–200 mM)	Reduction in dry matter content in shoots and roots, decline in phenolic compounds and photosynthetic pigment	Taïbi et al. (2016)
European sea rocket	Medium (100–400 mM)	Reduction in osmotic potential, enhancement in proline and amino acids	Hmidi et al. (2018)
Water mint	Low (50–150 mM)	Reduction in water content and growth, higher proline, protein levels, polyphenol oxidase activity, activities of superoxide dismutase, peroxidase, and ascorbate peroxidize	Haddadi et al. (2016)
Tomato	Medium (150–450 mM)	Lowered carotenoid content and chlorophyll, enhanced oxidative stress and proline content	Al Hassan et al. (2015)
Mustard	–	Reduced photosynthetic rate and electron transport	Mittal et al. (2012)
Potato	Low (25–50 mM)	Decline in root length, leaf water potential, shoot biomass, tuber yield but enhanced ABA content in plant	Akhtar et al. (2015)
Rice	Moderate (200 mM)	Reduction in starch content, chlorophyll a, leaf weight, relative water content, chlorophyll b and Fv/Fm	Amirjani (2011)
Perennial ryegrass	Medium (250 mM)	Enhanced hydrogen peroxide, higher activity of ascorbate (APX), superoxide dismutase (SOD) and peroxidase (POD)	Hu et al. (2012)

2.2.3 *Drought*

Water stress significantly impairs crop growth, reducing leaf area, causing premature senescence, and altering root-to-shoot ratios. Prolonged drought disrupts key plant processes such as nutrient uptake, photosynthesis, and protein synthesis, leading to the accumulation of ROS (Qiao et al. 2024). Drought affects plant growth at all stages, influencing both morphological and molecular aspects. It impedes seed germination, which requires adequate moisture across various crops (Tholen et al. 2012). Since cell division and elongation drive plant growth, sufficient water is critical for seed germination and subsequent developmental stages (Carbone et al. 2021). Prolonged anthesis and delayed grain formation further reduce yield in field crops, primarily due to diminished biosynthetic enzyme activity and decreased photosynthesis (Andreo-Jimenez et al. 2019). Globally, drought adversely affects the

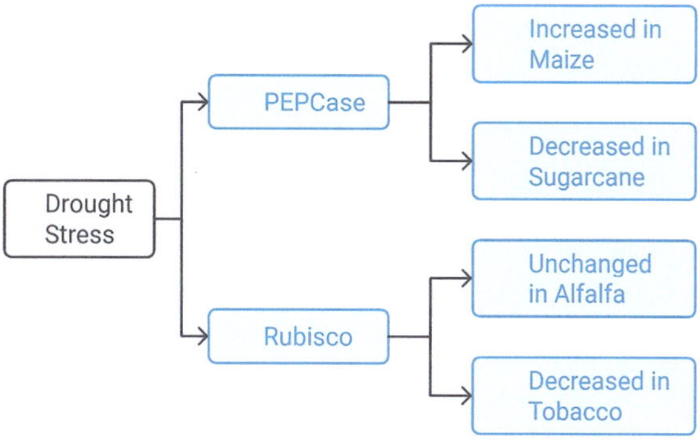

Fig. 2.2 Effects of drought stress on photosynthetic enzyme activity in different field crops

production of barley, maize, soybeans, wheat, and other major crops (EL Sabagh et al. 2020; Debray et al. 2022).

Water scarcity reduces root nutrient translocation and soil nutrient availability, leading to altered ion content in plant tissues. This results in decreased potassium uptake, reduced transpiration, and impaired root membrane transporter activity (Cakmak and Rengel 2024). Water scarcity significantly impairs photosynthesis in plants by reducing leaf area and photosynthetic rate (Fig. 2.2). This reduction is primarily attributed to decreased CO_2 conductance, which hampers Rubisco activity and lowers the efficiency of nitrate reductase and sucrose phosphate synthase (Zhang et al. 2017). Drought stress during the reproductive stage resulted in a 43% reduction in photosynthetic chlorophyll fluorescence, indicating damage to photosystem II, which is crucial for photosynthesis (Yasin et al. 2024). Post-anthesis water stress led to a 34% decrease in grain yield and a 27% reduction in 1000-grain weight, primarily due to diminished photo-assimilate production (Tamrazov 2022). On the other side, drought-tolerant varieties exhibited smaller stomata and higher water use efficiency, maintaining better physiological balance under water deficit (Xu et al. 2023). Varieties showed increased antioxidant activity and proline accumulation, with some demonstrating enhanced yield potential despite drought stress (Rehman et al. 2022).

2.3 Elicitors and Classification of Elicitors

Elicitors, which are chemical agents, trigger plant defense mechanisms when applied at low concentrations (Gaddam et al. 2024). This approach is widely employed to enhance the in vitro production of targeted SMs (Ghazal et al. 2024).

Elicitors are categorized as biotic or abiotic based on their origin (Fig. 2.3). Biotic elicitors are molecules derived from biological origins, including microorganisms or components of plant cell walls like pectin, cellulose, and chitin. In contrast, abiotic elicitors encompass hormonal, physical, and chemical factors that arise from non-biological sources.

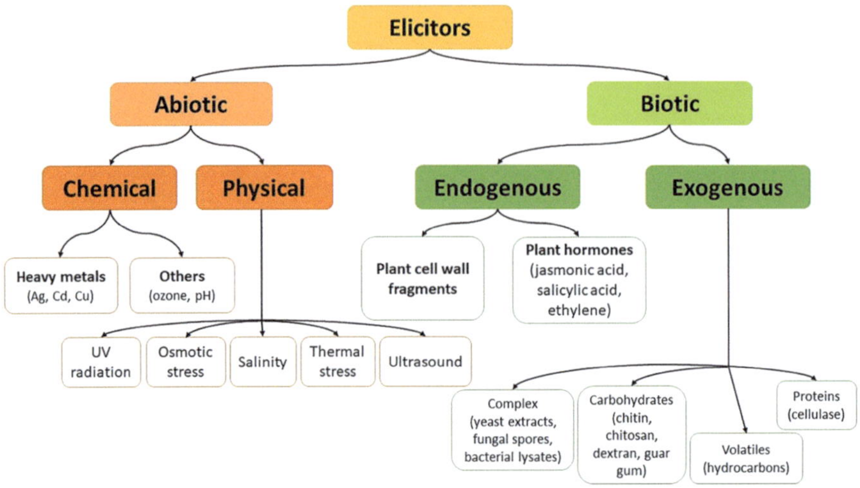

Fig. 2.3 Elicitors classification based on their nature. (Rogowska and Szakiel 2021)

2.3.1 Chemical Elicitors

The significance of metals as chemical elicitors in plant defense mechanisms has gained considerable focus because of their capacity to affect the biosynthesis of SMs, which play a key role in helping plants adapt to different abiotic stresses (Inam et al. 2023). Metals can significantly alter the production of SMs by influencing key aspects of secondary metabolism. Due to their increasing use in industries like agriculture and manufacturing, coupled with their high toxicity and tendency to bioaccumulate, metals have become major abiotic stress factors for living organisms (Alengebawy et al. 2021). Studies have shown that plants respond to metals such as cobalt (Co), silver (Ag), iron (Fe), and nickel (Ni) by producing SMs. At low concentrations (5 μM), cadmium (Cd) and Ag were particularly effective in promoting the synthesis of phenolic acids in *Vitis vinifera* cell cultures, with Co showing the same effect at concentrations of 5, 25, and 50 μM. After 4 h of treatment, 3-O-glucosyl-resveratrol levels increased by 1.6 times compared to the control (Cai et al. 2013). Similarly, *Ambrosia artemisiifolia* hairy root cultures exhibited an eightfold increase in thiarubrine production after 72 h of exposure to 50 mg/L vanadyl sulfate ($VOSO_4$) in 16-day-old cultures (Bhagwath and Hjortsø 2000). Metal ions have also been employed to stimulate betalain production (Rudrappa et al. 2004). Chemical elicitors activate signaling pathways that lead to the expression of defense-related genes and accumulation of phytoalexins, which help plants withstand stress. Various elicitors, such as Acibenzolar-S-Methyl and carrageenan, have been shown to improve plant health and yield in crops like wheat (Ducatti et al. 2023). These insights into metal elicitors reinforce their potential application in both improving stress tolerance in plants and promoting the production of valuable SMs for pharmaceutical and agricultural uses.

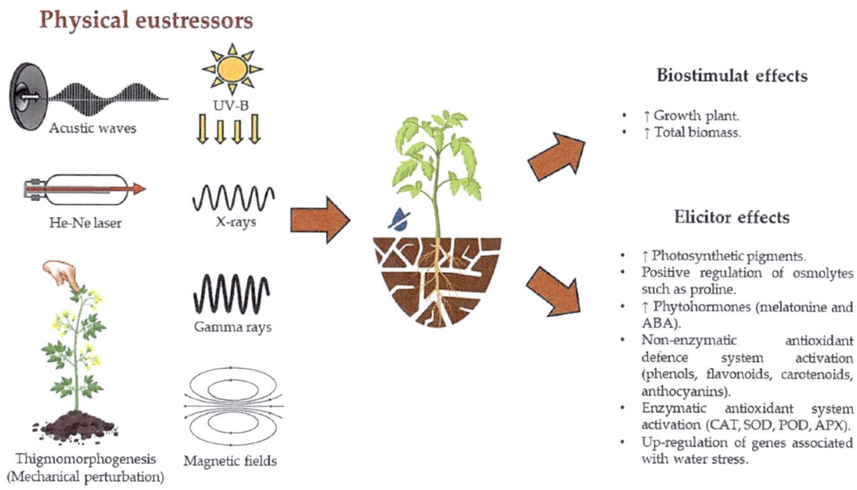

Fig. 2.4 The overall impact of biostimulants and physical elicitors in reducing water stress. (Melo-Sabogal and Contreras-Medina 2024)

2.3.2 *Physical Elicitors*

Physical elicitors encompass a range of environmental stimuli that trigger plant defense mechanisms and stress responses, playing a crucial role in plant adaptation to abiotic stress conditions (Inam et al. 2023). These stimuli include factors such as heat, salinity, drought, and light stress, all of which can significantly influence plant physiology and metabolic processes. The application of these physical stresses induces a cascade of molecular and biochemical changes that enhance plant resilience and promote the synthesis of SMs, which are vital for plant survival under adverse environmental conditions (Selwal et al. 2024). In the context of drought, salinity, and temperature stress, understanding the mechanisms by which physical elicitors modulate plant responses is key to developing strategies for improving crop tolerance and productivity (Fig. 2.4).

2.3.2.1 Light

Light is one of the most significant physical elicitors (Fig. 2.5) and plays a crucial role in plant metabolite production, promoting the synthesis of SMs like zingiberene and gingerol. Light exposure affects growth and metabolite production in hairy roots, such as Hyoscyamus albus and P. ginseng. UV radiation, an abiotic factor, also stimulates SMs production. UV-C, the most harmful form, is absorbed by the stratosphere, while UV-B is a stress factor that induces the synthesis of phenolic compounds (Liaqat et al. 2024c). Excessive UV-B radiation alters the antioxidant

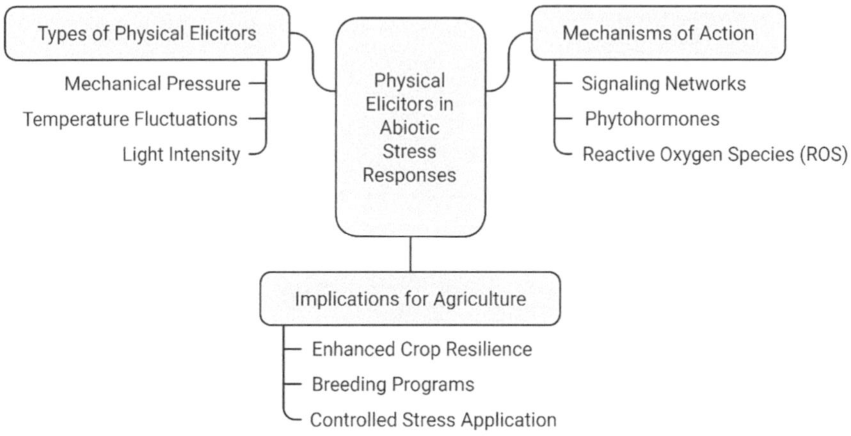

Fig. 2.5 Physical elicitors in abiotic stress responses

profile of plants, increasing vitamin C content and reducing phenolic levels. UV-C radiation enhances resveratrol and piceatannol production in peanut callus cultures and boosts stilbene synthesis in grape calli (Ghozlen et al. 2023).

2.3.2.2 Salt and Drought Stress

Osmotic stress, an abiotic elicitor, impacts plant growth, development, and SMs production. Sucrose, a crucial carbon and energy source, induces water stress in plants, enhancing capsaicin accumulation and steviol glycoside production. Salt stress stimulates SMs like alkaloids, phenols, and terpenes, but decreases anthocyanin levels in salt-sensitive species (Sytar et al. 2018). Sodium chloride enhances vincristine and vinblastine production in plants, while canavanine production does not significantly hinder it. Drought stress affects plant growth, reproductive development, and survival, with tolerance levels varying between species. Studies show that drought stress influences SMs production and accumulation, with higher concentrations of Polyethylene glycol (PEG) causing physiological instability (Zhao et al. 2021).

2.3.2.3 Thermal Stress

Intense heat is a major environmental stress that impedes plant growth and productivity by interfering with several metabolic processes, including the synthesis and breakdown of primary metabolites. Callus tissue induction and cell proliferation in cultures generally occur optimally within a temperature range of 17–25 °C. For example, *Melastoma malabathricum* cells cultured at lower temperatures (20 ± 2 °C) exhibited superior growth and higher anthocyanin production compared to cultures maintained at 26 ± 2 °C and 29 ± 2 °C (Chan et al. 2010). Similarly, anthocyanin production in cell cultures of strawberry and *Perilla frutescens* was found to be maximized at an optimal temperature of 25 °C. While hairy root cultures are typically cultivated at around 25 °C, lowering the temperature to 19.5 °C enhanced both linolenic acid content and total indole alkaloid production in the hairy roots of Catharanthus roseus. Additionally, *Panax quinquefolius* roots exhibited a significant increase in ginsenoside concentration with a 5 °C increase in temperature (Jochum et al. 2007).

2.3.3 Hormonal Elicitors

2.3.3.1 Salicylic Acid

Salicylic acid (SA) is a widely distributed phenolic plant growth regulator present throughout the plant kingdom. It plays a key role in regulating various developmental processes and growth, particularly in response to stress (Fig. 2.6) (Song et al. 2023). This endogenous signaling molecule enables plants to modulate their defense mechanisms against a range of stresses, which negatively impact several crop species. SA is also pivotal in controlling several physiological functions, such as photosynthesis, protein synthesis, transpiration, stomatal regulation, and antioxidant activity (Parveen et al. 2021; Khan et al. 2015). Exogenous application of SA has been associated with enhanced transpiration, increased net photosynthetic rates, improved wateruse efficiency, higher internal CO_2 concentrations, and more efficient sugar translocation. It also helps prevent degradation of the CO_2-fixing enzyme

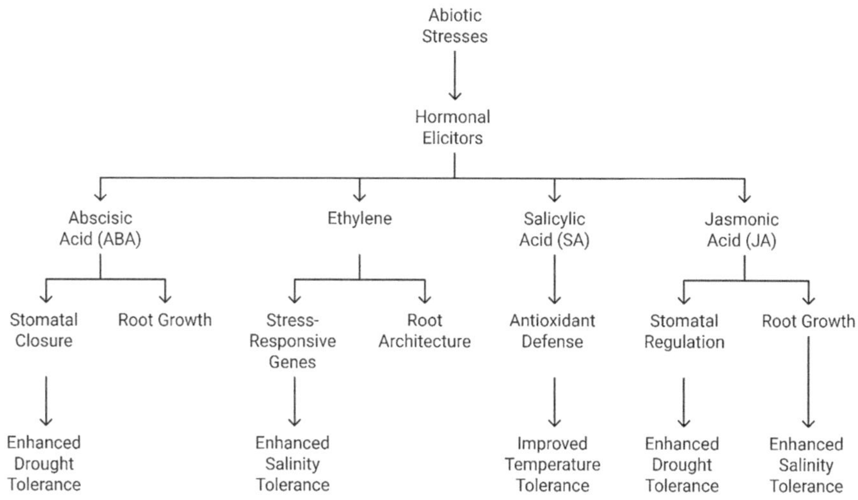

Fig. 2.6 Hormonal elicitors in the context of abiotic stress elicitors

Rubisco (Khalvandi et al. 2021; Garg and Bharti 2018; Gondor et al. 2016). SA promotes root architecture and nutrient uptake, which are essential for maintaining plant health under stress (Khan et al. 2021). In a case study on *Brassica napus* exposed to drought (D), salinity (S), and their combination (D + S), the application of salicylic acid (SA) significantly alleviated the stress effects, reducing leaf area loss by 13.65–21.87%, lowering chlorophyll content reduction by 30–50%, and enhancing antioxidant levels while decreasing malondialdehyde (MDA) content by 47% (Ali et al. 2023). SA enhances the antioxidant defense system, reducing oxidative damage by regulating reactive oxygen species (ROS) production (Yang et al. 2023). It induces the expression of stress-responsive genes, improving overall plant resilience (Song et al. 2023). While SA shows promise in improving abiotic stress tolerance, the underlying mechanisms remain complex and not fully understood. Further research is essential to optimize its application in agricultural practices and enhance crop resilience under changing environmental conditions.

2.3.3.2 Jasmonic Acid

Jasmonic acid (JA) and its methyl ester, methyl jasmonate (MeJA), are widely distributed across the plant kingdom. Both JA and MeJA are recognized as natural regulators of plant growth, influencing a range of physiological processes (Hewedy et al. 2023). While MeJA is generally known to inhibit plant growth, reduce photosynthetic activity, and limit stomatal opening, its role in protecting plants from abiotic stress remains inconsistent (Yan et al. 2015; Anjum et al. 2011). Nonetheless, at low concentrations, MeJA has been shown to act as a stress modulator, promoting growth and enhancing plant resistance to abiotic stress (Santisree et al. 2020). The pivotal functions of jasmonates in plant reactions to various stresses, including heat, heavy metals, drought, and salinity, are well-established (Yang et al. 2021; Raza et al. 2021; Rahman et al. 2024; Lopes et al. 2024). Through modulation of gene expression, MeJA significantly impacts photosynthesis and the antioxidant defense system.

2.4 Plant Hormones and Abiotic Stress Tolerance

Phytohormones, produced in small quantities, regulate signal transduction pathways and play a crucial role in plant response to abiotic stress (Mearaji et al. 2021). External application of plant growth regulators (PGRs), like foliar spraying or seed soaking, improves tolerance (Shah et al. 2021; Banerjee and Roychoudhury 2018). Plant extracts are sometimes used as PGRs to promote sustainable agriculture and reduce the use of hazardous pesticides (Shahid et al. 2024). These extracts, containing biologically active compounds such as zeatin in Moringa oleifera and brassinolides in Brassica species, serve as cost-effective plant growth stimulants and

Table 2.3 Effectiveness of different elicitors in mitigating abiotic stresses in plants

Crops/ plants	Elicitors	Findings	References
Rice	Polyamines	Resistance to salt stress through control of antioxidant and osmolyte levels as well as the regulation of several metabolic genes	Paul and Roychoudhury (2017)
Tomato	Calcium chloride	Higher levels of antioxidant enzymes and less reactive stress	Chakraborty and Acharya (2016)
Radish	Sorghum, brassica, and Moringa leaf extracts	Extracts applied as a growth regulator demonstrated a positive impact on plant antioxidant and biochemical activity as well as growth enhancement	Ashraf et al. (2016)
Pepper	Silicon	Modulated expression of protein, antioxidant metabolism, and physiology under salinity	Manivannan et al. (2016)
Common basil	Chitosan	Improved levels of carotenoid and chlorophyll as well as improved root and shoot growth, plant height, and leaf area under water stress	Malekpoor et al. (2016)
Wheat	aminocyclopropane1-carboxylic acid, abscisic acid and1-ethylene precursor	Enhanced shoot growth under slight drought circumstances	Valluru et al. (2016)
Rice	100 mg l^{-1} (SA)	Resistance to drought stress	Kareem Rihan and Fuller (2017)
Rice	Accumulation of protective compounds	Heat stress resistance	Fahad et al. (2016)
Mung bean	0.5 mM (SA)	Cadmium chloride resistance	Roychoudhury et al. (2016)
Pepper	Hydrogen peroxide (H_2O_2), chitosan, and SA	Oxidative stress response, increases endogenous H_2O_2, expression of pr1, cat1, pal genes, and enzymatic activities	Mejía-Teniente et al. (2013)

protectants against a range of abiotic stresses (EL Sabagh et al. 2022; Ashraf et al. 2016). Table 2.3 shows a full breakdown of the different phytohormones and elicitors that are very important for lowering abiotic stress.

2.5 Phytohormone Crosstalk in Abiotic Stress Responses

Phytohormones are essential in regulating plant processes and stress responses by interacting in complex ways with one another and other signaling molecules. These interactions, which may be either synergistic or antagonistic, profoundly influence plant growth, development, and stress tolerance (Zaid et al. 2021; Iqbal et al. 2022). A deeper understanding of these mechanisms sheds light on how signaling pathways modulate plant responses to abiotic stresses. Abscisic acid is crucial for coping with abiotic stresses like drought and salinity, as it regulates physiological processes such as stomatal closure and enhances overall stress tolerance (Rehman et al. 2021). Nitric oxide (NO) contributes to these processes through intricate interactions with ABA. Research indicates that NO and ABA can have both synergistic and antagonistic effects. For instance, NO plays a key role in ABA-induced stomatal closure, while ABA also controls NO production in guard cells. Arabidopsis mutants lacking NO production display defective ABA signaling, highlighting the importance of NO in ABA pathways (Hancock et al. 2011). Furthermore, the enzyme nitric oxide synthase (NOA1), which is regulated by both ABA and environmental stresses such as NaCl, is essential for NO production (Sun et al. 2010).

Under stress conditions such as drought and UV-B radiation, ABA and NO interact to mediate adaptive responses in plants, including stomatal closure and antioxidant defenses. For example, UV-B exposure elevates levels of ABA, NO, and hydrogen peroxide in maize leaves. In mutants deficient in ABA synthesis, UV-B-induced damage is exacerbated, though ABA application can restore NO and ABA levels, mitigating damage (Wilkinson and Davies 2010). ABA-induced H_2O_2 production facilitates NO accumulation via mitogen-activated protein kinase activity and the upregulation of antioxidant enzymes. Overexpression of the SgNCED1 gene in tobacco enhances ABA levels, resulting in increased H_2O_2 and NO levels, and heightened antioxidant enzyme activity (Zhang et al. 2009) (Table 2.4).

2.6 Mechanisms of Abiotic Stress Elicitors

Abiotic stress elicitors, such as drought, salinity, and temperature extremes, profoundly affect plant physiology by initiating complex biochemical and physiological responses. These stresses activate a range of adaptive mechanisms, including osmotic adjustment, ion homeostasis, stress protein synthesis, and antioxidative defenses, which enable plants to survive and function under adverse conditions. Understanding these mechanisms provides valuable insights into plant adaptation and offers potential strategies for enhancing stress resilience in crops.

2.6.1 Drought Stress Elicitors

Drought stress is primarily caused by reduced water availability, which leads to decreased turgor pressure and cellular function. One of the first responses to drought is the activation of osmotic stress mechanisms. Plants accumulate osmoprotectants like proline, trehalose, and other compatible solutes to preserve cellular turgidity, stabilize proteins, and safeguard cellular structures from damage caused by

Table 2.4 Interaction of various elicitors with different plant species under optimal and stress conditions

Elicitor 1	Elicitor 2	Crop	Condition	Effects	Reference
Methyl jasmonate	Salicylic acid	*Citrus limon*	Chilling stress	SA and MeJ interaction enhance phenolic synthesis and phenylalanine ammonia lyase activity, reduces polyphenol oxidase and peroxidase activities, and decreases chilling-induced membrane permeability and lipid peroxidation	Siboza et al. (2014)
Hydrogen peroxide	Salicylic acid	*Triticum aestivum*	Freezing stress	SA-induced freezing is mediated by cross talk between H_2O_2 and ABA signaling, resulting in reduced cell membrane damage, increased antioxidant capacity, and cold-responsive gene expression	Wang et al. (2018)
Nitric oxide	Salicylic acid	*Zea mays*	Water deficit	Salicylic acid (SA) and nitric oxide (NO) interact to enhance photosynthesis by increasing Rubisco and Rubisco-activase activities, along with elevating transcription of RbcL, ZmRCAα, and ZmRCAβ mRNA	Shao et al. (2018)
28-homobrassinolide	Salicylic acid	*Brassica juncea*	Salt stress	The combined supplementation significantly improved SPAD chlorophyll, photosystem II quantum yield, photosynthetic traits, growth attributes, and nitrate reductase	Hayat et al. (2012)
Nitric oxide	Salicylic acid	*Oryza sativa*	Metalloid phytotoxicity	Salicylic acid and nitric oxide signaling protect plants against arsenite stress by decreasing arsenic translocation and accumulation in roots and shoots, demonstrating a close relationship	Singh et al. (2017a, 2017b)

(continued)

Table 2.4 (continued)

Elicitor 1	Elicitor 2	Crop	Condition	Effects	Reference
Nitric oxide	Salicylic acid	*Lablab purpureus*	Heat stress	Both significantly influenced biophysical and growth processes under stress, by regulating antioxidant enzyme activities, improving plant–water relations, and recovering photosynthesis under stress	Rai et al. (2018)
Salicylic acid	Jasmonic acid	*Noccaea praecox*	Wounding and fungal stress	Disrupted JA/SA levels and SA signaling increase stress vulnerability	Llugany et al. (2013)
Nitric oxide	Salicylic acid	*Arabidopsis thaliana*	Pathogen stress	SA and NO synergistically activate defense against pathogens	Klessig et al. (2000)

dehydration (Taiwo et al. 2020). On a molecular level, drought stress activates ABA signaling pathways. ABA serves as a key regulator in drought responses by promoting stomatal closure to reduce water loss and influencing the expression of genes associated with stress tolerance (Liu et al. 2022). This involves the upregulation of genes encoding dehydration-responsive proteins, including dehydrins and late embryogenesis abundant (LEA) proteins, which help in protein stabilization and cellular repair. Drought stress also activates various transcription factors, such as dehydration-responsive element binding (DREB) and ABA-responsive element binding (AREB), which enhance the expression of stress-responsive genes (Riyazuddin et al. 2022; Abdul Aziz et al. 2021). These factors play a crucial role in orchestrating the plant's adaptation to water-deficient conditions by regulating stress-related pathways and enhancing the synthesis of protective proteins..

2.6.2 *Salinity Stress Elicitors*

The primary response to salinity stress involves the disruption of ionic balance, particularly the accumulation of sodium (Na^+) and chloride (Cl^-) ions in the plant cells. Plants counteract this by employing ion transport systems that include Na^+/H^+ antiporters and salt overly sensitive (SOS) pathways, which help to exclude or compartmentalize excess ions, thereby maintaining ionic homeostasis (Chakraborty et al. 2018). In response to salinity stress, plants also enhance the synthesis of osmoprotectants and upregulate genes related to stress tolerance (Omari Alzahrani 2021). Key components of this response include the activation of antioxidant defense systems to mitigate oxidative stress caused by the accumulation of ROS. Enzymes such as catalase, peroxidase, and superoxide dismutase (SOD) play pivotal roles in detoxifying ROS and preventing oxidative damage (Hasanuzzaman et al. 2021). Salinity stress induces changes in the expression of several genes involved in ion transport, osmotic regulation, and stress protection. For example, the production of stress-responsive proteins, such as osmotic regulators and ion transporters, helps in maintaining cellular integrity and functionality under high salinity conditions (Joshi et al. 2022; Mansour and Hassan 2022).

2.6.3 Temperature Stress Elicitors

Temperature stress encompasses both heat and cold stress, each with distinct impacts on plant physiology. Heat stress leads to the denaturation of proteins and enhanced respiration rates, resulting in increased ROS production (Suzuki 2023). In response, plants synthesize heat shock proteins (HSPs), which function as molecular chaperones to stabilize and refold denatured proteins, thereby mitigating heat-induced damage (Khan and Shahwar 2020). Cold stress induces different responses, including the formation of ice crystals within plant tissues, which can cause cellular damage and inhibit enzymatic activity. To combat this, plants produce cryoprotectants such as soluble sugars and polyols, which help to lower the freezing point of intracellular fluids and stabilize cellular membranes (Román-Figueroa et al. 2021). Cold stress also triggers the expression of cold-responsive genes that contribute to cold acclimation by enhancing cellular protection and adaptation mechanisms (Soualiou et al. 2022). Both heat and cold stress lead to alterations in cellular metabolism and gene expression, involving a network of stress-responsive transcription factors and signaling pathways. These include the C-repeat binding factor pathway (CBF) for cold stress and the heat shock factor (HSF) pathway for heat stress, which regulate the expression of genes involved in stress tolerance and adaptation..

2.7 Conclusion and Future Prospects

This chapter concludes that abiotic stress elicitors present a valuable approach for reducing the harmful effects of drought, salinity, and extreme temperatures on agricultural crops. By triggering specific physiological and molecular pathways, these elicitors bolster plant defense mechanisms and enhance tolerance to stress. Incorporating elicitors into modern agricultural practices emerges as a sustainable and efficient strategy for developing climate-resilient crops, ensuring consistent yields in the face of increasingly unpredictable environmental conditions.

Looking ahead, further research into optimizing elicitor use, particularly across diverse crop species, will be essential. Understanding the molecular mechanisms of elicitors and their application in different crops will help refine their effectiveness. The future of agriculture may see the widespread integration of these elicitors, offering an environmentally friendly and innovative strategy to develop climate-resilient crops, contributing to global efforts to ensure food security amidst changing global conditions.

References

AbdElgawad H, Zinta G, Hegab MM, Pandey R, Asard H, Abuelsoud W (2016) High salinity induces different oxidative stress and antioxidant responses in maize seedlings' organs. Front Plant Sci 7:276

Abdul Aziz M, Sabeem M, Mullath SK, Brini F, Masmoudi K (2021) Plant group II LEA proteins: intrinsically disordered structure for multiple functions in response to environmental stresses. Biomolecules 11(11):1662

Acosta-Motos JR, Ortuño MF, Bernal-Vicente A, Diaz-Vivancos P, Sanchez-Blanco MJ, Hernandez JA (2017) Plant responses to salt stress: adaptive mechanisms. Agronomy 7(1):18

Ahmad B, Zaid A, Sadiq Y, Bashir S, Wani SH (2019) Role of selective exogenous elicitors in plant responses to abiotic stress tolerance. In: Plant abiotic stress tolerance: agronomic, molecular and biotechnological approaches, pp 273–290

Ahmed M, Tóth Z, Decsi K (2024) The impact of salinity on crop yields and the confrontational behavior of transcriptional regulators, nanoparticles, and antioxidant defensive mechanisms under stressful conditions: a review. Int J Mol Sci 25(5):2654

Akhtar SS, Andersen MN, Liu F (2015) Biochar mitigates salinity stress in potato. J Agron Crop Sci 201(5):368–378

Al Hassan M, Fuertes MM, Sánchez FJR, Vicente O, Boscaiu M (2015) Effects of salt and water stress on plant growth and on accumulation of osmolytes and antioxidant compounds in cherry tomato. Not Bot Horti Agrobot Cluj-Napoca 43(1):1–11

Alengebawy A, Abdelkhalek ST, Qureshi SR, Wang MQ (2021) Heavy metals and pesticides toxicity in agricultural soil and plants: ecological risks and human health implications. Toxics 9(3):42

Ali E, Hussain S, Jalal F, Khan MA, Imtiaz M, Said F, Shah F (2023) Salicylic acid mitigates abiotic stress tolerance via altering defense mechanisms in Brassica napus (L.). Front Plant Sci 14:1187260

Amirjani MR (2011) Effect of salinity stress on growth, sugar content, pigments and enzyme activity of rice. Int J Bot 7(1):73–81

Andreo-Jimenez B, Vandenkoornhuyse P, Lê Van A, Heutinck A, Duhamel M, Kadam N, Bouwmeester H (2019) Plant host and drought shape the root associated fungal microbiota in rice. PeerJ 7:e7463

Anjum SA, Xie XY, Farooq M, Wang LC, Xue LI, Shahbaz M, Salhab J (2011) Effect of exogenous methyl jasmonate on growth, gas exchange and chlorophyll contents of soybean subjected to drought. Afr J Biotechnol 10(47):9647–9656

Apel K, Hirt H (2004) Reactive oxygen species: metabolism, oxidative stress, and signal transduction. Annu Rev Plant Biol 55(1):373–399

Ashraf R, Sultana B, Iqbal M, Mushtaq M (2016) Variation in biochemical and antioxidant attributes of Raphanus sativus in response to foliar application of plant leaf extracts as plant growth regulator. J Genet Eng Biotechnol 14(1):1–8

Ashraf MA, Akbar A, Parveen A, Rasheed R, Hussain I, Iqbal M (2018) Phenological application of selenium differentially improves growth, oxidative defense and ion homeostasis in maize under salinity stress. Plant Physiol Biochem 123:268–280

Balasubramaniam T, Shen G, Esmaeili N, Zhang H (2023) Plants' response mechanisms to salinity stress. Plants 12(12):2253

Banerjee A, Roychoudhury A (2018) Strigolactones: multi-level regulation of biosynthesis and diverse responses in plant abiotic stresses. Acta Physiol Plant 40:1–10
Bhagwath SG, Hjortsø MA (2000) Statistical analysis of elicitation strategies for thiarubrine a production in hairy root cultures of Ambrosia artemisiifolia. J Biotechnol 80(2):159–167
Bistgani ZE, Hashemi M, DaCosta M, Craker L, Maggi F, Morshedloo MR (2019) Effect of salinity stress on the physiological characteristics, phenolic compounds and antioxidant activity of Thymus vulgaris L. and thymus daenensis Celak. Ind Crop Prod 135:311–320
Bita CE, Gerats T (2013) Plant tolerance to high temperature in a changing environment: scientific fundamentals and production of heat stress-tolerant crops. Front Plant Sci 4:273
Cai Z, Kastell A, Speiser C, Smetanska I (2013) Enhanced resveratrol production in Vitis vinifera cell suspension cultures by heavy metals without loss of cell viability. Appl Biochem Biotechnol 171:330–340
Cakmak I, Rengel Z (2024) Potassium may mitigate drought stress by increasing stem carbohydrates and their mobilization into grains. J Plant Physiol 303:154325
Cao YY, Duan H, Yang LN, Wang ZQ, Liu LJ, Yang JC (2009) Effect of high temperature during heading and early filling on grain yield and physiological characteristics in indica rice. Acta Agron Sin 35:512–521
Carbone MJ, Alaniz S, Mondino P, Gelabert M, Eichmeier A, Tekielska D, Gramaje D (2021) Drought influences fungal community dynamics in the grapevine rhizosphere and root microbiome. J Fungi 7(9):686
Cardoso AA, Gori A, Da-Silva CJ, Brunetti C (2020) Abscisic acid biosynthesis and signaling in plants: key targets to improve water use efficiency and drought tolerance. Appl Sci 10(18):6322
Carmody M, Waszczak C, Idänheimo N, Saarinen T, Kangasjärvi J (2016) ROS signalling in a destabilised world: a molecular understanding of climate change. J Plant Physiol 203:69–83
Chaffai R, Ganesan M, Cherif A (2024) Plant adaptations and molecular responses to salt stress. In: Plant adaptation to abiotic stress: from signaling pathways and microbiomes to molecular mechanisms. Springer, Singapore, pp 137–176
Chakraborty N, Acharya K (2016) Ex vivo analyses of formulated bio-elicitors from a phytopathogen in the improvement of innate immunity in host. Arch Phytopathol Plant Protect 49(17–18):485–505
Chakraborty K, Basak N, Bhaduri D, Ray S, Vijayan J, Chattopadhyay K, Sarkar RK (2018) Ionic basis of salt tolerance in plants: nutrient homeostasis and oxidative stress tolerance. In: Plant nutrients and abiotic stress tolerance, pp 325–362
Chan LK, Koay SS, Boey PL, Bhatt A (2010) Effects of abiotic stress on biomass and anthocyanin production in cell cultures of Melastoma malabathricum. Biol Res 43(1):127–135
Chaudhary SB, Gujjar D, Kumar B, Patel S (2024) Salinity stress in legume crops: a comprehensive review of effects, mechanisms, and mitigation strategies. J Sci Res Rep 30(7):783–799
Chaudhry S, Sidhu GPS (2022) Climate change regulated abiotic stress mechanisms in plants: a comprehensive review. Plant Cell Rep 41(1):1–31
Devireddy AR, Tschaplinski TJ, Tuskan GA, Muchero W, Chen JG (2021) Role of reactive oxygen species and hormones in plant responses to temperature changes. Int J Mol Sci 22(16):8843
Debray R, Socolar Y, Kaulbach G, Guzman A, Hernandez CA, Curley R, Koskella B (2022) Water stress and disruption of mycorrhizas induce parallel shifts in phyllosphere microbiome composition. New Phytol 234(6):2018–2031
Djanaguiraman M, Prasad PV, Seppanen M (2010) Selenium protects sorghum leaves from oxidative damage under high temperature stress by enhancing antioxidant defense system. Plant Physiol Biochem 48(12):999–1007
Djanaguiraman M, Prasad PVV, Al-Khatib K (2011) Ethylene perception inhibitor 1-MCP decreases oxidative damage of leaves through enhanced antioxidant defense mechanisms in soybean plants grown under high temperature stress. Environ Exp Bot 71:215–223
Dolatabadian A, Sanavy SAMM, Ghanati F (2011) Effect of salinity on growth, xylem structure and anatomical characteristics of soybean. Not Sci Biol 3(1):41–45

Dos Santos TB, Ribas AF, de Souza SGH, Budzinski IGF, Domingues DS (2022) Physiological responses to drought, salinity, and heat stress in plants: a review. Stress 2(1):113–135
Ducatti RD, Wordell Filho JA, Tironi SP, Tramontin MA, Radünz AL, da Silva SM, Mazaro SM (2023) Wheat elicitation performed in open field with the use of different elicitors. Contribuciones a las Ciencias Sociales 16(7):6165–6185
Edreira JIR, Mayer LI, Otegui ME (2014) Heat stress in temperate and tropical maize hybrids. Kernel growth, water relations and assimilate availability for grain filling. Field Crop Res 166:162–172
El Sabagh A, Hossain A, Barutçular C, Islam MS, Ahmad Z, Wasaya A, Hasanuzzaman M (2020) Adverse effect of drought on quality of major cereal crops: implications and their possible mitigation strategies. In: Agron crops vol. 3: stress responses and tolerance, pp 635–658
El Sabagh A, Islam MS, Hossain A, Iqbal MA, Mubeen M, Waleed M, Abdelhamid MT (2022) Phytohormones as growth regulators during abiotic stress tolerance in plants. Front Agron 4:765068
Fahad S, Hussain S, Saud S, Hassan S, Ihsan Z, Shah AN, Huang J (2016) Exogenously applied plant growth regulators enhance the morpho-physiological growth and yield of rice under high temperature. Front Plant Sci 7:1250
Flowers TJ, Colmer TD (2015) Plant salt tolerance: adaptations in halophytes. Ann Bot 115(3):327–331
Gaddam NR, Devi TM, Rupali JS, Darapureddy NSSS, Reddy GR (2024) Exploiting induced plant resistance for sustainable pest management: mechanisms, elicitors, and applications: a review. J Exp Agric Int 46(9):586–599
Garg N, Bharti A (2018) Salicylic acid improves arbuscular mycorrhizal symbiosis, and chickpea growth and yield by modulating carbohydrate metabolism under salt stress. Mycorrhiza 28(8):727–746
Ghazal B, Fareed A, Ahmad N, Azra, Salmen SH, Ansari MJ, Qayyum A (2024) Elicitors directed in vitro growth and production of stevioside and other secondary metabolites in Stevia rebaudiana (Bertoni) Bertoni. Sci Rep 14(1):14714
Ghozlen HB, Mangelinckx S, Werbrouck SP (2023) Mycorrhizal and pathogenic fungi as bioelicitors for in vitro production of natural stilbenes in callus and cell suspension cultures of peanut (Arachis hypogaea L.). Process Biochem 135:1–11
Gondor OK, Janda T, Soós V, Pál M, Majláth I, Adak MK, Szalai G (2016) Salicylic acid induction of flavonoid biosynthesis pathways in wheat varies by treatment. Front Plant Sci 7:1447
Gunawardhana MDM, de Silva CS (2011) Impact of temperature and water stress on growth yield and related biochemical parameters of okra. Trop Agric Res 23:77–83
Haddadi BS, Hassanpour H, Niknam V (2016) Effect of salinity and waterlogging on growth, anatomical and antioxidative responses in Mentha aquatica L. Acta Physiol Plant 38:1–11
Hancock JT, Neill SJ, Wilson ID (2011) Nitric oxide and ABA in the control of plant function. Plant Sci 181(5):555–559
Hasanuzzaman M, Raihan MRH, Masud AAC, Rahman K, Nowroz F, Rahman M, Fujita M (2021) Regulation of reactive oxygen species and antioxidant defense in plants under salinity. Int J Mol Sci 22(17):9326
Hatfield JL, Prueger JH (2015) Temperature extremes: effect on plant growth and development. Weather Clim Extremes 10:4–10
Hayat S, Maheshwari P, Wani AS, Irfan M, Alyemeni MN, Ahmad A (2012) Comparative effect of 28 homobrassinolide and salicylic acid in the amelioration of NaCl stress in Brassica juncea L. Plant Physiol Biochem 53:61–68
Hewedy OA, Elsheery NI, Karkour AM, Elhamouly N, Arafa RA, Mahmoud GAE, Brestic M (2023) Jasmonic acid regulates plant development and orchestrates stress response during tough times. Environ Exp Bot 208:105260
Hmidi D, Abdelly C, Athar HUR, Ashraf M, Messedi D (2018) Effect of salinity on osmotic adjustment, proline accumulation and possible role of ornithine-δ-aminotransferase in proline biosynthesis in Cakile maritima. Physiol Mol Biol Plants 24:1017–1033

Hu L, Li H, Pang H, Fu J (2012) Responses of antioxidant gene, protein and enzymes to salinity stress in two genotypes of perennial ryegrass (Lolium perenne) differing in salt tolerance. J Plant Physiol 169(2):146–156

Hu Y, Yue J, Nie J, Luo D, Cao S, Wang C, Chen P (2023) Salicylic acid alleviates the salt toxicity in kenaf by activating antioxidant system and regulating crucial pathways and genes. J Ind Crops Prod 199:116691

Hualpa-Ramirez E, Carrasco-Lozano EC, Madrid-Espinoza J, Tejos R, Ruiz-Lara S, Stange C, Norambuena L (2024) Stress salinity in plants: new strategies to cope with in the foreseeable scenario. Plant Physiol Biochem 208:108507

Humbal A, Pathak B (2023) Influence of exogenous elicitors on the production of secondary metabolite in plants: a review ("VSI: secondary metabolites"). Plant Stress 8:100166

Hurkman WJ, Vensel WH, Tanaka CK, Whitehand L, Altenbach SB (2009) Effect of high temperature on albumin and globulin accumulation in the endosperm proteome of the developing wheat grain. J Cereal Sci 49:12–23

Hussain S, Khan M, Altaf MT, Shah MN, Alfagham AT (2025) Deciphering the morpho-physiological and biochemical response of sunflower hybrids with the application of biochar and slow-release nitrogen fertilizers under drought stress for sustainable crop production. Front Plant Sci 16:1541123

Ifeduba AM, Zhen S, Pandey J, Vales MI (2024) Leaf membrane stability under high temperatures as an indicator of heat tolerance in potatoes and genome-wide association studies to understand the underlying genetics. Plants 13(16):2175

Inam M, Attique I, Zahra M, Khan AK, Hahim M, Hano C, Anjum S (2023) Metal oxide nanoparticles and plant secondary metabolism: unraveling the game-changer nano-elicitors. Plant Cell Tissue Organ Cult 155(2):327–344

Iqbal S, Wang X, Mubeen I, Kamran M, Kanwal I, Díaz GA, Fahad S (2022) Phytohormones trigger drought tolerance in crop plants: outlook and future perspectives. Front Plant Sci 12:799318

Jha Y, Subramanian RB (2016) Regulation of plant physiology and antioxidant enzymes for alleviating salinity stress by potassium-mobilizing bacteria. In: Potassium solubilizing microorganisms for sustainable agriculture, pp 149–162

Jochum GM, Mudge KW, Thomas RB (2007) Elevated temperatures increase leaf senescence and root secondary metabolite concentrations in the understory herb Panax quinquefolius. Am J Bot 94(5):819–826

Joshi S, Nath J, Singh AK, Pareek A, Joshi R (2022) Ion transporters and their regulatory signal transduction mechanisms for salinity tolerance in plants. Physiol Plant 174(3):e13702

Joshi MK, Marviya GV, Jacob F, Kandoliya UK, Pandya PM, Vala AG (2024) System-wide analysis of groundnut's salinity resilience: integrating plant-cell interactions with environmental stress dynamics through cutting-edge transcriptomics. J Biotechnol 394:34–47

Jumrani K, Bhatia VS, Kataria S, Rastogi A (2024) The interactive effect of high temperature and water deficit stress on nitrogen fixation, photosynthesis, chlorophyll fluorescence, seed yield and quality in soybean. Plant Physiol. Rep. 29(1):125–140

Kamatchi KM, Anitha K, Kumar KA, Senthil A, Kalarani MK, Djanaguiraman M (2024) Impacts of combined drought and high-temperature stress on growth, physiology, and yield of crops. Plant Physiol Rep 29(1):28–36

Kang SM, Khan AL, Waqas M, You YH, Kim JH, Kim JG, Lee IJ (2014) Plant growth-promoting rhizobacteria reduce adverse effects of salinity and osmotic stress by regulating phytohormones and antioxidants in Cucumis sativus. J Plant Interact 9(1):673–682

Kareem Rihan H, Fuller M (2017) The effect of exogenous applications of salicylic acid and molybdenum on the tolerance of drought in wheat. Agric. Res. Technol. Open Access J 9(4). https://doi.org/10.19080/ARTOAJ.2017.09.555768

Khalvandi M, Siosemardeh A, Roohi E, Keramati S (2021) Salicylic acid alleviated the effect of drought stress on photosynthetic characteristics and leaf protein pattern in winter wheat. Heliyon 7(1):e05908

Khan Z, Shahwar D (2020) Role of heat shock proteins (HSPs) and heat stress tolerance in crop plants. In: Sustainable agriculture in the era of climate change, pp 211–234
Khan MIR, Fatma M, Per TS, Anjum NA, Khan NA (2015) Salicylic acid-induced abiotic stress tolerance and underlying mechanisms in plants. Front Plant Sci 6:462
Khan N, Ali S, Shahid MA, Mustafa A, Sayyed RZ, Curá JA (2021) Insights into the interactions among roots, rhizosphere, and rhizobacteria for improving plant growth and tolerance to abiotic stresses: a review. Cells 10(6):1551
Khondoker M, Mandal S, Gurav R, Hwang S (2023) Freshwater shortage, salinity increase, and global food production: a need for sustainable irrigation water desalination–a scoping review. Earth 4(2):223–240
Klessig DF, Durner J, Noad R, Navarre DA, Wendehenne D, Kumar D, Silva H (2000) Nitric oxide and salicylic acid signaling in plant defense. Proc Natl Acad Sci USA 97(16):8849–8855
Kumar S, Bhushan B, Wakchaure GC, Dutta R, Jat BS, Meena KK, Pathak H (2023) Unveiling the impact of heat stress on seed biochemical composition of major cereal crops: implications for crop resilience and nutritional value. Plant Stress 9:100183
Laghmouchi Y, Belmehdi O, Bouyahya A, Senhaji NS, Abrini J (2017) Effect of temperature, salt stress and pH on seed germination of medicinal plant Origanum compactum. Biocatal Agric Biotechnol 10:156–160
Lal MK, Tiwari RK, Gahlaut V, Mangal V, Kumar A, Singh MP, Zinta G (2021) Physiological and molecular insights on wheat responses to heat stress. Plant Cell Rep 41:501–518
Liaqat W, Barutcular C, Farooq M, Ahmad H, Jan M, Ahmad Z, Li M (2022) Climate change in relation to agriculture: a review. Span J Agric Res 20(2):e03R01
Liaqat W, Altaf MT, Barutçular C, Mohamed HI, Ahmad H, Jan MF, Khan EH (2024a) Sorghum: a star crop to combat abiotic stresses, food insecurity, and hunger under a changing climate: a review. J Soil Sci Plant Nutr 24(1):74–101
Liaqat W, Altaf MT, Barutçular C, Mohamed HI, Ali Z, Khan MO (2024b) Drought stress in sorghum: physiological tools, breeding technology, omics approaches and genomic-assisted breeding—a review. J. Soil Sci. Plant Nutr. 24:1665–1691
Liaqat W, Altaf MT, Barutçular C, Nawaz H, Ullah I, Basit A, Mohamed HI (2024c) Ultraviolet-B radiation in relation to agriculture in the context of climate change: a review. Cereal Res Commun 52(1):1–24
Liu L, Huang L, Sun C, Wang L, Jin C, Lin X (2021) Cross-talk between hydrogen peroxide and nitric oxide during plant development and responses to stress. J Agric Food Chem 69(33):9485–9497
Liu H, Song S, Zhang H, Li Y, Niu L, Zhang J, Wang W (2022) Signaling transduction of ABA, ROS, and Ca2+ in plant stomatal closure in response to drought. Int J Mol Sci 23(23):14824
Liu X, Chang X, Wang D, Wang Y, Wang X, Meng Q, Wang P (2024) Impact of individual and combined water deficit and high temperature on maize seed set: a field study. Plant Growth Regul 104:789–802
Llugany M, Martin SR, Barceló J, Poschenrieder C (2013) Endogenous jasmonic and salicylic acids levels in the Cd-hyperaccumulator Noccaea (Thlaspi) praecox exposed to fungal infection and/or mechanical stress. Plant Cell Rep 32:1243–1249
Lone R, Hassan N, Bashir B, Rohela GK, Malla NA (2023) Role of growth elicitors and microbes in stress management and sustainable production of Sorghum. Plant Stress 9:100179
Lopes AS, Dias TJ, Henschel JM, da Silva TI, de Moura VS, Silva AMO, Batista DS (2024) Methyl jasmonate mitigates salt stress and increases quality of purple basil (Ocimum basilicum L.). S Afr J Bot 171:710–718
Maity S, Shrivastav SP (2024) Understanding heat stress and tolerance mechanisms in wheat (Triticum aestivum L.): a comprehensive review. J. Adv. Biol Biotechnol 27(7):1196–1211
Malekpoor F, Pirbalouti AG, Salimi A (2016) Effect of foliar application of chitosan on morphological and physiological characteristics of basil under reduced irrigation. Res Crops 17(2):354–359

Malik Z, Afzal S, Danish M, Abbasi GH, Bukhari SAH, Khan MI, Ali S (2020) Role of nitric oxide and calcium signaling in abiotic stress tolerance in plants. In: Protective Chemical Agents in the Amelioration of plant abiotic stress: Biochemical Molecular Perspectives, pp 563–581

Manivannan A, Soundararajan P, Muneer S, Ko CH, Jeong BR (2016) Silicon mitigates salinity stress by regulating the physiology, antioxidant enzyme activities, and protein expression in Capsicum annuum 'Bugwang'. Biomed Res Int 2016(1):3076357

Mansour MMF, Hassan FA (2022) How salt stress-responsive proteins regulate plant adaptation to saline conditions. Plant Mol Biol 108(3):175–224

Marsic NK, Vodnik D, Mikulic-Petkovsek M, Veberic R, Sircelj H (2018) Photosynthetic traits of plants and the biochemical profile of tomato fruits are influenced by grafting, salinity stress, and growing season. J Agric Food Chem 66(22):5439–5450

Mearaji HS, Ansari A, Igdelou NKM, Lajayer BA, Pessarakli M (2021) Phytohormones and abiotic stresses: roles of phytohormones in plants under abiotic stresses. In: Handbook of plant and crop physiology. CRC Press, pp 175–213

Meena M, Yadav G, Sonigra P, Nagda A, Mehta T, Swapnil P, Marwal A (2022) Role of elicitors to initiate the induction of systemic resistance in plants to biotic stress. Plant Stress 5:100103

Mejía-Teniente L, de Dalia Durán-Flores F, Chapa-Oliver AM, Torres-Pacheco I, Cruz-Hernández A, González-Chavira MM, Guevara-González RG (2013) Oxidative and molecular responses in Capsicum annuum L. after hydrogen peroxide, salicylic acid and chitosan foliar applications. Int J Mol Sci 14(5):10178–10196

Melo-Sabogal DV, Contreras-Medina LM (2024) Elicitors and biostimulants to mitigate water stress in vegetables. Horticulturae 10(8):837

Mittal S, Kumari N, Sharma V (2012) Differential response of salt stress on Brassica juncea: photosynthetic performance, pigment, proline, D1 and antioxidant enzymes. Plant Physiol Biochem 54:17–26

Mohammed AR, Tarpley L (2010) Effects of high night temperature and spikelet position on yield-related parameters of rice (Oryza sativa L.) plants. Eur J Agron 33:117–123

Nxele X, Klein A, Ndimba BK (2017) Drought and salinity stress alters ROS accumulation, water retention, and osmolyte content in sorghum plants. S Afr J Bot 108:261–266

Omari Alzahrani F (2021) Metabolic engineering of osmoprotectants to elucidate the mechanism(s) of salt stress tolerance in crop plants. Planta 253(1):24

Parihar P, Singh S, Singh R, Singh VP, Prasad SM (2015) Effect of salinity stress on plants and its tolerance strategies: a review. Environ Sci Pollut Res 22:4056–4075

Parveen A, Arslan Ashraf M, Hussain I, Perveen S, Rasheed R, Mahmood Q, Abd Allah EF (2021) Promotion of growth and physiological characteristics in water-stressed Triticum aestivum in relation to foliar-application of salicylic acid. Water 13(9):1316

Paul S, Roychoudhury A (2017) Seed priming with spermine and spermidine regulates the expression of diverse groups of abiotic stress-responsive genes during salinity stress in the seedlings of indica rice varieties. Plant Gene 11:124–132

Qados AMA (2011) Effect of salt stress on plant growth and metabolism of bean plant Vicia faba (L.). J. Saudi Soc. Agric. Sci. 10(1):7–15

Qiao M, Hong C, Jiao Y, Hou S, Gao H (2024) Impacts of drought on photosynthesis in major food crops and the related mechanisms of plant responses to drought. Plants 13(13):1808

Rahman MA, Chikushi J, Yoshida S, Karim AJMS (2009) Growth and yield components of wheat genotypes exposed to high temperature stress under control environment. Bangl J Agric Res 34(3):360–372

Rahman MM, Mostofa MG, Keya SS, Ghosh PK, Abdelrahman M, Anik TR, LSP T (2024) Jasmonic acid priming augments antioxidant defense and photosynthesis in soybean to alleviate combined heat and drought stress effects. Plant Physiol Biochem 206:108193

Rahneshan Z, Nasibi F, Moghadam AA (2018) Effects of salinity stress on some growth, physiological, biochemical parameters and nutrients in two pistachio (Pistacia vera L.) rootstocks. J Plant Interact 13(1):73–82

Rai KK, Rai N, Rai SP (2018) Salicylic acid and nitric oxide alleviate high temperature induced oxidative damage in Lablab purpureus L plants by regulating bio-physical processes and DNA methylation. Plant Physiol Biochem 128:72–88

Rajput J, Singh AK, Singh S, Singh S, Singh S, Singh AK, Yadav RK (2023) Physiological impact of heat stress on wheat varities at different growth stages. Plant Arch 23(2)

Rani A, Guleria M, Sharma Y, Sharma S, Chaudhary A, Sharma R, Kumar P (2023) Insights into elicitor's role in augmenting secondary metabolites production and climate resilience in genus Ocimum–a globally important medicinal and aromatic crop. Ind Crops Prod 202:117078

Rattalino-Edreira J, Otegui M (2013) Heat stress in temperate and tropical maize hybrids. A novel approach for assessing sources of kernel loss in field conditions. J Agron Crop Sci 142:58–67

Raza A, Charagh S, Najafi-Kakavand S, Siddiqui MH (2021) The crucial role of jasmonates in enhancing heavy metals tolerance in plants. In: Jasmonates and Salicylates Signaling in Plants, pp 159–183

Rehman A, Azhar MT, Hinze L, Qayyum A, Li H, Peng Z, Du X (2021) Insight into abscisic acid perception and signaling to increase plant tolerance to abiotic stress. J Plant Interact 16(1):222–237

Rehman T, Tabassum B, Yousaf S, Sarwar G, Qaisar U (2022) Consequences of drought stress encountered during seedling stage on physiology and yield of cultivated cotton. Front Plant Sci 13:906444

Riyazuddin R, Nisha N, Singh K, Verma R, Gupta R (2022) Involvement of dehydrin proteins in mitigating the negative effects of drought stress in plants. Plant Cell Rep 41(3):519–533

Rogowska A, Szakiel A (2021) Enhancement of phytosterol and triterpenoid production in plant hairy root cultures—simultaneous stimulation or competition? Plants 10(10):2028

Román-Figueroa C, Bravo L, Paneque M, Navia R, Cea M (2021) Chemical products for crop protection against freezing stress: a review. J Agron Crop Sci 207(3):391–403

Roy D, Biswakarma N, Ghosh T, Bag K, Sarkar A, Paul K, Hari Krishna B (2024) Climate change on seeds physiology. In: Climate change impacts on soil-plant-atmosphere continuum. Springer, Singapore, pp 347–382

Roychoudhury A, Ghosh S, Paul S, Mazumdar S, Das G, Das S (2016) Pre-treatment of seeds with salicylic acid attenuates cadmium chloride-induced oxidative damages in the seedlings of mungbean (Vigna radiata L. Wilczek). Acta Physiol Plant 38:1–18

Rudrappa T, Neelwarne B, Aswathanarayana RG (2004) In situ and ex situ adsorption and recovery of betalains from hairy root cultures of Beta vulgaris. Biotechnol Prog 20(3):777–785

Salmanpour V, Sharafi S, Rezaei M, Khalili M, Abbaspour N (2024) Evaluation of morphophysiological characteristics of wheat genotypes in different foliar spraying treatments under normal conditions and salinity stress. Plant Sci Today 11(3):159–171

Santisree P, Jalli LCL, Bhatnagar-Mathur P, Sharma KK (2020) Emerging roles of salicylic acid and jasmonates in plant abiotic stress responses. In: Protective Chemical Agents in the Amelioration of Plant Abiotic Stress: Biochemical Molecular Perspectives, pp 342–373

Selwal N, Goutam U, Akhtar N, Sood M, Kukreja S (2024) Elicitation: "a trump card" for enhancing secondary metabolites in plants. J Plant Growth Regul 43:3027–3047

Shabala S, Munns R (2017) Salinity stress: physiological constraints and adaptive mechanisms. In: Plant stress physiology. CABI, Wallingford, pp 24–63

Shah SH, Islam S, Parrey ZA, Mohammad F (2021) Role of exogenously applied plant growth regulators in growth and development of edible oilseed crops under variable environmental conditions: a review. J Soil Sci Plant Nutr 21(4):3284–3308

Shahid M, Shafi Z, Ilyas T, Singh UB, Pichtel J (2024) Crosstalk between phytohormones and pesticides: insights into unravelling the crucial roles of plant growth regulators in improving crop resilience to pesticide stress. Sci Hortic 338:113663

Shao RX, Xin LF, Guo JM, Zheng HF, Mao J, Han XP, Elmore RW (2018) Salicylic acid-induced photosynthetic adaptability of Zea mays L. to polyethylene glycol-simulated water deficit is associated with nitric oxide signaling. Photosynthetica 56:1370–1377

Shekari F, Abbasi A, Mustafavi SH (2017) Effect of silicon and selenium on enzymatic changes and productivity of dill in saline condition. J Saudi Soc Agric Sci 16(4):367–374

Siboza XI, Bertling I, Odindo AO (2014) Salicylic acid and methyl jasmonate improve chilling tolerance in cold-stored lemon fruit (Citrus limon). J Plant Physiol 171(18):1722–1731
Singh AP, Dixit G, Kumar A, Mishra S, Kumar N, Dixit S, Tripathi RD (2017a) A protective role for nitric oxide and salicylic acid for arsenite phytotoxicity in rice (Oryza sativa L.). Plant Physiol Biochem 115:163–173
Singh M, Khan MM, Uddin M, Naeem M, Qureshi MI (2017b) Proliferating effect of radiolytically depolymerized carrageenan on physiological attributes, plant water relation parameters, essential oil production and active constituents of Cymbopogon flexuosus Steud. Under drought stress. PLoS One 12(7):e0180129
Song W, Shao H, Zheng A, Zhao L, Xu Y (2023) Advances in roles of salicylic acid in plant tolerance responses to biotic and abiotic stresses. Plants 12(19):3475
Soualiou S, Duan F, Li X, Zhou W (2022) Crop production under cold stress: an understanding of plant responses, acclimation processes, and management strategies. Plant Physiol Biochem 190:47–61
Sun LR, Hao FS, Lu BS, Ma LY (2010) AtNOA1 modulates nitric oxide accumulation and stomatal closure induced by salicylic acid in Arabidopsis. Plant Signal Behav 5(8):1022–1024
Surówka E, Rapacz M, Janowiak F (2020) Climate change influences the interactive effects of simultaneous impact of abiotic and biotic stresses on plants. In: Hasanuzzaman M (ed) Plant ecophysiology and adaptation under climate change: mechanisms and perspectives I. Springer, Singapore
Suzuki N (2023) Fine tuning of ROS, redox and energy regulatory systems associated with the functions of chloroplasts and mitochondria in plants under heat stress. Int J Mol Sci 24(2):1356
Sytar O, Mbarki S, Zivcak M, Brestic M (2018) The involvement of different secondary metabolites in salinity tolerance of crops. In: Salinity responses and tolerance in plants, volume 2: exploring RNAi, genome editing and systems biology, pp 21–48
Taïbi K, Taïbi F, Abderrahim LA, Ennajah A, Belkhodja M, Mulet JM (2016) Effect of salt stress on growth, chlorophyll content, lipid peroxidation and antioxidant defence systems in Phaseolus vulgaris L. S Afr J Bot 105:306–312
Taiwo AF, Daramola O, Sow M, Semwal VK (2020) Ecophysiology and responses of plants under drought. In: Plant ecophysiology and adaptation under climate change: mechanisms and perspectives I: general consequences and plant responses, pp 231–268
Talukder A, McDonald GK, Gill GS (2014) Effect of short-term heat stress prior to flowering and early grain set on the grain yield of wheat. Field Crop Res 160:54–63
Tamrazov TH (2022) Drought effects on physiological parameters of durum and bread wheat. SABRAO J Breed Genet 54(4):876–884
Tan W, Meng QW, Brestic M, Olsovska K, Yang X (2011) Photosynthesis is improved by exogenous calcium in heat-stressed tobacco plants. J Plant Physiol 168:2063–2071
Tang X, Mu X, Shao H, Wang H, Brestic M (2015) Global plant-responding mechanisms to salt stress: physiological and molecular levels and implications in biotechnology. Crit Rev Biotechnol 35(4):425–437
Tholen D, Boom C, Zhu XG (2012) Opinion: prospects for improving photosynthesis by altering leaf anatomy. Plant Sci 197:92–101
Valluru R, Davies WJ, Reynolds MP, Dodd IC (2016) Foliar abscisic acid-to-ethylene accumulation and response regulate shoot growth sensitivity to mild drought in wheat. Front Plant Sci 7:461
Vanaja M, Sarkar B, Sathish P, Jyothi Lakshmi N, Yadav SK, Mohan C, Singh VK (2024) Elevated CO2 ameliorates the high temperature stress effects on physio-biochemical, growth, yield traits of maize hybrids. Sci Rep 14(1):2928
Wang W, Wang X, Huang M, Cai J, Zhou Q, Dai T, Jiang D (2018) Hydrogen peroxide and abscisic acid mediate salicylic acid-induced freezing tolerance in wheat. Front Plant Sci 9:1137
Wang J, Song L, Gong X, Xu J, Li M (2020) Functions of jasmonic acid in plant regulation and response to abiotic stress. Int J Mol Sci 21(4):1446
Wilkinson S, Davies WJ (2010) Drought, ozone, ABA and ethylene: new insights from cell to plant to community. Plant Cell Environ 33(4):510–525

Xu Y, Song D, Qi X, Asad M, Wang S, Tong X, Wang S (2023) Physiological responses and transcriptome analysis of soybean under gradual water deficit. Front Plant Sci 14:1269884

Yamori W, Hikosaka K, Way DA (2014) Temperature response of photosynthesis in C3, C4, and CAM plants: temperature acclimation and temperature adaptation. Photosynth Res 119:101–117

Yan Z, Zhang W, Chen J, Li X (2015) Methyl jasmonate alleviates cadmium toxicity in Solanum nigrum by regulating metal uptake and antioxidative capacity. Biol Plant 59(2):373–381

Yang J, Miao W, Chen J (2021) Roles of jasmonates and brassinosteroids in rice responses to high temperature stress–a review. The Crop J 9(5):977–985

Yang H, Fang R, Luo L, Yang W, Huang Q, Yang C, Wang J (2023) Uncovering the mechanisms of salicylic acid-mediated abiotic stress tolerance in horticultural crops. Front Plant Sci 14:1226041

Yasin S, Zavala-García F, Niño-Medina G, Rodríguez-Salinas PA, Gutiérrez-Diez A, Sinagawa-García SR, Lugo-Cruz E (2024) Morphological and physiological response of maize (Zea mays L.) to drought stress during reproductive stage. Agronomy 14(8):1718

Yin Y, Li S, Liao W, Lu Q, Wen X, Lu C (2010) Photosystem II photochemistry, photoinhibition, and the xanthophyll cycle in heat-stressed rice leaves. J Plant Physiol 167(12):959–966

Zaid A, Mushtaq M, Wani SH (2021) Interactions of phytohormones with abiotic stress factors under changing climate. In: Frontiers in plant-soil interaction. Academic, pp 221–236

Zhang Y, Tan J, Guo Z, Lu S, He S, Shu W, Zhou B (2009) Increased abscisic acid levels in transgenic tobacco over-expressing 9 cis-epoxycarotenoid dioxygenase influence H2O2 and NO production and antioxidant defences. Plant Cell Environ 32(5):509–519

Zhang X, Cai J, Wollenweber B, Liu F, Dai T, Cao W, Jiang D (2013) Multiple heat and drought events affect grain yield and accumulations of high molecular weight glutenin subunits and glutenin macropolymers in wheat. J Cereal Sci 57:134–140

Zhang Q, Gong M, Yuan J, Hou Y, Zhang H, Wang Y, Hou X (2017) Dark septate endophyte improves drought tolerance in Sorghum. Int J Agric Biol 19(1)

Zhao M, Ren Y, Wei W, Yang J, Zhong Q, Li Z (2021) Metabolite analysis of Jerusalem artichoke (Helianthus tuberosus L.) seedlings in response to polyethylene glycol-simulated drought stress. Int J Mol Sci 22(7):3294

Chapter 3
Role of Elicitors on Crop Physiology and Morphology Under Abiotic Stress

Tony Kevork Sajyan, Keyanna Radi, Saleh M. AlTurki, Mohamed M. El-Mogy, and Rania Francis

Abstract Abiotic stresses such as salinity and drought significantly affect crop physiology and morphology, leading to reduced yield and productivity. Elicitors, which are bioactive compounds derived from biotic or abiotic sources, have emerged as beneficial agents in enhancing crop tolerance against these stresses. This chapter explores the role of elicitors in modulating crop physiology and morphology under adverse environmental conditions.

Elicitors can be categorized into abiotic (chemical and physical) and biotic (microbial and plant-derived) groups. Their mechanisms of action involve stress sensing, activation of plant defense pathways, signal transduction, and regulation of stress-responsive genes. These processes contribute to enhanced photosynthesis, improved water use efficiency, and better nutrient uptake. Furthermore, elicitors promote physiological adaptations such as osmoregulation and hormonal balance, which are critical for plant survival under stress.

Morphologically, elicitors influence root architecture, leading to deeper and more efficient root systems that facilitate water and nutrient acquisition. They also reinforce the structural integrity of leaves and stems, enhancing the plant's ability to withstand environmental stressors. Additionally, elicitor-mediated responses

T. K. Sajyan (✉)
Faculty of Agricultural and Food Sciences, American University of Beirut, Beirut, Lebanon

Department of Agriculture and Food Engineering, School of Engineering, Holy Spirit University of Kaslik, Jounieh, Lebanon
e-mail: ts71@aub.edu.lb

K. Radi
Faculty of Agricultural and Food Sciences, American University of Beirut, Beirut, Lebanon

S. M. AlTurki · M. M. El-Mogy
Department of Arid Land Agriculture, College of Agricultural and Food Science, King Faisal University, Al-Ahsa, Saudi Arabia

R. Francis
Department of Agriculture and Food Engineering, School of Engineering, Holy Spirit University of Kaslik, Jounieh, Lebanon

H. I. Mohamed (ed.), *Elicitors for Sustainable Crop Production*, Sustainability Sciences in Asia and Africa, https://doi.org/10.1007/978-981-95-3533-0_3

support reproductive development, ensuring sustained crop productivity even under unfavorable conditions.

The application of elicitors in agricultural practices is gaining attention due to their potential in sustainable crop management. Methods such as foliar sprays, soil treatments, and seed priming have demonstrated significant improvements in stress tolerance. Elicitors have shown potential in alleviating the effects of drought, salinity, temperature extremes, and oxidative stress, making them valuable methods for climate-resilient agriculture.

In conclusion, elicitors play a vital role in enhancing crop physiology and morphology under abiotic stress conditions. Further research into their mechanisms and large-scale applications could provide innovative solutions for improving crop tolerance and production facing global climate challenges.

Keywords Elicitors · Abiotic stress · Crop physiology · Plant morphology · Stress tolerance · Plant small peptides

3.1 Introduction

The global agricultural landscape is undergoing unprecedented pressure as the need to enhance food production increases in tandem with rapid population growth and the mounting effects of climate change. This challenge is further compounded by the demand for environmentally sustainable practices in farming systems (Ronga et al. 2020). Farmers must now improve crop yields under increasingly unpredictable environmental conditions, while researchers are urgently working to develop new technologies that not only promote productivity but also minimize ecological impact. A key limitation to crop growth and productivity is the exposure of plants to abiotic stress factors such as nutrient deficiency, drought, and salinity. Plants, as sessile organisms, cannot escape environmental stressors and therefore rely on intricate internal defense mechanisms to survive. Central to this adaptive capacity is the concept of the growth–defense tradeoff; plants must allocate limited resources between physiological processes like photosynthesis and reproduction, and defense strategies that mitigate damage from pathogens or environmental extremes (He et al. 2022).

Plants have evolved sophisticated immune systems that detect and respond to stress signals through the recognition of conserved molecular patterns. These include microbe-associated molecular patterns (MAMPs) and others. These molecules are perceived by pattern recognition receptors (PRRs), triggering a suite of defense reactions known as pattern-triggered immunity (PTI) (Boller and Felix 2009). Elicitor molecules bind to unique receptor proteins found on plant cell membranes. The receptors can recognize the chemical pattern of elicitors, which activates intracellular defense signaling and causes substantial synthesis of secondary metabolites, phytoalexins, which reduce damage and boost tolerance to pests, illnesses, and environmental challenges (Gowthami 2018).

Among the most promising areas in the study of plant immunity are peptide elicitors—small peptide molecules that play key roles in plant defense signaling. These peptides are often derived from precursor proteins known as plant elicitor peptides (PROPEPs) and serve as endogenous signals to amplify immune responses. The PROPEP–plant elicitor peptide receptor (PEPR) signaling system is well-documented in several crop species, where different *PROPEP* genes are expressed in response to specific biotic and abiotic stressors. For example, *Arabidopsis thaliana PROPEP1* (*AtPROPEP1*) is activated by fungal infection, while *AtPROPEP3* responds to herbivory (Bartels et al. 2013).

Despite the increasing recognition of peptide signaling, the molecular mechanisms governing the release and perception of PROPEPs remain poorly understood. Two main models attempt to explain peptide activity: the "damage model," where peptides are released upon cellular disruption to alert neighboring cells, and the "amplification model," in which peptide secretion serves to maintain or propagate the immune response via autocrine or paracrine signaling. Notably, PROPEPs lack conventional signal sequences for entry into the secretory pathway, suggesting they are secreted through noncanonical routes, similar to leaderless secretory proteins such as interleukin-1β in animals (Ding et al. 2012).

Beyond PROPEPs, a diverse group of small peptides_usually fewer than 100 amino acids—contribute to plant development and environmental adaptation. These include peptides derived from small open reading frames (sORFs) and larger precursor proteins that undergo posttranslational modifications such as proline hydroxylation. Once secreted, these peptides bind to leucine-rich repeat receptor-like kinases (LRR-RLKs) on the plasma membrane, activating intracellular signaling cascades that regulate responses to abiotic stresses (Zhu et al. 2023; Xiao et al. 2025).

Abiotic stressors like nitrogen, phosphorus, and sulfur deficiency disrupt key metabolic processes in plants, but specialized peptides help restore balance. For instance, C-terminally encoded peptides (CEPs), CLAVATA3/embryo, and root meristem growth factors (RGFs) regulate nutrient uptake and root architecture during deficiency (Xiao et al. 2025). Similarly, drought and salinity stresses are counteracted by peptide signaling systems involving peptides such as CLE25, CLE9, and CEP5, which mediate stomatal closure, root growth modulation, and abscisic acid (ABA) biosynthesis (Takahashi et al. 2018).

These findings underscore the importance of peptide signaling networks in enhancing plant resilience. Understanding the mechanisms by which these peptides function offers a pathway to more sustainable agricultural practices by reducing the use of chemicals and instead leveraging the plant's molecular toolkit. The aim of this chapter is to explore the role of elicitors—particularly small peptides—in enhancing crop tolerance to abiotic stresses by examining their sources, classifications, and mechanisms of action. This chapter focuses on how these bioactive compounds modulate plant physiology and morphology, activate signaling pathways, and contribute to sustainable agricultural practices by improving stress tolerance.

3.2 Types and Sources of Elicitors

The term "elicitor" referred to molecules that trigger the production of phytoalexins, which are defense compounds accumulated in response to pathogens or stress. This term was later applied to compounds stimulating any type of plant defense. Elicitors can be classified based on their origin and molecular composition. They include biotic or abiotic elicitors and complex or defined elicitors (Thakur and Sohal 2013).

The use of abiotic elicitors to increase plant tolerance to pathogens is considered a promising way for disease management and control (Gowtham et al. 2024). In addition, abiotic elicitors trigger the production of secondary metabolites and various compounds in plants (Thakur et al. 2020). For example, the application of abiotic elicitors in *Withania callus* cultures substantially enhances the production of bioactive compounds (Abed and Taha 2024).

Abiotic elicitors are classified into physical, chemical, and hormonal elicitors. Physical elicitors include light, ultraviolet radiation, drought, salinity, osmotic stress, and thermal stress. Chemical elicitors contain a range of substances, including heavy metals. Plant hormones have been extensively employed as elicitors in numerous research studies. Jasmonic acid (JA), salicylic acid (SA), and gibberellic acid are among the most extensively researched elicitors, owing to their pivotal roles in triggering plant defense responses. Upon exposure to abiotic stress, plants produce these elicitors, which in turn stimulate the accumulation of valuable secondary metabolites, including phenolics, alkaloids, carotenoids, and chlorophyll (Petrova et al. 2024).

3.3 Mechanisms of Elicitor-Mediated Abiotic Stress Tolerance

3.3.1 Stress Sensing and Signal Transduction

Plants utilize complex signaling mechanisms to detect and respond to abiotic stresses, including drought, salinity, extreme temperatures, and exposure to heavy metals. These signaling pathways coordinate biochemical and molecular responses to preserve cellular homeostasis and ensure survival under adverse conditions. Plants initiate abiotic stress responses through a coordinated signal transduction pathway involving receptor activation, secondary messengers, calcium signaling, and phosphorylation cascades. This pathway enables precise regulation of stress-responsive genes and physiological adaptations.

Exogenous elicitors often interact with plant receptors to initiate pathways of signal transduction, leading to the stimulation of genes and the production of diverse chemicals involved in plant protection or resistance (Singh and Dwivedi 2018).

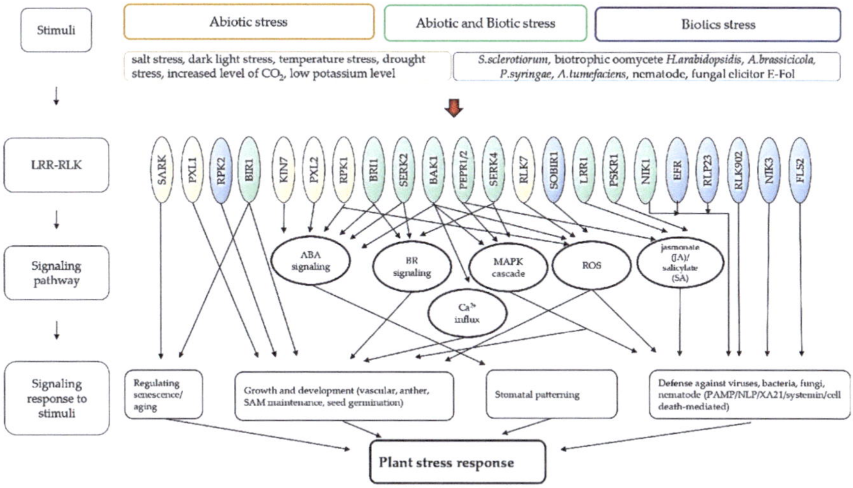

Fig. 3.1 Biotic- and abiotic-tested LRR-RLKs triggering signaling pathways in plant stress response (Soltabayeva et al. 2022)

Abiotic stress response pathways are initiated by the activation of membrane-localized LRR-RLKs, which are triggered by various endogenous peptide elicitors (Kim et al. 2021a, b). RLKs sense cell wall integrity (CWI) through their extracellular domains, which interact with cell wall components or danger-associated molecular patterns (DAMPs) to initiate downstream signaling pathways (Fig. 3.1). These RLKs are embedded in the plasma membrane, with their extracellular domains acting as sensors for physical or chemical changes in the cell wall (Soltabayeva et al. 2022). One example of an RLK is the LecRLK, which was found to contribute to salt tolerance in plants. This receptor can be activated by NaCl and mediates salt stress responses (Sun et al. 2020). Plant elicitor peptides such as RALF peptides, which are released during cell wall damage, are recognized by the LRR-RLK FERONIA (FER). The RALF22/23-FER module then regulates salt tolerance through interactions with cell wall LRR extensins (LRX3/4/5) (Zhao et al. 2018). Table 3.1 summarizes the main RLKs involved in abiotic stress response.

DAMPs, such as Pep1 and Pep3, are released during cell wall damage or stress and are perceived by RLKs PEPR1 and PEPR2 (Jing et al. 2020). Specifically, AtPEP3 was found to confer plant resistance to salt stress upon binding to PEPR1 and 2, triggering downstream signaling cascades (Nakaminami et al. 2018). Other peptides like CLE9/25 play a crucial role in the dehydration stress response. The CLE25 peptide transmits water-deficiency signals in *Arabidopsis* through its interaction with BARELY ANY MERISTEM (BAM) receptors in leaves (Takahashi et al. 2018). Peptides like IDA and PSK are involved in the osmotic stress response, with distinct but complementary pathways that regulate cellular responses to osmotic challenges. PSK also contributes to flower abscission in tomato under

Table 3.1 Examples of RLKs and their function in abiotic stress response

Plant species	RLKs	Subfamily	Function in abiotic stress	Reference
Arabidopsis thaliana	RLK7	LRR-RLK	Mitigates drought stress through stomatal closure	Kim et al. (2021a, b)
	RPK1/BAK1	LRR-RLK	Mitigates drought stress through ABA-induced stomatal closure	Collin et al. (2021)
Oryza sativa (rice)	OsSIT1	LecRLK	Negatively regulates salt stress	Shen et al. (2020)
	OsSTLK	LRR-RLK	Positively regulates salt stress tolerance	Chowdhury and Mubassir (2022)
	OsWAK11	WAK-RLK	Regulates metal stress (resistance to Al and Co)	Ali et al. (2019)
	LRK10-L	PR5K-RLK	Regulates metal stress (Cr)	Shang et al. (2020)
Glycine soja (soybean)	GsLRPK	LRR-RLK	Positively regulates cold stress	Chong et al. (2022)

drought stress, suggesting its role in both developmental processes and stress adaptation (Reichardt et al. 2020).

Although the precise mechanisms of elicitor-mediated abiotic stress responses remain to be fully understood, exogenous elicitors, such as chitosan, SA, JA, and nitric oxide (NO), also interact with membrane receptors to trigger signaling pathways that promote the expression of genes involved in stress tolerance and the synthesis of protective molecules like antioxidants, osmolytes, and secondary metabolites. For instance, the application of JA or SA enhances drought tolerance by modulating stomatal closure and increasing proline accumulation. Additionally, plants use multi-layered defenses against pathogens. Research has revealed substantial crosstalk between biotic and abiotic stress responses in plants, suggesting that pathways activated by one type of stress can influence the plant's ability to respond to the other. For instance, Yang et al. demonstrated that the exogenous application of the PAMP-induced PIP1 peptide enhances tomato resistance to both salt stress and *Phytophthora infestans*. This peptide activates the ABA signaling pathway, which is crucial for managing salt stress, while simultaneously triggering the accumulation of pathogenesis-related proteins, reactive oxygen species (ROS), and JA, which are involved in plant defense against pathogens. This interplay between the abiotic stress response (salt stress) and the biotic stress response (pathogen attack) underscores the complex and integrated nature of plant stress signaling, where mechanisms aimed at mitigating one type of stress can also provide a defense against another (Yang et al. 2023).

Upon stress perception, signaling molecules such as Ca2+ ions, ROS, and hormones like ABA accumulate and activate downstream effectors, mainly protein kinases and transcription factors, which then modify gene expression and the activities of proteins and enzymes, triggering the defense mechanisms (He et al. 2018).

3.3.2 Regulation of Stress-Responsive Genes

Upon sensing, RLKs transmit the signals through secondary messengers and protein kinases such as Ca^{2+} ions, ROS, and MAPKs, leading to transcriptional reprogramming and defense responses (Zhang et al. 2020). This mechanism allows plants to rapidly detect and respond to cell wall disruptions, reinforcing structural integrity and activating defense responses to enhance stress resilience (Manepalli et al. 2022).

MAPK cascades are central signaling modules that enable plants to detect and respond to environmental stresses such as drought, salinity, extreme temperatures, and pathogen attacks.

For example, the MEKK1-MKK2-MPK4 cascade is known to promote plant freezing tolerance by activating the ICE1 transcription factors (TFs) that regulate the *CBF* gene, which confers plant cold tolerance (Zhao et al. 2017). Once activated, MAPKs phosphorylate downstream targets, including transcription factors like dehydration-responsive element-binding proteins (DREBs), AREBs, WRKY, and NACs, which bind to stress-responsive gene promoters to initiate the production of osmoprotectants, chaperones, and detoxification enzymes (Lozano-Juste et al. 2020). For instance, upon sensing osmotic stress, MAPK cascades lead to transcriptional reprogramming, where phosphorylated WRKY transcription factors enhance stress gene expression (e.g., the *RD29A* gene for osmolyte synthesis) and MYB transcription factors upregulate genes like *COR15* to stabilize membranes during dehydration (Lin et al. 2021). While *COR15* stabilizes membranes during dehydration, its regulation overlaps with *RD29A* through CBF/DREB1 transcription factors. However, studies highlight that *RD29A* promoters are more responsive to drought and cold, whereas *RD29B* is salt-specific (Liu et al. 2023). Additionally, the same MAPK pathway can respond to various stresses, as seen in strawberries, where MAPK5 is activated by both drought and salinity (He et al. 2020).

Additionally, RLKs activate calcium influx, enhancing the signal through calcium-dependent protein kinases (CDPKs), and induce ROS bursts that act as secondary messengers to potentiate immune responses and activate antioxidant defense mechanisms (Xu et al. 2022). In fact, although stress is typically perceived at the cell surface, it is subsequently relayed to various organelles, including the chloroplast, mitochondrion, peroxisome, nucleus, and cell wall. The signals generated by these organelles are integrated to regulate stress-responsive gene expression and other cellular processes, thereby restoring cellular homeostasis (Zhu 2016). For instance, under unfavorable environmental conditions, the chloroplast acts as an environmental sensor, generating ROS molecules that act as retrograde signals (Song et al. 2021). Chloroplast-derived ROS, particularly hydrogen peroxide (H_2O_2), can then regulate nuclear gene expression through direct signaling. H_2O_2 diffuses into the nucleus, where it oxidizes transcription factors (TF) such as NRP1 and heat shock factors (HSFs), thereby activating the expression of stress-responsive genes, including antioxidant enzymes and heat shock proteins (Kuźniak and

Table 3.2 Role of chloroplasts in plant stress responses

Species	Components	Process/stimulus	Reference
Heat stress			
Triticum aestivum	Rca1	Modulate Rubisco activity	Kumar et al. (2016)
Triticum aestivum	Ub2	Improve antioxidant capacity	Tian et al. (2014)
Arabidopsis thaliana	Hsp21	Associate with the thylakoid membranes	Bernfur et al. (2017)
Chilling stress			
Oryza sativa	DUA1	Regulate chloroplast development	Cui et al. (2019)
Lycopersicon esculentum	CDJ1	Maintain PSII activity	Kong et al. (2014)
Arabidopsis thaliana	RBD1	Regulate chloroplast protein translation by influencing 23S rRNA processing	Wang et al. (2016)
Salt stress			
Arabidopsis thaliana	Hsp70,Clp protease	Protect plants from salinity-triggered oxidative stress	Pulido et al. (2017)
Oryza sativa	CspA, CspB	Impart SGR phenotype	Guddimalli et al. (2021)
High-light stress			
Chlamydomonas reinhardtii	FtSH	Control the quality of thylakoid membrane proteins	Malnoë et al. (2014)
Drought stress			
Zea mays	LOX6	Additional storage of nitrogen	Abbaraju et al. (2022)
Salsola laricifolia	NADP-ME4	Alleviate chlorophyll content decrease and PSII photochemical efficiency	Wen et al. (2021)
Solanum tuberosum	BBX21	Reduce chloroplast electron transport capacity	Gómez-Ocampo et al. (2021)

Kopczewski 2020). Table 3.2 summarizes known chloroplast biological processes involved in plant stress response.

Lastly, MAPKs regulate plant hormone signaling (e.g., ABA, JA) to coordinate physiological responses, such as stomatal closure during drought or antioxidant synthesis to mitigate oxidative damage. For instance, ABA plays a crucial role in drought and salinity tolerance by regulating stomatal closure, reducing transpiration, and activating stress-responsive genes, including dehydration-responsive element-binding proteins (DREB) (Zhang et al. 2023). Research indicates that binding the CLE25 peptide to the BAM receptor facilitates the transmission of drought signals from the roots to the leaves. This interaction triggers the accumulation of ABA by upregulating the *NCED3* gene, which is responsible for de novo ABA synthesis, thereby improving drought resistance (Tsintzou and Madesis 2024).

3.3.3 Activation of Plant Defense Responses

Elicitors can catalyze stress responses in plants, leading to the formation of biologically active compounds or the induction of unprecedented secondary metabolites. The use of elicitors causes stress in the plant system, which activates plant defense mechanisms and promotes secondary metabolite synthesis (Bhaskar et al. 2022).

Upon sensing abiotic stress, plants activate a sophisticated array of defense mechanisms spanning molecular, biochemical, and physiological responses. These responses are coordinated through an intricate regulatory network involving transcription factors like WRKY and NAC, calcium ions as signaling intermediates, and cross talk between different stress pathways, ensuring plant survival under diverse environmental stressors (Fig. 3.2).

Extreme temperatures can disrupt the plasma membrane by causing cold-induced rigidification or heat-induced fluidization, both of which result in protein deactivation and ion leakage. Membrane modulation is achieved through increased unsaturated fatty acids (UFAs) production, which enhances membrane fluidity, preventing rigidity in cold or leakage in heat (He et al. 2018). Additionally, heat shock proteins (HSPs), like HSP70 and HSP16.4, play a key role in stabilizing and refolding denatured proteins under extreme temperature conditions. Under abiotic stress conditions, these molecular chaperones confer plant resistance. For instance, in transgenic *Arabidopsis*, overexpression of HSP16.4 enhanced resistance to both drought and heat stress, compared to the nontransformed control line (Huang et al. 2019).

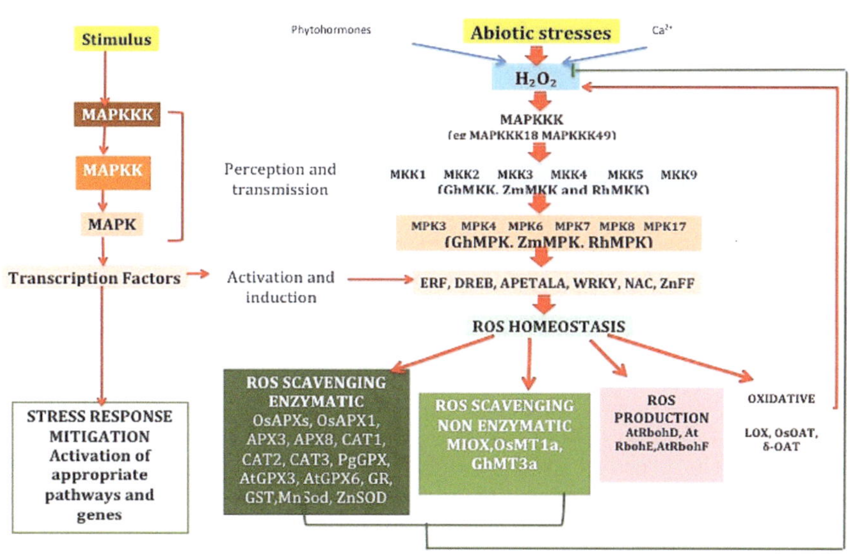

Fig. 3.2 The pathway from stimulus perception to downstream activation of genes and the genes involved in their regulation (Nadarajah 2020)

ROS, including H_2O_2, singlet oxygen ($^1O_2$), and hydroxyl radicals (OH), are abundantly produced in plants under abiotic stress conditions, where they contribute to cellular damage. However, during oxidative stress induced by abiotic factors, certain ROS, particularly H_2O_2, can also act as signaling molecules. This triggers the activation of ROS-scavenging mechanisms, such as superoxide dismutase (SOD) and catalase (CAT), which help to protect cellular integrity and mitigate further oxidative damage (Nadarajah 2020).

Osmotic adjustment is a critical physiological mechanism enabling plants to withstand drought and salinity by accumulating compatible solutes like proline and glycine betaine (GB), which maintain cell turgor, stabilize cellular structures, and mitigate oxidative damage (Nahar et al. 2016). For example, proline serves multifaceted roles as an osmoprotectant by lowering cellular water potential, scavenging ROS to prevent oxidative stress, and stabilizing membranes and proteins under dehydration (Saibi and Brini 2020). For instance, salt-tolerant rice lines (CSSL8–94 and CSSL8–95) exhibited enhanced osmotic adjustment (up to 1.16 MPa) and accumulated 8.4 times more proline under salinity compared to sensitive varieties, correlating with reduced electrolyte leakage and oxidative damage (Nounjan et al. 2018). Similarly, GB preserves photosynthetic machinery by lowering the effect of photoinhibition caused by excessive light and protects subcellular structures by acting as a molecular chaperone (Jarin et al. 2024).

3.4 Elicitors in Abiotic Stress Tolerance: Role and Applications

3.4.1 Effects on Crop Physiology

3.4.1.1 Enhanced Photosynthesis and Water Use Efficiency

Abiotic stress, particularly drought stress, significantly impacts various physiological parameters in plants, including photosynthetic efficiency, biomass accumulation, and overall yield, thereby compromising plant growth and productivity. In fact, under drought conditions, plants close their stomata to conserve water, reducing CO_2 intake and thus inhibiting photosynthesis. This stomatal closure is a primary mechanism by which drought stress affects photosynthesis, particularly during mild-to-moderate drought (Lawson and Vialet-Chabrand 2018). For instance, in wheat under moderate-to-severe water stress, photosynthetic parameters (such as net photosynthetic rate, intercellular carbon concentration, stomatal conductance, and transpiration) decrease, leading to a reduction in wheat height, biomass, and grain size. This ultimately results in lower water use efficiency and irrigation water productivity (Zhao et al. 2020a, b). Elicitors, such as methyl jasmonate (MeJA) and yeast extract, serve as valuable tools for enhancing photosynthesis and water use efficiency in plants. By stimulating the production of photosynthetic pigments like chlorophyll and carotenoids, these elicitors can improve photosynthetic rates, as observed in aeroponically cultivated *V. jatamansi, where* yeast extract increased photosynthesis. Additionally, these elicitors can influence stomatal conductance, which is crucial for balancing CO_2 uptake and water loss, thereby optimizing water use efficiency. MeJA and yeast extract treatments have been shown to modulate stomatal behavior, enhancing photosynthesis while potentially improving water conservation under stress conditions (Partap et al. 2020). Similarly, the exogenous application of the PpT-2 peptide significantly enhances water use efficiency (WUE) and photosynthesis in tomato plants, particularly under water-limited conditions. Application of PpT-2 leads to improved biomass production per unit of water, indicating better drought tolerance. Higher doses (150 mg L^{-1}) were especially effective in boosting WUE. Additionally, PpT-2 treatment enhanced key photosynthetic parameters, including CO_2 assimilation, chlorophyll fluorescence, and the efficiency of photosystem II (PSII). These improvements also extended to both photochemical and non-photochemical processes, essential for energy conversion under stress (Silva et al. 2025). This makes elicitors promising for improving crop resilience and productivity, particularly in environments where water is limited.

3.4.1.2 Regulation of Nutrient Uptake

Abiotic stresses such as drought, salinity, and extreme temperatures disrupt plant nutrient uptake by altering root architecture, reducing nutrient transport efficiency, and causing ionic imbalances. Drought limits soil moisture, impairing the mobility and absorption of essential nutrients such as nitrogen and potassium. Salinity induces sodium toxicity, which competes with potassium and calcium uptake, while heavy metals like cadmium block nutrient transporters (Wang et al. 2025). Collectively, these effects compromise plant growth, productivity, and resilience under adverse conditions. Elicitors regulate nutrient uptake under abiotic stress by modulating molecular and physiological pathways. They bind to receptors, triggering signaling cascades that modulate ion transporters, such as downregulating toxic metal uptake while preserving essential nutrient transporters. In lead (Pb)-stressed rice, elicitors like MeJA and SA reduced Pb accumulation by downregulating metal transporters (IRT1 and HMA3) while enhancing antioxidant defenses and osmolyte synthesis. This preserved ion homeostasis and reduced competition between Pb and essential nutrients (Altun and Orcan 2024). Menadione sodium bisulfate (MSB) at an optimal concentration (100 μM) alleviated cadmium-induced nutrient imbalances in okra by enhancing the uptake of essential nutrients such as calcium, potassium, and magnesium, while simultaneously reducing Cd accumulation and translocation to aerial plant parts. This suggests MSB supports nutrient homeostasis under heavy metal stress (Rasheed et al. 2018). In a study by Roy et al., the C-terminally encoded peptide (CEP1) was shown to enhance nutrient uptake in plants. The researchers discovered that CEP1 increased the absorption of nitrate, phosphate, and sulfate in *Medicago truncatula*, particularly under nitrogen-deficient conditions. Transcriptome analysis revealed that synthetic CEP1 activated several genes involved in nutrient transport, including those for nitrate and sulfate, and also stimulated genes related to signaling pathways such as RLKs and transcription factors (Roy et al. 2022).

3.4.1.3 Improvement of Osmoregulation

Osmoregulation is a vital process in plants that involves regulating the balance of water and solutes within cells to maintain homeostasis, especially under abiotic stress conditions such as drought and salinity. When plants face low water potential, they accumulate small molecules known as osmolytes, such as free amino acids (proline and GB), soluble sugars, and proteins, which play a critical role in mitigating the negative effects of these stresses (Ozturk et al. 2021). These molecules help plants maintain cellular osmotic pressure and water retention, thereby preserving turgor pressure, preventing dehydration, and enhancing the plant's ability to survive water-deficient conditions. By increasing cellular osmotic pressure, osmolytes contribute significantly to plant resilience under challenging environmental conditions (Forlani et al. 2019). Several studies have demonstrated that the exogenous application of elicitors can enhance the content of osmoprotective molecules in plants. For example, a study on salt-stressed soybean demonstrated that the application of MeJA and SA elicitors enhanced the accumulation of osmolytes, such as GB, soluble sugars, and proteins, thereby improving osmotic adjustment and helping the plants better tolerate salt stress (Farhangi-Abriz and Ghassemi-Golezani 2018). Similarly, under drought stress, the application of SA in *Brassica rapa* (La et al. 2019) and quinoa (Forouzandeh et al. 2024) induced proline accumulation through the increased expression of genes involved in proline synthesis.

3.4.2 Effects on Crop Morphology

3.4.2.1 Root Architecture Modification

The root system architecture (RSA) consists of several features, including the root length, number, spread, and length of lateral roots. The root system plays a crucial role in nutrient uptake, storage, and anchorage in plants. Abiotic stresses, such as drought, salinity, and nutrient deficiency significantly influence RSA by inducing structural and functional changes. For instance, heat stress modifies lateral root elongation and reduces overall root biomass. It also disrupts cellular processes, leading to impaired nutrient uptake and reduced growth (Khan et al. 2016). Elicitors play a vital role in modifying root architecture during abiotic stress by activating signaling pathways that enhance root plasticity and resilience. The application of a tannin-based elicitor significantly enhances root architecture in tomato plants, particularly under salt stress. It promotes substantial increases in root length and weight by 23–24%, respectively, supporting better anchorage and nutrient acquisition. At the molecular level, the elicitor upregulates over 280 genes, many associated with root growth and salt stress tolerance, indicating a strong influence on stress-responsive genetic pathways. Morphological adaptations in root structure were also observed, aiding plant survival under adverse conditions. Additionally, the elicitor modulates genes related to nutrient uptake, highlighting its role in improving both structural and functional root traits (Campobenedetto et al. 2021). A study conducted by Kirubakaran et al. found that the foliar application of elicitors such as seaweed extract and chitosan significantly improved root architecture in grapevines under drought stress. These treatments enhanced root biomass, particularly in the 1103 rootstock, which exhibited deeper and thicker roots. Root imaging analysis revealed notable improvements in structural traits, including increased convex area and favorable width-to-depth ratios, contributing to enhanced water uptake capacity. The biostimulants also appeared to support below-ground moisture retention, possibly through mechanisms such as hydraulic lift. Additionally, the study highlighted the role of biostimulants in modulating root architectural plasticity by influencing shoot–root communication pathways, thereby promoting physiological adaptation and resilience under water-limited conditions (Kirubakaran et al. 2024).

3.4.2.2 Structural Integrity of Leaves and Stems

Elicitors play a significant role in enhancing the structural integrity of leaves and stems during abiotic stress by activating defense mechanisms and reinforcing cellular structures.

For instance, JA and SA function as elicitors that enhance leaf resilience to oxidative damage by activating antioxidant defense pathways. These compounds stimulate the transcription of antioxidant enzymes such as SOD and CAT, as well as stress-responsive transcription factors like those in the WRKY family. Additionally, JA and SA promote the accumulation of nonenzymatic antioxidants, including glutathione and ascorbate, which contribute to ROS scavenging. This coordinated response improves leaf cell membrane stability and reduces electrolyte leakage under salinity stress (Melo et al. 2024). Moreover, the fungal elicitor *Piriformospora indica* plays a crucial role in maintaining leaf integrity during abiotic stress, such as drought and salinity, by enhancing antioxidative defenses and osmotic regulation. It induces the activity of key antioxidative enzymes, including SOD, CAT, ascorbate peroxidase (APX), glutathione reductase (GR), and peroxidase (POD), which detoxify ROS generated under stress conditions. This prevents oxidative damage to cellular structures, ensuring the stability of plastids and maintaining photosynthetic efficiency. Additionally, *P. indica* promotes the accumulation of osmolytes like polyamines and proline, which help maintain cellular water balance and protect membrane integrity during stress. The combined action of antioxidative enzymes and osmolytes creates a stable environment within leaf cells, preserving their structural integrity. By mitigating oxidative stress and stabilizing plastids, *P. indica* enhances leaf resilience, ensuring sustained photosynthesis and overall plant health under adverse environmental conditions (Li et al. 2023).

3.4.2.3 Enhanced Reproductive Development

Different types of abiotic stress, such as drought and heat, impact plant reproductive processes in distinct ways. For instance, extreme temperatures disrupt key stages of pollen development, including production, germination, and tube elongation (Gupta et al. 2024). In *Pennisetum glaucum* (pearl millet), exposure to high temperatures led to a decrease in pollen germination (Djanaguiraman et al. 2018). These temperature-induced failures in pollen functionality directly reduce seed set and crop yields, highlighting the vulnerability of reproductive phases to climate extremes. On the other hand, drought reduces water availability, causing cellular dehydration and oxidative stress. This leads to abnormal floral development, including anther defects, reduced pollen viability, ovule abortion, and failure of flowers to open (Ma et al. 2020). As environmental stresses increasingly impact crop productivity, elicitors have emerged as a promising tool to enhance plant resilience, particularly in improving pollen performance and pollen–pistil interactions under adverse conditions. For example, the application of a protein hydrolysate-based elicitor improved pollen viability in tomatoes under drought stress, contributing to enhanced yield performance (Francesca et al. 2021). Similarly, a study was conducted in which a 10^{-5} M solution of indole-3-acetic acid (IAA) was applied as a foliar spray to various rice genotypes following irrigation withdrawal during the anthesis stage. The findings indicated that drought stress negatively impacted pollen viability and spikelets. However, the external application of IAA enhanced both pollen viability and spikelet fertility, ultimately contributing to improved yield under water-limited conditions fertility (Sharma et al. 2018). Melatonin application has been shown to enhance pollen thermotolerance in tomatoes by mitigating high-temperature-induced damage. It reduces ROS in anthers through the upregulation of antioxidant enzymes and protects pollen development by preserving tapetum integrity and preventing organelle degradation. This protection is supported by increased expression of heat shock proteins and autophagy-related genes, ultimately maintaining pollen viability and function under heat stress (Qi et al. 2018). Moreover, another study revealed that microalgae-based elicitor treatment influenced the reproductive development of tomato plants primarily by enhancing stress response pathways rather than directly targeting flowering or pollen development. Although few genes associated with flowering were differentially expressed, the treatment upregulated key stress-responsive genes—such as those involved in JA, ABA, and ethylene signaling—along with protective proteins like dehydrins and oleosins. This priming effect may indirectly support reproductive success under abiotic stress conditions by improving the resilience of flower buds and promoting overall plant health (Gitau et al. 2023).

3.4.3 Abiotic Stress-Specific Applications

3.4.3.1 Drought and Water Deficit Stress

Drought and water deficit stress present major obstacles to plant growth and agricultural productivity, often leading to reduced crop yields and compromised plant health. Moreover, drought stress not only impairs plant growth but also triggers an increase in the production of secondary metabolites (Abdeldaym et al. 2024). For example, moderate water stress on *Bupleurum chinense* showed noticeably greater amounts of anti-inflammatory saikosaponins (Zhu et al. 2009). In addition, drought stress increased the production of rosmarinic, ursolic, and oleanolic acid in *Prunella vulgaris* (Chen et al. 2011). Similarly, water deficit resulted in the accumulation of glycyrrhizic acid in the roots of *Glycyrrhiza uralensis* (Li et al. 2011). Osmotic agents like sucrose, glucose, and polyols can be used to induce osmotic stress and boost the production of secondary metabolites in plant cell cultures (Salehi et al. 2019).

As these environmental conditions become more frequent and severe, researchers have turned to innovative approaches to mitigate their effects. Among these strategies, the application of elicitors has emerged as a promising method to improve drought tolerance in plants. An example of an elicitor application under drought stress is the use of chitosan on thyme plants. When subjected to drought stress, plants have shown significant changes in various physiological parameters, including photosynthetic pigments, proline content, and lipid peroxidation. The foliar application of chitosan played a crucial role in mitigating the negative effects of drought. Chitosan enhanced proline accumulation, which helped with osmotic adjustment, and it also reduced lipid peroxidation and improved membrane integrity, thus protecting the plants from severe drought-induced damage (Bistgani et al. 2017). SA is another widely recognized elicitor molecule utilized to alleviate the effects of drought stress in plants. Melo et al. investigated the effects of SA on two cowpea genotypes under drought conditions. The exogenous application of SA enhanced antioxidant enzyme activity and improved grain yield (Melo et al. 2024). Similarly, in quinoa, it was found that the exogenous application of SA along with MeJA improved drought tolerance by enhancing antioxidant enzyme activity and increasing carotenoid content under water deficit stress. The treatment also positively influenced the expression of drought-related genes, particularly *MYB62*, and increased proline and carbohydrate content (Forouzandeh et al. 2024). Endophyte-derived elicitors, such as polysaccharides, have also been shown to enhance plant tolerance to abiotic stresses. For instance, the application of polysaccharides from the endophyte *Bionectria* sp. Fat6 significantly improved the growth of tartary buckwheat under drought stress. It enhanced the net photosynthetic rate, stomatal conductance, and transpiration rate, while reducing intercellular CO_2 concentration. The treatment also increased chlorophyll content, soluble protein, and proline levels while decreasing malondialdehyde content, indicating improved photosynthetic capacity and stress tolerance (Xiang et al. 2021). These findings suggest that the polysaccharide acts as an effective abiotic elicitor to mitigate the negative impacts of drought on tartary buckwheat. Lastly, some microbial-derived peptides have been shown to promote drought stress responses in plants. AMEP412 is a protein elicitor taken from *Bacillus subtilis*. AMEP412 application was shown to enhance drought tolerance in soybean seedlings by improving physiological performance and boosting antioxidant activity. It helped regulate ROS accumulation, improved stress recovery, and reduced lipid peroxidation, thereby alleviating the negative effects of drought stress (Liu et al. 2022). Another bacterial-derived peptide is the bacteriocin Thuricin-17. Research showed that the application of Thuricin-17 enhanced soybean tolerance to water stress by improving root architecture, biomass accumulation, and nitrogen fixation. This intervention helped maintain balanced water and nitrogen status, thereby mitigating the detrimental effects of drought stress on plant growth (Prudent et al. 2015).

3.4.3.2 Salinity Stress

Salt stress is another major abiotic stressor that disrupts ion homeostasis, induces oxidative damage, and reduces crop productivity. Salinity has a massive effect on various plant physiological and metabolic processes, including photosynthesis, protein synthesis, energy production, and lipid metabolism. Plants experiencing water deficit due to salinity are vulnerable to oxidative stress, triggered by an overproduction of ROS, which can cause damage to nucleic acids, proteins, and lipids. However, exposure to salinity has been found to stimulate the accumulation of secondary metabolites, such as phenolics, terpenes, and alkaloids, which can have valuable applications (Petrova et al. 2024).

Several types of elicitors have been used to help mitigate the effects of salt-induced stress. In *Salvia virgata*, a combination of SA and zinc oxide nanoparticles (ZnONP) under salinity stress improved chlorophyll content and reduced oxidative damage while increasing proline and sugar levels (Bozaba and Kuru 2024). In addition to being used in water-deficient conditions, chitosan has demonstrated significant potential as an elicitor to alleviate the adverse effects of salt stress on plants. Chitosan helps mitigate salinity-induced oxidative damage by enhancing antioxidant enzyme activities like superoxide dismutase (SOD), CAT, and peroxidase (POX). In *Phaseolus vulgaris*, chitosan nanoparticles significantly reduced H_2O_2 levels and lipid peroxidation (Alenazi et al. 2024).

Additionally, this biopolymer regulates ion leakage and improves membrane stability under salt stress. In tomato plants treated with chitosan foliar sprays, ion leakage was reduced, indicating improved cellular integrity under saline conditions (Özkurt and Bektaş 2022).

Microalgal exopolysaccharides (EPS) have been identified as effective biostimulants for enhancing plant resilience under salt stress. These polymers contribute to osmotic balance by promoting the accumulation of osmoprotectants, which help maintain cellular ion homeostasis. Additionally, EPS mitigate oxidative stress by boosting antioxidant enzyme activities, thereby reducing ROS accumulation. In practical applications, EPS derived from microalgae like *Dunaliella salina* have been shown to improve growth parameters in tomatoes under salt stress, such as shoot and root length, dry weight, and protein content (El Arroussi et al. 2018). Another study investigated the efficacy of Spirulina-derived EPS in alleviating salt stress effects on hydroponically grown basil and spinach. The results indicated that EPS treatment significantly enhanced biomass productivity in both basil and spinach under moderate salt stress. Even under high salinity, EPS-treated plants showed improved productivity compared to untreated ones. The treated plants exhibited increased levels of proline, phenolic compounds, and soluble sugars, along with improved potassium-to-sodium (K^+/Na^+) ratio, suggesting enhanced stress tolerance mechanisms (Chaudhuri and Balasubramanian 2025).

3.4.3.3 Temperature Extremes: Heat and Cold Stress

Extreme temperature stress, encompassing both heat and cold, significantly impacts plant growth, development, and productivity by disrupting vital physiological and biochemical processes. Heat stress can induce premature aging of leaves, accelerate leaf senescence, and simultaneously trigger the accumulation of secondary metabolites. Furthermore, temperature fluctuations have a significant impact on various cellular processes in plant cell cultures, including metabolic regulation, membrane permeability, and the rate of intracellular responses (Asad et al. 2019). High temperatures can damage cellular structures, such as membranes and proteins, impair photosynthesis and respiration, and reduce water and nutrient uptake. This often leads to oxidative stress, stunted growth, and diminished crop yields (Zhao et al. 2020a, b). On the other hand, cold stress—ranging from chilling (0–12 °C) to freezing (below 0 °C)—induces cellular dehydration, membrane damage due to ice crystal formation and metabolic disruptions (Qian et al. 2024) (Fig. 3.3). Heat stress, especially during the reproductive phase, negatively impacts grain filling and seed setting in spring maize, reducing its yield. It was found that elevated temperatures led to membrane damage, decreased photosynthetic pigments, and lowered net assimilation, all contributing to reducing grain yield. Iqbal et al. assessed the effectiveness of different elicitors (SA, $CaCl_2$, and *Moringa oleifera* leaf extract) in improving the performance of heat-stressed maize. The findings indicated that the application of elicitors enhanced seedling emergence, increased chlorophyll content, and promoted better growth (Iqbal et al. 2020). Melatonin applications have been observed to have a beneficial effect in increasing heat-stress tolerance in plants. The exogenous application of melatonin effectively alleviated heat-induced leaf senescence in perennial ryegrass by suppressing senescence-associated genes (Zhang et al. 2017). Under heat stress, melatonin-treated plants exhibited improved growth, including plant height, chlorophyll content, and improved photochemical efficiency and cell membrane stability compared to untreated plants. In a 2-year greenhouse experiment, the effects of three elicitors, SA, MeJA, and Sodium nitroprusside (SNP), on cold stress tolerance in grapevine saplings were tested. While some elicitors were more effective than others, in general, the treatment with

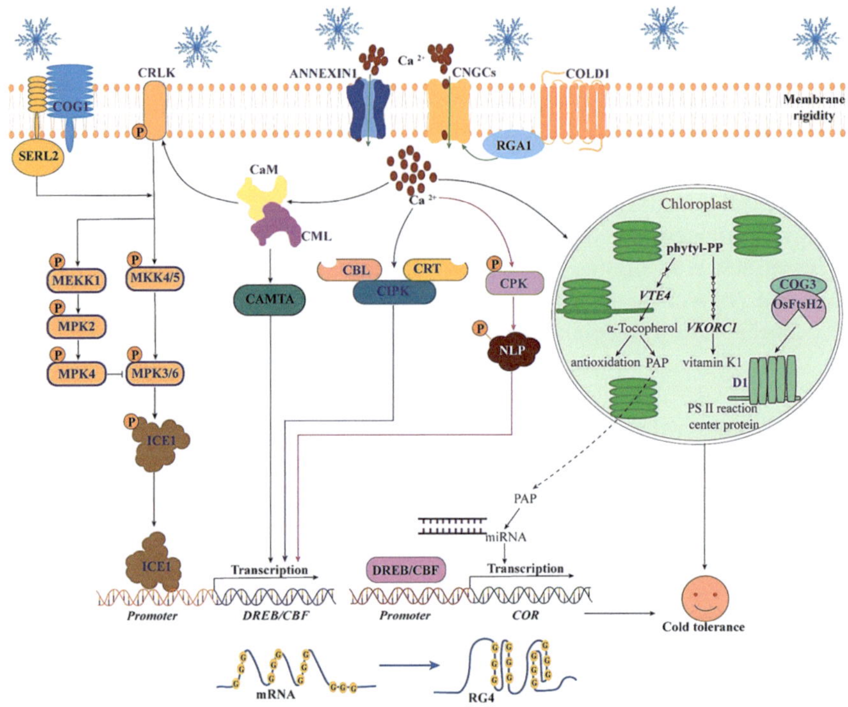

Fig. 3.3 Perception and transduction of cold signals in plants (Qian et al. 2024)

elicitors resulted in enhanced cold tolerance through increased antioxidant enzyme activities, reduced electrolyte leakage and lipid peroxidation, and improved osmotic adjustment capacity (Daler et al. 2025).

Another study examined the effectiveness of the cytokinin elicitor "kinetin" on coffee plants under cold stress. The exogenous application of the elicitor led to an improved ROS-scavenging capacity by stimulating the accumulation of nonenzymatic antioxidant compounds such as anthocyanins, caffeine, and phenolic compounds (Acidri et al. 2020).

3.4.3.4 Heavy Metal and Oxidative Stress

Heavy metal (HM) stress is a significant environmental challenge affecting plant growth and productivity worldwide. Heavy metals, such as lead (Pb), cadmium (Cd), mercury (Hg), and arsenic (As), are toxic to plants even at low concentrations, causing oxidative stress, disrupting cellular homeostasis, and impairing physiological processes like photosynthesis and nutrient uptake (Emamverdian et al. 2015). In a study conducted by Altun and Orcan, the authors examined the effect of elicitors MeJA and SA on Karacadağ rice exposed to lead (Pb) stress. The results showed that both elicitors improved various physiological and biochemical parameters, including photosynthetic pigments, proline, total protein, and secondary metabolites. SA was more effective at increasing flavonoid content, while MeJA better enhanced photosynthetic pigments, protein, and proline levels, and reduced malondialdehyde (MDA) and Pb accumulation. The combined treatment of 2 mM SA and 20 μM MeJA at 400 ppm Pb provided the best overall improvements. This study demonstrates that elicitor treatments, particularly in combination, can effectively alleviate the harmful effects of Pb stress on rice plants (Altun and Orcan 2024). Kaushik et al. tested the efficacy of MeJA in pigeon peas under Cd stress. The authors found that the application of MeJA helped mitigate cadmium toxicity in plants by enhancing growth and reducing oxidative stress. The treatment improved root development and boosted antioxidant defenses, while also modulating gene expression to decrease Cd uptake (Kaushik et al. 2024). Brassinosteroids (BRs) represent a class of plant hormones that function as elicitors and have been utilized to mitigate the adverse effects of HM toxicity. Their protective role is primarily attributed to the activation of the plant's antioxidant defense mechanisms, as well as the enhanced accumulation of stress-related metabolites such as phytochelatins and proline (Faseela et al. 2025). For instance, in rice subjected to Chromium (Cr) stress, the exogenous application of epibrassinolide (EBL) helped the seedlings overcome the HM stress by activating the antioxidative defense system by upregulating genes encoding ROS-scavenging enzymes. Overall, EBL treatment was effective in reducing Cr uptake by the plants and improving their growth (Sharma et al. 2016).

3.5 Conclusion

Elicitors, especially small peptides, have emerged as powerful modulators of plant defense and stress adaptation mechanisms. Through intricate signaling pathways involving receptor-like kinases and hormone-mediated responses, these molecules enable plants to sense and respond to abiotic stressors such as drought, salinity, and nutrient deficiencies. By enhancing both physiological functions and structural integrity, elicitors contribute to improved plant survival, growth, and productivity under challenging environmental conditions. Their integration into crop management practices holds great promise for sustainable agriculture, especially in the context of climate change. Future research should focus on unraveling the detailed molecular mechanisms of peptide signaling and translating this knowledge into practical applications for crop improvement.

References

Abbaraju HKR, Gupta R, Appenzeller LM, Fallis LP, Hazebroek J, Zhu G, Bourett TM, Howard RJ, Weers B, Lafitte RH, Hakimi SM, Schussler JR, Loussaert DF, Habben JE, Dhugga KS (2022) A vegetative storage protein improves drought tolerance in maize. Plant Biotechnol J 20:374–389

Abdeldaym EA, Hassan HA, El-Mogy MM, Mohamed MS, Abuarab ME, Omar HS (2024) Elevated concentrations of soil carbon dioxide with partial root-zone drying enhance drought tolerance and agro-physiological characteristics by regulating the expression of genes related to aquaporin and stress response in cucumber plants. BMC Plant Biol 24:917

Abed AS, Taha AJ (2024) Biotic and abiotic elicitors induce biosynthesis and accumulation of some withanolide compounds in the callus culture of *Withania somnifera* L. Egypt J Bot 64:245–256

Acidri R, Sawai Y, Sugimoto Y, Handa T, Sasagawa D, Masunaga T, Yamamoto S, Nishihara E (2020) Exogenous kinetin promotes the nonenzymatic antioxidant system and photosynthetic activity of coffee (*Coffea arabica* L.) plants under cold stress conditions. Plants 9:281

Alenazi MM, El-Ebidy AM, El-Shehaby OA, Seleiman MF, Aldhuwaib KJ, Abdel-Aziz HMM (2024) Chitosan and chitosan nanoparticles differentially alleviate salinity stress in *Phaseolus vulgaris* L. plants. Plants 13:398

Ali A, Kim JK, Jan M, Khan HA, Khan IU, Shen M, Park J, Lim CJ, Hussain S, Baek D, Wang K, Chung WS, Rubio V, Lee SY, Gong Z, Kim WY, Bressan RA, Pardo JM, Yun DJ (2019) Rheostatic control of ABA signaling through HOS15-mediated OST1 degradation. Mol Plant 12:1447–1462

Altun H, Orcan P (2024) Responses to exogenous elicitor treatment in lead-stressed *Oryza sativa* L. BMC Plant Biol 24:897

Asad MAU, Zakari SA, Zhao Q, Zhou L, Ye Y, Cheng F (2019) Abiotic stresses intervene with ABA signaling to induce destructive metabolic pathways leading to death: premature leaf senescence in plants. Int J Mol Sci 20:256

Bartels S, Lori M, Mbengue M, van Verk M, Klauser D, Hander T, Böni R, Robatzek S, Boller T (2013) The family of Peps and their precursors in *Arabidopsis*: differential expression and localization but similar induction of pattern-triggered immune responses. J Exp Bot 64:5309–5321

Bernfur K, Rutsdottir G, Emanuelsson C (2017) The chloroplast-localized small heat shock protein Hsp21 associates with the thylakoid membranes in heat-stressed plants. Protein Sci 26:1773–1784

Bhaskar R, Xavier LSE, Udayakumaran G, Kumar DS, Venkatesh R, Nagella P (2022) Biotic elicitors: a boon for the *in-vitro* production of plant secondary metabolites. Plant Cell Tissue Organ Cult 149:7–24

Bistgani ZE, Siadat SA, Bakhshandeh A, Pirbalouti AG, Hashemi M (2017) Interactive effects of drought stress and chitosan application on physiological characteristics and essential oil yield of *Thymus daenensis Celak*. Crop J 5:407–415

Boller T, Felix G (2009) A renaissance of elicitors: perception of microbe-associated molecular patterns and danger signals by pattern-recognition receptors. Annu Rev Plant Biol 60:379–406

Bozaba TO, Kuru İS (2024) The effect of the combined application of elicitors to *Salvia virgata Jacq*. under salinity stress on physiological and antioxidant defense. BMC Plant Biol 24:788

Campobenedetto C, Mannino G, Beekwilder J, Contartese V, Karlova R, Bertea CM (2021) The application of a biostimulant based on tannins affects root architecture and improves tolerance to salinity in tomato plants. Sci Rep 11:354

Chaudhuri R, Balasubramanian P (2025) Enhancing salt tolerance in leafy vegetables using microalgal exopolysaccharides. J Crop Health 77:70

Chen Y, Guo Q, Liu L, Liao L, Zhu Z (2011) Influence of fertilization and drought stress on the growth and production of secondary metabolites in *Prunella vulgaris* L. J Med Plants Res 5:1749–1755

Chong L, Xu R, Ku L, Zhu Y (2022) Beyond stress response: OST1 opening doors for plants to grow. Stress Biol 2:44

Chowdhury R, Mubassir MHM (2022) How *Arabidopsis* receptor-like kinase 7 (RLK7) manifests: delineating its structure and function. Adv Agric 2022:4715110

Collin A, Daszkowska-Golec A, Szarejko I (2021) Updates on the role of abscisic acid insensitive 5 (ABI5) and abscisic acid-responsive element binding factors (ABFs) in ABA signaling in different developmental stages in plants. Cells 10:1996

Cui X, Wang Y, Wu J, Han X, Gu X, Lu T, Zhang Z (2019) The RNA editing factor DUA1 is crucial to chloroplast development at low temperature in rice. New Phytol 221:834–849
Daler S, Yağci A, Cangí R (2025) The effect of different elicitor treatments on cold stress in *V. vinifera L.* Cv. "Victoria". Res Sq 47(8):78
Ding Y, Wang J, Wang J, Stierhof YD, Robinson DG, Jiang L (2012) Unconventional protein secretion. Trends Plant Sci 17:606–615
Djanaguiraman M, Perumal R, Ciampitti I, Gupta S, Prasad P (2018) Quantifying pearl millet response to high temperature stress: thresholds, sensitive stages, genetic variability and relative sensitivity of pollen and pistil. Plant Cell Environ 41:993–1007
El Arroussi H, Benhima R, Elbaouchi A, Sijilmassi B, El Mernissi N, Aafsar A, Meftah-Kadmiri I, Bendaou N, Smouni A (2018) *Dunaliella salina* exopolysaccharides: a promising biostimulant for salt stress tolerance in tomato (*Solanum lycopersicum*). J Appl Phycol 30:2929–2941
Emamverdian A, Ding Y, Mokhberdoran F, Xie Y (2015) Heavy metal stress and some mechanisms of plant defense response. Sci World J 2015:756120
Farhangi-Abriz S, Ghassemi-Golezani K (2018) How can salicylic acid and jasmonic acid mitigate salt toxicity in soybean plants? Ecotoxicol Environ Saf 147:1010–1016
Faseela P, Veena M, Sen A, Anjitha KS, Aswathi KPR, Sruthi P, Puthur JT (2025) Elicitors fortifies the plant resilience against metal and metalloid stress. Int J Phytoremediation 27:372–389
Forlani G, Trovato M, Funck D, Signorelli S (2019) Chapter 3 - Regulation of proline accumulation and its molecular and physiological functions in stress defence. In: Hossain M, Kumar V, Burritt D, Fujita M, Mäkelä P (eds) Osmoprotectant-mediated abiotic stress tolerance in plants: recent advances and future perspectives, 1st edn. Springer, Cham, pp 73–97
Forouzandeh M, Parsa S, Mahmoodi S, Izanloo A (2024) Physiological, biochemical, and molecular responses of quinoa (*Chenopodium quinoa Willd.*) to elicitors under drought stress. Plant Mol Biol Report 42:515–531
Francesca S, Cirillo V, Raimondi G, Maggio A, Barone A, Rigano MM (2021) A novel protein hydrolysate-based biostimulant improves tomato performances under drought stress. Plants 10:783
Gitau MM, Shetty P, Maróti G (2023) Transcriptional analysis reveals induction of systemic resistance in tomato treated with *Chlorella* microalgae. Algal Res 72:103106
Gómez-Ocampo G, Ploschuk EL, Mantese A, Crocco CD, Botto JF (2021) BBX21 reduces abscisic acid sensitivity, mesophyll conductance and chloroplast electron transport capacity to increase photosynthesis and water use efficiency in potato plants cultivated under moderated drought. Plant J 108:1131–1144
Gowtham HG, Murali M, Shilpa N, Amruthesh KN, Gafur A, Antonius S, Sayyed RZ (2024) Harnessing abiotic elicitors to bolster plant's resistance against bacterial pathogens. Plant Stress 11:100371
Gowthami L (2018) Role of elicitors in plant defense mechanism. J Pharmacogn Phytochem 7:2806–2812
Guddimalli R, Somanaboina AK, Palle SR, Edupuganti S, Kummari D, Palakolanu SR, Naravula J, Gandra J, Qureshi IA, Marka N, Polavarapu R, Kishor OBK (2021) Overexpression of RNA-binding bacterial chaperones in rice leads to stay-green phenotype, improved yield and tolerance to salt and drought stresses. Physiol Plant 173:1351–1368
Gupta S, Novák O, Kulkarni MG, Doležalova I, Van Staden J, Doležal K (2024) Unleashing the potential of biostimulants in stimulating pollen germination and tube growth. J Plant Growth Regul 43:3392–3423
He M, He CQ, Ding NZ (2018) Abiotic stresses: general defenses of land plants and chances for engineering multistress tolerance. Front Plant Sci 9:1771
He X, Wang C, Wang H, Li L, Wang C (2020) The function of MAPK cascades in response to various stresses in horticultural plants. Front Plant Sci 11:952
He Z, Webster S, He SY (2022) Growth-defense trade-offs in plants. Curr Biol 32:R634–R639

Huang LJ, Cheng GX, Khan A, Wei AM, Yu QH, Yang SB, Luo DX, Gong ZH (2019) CaHSP16.4, a small heat shock protein gene in pepper, is involved in heat and drought tolerance. Protoplasma 256:39–51

Iqbal H, Yaning C, Rehman HU, Waqas M, Ahmed Z, Raza ST, Shareef M (2020) Improving heat stress tolerance in late planted spring maize by using different exogenous elicitors. Chil J Agric Res 80:30–40

Jarin A, Ghosh UK, Hossain MS, Mahmud A, Khan MAR (2024) Glycine betaine in plant responses and tolerance to abiotic stresses. Discov Agric 2:127

Jing Y, Shen N, Zheng X, Fu A, Zhao F, Lan W, Luan S (2020) Danger-associated peptide regulates root immune responses and root growth by affecting ROS formation in *Arabidopsis*. Int J Mol Sci 21:4590

Kaushik S, Ranjan A, Singh AK, Sirhindi G (2024) Methyl jasmonate reduces cadmium toxicity by enhancing phenol and flavonoid metabolism and activating the antioxidant defense system in pigeon pea (*Cajanus cajan*). Chemosphere 346:140681

Khan MA, Gemenet DC, Villordon A (2016) Root system architecture and abiotic stress tolerance: current knowledge in root and tuber crops. Front Plant Sci 7:1584

Kim H, Seomun S, Yoon Y, Jang G (2021a) Jasmonic acid in plant abiotic stress tolerance and interaction with abscisic acid. Agronomy 11:1886

Kim JS, Jeon BW, Kim J (2021b) Signaling peptides regulating abiotic stress responses in plants. Front Plant Sci 12:704490

Kirubakaran S, Sehgal A, Deys K, Szewc-McFadden A, Duplais C, Gutierrez B, Meakem V, Galarneau E, Londo J, Turner BN, Cadle-Davidson LE, Zhong G, Gouker F (2024) Effect of grapevine rootstock and foliar biostimulants in regulating scion physiology, secondary metabolites, and root architectural adaptation to drought stress. Soc Sci Res. https://doi.org/10.2139/ssrn.4897138

Kong F, Deng Y, Zhou B, Wang G, Wang Y, Meng Q (2014) A chloroplast-targeted DnaJ protein contributes to maintenance of photosystem II under chilling stress. J Exp Bot 65:143–158

Kumar RR, Goswami S, Singh K, Dubey K, Singh S, Sharma R, Verma N, Kala YK, Rai GK, Grover M, Mishra DC, Singh B, Pathak H, Chinnusamy V, Rai A, Praveen S (2016) Identification of putative RuBisCo Activase (TaRca1)-the catalytic chaperone regulating carbon assimilatory pathway in wheat (Triticum aestivum) under the heat stress. Front Plant Sci 7:986

Kuźniak E, Kopczewski T (2020) The chloroplast reactive oxygen species-redox system in plant immunity and disease. Front Plant Sci 11:572686

La VH, Lee BR, Zhang Q, Park SH, Tabibul Islam Md, Kim TH (2019) Salicylic acid improves drought-stress tolerance by regulating the redox status and proline metabolism in *Brassica rapa*. Hortic Environ Biotechnol 60:31–40

Lawson T, Vialet-Chabrand S (2018) Speedy stomata, photosynthesis and plant water use efficiency. New Phytol 221:93–98

Li W, Hou J, Wang W, Tang X, Liu C, Xing D (2011) Effect of water deficit on biomass production and accumulation of secondary metabolites in roots of *Glycyrrhiza uralensis*. Russ J Plant Physiol 58:538–542

Li L, Feng Y, Qi F, Hao R (2023) Research progress of *Piriformospora indica* in improving plant growth and stress resistance to plant. J Fungi 9:965

Lin L, Wu J, Jiang M, Wang Y (2021) Plant mitogen-activated protein kinase cascades in environmental stresses. Int J Mol Sci 22:1543

Liu Q, Wang S, Du Y, Yin K (2022) Improved drought tolerance in soybean by protein elicitor AMEP412 induced ROS accumulation and scavenging. Biotechnol Biotechnol Equip 36:401–412

Liu W, Thapa P, Park SW (2023) RD29A and RD29B rearrange genetic and epigenetic markers in priming systemic defense responses against drought and salinity. Plant Sci 337:111895

Lozano-Juste J, Alrefaei AF, Rodriguez PL (2020) Plant osmotic stress signaling: MAPKKKs meet SnRK2s. Trends Plant Sci 25:1179–1182

Ma X, Su Z, Ma H (2020) Molecular genetic analyses of abiotic stress responses during plant reproductive development. J Exp Bot 71:2870–2885

Malnoë A, Wang F, Girard-Bascou J, Wollman FA, de Vitry C (2014) Thylakoid FtsH protease contributes to photosystem II and cytochrome B6f remodeling in *Chlamydomonas reinhardtii* under stress conditions. Plant Cell 26:373–390

Manepalli SB, Tomar S, Gaikwad DJ, Maitra S (2022) Abiotic stress signaling in plants and transgenic technology as a triumph: a review. J App Biol Biotech 10:5–13

Melo AS, Costa RRD, Sá FVDS, Dias GF, Alencar RS, Viana PMO, Peixoto TDC, Suassuna JF, Brito MEB, Ferraz RLS, Costa PDS, Melo YL, Corrêa ÉB, Lacerda CF, Dantas Neto J (2024) Modulation of drought-induced stress in cowpea genotypes using exogenous salicylic acid. Plants 13:634

Nadarajah KK (2020) ROS homeostasis in abiotic stress tolerance in plants. Int J Mol Sci 21:5208

Nahar K, Hasanuzzaman M, Fujita M (2016) Chapter 4 - Roles of osmolytes in plant adaptation to drought and salinity. In: Iqbal N, Nazar R, Khan N (eds) Osmolytes and plants acclimation to changing environment: emerging omics technologies, 1st edn. Springer, pp 37–68

Nakaminami K, Okamoto M, Higuchi-Takeuchi M, Yoshizumi T, Yamaguchi Y, Fukao Y, Shimizu M, Ohashi C, Tanaka M, Matsui M, Shinozaki K, Seki M, Hanada K (2018) AtPep3 is a hormone-like peptide that plays a role in the salinity stress tolerance of plants. Proc Natl Acad Sci USA 115:5810–5815

Nounjan N, Chansongkrow P, Charoensawan V, Siangliw JL, Toojinda T, Chadchawan S, Theerakulpisut P (2018) High performance of photosynthesis and osmotic adjustment are associated with salt tolerance ability in rice carrying drought tolerance QTL: physiological and co-expression network analysis. Front Plant Sci 9:1135

Özkurt N, Bektaş Y (2022) Alleviation of salt stress with chitosan foliar application and its effects on growth and development in tomato (*Solanum lycopersicum* L.). Turk J Agric Res 9:342–351

Ozturk M, Turkyilmaz Unal B, García-Caparrós P, Khursheed A, Gul A, Hasanuzzaman M (2021) Osmoregulation and its actions during the drought stress in plants. Physiol Plant 172:1321–1335

Partap M, Kumar P, Kumar A, Joshi R, Kumar D, Warghat AR (2020) Effect of elicitors on morpho-physiological performance and metabolites enrichment in *Valeriana jatamansi* cultivated under aeroponic conditions. Front Plant Sci 11:01263

Petrova M, Miladinova-Georgieva K, Geneva M (2024) Influence of abiotic and biotic elicitors on organogenesis, biomass accumulation, and production of key secondary metabolites in *Asteraceae* plants. Int J Mol Sci 25:4197

Prudent M, Salon C, Souleimanov A, Neil Emery RJ, Smith DL (2015) Soybean is less impacted by water stress using *Bradyrhizobium japonicum* and thuricin-17 from *Bacillus thuringiensis*. Agron Sustain Dev 35:749–757

Pulido P, Llamas E, Rodriguez-Concepcion M (2017) Both Hsp70 chaperone and Clp protease plastidial systems are required for protection against oxidative stress. Plant Signal Behav 12:e1290039

Qi ZY, Wang KX, Yan MY, Kanwar MK, Li DY, Wijaya L, Alyemeni MN, Ahmad P, Zhou J (2018) Melatonin alleviates high temperature-induced pollen abortion in *Solanum lycopersicum*. Molecules 23:386

Qian Z, He L, Li F (2024) Understanding cold stress response mechanisms in plants: an overview. Front Plant Sci 15:1443317

Rasheed R, Arslan Ashraf M, Kamran S, Iqbal M, Hussain I (2018) Menadione sodium bisulphite mediated growth, secondary metabolism, nutrient uptake and oxidative defense in okra (*Abelmoschus esculentus Moench*) under cadmium stress. J Hazard Mater 360:604–614

Reichardt S, Piepho HP, Stintzi A, Schaller A (2020) Peptide signaling for drought-induced tomato flower drop. Science 367:1482–1485

Ronga D, Caradonia F, Parisi M, Bezzi G, Parisi B, Allesina G, Pedrazzi S, Francia E (2020) Using digestate and biochar as fertilizers to improve processing tomato production sustainability. Agronomy 10:138

Roy S, Griffiths M, Torres-Jerez I, Sanchez B, Antonelli E, Jain D, Krom N, Zhang S, York LM, Scheible WR, Udvardi M (2022) Application of synthetic peptide CEP1 increases nutrient uptake rates along plant roots. Front Plant Sci 12:793145

Saibi W, Brini F (2020) Proline, a peculiar amino acid with astucious functions in development and salt tolerance process in plants. J Food Nutr Metabol 3:1–8

Salehi M, Karimzadeh G, Naghavi MR (2019) Synergistic effect of coronatine and sorbitol on artemisinin production in cell suspension culture of *Artemisia annua* L. cv. Anamed. Plant Cell Tissue Organ Cult 137:587–597

Shang Y, Yang DM, Ha YM, Shin HY, Nam KH (2020) Receptor-like protein kinases RPK1 and BAK1 sequentially form complexes with the cytoplasmic kinase OST1 to regulate ABA-induced stomatal closure. J Exp Bot 71:1491–1502

Sharma P, Kumar A, Bhardwaj R (2016) Plant steroidal hormone epibrassinolide regulate – heavy metal stress tolerance in *Oryza sativa L.* by modulating antioxidant defense expression. Environ Exp Bot 122:1–9

Sharma L, Dalal M, Verma RK, Kumar SV, Yadav SK, Pushkar S, Kushwaha SR, Bhowmik A, Chinnusamy V (2018) Auxin protects spikelet fertility and grain yield under drought and heat stresses in rice. Environ Exp Bot 150:9–24

Shen JL, Diao WZ, Zhang LF, Acharya BR, Wang M, Zhao XY, Chen DH, Zhang W (2020) Secreted peptide PIP1 induces stomatal closure by activation of guard cell anion channels in *Arabidopsis*. Front Plant Sci 11:1029

Silva S, Teixeira C, Dias MC, Pereira BR, Gomes P, Silva MSA (2025) Bioprospecting a natural antioxidant peptide for tomato biostimulation under water limitations: physiological approach. J Plant Growth Regul 44:2569–2582

Singh A, Dwivedi P (2018) Methyl-jasmonate and salicylic acid as potent elicitors for secondary metabolite production in medicinal plants: a review. J Pharmacogn Phytochem 7:750–757

Soltabayeva A, Dauletova N, Serik S, Sandybek M, Omondi JO, Kurmanbayeva A, Srivastava S (2022) Receptor-like kinases (LRR-RLKs) in response of plants to biotic and abiotic stresses. Plants 11:2660

Song Y, Feng L, Alyafei MAM, Jaleel A, Ren M (2021) Function of chloroplasts in plant stress responses. Int J Mol Sci 22:13464

Sun Y, Qiao Z, Muchero W, Chen J (2020) Lectin receptor-like kinases: the sensor and mediator at the plant cell surface. Front Plant Sci 11:596301

Takahashi F, Suzuki T, Osakabe Y, Betsuyaku S, Kondo Y, Dohmae N, Fukuda H, Yamaguchi-Shinozaki K, Shinozaki K (2018) A small peptide modulates stomatal control via abscisic acid in long-distance signaling. Nature 556:235–238

Thakur M, Sohal BS (2013) Role of elicitors in inducing resistance in plants against pathogen infection: a review. ISRN Biochem 2013:762412

Thakur M, Bhattacharya S, Khosla PK, Puri S (2020) Improving production of plant secondary metabolites through biotic and abiotic elicitation. J Appl Res Med Aromat Plants 12:1–12

Tian F, Gong J, Zhang J, Feng Y, Wang G, Guo Q, Wang W (2014) Overexpression of monoubiquitin improves photosynthesis in transgenic tobacco plants following high temperature stress. Plant Sci 226:92–100

Tsintzou G, Madesis P (2024) Chapter 2 - Peptide elicitors for defense against abiotic stress. In: Singh S, Mehrotra R (eds) Plant elicitor peptides: new tool for sustainable agriculture, 1st edn. Springer, pp 17–47

Wang S, Bai G, Wang S, Yang L, Yang F, Wang Y, Zhu JK, Hua J (2016) Chloroplast RNA-binding protein RBD1 promotes chilling tolerance through 23S rRNA processing in *Arabidopsis*. PLoS Genet 12:e1006027

Wang L, Ju C, Han C, Yu Z, Bai MY, Wang C (2025) The interaction of nutrient uptake with biotic and abiotic stresses in plants. J Integr Plant Biol 67:455–487

Wen Z, Wang Y, Xia C, Zhang Y, Zhang H (2021) Chloroplastic SaNADP-ME4 of C(3)-C(4) woody desert species *Salsola laricifolia* confers drought and salt stress resistance to *Arabidopsis*. Plants 10:1827

Xiang D, Wei W, Wan Y, Wu X, Ye X, Peng L, Zhong L, Wu Q, Zou L, Zhao G, Zhao J (2021) Polysaccharide elicitor from the endophyte *Bionectria sp.* Fat6 improves growth of tartary buckwheat under drought stress. Phyton 90:461–473

Xiao F, Zhou H, Lin H (2025) Decoding small peptides: regulators of plant growth and stress resilience. J Integr Plant Biol 67:596–631

Xu T, Niu J, Jiang Z (2022) Sensing mechanisms: calcium signaling mediated abiotic stress in plants. Front Plant Sci 13:925863

Yang R, Wang Z, Zhao L, Liu J, Meng J, Luan Y (2023) Secreted peptide SpPIP1 modulates disease resistance and salt tolerance in tomato. J Agric Food Chem 71:12264–12279

Zhang J, Shi Y, Zhang X, Du H, Xu B, Huang B (2017) Melatonin suppression of heat-induced leaf senescence involves changes in abscisic acid and cytokinin biosynthesis and signaling pathways in perennial ryegrass (*Lolium perenne L.*). Environ Exp Bot 138:36–45

Zhang H, Zhao Y, Zhu JK (2020) Thriving under stress: how plants balance growth and the stress response. Dev Cell 55:529–543

Zhang Y, Xu J, Li R, Ge Y, Li Y, Li R (2023) Plants' response to abiotic stress: mechanisms and strategies. Int J Mol Sci 24:10915

Zhao C, Wang P, Si T, Hsu CC, Wang L, Zayed O, Yu Z, Zhu Y, Dong J, Tao WA, Zhu JK (2017) MAP kinase cascades regulate the cold response by modulating ICE1 protein stability. Dev Cell 43:618–629.e5

Zhao C, Zayed O, Yu Z, Jiang W, Zhu P, Hsu CC, Zhang L, Tao WA, Lozano-Durán R, Zhu JK (2018) Leucine-rich repeat extensin proteins regulate plant salt tolerance in *Arabidopsis*. Proc Natl Acad Sci USA 115:13123–13128

Zhao J, Lu Z, Wang L, Jin B (2020a) Plant responses to heat stress: physiology, transcription, noncoding RNAs, and epigenetics. Int J Mol Sci 22:117

Zhao W, Liu L, Shen Q, Yang J, Han X, Tian F, Wu J (2020b) Effects of water stress on photosynthesis, yield, and water use efficiency in winter wheat. Water 12:2127

Zhu JK (2016) Abiotic stress signaling and responses in plants. Cell 167:313–324

Zhu Z, Liang Z, Han R, Wang X (2009) Impact of fertilization on drought response in the medicinal herb *Bupleurum chinense* DC.: growth and saikosaponin production. Ind Crop Prod 29:629–633

Zhu Q, Feng Y, Xue J, Chen P, Zhang A, Yu Y (2023) Advances in receptor-like protein kinases in balancing plant growth and stress responses. Plants 12:427

Chapter 4
Harnessing Chitosan for Sustainable Agriculture: Strengthening Plant Defense Mechanisms Under a Changing Climate

Muhammad Faheem Jan, Waqas Liaqat, Muhammad Tanveer Altaf, Wang Shuai, Changzhuang Liu, Miaoli Zhang, Ehtisham Hassan Khan, Haseeb Ahmad, Unaiza, and Ming Li

Abstract Chitosan, a natural biopolymer derived from chitin, has become a significant factor in contemporary agriculture, providing potential avenues for improving crop yield in response to environmental stresses. Chitosan's distinctive characteristics enable plants to exhibit enhanced resistance against abiotic stressors, including drought, salinity, heat, and heavy metal toxicity, as well as biotic hazards such as pests and diseases. Chitosan nanoparticles (NPs) have garnered significant interest due to their capacity to stimulate plant defense mechanisms and enhance stress tolerance. The incorporation of these NPs into plant systems enhances the bioavailability and absorption efficiency of chitosan, resulting in more potent and resilient responses to stresses. Moreover, the sustainable and environmentally friendly characteristics of chitosan make it a compelling option for decreasing dependence on chemical fertilizers and pesticides, aligning with the increasing demand for more eco-friendly agricultural practices. Given the escalating uncertainties posed by climate change and resource constraints, chitosan emerges as a valuable tool for enhancing crop health, productivity, and general resilience in global agricultural

M. F. Jan (✉) · H. Ahmad
Guangxi Key Laboratory of Agro-environment and Agro-products Safety, College of Agriculture, Guangxi University, Nanning, China

W. Liaqat (✉) · M. T. Altaf
Department of Field Crops, Faculty of Agriculture, Recep Tayyip Erdoğan University, Rize/Pazar, Turkey

W. Shuai · C. Liu · M. Zhang · M. Li
College of Agriculture, Northeast Agricultural University, Harbin, China

E. H. Khan
MOA Key Laboratory of Crop Ecophysiology and Farming System in the Middle Reaches of the Yangtze River, College of Plant Science and Technology, Huazhong Agricultural University, Wuhan, China

Unaiza
Department of Field Crops, Faculty of Agriculture,Institute of Natural and Applied Sciences, Çukurova University, Adana, Turkey

H. I. Mohamed (ed.), *Elicitors for Sustainable Crop Production*, Sustainability Sciences in Asia and Africa, https://doi.org/10.1007/978-981-95-3533-0_4

systems. This chapter emphasizes the significance of chitosan in shaping a more environmentally sustainable future for agriculture, providing a natural and efficient solution to the increasing demands of global food production.

Keywords Chitosan · Drought · Salinity · Heavy metals · Nanoparticles

4.1 Introduction

Agricultural production is increasingly facing challenges associated with climate change, including the prevalence of pests, diseases, nutrient depletion, and reduced productivity, driven by environmental stressors such as salinity, drought, temperature fluctuations, and heavy metal toxicity (Fig. 4.1). Additionally, the rising global population, urbanization, and depletion of arable land have escalated the demand for food. Global food production is expected to increase by 70–100% by 2050 (Van Dijk et al. 2021). These challenges have prompted researchers to seek natural and environmentally friendly alternatives for agricultural sustainability. Chitosan has shown promise in addressing these issues because it provides advantages without endangering the agroecosystem and health of the soil. Due to its beneficial qualities, including biodegradability, biocompatibility, and nontoxicity, as well as its capacity to encourage plant development, improve yield, and increase resilience to a variety of biotic and abiotic challenges, its application in agriculture is increasing (Shahrajabian et al. 2021). Chitosan has been shown in several experiments to positively affect crop growth and yield, such as rice (Moolphuerk et al. 2022), soybean (Abou El-Enin et al. 2023), and wheat (Kolesnikov et al. 2022). These advantages are associated with improved physiological processes, enhanced nutrient uptake, increased cell division, and protein synthesis, among others.

Chitosan exhibits strong antimicrobial properties, suppressing various fungal, bacterial, viral, and nematode pathogens. For example, fungal pathogens like

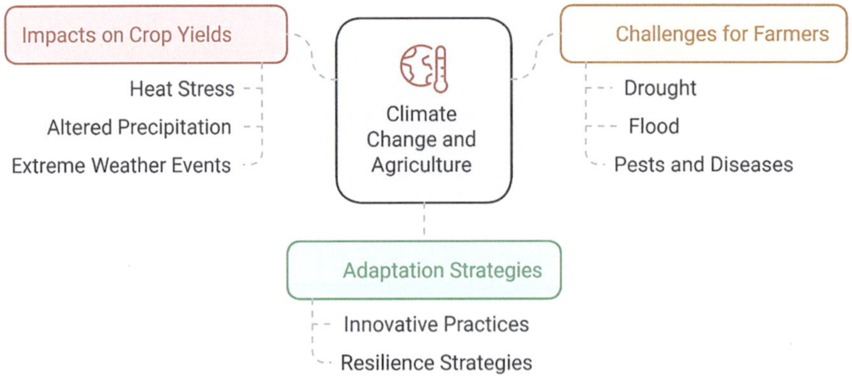

Fig. 4.1 Climate change impacts on agriculture and adaptation strategies

Alternaria, Rhizophus, Botrytis, and *Fusarium* have been significantly suppressed by chitosan treatment (Karamchandani et al. 2022). Chitosan has exhibited antibacterial activity against plant pathogens such as *Xanthomonas* spp. and *Pseudomonas* spp. (Narasimhamurthy et al. 2022). In tomatoes, it enhances resistance to the tomato mosaic virus while improving growth and yield (El-Gendi et al. 2022). Additionally, chitosan is effective against nematodes like *Meloidogyne* spp. in tomatoes (Krif et al. 2022). The production of defense-related genes and protective macromolecules is two of its resistance mechanisms (Selvaraj et al. 2022). Chitosan significantly improved crop yield and quality across various plants, including vegetables, fruits, and ornamental species (Boamah et al. 2023).

Chitosan's role as an elicitor of plant defense systems makes it a promising tool for addressing the challenges posed by both biotic and abiotic stresses in agriculture (Sree Rayanoothala et al. 2024). With increasing food demand and unsustainable use of synthetic chemicals, chitosan presents a viable solution for stress adaptation. Plant resistance to abiotic stressors such as heat, salinity, drought, and heavy metal toxicity is increased by chitosan and its oligomers. Its capacity to improve plant performance in stressful conditions and scavenge reactive oxygen species (ROS) has spurred further investigation into its possible uses. Chitosan's ability to stimulate the production of antioxidant enzymes, enhance root growth, and boost photosynthetic activity contributes to its effectiveness in mitigating drought stress (Dawood et al. 2024). As a growth elicitor, it stimulates the manufacture of other regulators such as ethylene and may mimic the actions of auxins and other plant growth regulators (Mukarram et al. 2023). Strengthening the antioxidant defense system also increases plants' tolerance to environmental challenges (Hidangmayum et al. 2019). Chitosan's chemical makeup, particularly the patterns of polymerization and acetylation, determines how efficient it is as a growth elicitor (Fan et al. 2023). Chitosan can boost plant height, root length, and biomass output when applied to the roots, leaves, or seeds of plants. This can result in increased crop yields. Additionally, it increases fruit size, vitamin and mineral content, and shelf life, all of which enhance crop quality (Saha et al. 2024). Furthermore, chitosan is employed in bioremediation to lessen the levels of heavy metals and other harmful compounds in contaminated soils. This chapter explores the role of chitosan in alleviating the negative effects of stress on agricultural crops.

4.2 Overview of Chitosan

Chitosan, a polycationic polymer formed by the deacetylation of chitin, is found in various organisms, including crustaceans, insects, fungi, mollusks, and marine species (Lalarukh et al. 2024). Chitin, the second most abundant polysaccharide after cellulose, is converted into chitosan, which introduces amino-based functional groups, enhancing its biological activity (Mohan et al. 2022). The molecular weight, purity, and properties of chitosan depend on its source and extraction process. Notable characteristics of chitosan include nontoxicity, mucoadhesiveness (Mura et al. 2022), hemostatic properties, film-forming capability (Ibrahim et al. 2023), and antioxidative activities (Yu et al. 2022). These characteristics make it valuable for applications in fields such as agriculture, food technology, and biomedicine. In agriculture, chitosan supports sustainability by mitigating biotic and abiotic stresses, such as drought, salinity, and heavy metal toxicity (Shinde et al. 2024). Studies have shown that chitosan combined with plant growth-promoting rhizobacteria (PGPR) enhances growth and plant metabolites more effectively than PGPR alone (Riseh et al. 2022). Encapsulation of agrochemicals and metal components within chitosan has demonstrated increased defense-related metabolites, antioxidant activity, and growth promotion, while its slow-release mechanism ensures sustained protection with minimal environmental impact.

4.3 Chitosan-Induced Mechanisms of Plant Defense

The mechanism of chitosan in plants is not fully understood, but many studies have suggested that it elicits various defense responses (Suwanchaikasem et al. 2023). Chitin-specific receptors in plant cell membranes trigger these responses, as chitosan treatments mimic compounds from chitin-containing organisms (Riseh et al. 2024a, b, c). Chitin elicitor binding proteins (CEBiP), identified in several crops, induce defense mechanisms by affecting chitosan-responsive gene expression (Arif et al. 2022). Chitosan and its derivatives act as elicitors in various plants (Dutta et al. 2022), enhancing defense responses under stress (Fig. 4.2). Phytoalexins, pathogenesis-related proteins (chitinase, β-glucanase), proteinase inhibitors, callose deposition, lignin biosynthesis, and stress-responsive gene activation are some of the reactions that occur throughout these processes. Treatments with chitosan have been shown to increase these defense-related molecules, making it an effective antibacterial agent and an elicitor of plant protection (Ji et al. 2022). Moreover, studies have shown that chitosan can induce the chitinase and glucanase enzymes in crops like tomatoes (Narasimhamurthy et al. 2022) and peaches (Zhao et al. 2020). The effectiveness of chitosan in inducing pathogenesis-related proteins, however, can vary depending on its molecular weight, as seen in rice seedlings (Fan et al. 2023).

4.4 Biological Activity of Chitosan

Chitosan is a naturally derived polysaccharide obtained from the deacetylation of chitin, the second most abundant natural biopolymer after cellulose (Ali et al. 2024). Its biological activities have received extensive attention for their potential application in plant health and disease resistance. The chemical structure of chitosan, featuring amino groups on D-glucosamine units, imparts it with polycationic properties, enabling interactions with negatively charged microbial cell walls and plant surfaces (Lingait et al. 2023). Chitosan has demonstrated potent antimicrobial

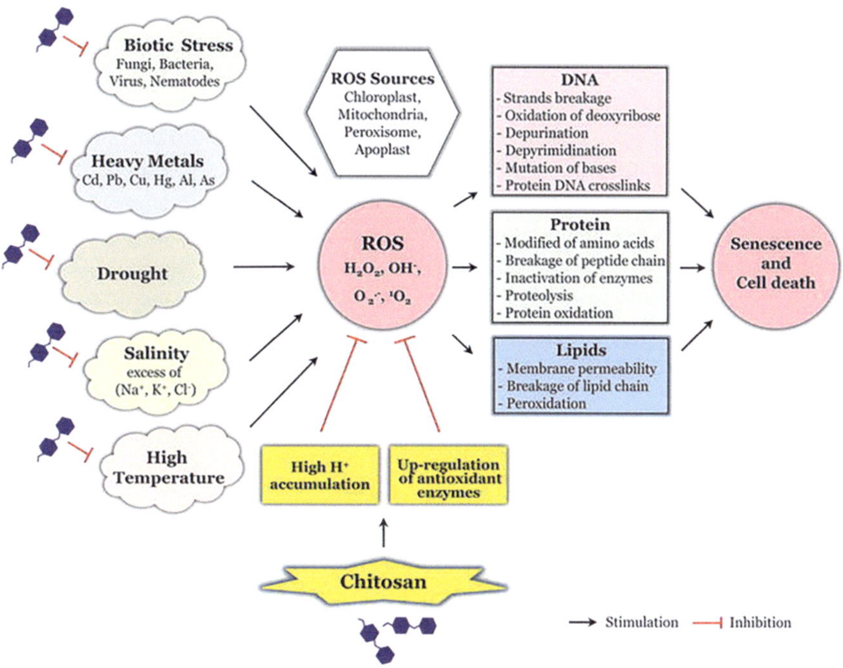

Fig. 4.2 An overview of chitosan's mechanism in combating various stresses

properties against a wide range of pathogens, including bacteria, fungi, viruses, and nematodes (Fan et al. 2023). Its mechanism of action involves the disruption of microbial cell membranes through electrostatic interactions, causing the leakage of intracellular components, which leads to microbial death. The polymer's effectiveness against fungi such as *Alternaria* and *Fusarium*, and bacterial pathogens like *Xanthomonas* and *Pseudomonas* has been well-documented (Biradar et al. 2022). Furthermore, chitosan enhances plant resistance to viral pathogens by triggering defense responses, reducing infection rates, and mitigating viral spread in crops such as maize (Wang et al. 2023), faba bean (Ji et al. 2023), tomato (Carmona et al. 2020), and potato (Al-Mokadem et al. 2022). Chitosan acts as an elicitor of plant immune responses, mimicking pathogen-associated molecular patterns (PAMPs) and activating the plant's innate immune system (Dutta et al. 2022). The application of chitosan induces the synthesis of defense enzymes, such as peroxidases (Pongprayoon et al. 2022a, b) and chitinases (Khaled and Alobaidi 2023), which strengthen plant cell walls and inhibit the growth of invading pathogens. It also promotes the accumulation of secondary metabolites like phenolic compounds, alkaloids, and flavonoids, which serve as chemical defenses against pests and diseases (Sun et al. 2023; Stasińska-Jakubas and Hawrylak-Nowak 2022). A notable biological function of chitosan is its ability to mitigate oxidative stress by scavenging reactive oxygen species (ROS). During biotic and abiotic stresses, ROS levels increase in plant tissues, leading to oxidative damage and cell death (Nadarajah 2020). Chitosan neutralizes these harmful radicals by enhancing the activity of antioxidant enzymes, such as SOD, CAT, and POD (Quitadamo et al. 2021; Mohammadi et al. 2021; Dawood et al. 2024). These enzymes not only prevent cellular damage but also stabilize physiological processes, ensuring optimal growth under stress conditions. Chitosan plays a significant role in improving soil health and nutrient availability. Its ability to chelate heavy metals helps detoxify contaminated soils and facilitates the uptake of essential minerals (Chouhan and Mandal 2021). In combination with fertilizers, chitosan aids in the slow and controlled release of nutrients, enhancing plant absorption while minimizing environmental contamination from fertilizer runoff.

4.5 Chitosan as a Biostimulant

Chitosan has emerged as a powerful biostimulant, promoting plant growth and enhancing resilience to environmental stresses. As a biopolymer, it is eco-friendly, biodegradable, and nontoxic, making it an excellent choice for sustainable agricultural practices. Its ability to regulate hormonal balance, enhance nutrient uptake, and support physiological processes contributes to increased crop yields and improved plant health (Shahrajabian et al. 2021; Riseh et al. 2023). The application of chitosan has been shown to improve seed germination rates, accelerate seedling emergence, and increase biomass production (Boamah et al. 2023). Studies demonstrate that chitosan-treated seeds display higher resistance to adverse conditions such as salinity and drought, ensuring better seedling establishment (Riseh et al. 2024a, b, c). This effect is attributed to chitosan's ability to regulate water uptake and activate enzymes involved in early growth phases. Additionally, foliar applications of chitosan improve photosynthetic activity by increasing chlorophyll content, thus promoting more efficient energy capture and utilization (Masoumi et al. 2024; Ahmed et al. 2020). The enhancement of photosynthesis is coupled with improved nutrient uptake through root systems, leading to increased root and shoot biomass. Chitosan plays a critical role in modulating plant hormone pathways, including the regulation of auxins, gibberellins, and abscisic acid (Khan et al. 2022a, b, c). This modulation helps plants adapt to abiotic stresses by regulating stomatal closure, improving water retention, and enhancing osmotic adjustment mechanisms. In addition to hormonal regulation, chitosan induces the production of compatible solutes, such as sugars and amino acids, which maintain cellular osmotic balance under drought and salinity conditions (Sadak et al. 2022).

Chitosan applications have consistently demonstrated yield improvements in various crops, including rice (Elshayb et al. 2022), soybean (Abou El-Enin et al. 2023), and wheat (Samdurkar et al. 2021). Beyond yield enhancement, chitosan-treated plants produce fruits and vegetables with higher antioxidant levels and better sensory qualities, including apple (Ackah et al. 2022), tomato (Hernández et al. 2022), and spinach (Meitha et al. 2022). Incorporating chitosan into agricultural systems helps reduce the reliance on chemical fertilizers and pesticides, aligning with environmentally sustainable practices. Its biodegradability ensures that residues do not accumulate in the environment, reducing pollution and promoting ecosystem health. Moreover, chitosan's ability to enhance soil microbial activity contributes to healthier soil ecosystems and improved crop resilience.

4.6 Advantages of Using Chitosan Nanoparticles in Plants

Chitosan nanoparticles (CNPs) are a promising carrier beyond biomedical applications. Their biodegradability, biocompatibility, adsorption capabilities, and low toxicity make them useful in generating stress resistance (Jiang et al. 2024; Khan et al. 2025). They are also more adaptable as drug delivery systems because of their simple production method (Jafernik et al. 2023). According to Kolbert et al. (2021), the regulated release of nitric oxide (NO) was demonstrated by CNPs containing S-nitroso-mercaptosuccinic acid (S-nitroso-MSA), a nitric oxide donor, to mitigate salt stress in maize. To sustain plant growth and development, NO is essential for plants to respond to abiotic challenges (Kumari et al. 2024). While the encapsulation shields the NO donor from heat and degradation, the gradual release of NO from the nanoformulation probably lessens the effects of salinity (Pande et al. 2022). Chitosan's ability to encapsulate active compounds promotes sustainable agriculture by enabling effective gene delivery systems for plant transformation and regulated release of these ingredients (Riseh et al. 2024a, b, c). Narasimhamurthy et al. (2022) found that CNPs significantly enhanced plant immune responses and defense-related enzymes, suggesting that they are more effective for phytosanitation and disease management compared to natural chitosan.

4.7 Uptake and Translocation of Chitosan-Based Nanoparticles

Plants take up nanoparticles (NPs) through foliar and root pathways. Foliar uptake occurs via cuticular and stomatal routes. The waxy cuticle poses challenges for NP entry due to its small pore size (0.6–4.8 nm) (Hidangmayum and Dwivedi 2022). In contrast, stomatal uptake is more evident, with NPs entering through stomatal apertures, which range from 3 to 10 μm in width and 25 μm in length (Eichert et al. 2008). Studies have shown NP movement through stomatal pathways in various plants, including *Citrullus lanatus, Cucurbita pepo, Lactuca sativa, Allium porrum, and Arabidopsis thaliana* (Khan et al. 2022a; Ahmad et al. 2022). Although the exact size exclusion limit for NP varies based on plant species and leaf morphology, translocation through the xylem and phloem has been observed, though the mechanisms are not fully understood. In watermelon, foliar-applied NP (24–47 nm) was shown to enter through stomata and translocate from shoots to roots via phloem

(Khan et al. 2022b). In wheat and beans, chitosan-loaded NPK nanoparticles were localized in xylem and phloem after foliar application (Hidangmayum and Dwivedi 2022). Root uptake is facilitated by the positive charge of NPs, which interacts with the negatively charged root hairs. The root epidermis allows entry through the apoplastic pathway, as supported by transmission electron microscopy (TEM) and confocal scanning laser microscopy (CLSM) studies, including the uptake of ZnO in wheat and Ag NP (20–80 nm) in *Arabidopsis thaliana* (Zhu et al. 2021a, b; He et al. 2022).

4.8 Chitosan in Mitigating the Effects of Abiotic Stresses

4.8.1 Drought Stress (DS)

Due to the production of ROS, which causes lipid peroxidation of membranes and interacts with other macromolecules to reduce plant growth and yield, drought stress (DS) severely reduces agricultural productivity and has several negative effects on plant health (Seleiman et al. 2021; Jan et al. 2025). On the other hand, chitosan application has been demonstrated to improve ROS scavenging capabilities by promoting plant development and boosting the availability and uptake of water and vital nutrients (AL Kahtani et al. 2020). Chitosan treatment reduced DS in white clover (*Trifolium repens*) and enhanced the synthesis of metabolites that protect against stress (Li et al. 2017). The negative impacts of drought circumstances in corn (Elshamly 2023), wheat (Jahanbani et al. 2023), soybeans (Methela et al. 2023), and rice (Moolphuerk et al. 2022) were also lessened by foliar application of chitosan (Table 4.1).

Drought stress leads to impaired chloroplast function, decreased chlorophyll content, and reduced photosynthetic enzyme activity (Jan et al. 2024). It disrupts CO_2 intake due to stomatal closure, resulting in diminished photosynthesis and plant growth (Qiao et al. 2024). Chitosan can cause stomatal closure via an abscisic acid (ABA)-dependent mechanism. It is well established that ABA increases stomatal closure and decreases transpiration, yet the precise mechanisms underlying stomatal closure mediated by chitosan remain unclear. Chitosan foliar application has been used to decrease water consumption in tomato (Hassnain et al. 2020), sugar beet (Elmasry and Al-Maracy 2023), and maize (Almeida et al. 2020). Furthermore, *Thymus daenensis* treated with chitosan reduced the DS effect without compromising dry matter content and essential oil production (Bistgani et al. 2023).

Plant development is stimulated by chitosan, which also boosts the production of several antioxidant enzymes. In maize (Wu et al. 2024), wheat (Liu et al. 2021), and alfalfa (Mustafa et al. 2022), it decreases lipid peroxidation and mitigates the DS effect by increasing the activity of SOD, CAT, and reducing Malondialdehyde (MDA). Furthermore, chitosan increases the amount of soluble sugar in maize (Behboud et al. 2020), tomatoes (Attia et al. 2021), and sugar beets (Makhlouf et al. 2022). Sugars like fructose and glucose influence signal transduction, modify stress responses, and promote growth and development, which help to mitigate the effects

Table 4.1 Some studies of chitosan related to drought stress

Crop	Application	Effects	References
Mungbean	Seed priming (0.15% chitosan solution) and foliar spray (100 mL plant^{-1})	Improved growth, photosynthetic pigments, and relative water content, and enhanced drought tolerance	Hidangmayum et al. (2023)
Rice	Seed priming and foliar spray (100 mg L^{-1})	Enhanced growth of the shoots and roots, increased relative water content, decreased levels of malondialdehyde and electrolyte leakage, and enhanced photosynthetic effectiveness	Moolphuerk et al. (2022)
Sunflower	Foliar spray of K (11.48 g/L) and chitosan (0.28 g/L)	Boosted the activity of antioxidant enzymes, maintained turgor, increased photosynthetic activity, and raised the concentration of suitable solutes such as free amino acids, soluble proteins, proline, and total phenolics	Shehzad et al. (2020)
Wheat	chitosan nanoparticle (NPs) concentrations (0, 30, 60, and 90 ppm) were sprayed topically and applied to the soil	In comparison to the control, NPs, especially 90 ppm, enhanced leaf area, relative water content, chlorophyll content, photosynthetic rate, the activities of catalase and superoxide dismutase, yield, and biomass	Behboudi et al. (2019)
Tomato	Chitosan nanoparticles (60–90 ppm) applied to the soil	Drought-responsive gene expression was significantly upregulated in both concentrations, specifically HsfA1a, SlAREB1, LeNCED1, and LePIP1	Mohamed and Abdel-Hakeem (2023)
Bermuda grass	Foliar application of chitosan (10 mg L^{-1})	Increased turf quality, chlorophyll content, and leaf relative water content, and decreased electrolyte leakage, malonaldehyde, and hydrogen peroxide content under DS	Hao et al. (2023)
Faba bean	Foliar application of chitosan at concentrations of 0.5, 1.0, and 2.0 g L^{-1} and chitosan NP at concentrations of 10, 20, and 30 mg L^{-1}	Plant resistance to DS was improved by all applied treatments, including chitosan and chitosan NP at different doses. Vegetative growth metrics, photosynthetic pigments, total soluble sugars, soluble proteins, proline, phenolic compounds, and antioxidant enzyme activity improved, and levels of lipoxygenase, malondialdehyde, and hydrogen peroxide reduced	Dawood et al. (2024)
Common bean	Foliar spray of chitosan NP (50 and 100 ppm)	Chitosan NP at 100 ppm enhanced seed yield, biological yield, relative water content, and SPAD, and lowered total soluble sugar and MDA under stress	Dolatkhah Dashtmian et al. (2023)
Sugar beet	Chitosan spray (200 mg L^{-1})	Plants treated with chitosan showed increased CAT and GPX activities, boosted sucrose%, extracted sugar%, and sugar yield ha^{-1}, as well as improved root length, fresh weight, leaf area, and root yield	Makhlouf et al. (2022)

(continued)

Table 4.1 (continued)

Crop	Application	Effects	References
White clover	Seed priming with chitosan (5 mg/L)	Activated the activities of superoxide dismutase, catalase, and peroxidase, and enhanced the expression levels of genes encoding dehydration-responsive transcription factors (DREB2, DREB4, and DREB5) and dehydrins (Y2K, Y2SK, and SK2)	Ling et al. (2022)
White clover	Foliar application on seedlings (1 mg/mL)	Increased synthesis of metabolites sensitive to stress	Li et al. (2017)

of drought. According to Li et al. (2017), white clover treated with chitosan has been demonstrated to upregulate genes associated with glucose transport and metabolism, which are essential for reducing DS.

Foliar application of chitosan nanoparticles has been shown to mitigate DS by enhancing proline accumulation, antioxidative activity, alkaloid content, and defense-related gene activation in periwinkle (*Catharanthus roseus*), improving root biomass and photosynthetic rate in soybean when encapsulating S-nitrosoglutathione (Hidangmayum and Dwivedi 2022; Gomes et al. 2022), and enhancing physiological and biochemical attributes in wheat under drought conditions (Behboudi et al. 2019). Chitosan application has been reported to induce DS resistance by enhancing antioxidant activity, increasing H_2O_2 content, and promoting root system development in maize (Younas et al. 2022) and rice (Moolphuerk et al. 2022). Moreover, chitosan enhances ABA activity, which is crucial for regulating stomatal aperture and reducing the rate of transpiration under stress conditions (Bharath et al. 2021).

4.8.2 *Salinity Stress (SS)*

It is anticipated that high salinity will affect more than 20% of agricultural land worldwide (Mukhopadhyay et al. 2021). Salinity harms crop growth and development. In extreme situations, low external osmotic potential or toxic effects from high Na and Cl ion accumulation can prevent plants from absorbing water and nutrients. Salinity stress causes biochemical alterations and induces ROS, leading to oxidative stress that disrupts cellular processes. Research has shown that low concentrations of chitosan can alleviate the adverse effects of salinity stress (SS) (Hidangmayum et al. 2019). Chitosan-treated safflower and sunflower seeds have demonstrated reduced oxidative stress by lowering enzyme activities (Bakhoum et al. 2020). Similarly, it has been demonstrated that pretreating rice, maize, and mungbean with chitosan during SS increases antioxidant enzyme activities and lowers MDA levels, hence reducing the impacts of SS (González et al. 2015; Ray et al. 2016; Al-Tawaha et al. 2018). Furthermore, it has been found that treating wheat, lentils, and sunflowers with chitosan and oligo-chitosan reduced SS effects (Al-Tawaha and Al-Ghzawi 2013; Bakhoum et al. 2020; Sadak and Talaat 2021).

In mungbean seedlings, solid matrix priming with nano-chitosan reduced the detrimental effects of salinity and increased plant development, chlorophyll content, and protein levels (Sen et al. 2020). Treatment with chitosan-polyvinyl alcohol hydrogels, either with or without copper nanoparticles, enhanced the expression of genes linked to superoxide dismutase and jasmonic acid in tomato plants that were stressed by salinity (Hernández-Hernández et al. 2018). It has been demonstrated that chitosan strengthens a plant's defense mechanisms against SS. Chitosan reduces lipid peroxidation and oxidative stress in plants, maintains membrane integrity, enhances total alkaloid content, and improves antioxidant enzyme activities under SS (Amer and Shoala 2020). Additionally, treatment with chitosan has been associated with higher levels of chlorophyll in plants under SS, suggesting that it plays a role in reducing negative impacts and enhancing growth characteristics (Ali et al. 2018).

A wide variety of plant species, including rice, wheat, barley, soybean, tomato, and *Arabidopsis thaliana*, have been shown to exhibit chitosan-induced salinity tolerance (Table 4.2), which is mediated by controlling ion transporters (Sree Rayanoothala et al. 2024). Chitosan application has been shown to increase biomass production, ion balance, and antioxidant defense in wheat, enhance root growth and nutrient uptake in soybean (Aazami et al. 2023), improve fruit yield and photosynthetic efficiency in tomato (El Amerany et al. 2023), and induce stress-responsive gene expression in *Arabidopsis thaliana* (Hassan et al. 2021). Mulaudzi et al. (2022) demonstrated a positive effect of chitosan on sorghum under SS by reducing salt-induced oxidative damage. In sorghum plants exposed to 300 mM NaCl, chitosan (0.25 and 0.5 mg/mL) mitigated the detrimental effects on growth and on fresh and dry weights. Chitosan significantly reduced oxidative stress markers like H_2O_2 and malondialdehyde, lowered proline and total soluble sugar levels, and decreased antioxidant enzyme activities by more than 50%.

Bakhoum et al. (2020) reported that chitosan significantly improved sunflower growth and productivity under SS. Soaking seeds of the sunflower plant with a 50 mg/L concentration of chitosan was the most effective for enhancing growth, yield, and quality by boosting photosynthetic pigments and physiological traits. Jiao et al. (2024) found that applying chitosan (75 mg/L) to maize seedlings under SS (150 mM) effectively restored biomass accumulation and root architecture, which were inhibited by SS. This recovery was attributed to increased chlorophyll content, enhanced photosynthetic assimilation, and a reduction in sodium content in both shoots and roots by 25.42% and 5.12%, respectively, compared to NaCl-treated plants. Additionally, chitosan alleviated salt-induced oxidative stress by boosting the activities of antioxidant enzymes, including superoxide dismutase, catalase, ascorbate peroxidase, and peroxidase, and increasing ascorbate levels. Foliar application of nano-chitosan improved water status and reduced stomatal conductance and transpiration in pearl millet under SS (Priyaadharshini et al. 2019).

Table 4.2 Some studies of chitosan related to salinity stress

Crop	Application	Effects	References
Maize	Foliar application (0.1 mol/L NaCL, 10 mg/L) Chitosan application and 0.1 mol/L NaCL + 10 mg/L Chitosan	Enhanced growth of root and shoot	Al-Tawaha et al. (2018)
Rice	Seed soaking (0, 100–500 mg L^{-1})	Enhanced catalase and peroxidase enzymes	González et al. (2015)
Tomato	Foliar spray (0.03–0.05%)	Improved plant growth and chlorophyll content	Özkurt and Bektaş (2022)
Sunflower	Seed soaking (25, 50, and 75 mg L^{-1})	Enhanced growth and productivity	Bakhoum et al. (2020)
Maize	Foliar spray (1000 mg L^{-1})	Increased plant height, root length, and total mitochondrial respiration rate	Turk (2019)
Wheat	Foliar spray (10–20 mg L^{-1})	Increased carotenoids, chlorophyll, and antioxidant enzymes	Abdallah et al. (2020)
Wheat	Seed soaking (250, 500, and 750 mg L^{-1})	Enhanced growth parameters, antioxidant enzymes, MDA, and proline	Shams Peykani (2018)
Mungbean	Seed priming (1, 2, and 5 mg L^{-1})	Increased root length, germination rate, and plant height	Sen and Mandal (2016)
Safflower	Tissue culture (25–50 mg L^{-1})	Enhanced callus growth, flavonoids, and anthocyanin	Golkar et al. (2019)
Wheat	Seed soaking (20–40 mg/L)	Increased plant growth, H_2O_2 and MDA, Enhanced antioxidant enzymes (SOD and POX) and yield	Sadak and Talaat (2021)

4.8.3 Heavy Metal Stress (HMS)

Studies have verified that the amino and hydroxyl groups in chitosan enable it to form compounds with heavy metals. By preventing cellular dysfunction in the shoots and roots, this capacity can lessen the phytotoxicity that heavy metals produce in plants (Sobral et al. 2022; Maryam et al. 2025). Several investigations have verified that chitosan's amino and hydroxyl groups enable it to create complexes with heavy metals. Through the foliar application of various molecular weights, Zong et al. (2017a) reported that it was efficient in reducing cadmium toxicity in rape (*Brassica rapa L.*); nevertheless, the same experiment conducted under greenhouse conditions also demonstrated protective benefits against cadmium toxicity (Zong et al. 2017b). Rasheed et al. (2020) observed that chitosan can bind metals such as Ag, Zn, Cd, and Pb, leading to their accumulation in peas, a process reviewed by Vasconcelos (2014) in the context of phytoremediation and biofortification.

These results suggest that chitosan may help mitigate phytotoxicity caused by HMS. It is also possible that chitosan and its oligomers, by binding higher concentrations of toxic elements, could enhance the uptake of essential minerals. Consequently, it may be beneficial to utilize both bulk chitosan and chitosan nanoparticles, either by themselves or in combination with other potent substances that are known to activate defense enzymes or lessen HMS. Additionally, because of their higher adsorption, solubility, translocation, and controlled release of active substances, chitosan nanoparticles may be more advantageous than the bulk form.

Sadeghipour (2021) reported that the exogenous application of chitosan reduced nickel toxicity and enhanced growth traits by reducing nickel uptake and oxidative damage, raising relative water content, stomatal conductance, and chlorophyll value, and further boosting proline content and antioxidant enzyme activity. According to Qu et al. (2019), in maize seedlings subjected to cadmium stress, chitosan enhanced photosynthetic and antioxidant enzyme activity. Moreover, it boosted the amount of chlorophyll, improved the activity of photosynthetic enzymes, and promoted the development of root systems. Furthermore, chitosan mainly facilitated the metabolism of ROS in seedlings by minimizing chlorophyll loss and boosting antioxidant enzyme activity, thus protecting photosynthetic function and enhancing the tolerance of maize seedlings. According to Abdelhameed et al. (2024), foliar spraying with 0.1% chitosan increases the growth of faba bean, pigment fractions, and protein. It also lowered lead concentrations under lead stress and decreased MDA and H_2O_2 contents. The reduction of lead by chitosan is likely associated with the action of proline and phenylalanine ammonia lyase (PAL), two antioxidant enzymes.

4.8.4 Heat Stress (HS)

Since heat and drought stress frequently coexist, it might be challenging to monitor these two stresses independently (Zahra et al. 2021); this makes HS a complex challenge. Research has demonstrated that dry beans, particularly those planted late in the growing season, may withstand HS when chitosan, zinc, and humic acid are applied topically (Ibrahim and Ramadan 2015). Nevertheless, the use of chitosan in HS situations has not been extensively studied (Hidangmayum and Dwivedi 2022). According to studies (Soma et al. 2021), heat shock-related genes such as ABF3 (abscisic acid-responsive-element-binding factor 3) may be activated by ABA, potentially improving resistance to HS. Therefore, by triggering defense-related ABA-responsive genes and promoting ABA activity, which is linked to stomatal closure (Sree Rayanoothala et al. 2024), chitosan may be able to lessen the effects of HS. According to Yang et al. (2020), overexpression of ABF3 confers tolerance to HS. Thus, by inducing ABA activity, which in turn promotes the expression of defense-related ABA-responsive genes, chitosan may be useful in combating HS (Kaushal et al. 2023).

Table 4.3 Some studies of chitosan related to heat and heavy metal stress

Crop	Stress	Application	Effects	References
Egg plant	Heat	Foliar spray (0.0125%, 0.0150%, and 0.02% chitosan)	Significantly improved physiological and biochemical attributes of eggplant genotypes	Liaqat et al. (2019)
Pepper	Heat	Foliar spray (0, 12.5, 25, and 37.5 mg L^{-1})	Enhanced growth and yield	Al-Hassani and Majid (2019)
Tomato	Heat	(0, 30, 60, 90, and 120 mg/L)	Enhanced growth and quality attributes of tomato	Hussain et al. (2019)
Wheat	Heat	Foliar application (0.01–0.16%)	Increased the activity of glutathione reductase, glutathione peroxidase, ascorbate peroxidase, catalase, peroxidase, and superoxide dismutase	Mohan et al. (2024)
Wheat	Heavy metal	Seed priming (0.1, 0.25, and 0.5 mg/L)	Enhanced root elongation and antioxidant enzyme activities	Kurtuluş and Vardar (2021)
Soybean	Heavy metal	Soil application (0.5–1%), foliar application (100–200 ppm), and seed treatment (0.25–0.5%)	Foliar application was the most effective treatment, which increased growth and yield	Sadeghipour (2021)
Maize	Heavy metal	Seedling soaking (0.01% chitosan)	Increased root growth, elevated activities of SOD, POD, CAT, APX, and GR, and higher AsA and GSH contents in the leaves	Qu et al. (2019)
Wheat	Heavy metal	Soil treatment (180 g pot^{-1} + Melatonin)	Decreased amounts of Pb, Cd, Cr, Ni, Cu, and Co in grain, roots, and shoots	Dradrach et al. (2022)

Through a variety of molecular mechanisms, chitosan strengthens antioxidant defense systems and is essential for reducing HS in plants (Table 4.3). Its capacity to control membrane stability and heat shock proteins, which are crucial for shielding plant cells from heat-induced damage, is one important method (Ahmed et al. 2024). Moreover, plants can preserve water and combat heat-induced dehydration thanks to chitosan's control of stomatal conductance. By modulating certain genes that regulate pathogenesis-related protein 1 (PR1), glutathione S-transferase (GST), and SOD, hormones such as salicylic acid, jasmonic acid, and ethylene can activate heat shock proteins and scavenge ROS (Das et al. 2021). According to Shakoor et al. (2023), chitosan applied topically improved HS cotton's glycine betaine levels, seed cotton yield, early maturity, and fiber quality. It also decreased the amount of malondialdehyde. Lower doses of chitosan, such as 0, 0.2, 0.4, and 0.6 g L^{-1}, were less effective than the optimal dose of 0.8 g L^{-1}. The study determined that the exogenous application of 0.8 g L^{-1} chitosan was advantageous for mitigating the effects of HS since it improved the quality of the fiber and the yield of seed cotton.

According to Huang et al. (2021), the administration of chitosan (0.1 g L^{-1}) dramatically reduced heat-induced damage in creeping bentgrass leaves. This was achieved through improvements in water balance, scavenging of ROS, metabolism of chlorophyll, and photosynthesis. When exposed to HS, creeping bentgrass treated with chitosan exhibited significantly reduced expression levels of genes linked to senescence (AsSAG12, AsSAG39, AsI20, and Ash36) and higher transcription levels of genes involved in chlorophyll biosynthesis (AsPBGD and AsCHLH). As a result, during HS, photosynthetic activity was improved and leaf senescence was decreased. Furthermore, chitosan applied topically greatly enhanced antioxidant enzyme activity (SOD, CAT, POD, and Ascorbate Peroxidase (APX), hence reducing heat-induced oxidative damage. According to Ahmed et al. (2024), chitosan increased the antioxidant activity of SOD, CAT, APX, and POD while lowering H_2O_2 to lessen the effects of high temperatures on rice. Reduced lipid peroxidation, enhanced leaf greenness, net photosynthetic rate, and proline content were other noteworthy mitigating responses.

4.9 Signaling Mechanisms in Plant Responses to Abiotic Stress

Depending on the kind of stress encountered, different main cues are involved in plant responses to abiotic stress. Reactive oxygen species and inositol phosphates are two examples of secondary signals that are set off by particular receptor molecules during these signaling pathways. These signals ultimately cause the release of Ca^{2+} from cells. This release causes specific proteins or transcription factors to undergo phosphorylation-driven changes, which control the expression of genes related to the stress response (Pongprayoon et al. 2022a, b). Various receptors subsequently relay these secondary signals through downstream pathways, allowing the primary stress signal to be interpreted at different locations and times (Sofi et al. 2023). These secondary signaling events interact with multiple stress response pathways, providing plants with cross-protection against diverse stressors (Sadak et al. 2022).

While several studies report chitosan's role in enhancing defense responses to stress (Fig. 4.3), the precise mechanism of action remains incompletely understood (Sree Rayanoothala et al. 2024). Plants respond to stress, both biotic and abiotic, by activating genes linked to stress protection and defense-related compounds. For example, oligo-chitosan has been demonstrated to upregulate auxin and gibberellin synthesis-related genes, as well as activate the jasmonate and ethylene signaling pathways in rapeseed (Hidangmayum and Dwivedi 2022), which are crucial for defense against various stresses. Grapevines treated with oligo-chitosan produce more phenolic compounds like anthocyanins, which offer defense against biotic and abiotic stress (Singh et al. 2020). Similar to this, oligo-chitosan treatment improves chitinase and glucanase activity in bell peppers (*Capsicum annuum* L.) and increases total soluble phenolic content in blackberries (*Rubus* spp.) (Samarah et al. 2020). Increasing resilience to stress is largely dependent on the production of these defensive secondary metabolites (Cheplick et al. 2017).

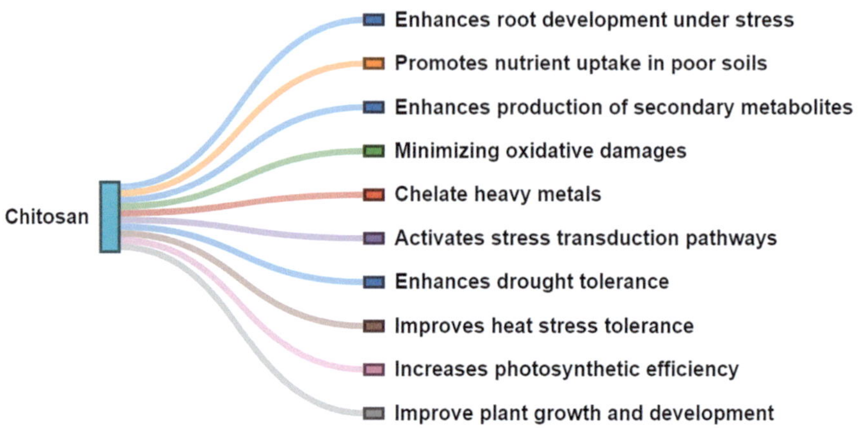

Fig. 4.3 Some advantages of chitosan against stress

Plants usually initiate stomatal closure and decrease the activity of photosynthetic enzymes, like ribulose-1,5-bisphosphate carboxylase/oxygenase, in response to DS (Feng et al. 2020). Plants also produce more drought-responsive metabolites, such as polyols, sugars, and amino acids, to assist in controlling turgor pressure in response to DS (Li et al. 2017). For example, higher levels of the osmoprotective amino acid proline were seen in barley plants treated with chitosan under conditions of water deficiency (Behboudi et al. 2018). Higher concentrations of organic molecules, such as aldehydes, ketones, and phenols, defensive metabolites against stress were seen in maize after treatment with nano-chitosan (Khati et al. 2017). Additionally, plants enhance antioxidant levels and antioxidative enzyme activities to protect against enhanced ROS levels, which are a result of disrupted electron transport chains. These plants also produce unique proteins, like late embryogenesis abundant proteins that play defensive roles against stress (Iriti et al. 2009; Behboudi et al. 2019). Antioxidants like superoxide dismutase and catalase have been demonstrated to exhibit increased activity when barley under DS is treated with nano-chitosan (Behboudi et al. 2018). Application of chitosan also increases proline levels, which are linked to increased activity of the proteinase enzyme (Attaran Dowom et al. 2022). Moreover, chitosan causes stomatal closure through ABA-dependent antitranspirant action. It is well known that ABA activates genes linked to defense and causes stomatal closure in response to stress (Bharath et al. 2021).

4.10 Conclusion and Future Prospects

The naturally occurring biopolymer, chitosan, has demonstrated remarkable versatility as a plant growth promoter, defense enhancer, and stress mitigation agent. Its significance lies in its ability to modulate physiological and molecular responses in plants, specifically under stress conditions. Chitosan is well-known for alleviating abiotic stresses such as heat, drought, salinity, and heavy metal toxicity. It supports plant survival in these challenging environments by enhancing root development, regulating stomatal conductance, and activating reactive oxygen species scavenging mechanisms. Chitosan nanoparticles and their derivatives have expanded their agricultural applications, providing enhanced bioactivity, controlled release, and improved stability, making them promising tools for plant protection and growth enhancement.

To fully unlock the potential of chitosan in agriculture, future research should prioritize understanding its molecular mechanisms, particularly its interaction with phytohormones, signaling molecules, and antioxidant systems. Optimizing chitosan nanoparticle formulations by refining variables like molecular weight and particle size is essential for improving their uptake and translocation in plants while minimizing environmental risks. Further studies are needed to elucidate the pathways through which chitosan mitigates abiotic stresses and expands its biotic stress applications in real-world settings. Additionally, rigorous environmental impact assessments are crucial to ensure the safe and sustainable use of chitosan-based technologies in agriculture.

References

Aazami MA, Maleki M, Rasouli F, Gohari G (2023) Protective effects of chitosan based salicylic acid nanocomposite (CS-SA NCs) in grape (*Vitis vinifera* cv. 'Sultana') under salinity stress. Sci Rep 13:883

Abdallah MMS, Ramadan AAEM, El-Bassiouny HMS, Bakry BA (2020) Regulation of antioxidant system in wheat cultivars by using chitosan or salicylic acid to improve growth and yield under salinity stress. Asian J Plant Sci 19:114–126

Abdelhameed RE, Abdalla H, Abdel-Haleem M (2024) Offsetting pb induced oxidative stress in Vicia faba plants by foliar spray of chitosan through adjustment of morpho-biochemical and molecular indices. BMC Plant Biol 24:557

Abou El-Enin MM, Sheha AM, El-Serafy RS, Ali OA, Saudy HS, Shaaban A (2023) Foliage-sprayed nano-chitosan-loaded nitrogen boosts yield potentials, competitive ability, and profitability of intercropped maize-soybean. Int J Plant Prod 17:517–542

Ackah S, Bi Y, Xue S, Yakubu S, Han Y, Zong Y, Prusky D (2022) Post-harvest chitosan treatment suppresses oxidative stress by regulating reactive oxygen species metabolism in wounded apples. Front Plant Sci 13:959762

Ahmad A, Hashmi SS, Palma JM, Corpas FJ (2022) Influence of metallic, metallic oxide, and organic nanoparticles on plant physiology. Chemosphere 290:133329

Ahmed KBM, Khan MMA, Siddiqui H, Jahan A (2020) Chitosan and its oligosaccharides, a promising option for sustainable crop production-a review. Carbohydr Polym 227:115331

Ahmed S, Ahmed SF, Biswas A, Sultana A, Issak M (2024) Salicylic acid and chitosan mitigate high temperature stress of rice via growth improvement, physio-biochemical adjustments and enhanced antioxidant activity. Plant Stress 11:100343

AL Kahtani MD, Attia KA, Hafez YM, Khan N, Eid AM, Ali MA, Abdelaal KA (2020) Chlorophyll fluorescence parameters and antioxidant defense system can display salt tolerance of salt acclimated sweet pepper plants treated with chitosan and plant growth promoting rhizobacteria. Agronomy 10:1180

Al-Hassani FA, Majid BH (2019) Effect of arginine, chitosan and agryl mulching on the growth and yield of pepper plant under the conditions of unheated greenhouses. Plant Arch 19:256–262

Ali EF, Hassan FAS, Elgimabi M (2018) Improving the growth, yield and volatile oil content of *Pelargonium graveolens* L. Herit by foliar application with moringa leaf extract through motivating physiological and biochemical parameters. S Afr J Bot 119:383–389

Ali G, Sharma M, Salama ES, Ling Z, Li X (2024) Applications of chitin and chitosan as natural biopolymer: potential sources, pretreatments, and degradation pathways. Biomass Convers Biorefinery 14:4567–4581

Almeida LG, Magalhães PC, Karam D, Silva EMD, Alvarenga AA (2020) Chitosan application in the induction of water deficit tolerance in maize plants. Acta Sci Agron 42:e42463

Al-Mokadem AZ, Alnaggar AEAM, Mancy AG, Sofy AR, Sofy MR, Mohamed AKS, Agha MS (2022) Foliar application of chitosan and phosphorus alleviate the potato virus Y-induced resistance by modulation of the reactive oxygen species, antioxidant defense system activity and gene expression in potato. Agronomy 12:3064

Al-Tawaha ARM, Al-Ghzawi ALA (2013) Effect of chitosan coating on seed germination and salt tolerance of lentil (Lens culinaris L.). Res Crops 14:489–491

Al-Tawaha AR, Turk MA, Al-Tawaha ARM, Alu'Datt MH, Wedyan M, Al-Ramamneh EADM, Hoang AT (2018) Using chitosan to improve growth of maize cultivars under salinity conditions. Bulgarian J Agric Sci 24(3):437–442

Amer A, Shoala T (2020) Physiological and phenotypic characters of sweet marjoram in response to pre-harvest application of hydrogen peroxide or chitosan nanoparticles. Sci Hortic 268:109374

Arif Y, Siddiqui H, Hayat S (2022) Role of chitosan nanoparticles in regulation of plant physiology under abiotic stress. Sustain Agric Rev 53:399–413

Attaran Dowom S, Karimian Z, Mostafaei Dehnavi M, Samiei L (2022) Chitosan nanoparticles improve physiological and biochemical responses of *Salvia abrotanoides* (Kar.) under drought stress. BMC Plant Biol 22:364

Attia MS, Osman MS, Mohamed AS, Mahgoub HA, Garada MO, Abdelmouty ES, Abdel Latef AAH (2021) Impact of foliar application of chitosan dissolved in different organic acids on isozymes, protein patterns and physio-biochemical characteristics of tomato grown under salinity stress. Plants 10:388

Bakhoum GS, Sadak MS, Badr EAEM (2020) Mitigation of adverse effects of salinity stress on sunflower plant (*Helianthus annuus* L.) by exogenous application of chitosan. Bull Natl Res Cent 44:1–11

Behboud R, Moradi A, Farajee H (2020) Effect of different chitosan concentrations on seed germination and some biochemical traits of sweet corn (*Zea mays* var. Saccharata) seedling under osmotic stress conditions. Iran J Seed Res 7:1–22

Behboudi F, Tahmasebi Sarvestani Z, Kassaee MZ, Modares Sanavi SAM, Sorooshzadeh A, Ahmadi SB (2018) Evaluation of chitosan nanoparticles effects on yield and yield components of barley (*Hordeum vulgare* L.) under late season drought stress. J Water Environ Nanotechnol 3:22–39

Behboudi F, Tahmasebi-Sarvestani Z, Kassaee MZ, Modarres-Sanavy SAM, Sorooshzadeh A, Mokhtassi-Bidgoli A (2019) Evaluation of chitosan nanoparticles effects with two application methods on wheat under drought stress. J Plant Nutr 42:1439–1451

Bharath P, Gahir S, Raghavendra AS (2021) Abscisic acid-induced stomatal closure: an important component of plant defense against abiotic and biotic stress. Front Plant Sci 12:615114

Biradar SP, Mehta MR, Mahajan HP, Bankhele RR, Hivrale AU (2022) Biological activities of chitosan-based nanomaterials. In: Role of chitosan and chitosan-based nanomaterials in plant sciences. Academic Press, London, pp 119–142

Bistgani ZE, Hashemi M, Akbari P, Mumivand H (2023) Influence of drought stress and foliar application of chitosan on nutrient accumulation and phenolic composition of Thymus daenensis Celak. Crop Sci 63:921–935

Boamah PO, Onumah J, Aduguba WO, Santo KG (2023) Application of depolymerized chitosan in crop production: a review. Int J Biol Macromol 235:123858

Carmona SL, del Pilar Villarreal-Navarrete A, Burbano-David D, Gómez-Marroquín M, Torres-Rojas E, Soto-Suárez M (2020) Boosting photosynthetic machinery and defense priming with chitosan application on tomato plants infected with *Fusarium oxysporum* f. sp. lycopersici. bioRxiv pp. 2020–08

Cheplick S, Sarkar D, Bhowmik PC, Shetty K (2017) Improved resilience and metabolic response of transplanted blackberry plugs using chitosan oligosaccharide elicitor treatment. Can J Plant Sci 98:717–731

Chouhan D, Mandal P (2021) Applications of chitosan and chitosan based metallic nanoparticles in agrosciences-a review. Int J Biol Macromol 166:1554–1569

Das D, Ullah H, Tisarum R, Cha-um S, Datta A (2021) Morpho-physiological responses of tropical rice to potassium and silicon fertilization under water-deficit stress. J Soil Sci Plant Nutr:1–18 23(1):220–237

Dawood MG, El-Awadi MES, Sadak MS (2024) Chitosan and its nanoform regulates physiological processes and antioxidant mechanisms to improve drought stress tolerance of *Vicia faba* plant. J Soil Sci Plant Nutr 24:5696–5709

Dolatkhah Dashtmian A, Hosseini Mazinani SM, Pazoki A (2023) Exogenous chitosan nanoparticles modulated drought stress through changing yield, biochemical attributes, and fatty acid profile of common bean (*Phaseolus Vulgaris* L.) cultivars. Gesunde Pflanzen 75:2463–2476

Dradrach A, Iqbal M, Lewińska K, Jędroszka N, Rana MAK, Tanzeem-ul-Haq HS (2022) Effects of soil application of chitosan and foliar melatonin on growth, photosynthesis, and heavy metals accumulation in wheat growing on wastewater polluted soil. Sustainability 14:8293

Dutta T, Nandy S, Singh J, Pandey DK, Dey A (2022) Chitin and chitosan as elicitors in sustainable production of medicinal crops. In: New and future developments in microbial biotechnology and bioengineering: sustainable agriculture: revisiting green chemicals. Elsevier, pp 413–426. https://doi.org/10.1016/B978-323-85581-5.00017-3

Eichert T, Kurtz A, Steiner U, Goldbach HE (2008) Size exclusion limits and lateral heterogeneity of the stomatal foliar uptake pathway for aqueous solutes and water-suspended nanoparticles. Physiol Plant 134:151–160

El Amerany F, Meddich A, Wahbi S, Taourirte M, Rhazi M (2023) Application of compost, as a peat substitute, and foliar spray of chitosan and/or roots inoculation with beneficial fungus increases tomato plants productivity and fruit size. Gesunde Pflanzen 75:1485–1495

El-Gendi H, Al-Askar AA, Király L, Samy MA, Moawad H, Abdelkhalek A (2022) Foliar applications of Bacillus subtilis HA1 culture filtrate enhance tomato growth and induce systemic resistance against tobacco mosaic virus infection. Horticulturae 8:301

Elmasry HM, Al-Maracy SH (2023) Efficacy of anti-transpiration on yield and quality of sugar beet subjected to water stress. J Cent Eur Agric 24:268–281

Elshamly AM (2023) Minimizing the adverse impact of drought on corn by applying foliar potassium humate combined with chitosan. J Soil Sci Plant Nutr 23:1913–1929

Elshayb OM, Nada AM, Farroh KY, Al-Huqail AA, Aljabri M, Binothman N, Seleiman MF (2022) Utilizing urea–chitosan nanohybrid for minimizing synthetic urea application and maximizing *Oryza sativa* L. productivity and N uptake. Agriculture 12:944

Fan Z, Wang L, Qin Y, Li P (2023) Activity of chitin/chitosan/chitosan oligosaccharide against plant pathogenic nematodes and potential modes of application in agriculture: a review. Carbohydr Polym 306:120592

Feng BH, Li GY, Islam M, Fu WM, Zhou YQ, Chen TT, Fu GF (2020) Strengthened antioxidant capacity improves photosynthesis by regulating stomatal aperture and ribulose-1, 5-bisphosphate carboxylase/oxygenase activity. Plant Sci 290:110245

Golkar P, Taghizadeh M, Yousefian Z (2019) The effects of chitosan and salicylic acid on elicitation of secondary metabolites and antioxidant activity of safflower under in vitro salinity stress. Plant Cell Tissue Organ Cult 137:575–585

Gomes DG, Debiasi TV, Pelegrino MT, Pereira RM, Ondrasek G, Batista BL, Oliveira HC (2022) Soil treatment with nitric oxide-releasing chitosan nanoparticles protects the root system and promotes the growth of soybean plants under copper stress. Plants 11:3245

González LM, Guerrero YR, Rodríguez AF, Vázquez MN (2015) Effect of seed treatment with chitosan on the growth of rice (*Oryza sativa* L.) seedlings cv. INCA LP-5 in saline medium. Cult Trop 36:136–142

Hao T, Yang Z, Liang J, Yu J, Liu J (2023) Foliar application of carnosine and chitosan improving drought tolerance in bermudagrass. Agronomy 13:442

Hassan FAS, Ali E, Gaber A, Fetouh MI, Mazrou R (2021) Chitosan nanoparticles effectively combat salinity stress by enhancing antioxidant activity and alkaloid biosynthesis in *Catharanthus roseus* (L.) G. Don. Plant Physiol Biochem 162:291–300

Hassnain K, Basit A, Alam M, Ahmad I, Ullah I, Alam N, Shair M (2020) Efficacy of chitosan on performance of tomato (*Lycopersicon esculentum* L.) plant under water stress condition. Pak J Agric Res 33(1):27–41

He J, Zhang L, He SY, Ryser ET, Li H, Zhang W (2022) Stomata facilitate foliar sorption of silver nanoparticles by Arabidopsis thaliana. Environ Pollut 292:118448

Hernández V, Botella MÁ, Hellín P, Fenoll J, Flores P (2022) Dose-dependent potential of chitosan to increase yield or bioactive compound content in tomatoes. Horticulturae 8:1152

Hernández-Hernández H, Juárez-Maldonado A, Benavides-Mendoza A, Ortega-Ortiz H, Cadenas-Pliego G, Sánchez-Aspeytia D, González-Morales S (2018) Chitosan-PVA and copper nanoparticles improve growth and overexpress the SOD and JA genes in tomato plants under salt stress. Agronomy 8:175

Hidangmayum A, Dwivedi P (2022) Chitosan based nanoformulation for sustainable agriculture with special reference to abiotic stress: a review. J Polym Environ 30:1264–1283

Hidangmayum A, Dwivedi P, Katiyar D, Hemantaranjan A (2019) Application of chitosan on plant responses with special reference to abiotic stress. Physiol Mol Biol Plants 25:313–326

Hidangmayum A, Dwivedi P, Kumar P, Upadhyay SK (2023) Seed priming and foliar application of chitosan ameliorate drought stress responses in mungbean genotypes through modulation of morpho-physiological attributes and increased antioxidative defense mechanism. J Plant Growth Regul 42:6137–6154

Huang C, Tian Y, Zhang B, Hassan MJ, Li Z, Zhu Y (2021) Chitosan (CTS) alleviates heat-induced leaf senescence in creeping Bentgrass by regulating chlorophyll metabolism, antioxidant defense, and the heat shock pathway. Molecules 26:5337

Hussain I, Ahmad S, Ullah I, Ahmad I, Alam M, Khan S, Ayaz S (2019) Foliar application of Chitosan modulates the morphological and biochemical characteristics of tomato. Asian J Agric Biol 7:365–372

Ibrahim EA, Ramadan WA (2015) Effect of zinc foliar spray alone and combined with humic acid or/and chitosan on growth, nutrient elements content and yield of dry bean (*Phaseolus vulgaris* L.) plants sown at different dates. Sci Hortic 184:101–105

Ibrahim MA, Alhalafi MH, Emam EAM, Ibrahim H, Mosaad RM (2023) A review of chitosan and chitosan nanofiber: preparation, characterization, and its potential applications. Polymers 15:2820

Iriti M, Picchi V, Rossoni M, Gomarasca S, Ludwig N, Gargano M, Faoro F (2009) Chitosan antitranspirant activity is due to abscisic acid-dependent stomatal closure. Environ Exp Bot 66:493–500

Jafernik K, Ładniak A, Blicharska E, Czarnek K, Ekiert H, Wiącek AE, Szopa A (2023) Chitosan-based nanoparticles as effective drug delivery systems—a review. Molecules 28:1963

Jahanbani A, Asghari Zakaria R, Ashrafi V, Ghasemi Kalkhoran M, Shahriary R (2023) Effect of foliar application of chitosan and humic acid on yield and yield components of bread wheat under end-season drought stress. Environ Stress Crop Sci 16:905–918

Jan MF, Altaf MT, Liaqat W, Ahmad S, Hussain M, Khan MA (2025) Approaches for the amelioration of adverse effects of drought stress on soybean plants: from physiological responses to agronomical, molecular, and cutting-edge technologies. Plant Soil 1:1–53

Jan MF, Li M, Liaqat W, Hussain M, Khan MA, Zhang Y (2024) Chlorophyll fluorescence: a smart tool for maize improvement. Cereal Res Commun 53:617–648

Ji H, Wang J, Chen F, Fan N, Wang X, Xiao Z, Wang Z (2022) Meta-analysis of chitosan-mediated effects on plant defense against oxidative stress. Sci Total Environ 851:158212
Ji H, Wang J, Xue A, Chen F, Guo H, Xiao Z, Wang Z (2023) Chitosan–silica nanocomposites induced resistance in faba bean plants against aphids (*Acyrthosiphon pisum*). Environ Sci Nano 10:1966–1977
Jiang X, Yang F, Jia W, Jiang Y, Wu X, Song S, Shen J (2024) Nanomaterials and nanotechnology in agricultural pesticide delivery: a review. Langmuir 40(36):18806–18820
Jiao Q, Shen F, Fan L, Song Z, Zhang J, Song J, Liu H (2024) Chitosan regulates the root architecture system, photosynthetic characteristics and antioxidant system contributing to salt tolerance in maize seedling. Agriculture 14:304
Karamchandani BM, Chakraborty S, Dalvi SG, Satpute SK (2022) Chitosan and its derivatives: promising biomaterial in averting fungal diseases of sugarcane and other crops. J Basic Microbiol 62:533–554
Kaushal P, Ali N, Saini S, Pati PK, Pati AM (2023) Physiological and molecular insight of microbial biostimulants for sustainable agriculture. Front Plant Sci 14:1041413
Khaled JM, Alobaidi AS (2023) Priming effect of chitosan on induces protection against tomato root wilt disease mediated through upregulation of defense enzymes. Physiol Mol Plant Pathol 127:102118
Khan I, Awan SA, Rizwan M, Hassan ZU, Akram MA, Tariq R, Xie W (2022a) Nanoparticle's uptake and translocation mechanisms in plants via seed priming, foliar treatment, and root exposure: a review. Environ Sci Pollut Res 29:89823–89833
Khan MAU, Arshad H, Majid A (2022b) Transport mechanism from quantum dots to plant systems. In: Quantum dots for plant systems. Springer, Cham, pp 103–136
Khan MN, Khan Z, Zhu L, Li Y, Wu H (2022c) Impact of chitosan and chitosan-based metallic nanoparticles on the regulation of plant hormones. In: Role of chitosan and chitosan-based nanomaterials in plant sciences. Academic Press, London, pp 343–356
Khan EH, Liaqat W, Jan MF, Altaf MT, Ali A, Ahmad H (2025) Elicitors in plant biology: types and defense mechanisms. In: Elicitors for sustainable crop production. Springer, Singapore, pp 41–61
Khati P, Chaudhary P, Gangola S, Bhatt P, Sharma A (2017) Nanochitosan supports growth of Zea mays and also maintains soil health following growth. 3 Biotech 7:1–9
Kolbert Z, Szőllősi R, Feigl G, Kónya Z, Rónavári A (2021) Nitric oxide signalling in plant nanobiology: current status and perspectives. J Exp Bot 72:928–940
Kolesnikov LE, Popova EV, Novikova II, Kolesnikova YR, Balagurova ED (2022) Application of chitosan to protect wheat from diseases and boost yields. Appl Biochem Microbiol 58:329–335
Krif G, Lahlali R, El Aissami A, Laasli SE, Mimouni A, Serderidis S, Mokrini F (2022) Efficacy of authentic bio-nematicides against the root-knot nematode, Meloidogyne javanica infecting tomato under greenhouse conditions. Physiol Mol Plant Pathol 118:101803
Kumari R, Kapoor P, Mir BA, Singh M, Parrey ZA, Rakhra G, Rakhra G (2024) Unlocking the versatility of Nitric Oxide in plants and insights into its molecular interplays under biotic and abiotic stress. Nitric Oxide 150:1–17
Kurtuluş G, Vardar F (2021) The effects of chitosan application against aluminum toxicity in wheat (*Triticum aestivum* L.) roots. Ann Bot 11:121–134
Lalarukh, Hussain SM, Ali S, Zahoor AF, Azmat H, Nazish N, Mahrukh (2024) Innovation of advanced polymers from seafood waste: applications of chitin and chitosan. Polym Adv Technol 35:e6471
Li Z, Zhang Y, Zhang X, Merewitz E, Peng Y, Ma X, Yan Y (2017) Metabolic pathways regulated by chitosan contributing to drought resistance in white clover. J Proteome Res 16:3039–3052
Liaqat A, Ihsan MZ, Rizwan MS, Mehmood A, Ijaz M, Alam M, Yaqub MS (2019) Inducing effect of chitosan on the physiological and biochemical indices of eggplant (*Solanum melongena* L.) genotypes under heat and high irradiance. Appl Ecol Environ Res 17(5):11273–11287
Ling Y, Zhao Y, Cheng B, Tan M, Zhang Y, Li Z (2022) Seed priming with chitosan improves germination characteristics associated with alterations in antioxidant defense and dehydration-responsive pathway in white clover under water stress. Plants 11:2015

Lingait D, Rahagude R, Gaharwar SS, Das RS, Verma MG, Srivastava N, Mandavgane S (2023) A review on versatile applications of biomaterial/polycationic chitosan: an insight into the structure–property relationship. Int J Biol Macromol 257:128676

Liu J, Gai L, Zong H (2021) Foliage application of chitosan alleviates the adverse effects of cadmium stress in wheat seedlings (*Triticum aestivum* L.). Plant Physiol Biochem 164:115–121

Makhlouf BSI, Khalil SRAE, Saudy HS (2022) Efficacy of humic acids and chitosan for enhancing yield and sugar quality of sugar beet under moderate and severe drought. J Soil Sci Plant Nutr 22:1676–1691

Maryam, El-Mogy MM, Jan MF, Naz I, Ahmad I, Ahmad R, Altaf MT (2025) Nanoparticle innovations for mitigating metal toxicity in plants. Phyton 94(3):623–640

Masoumi Z, Haghighi M, Mozafarian M (2024) Effects of foliar spraying with melatonin and chitosan nano-encapsulated melatonin on tomato (*Lycopersicon esculentum* L. cv. Falcato) plants under salinity stress. BMC Plant Biol 24:961

Meitha K, Pramesti Y, Signorelli S, Kriswantoro JA (2022) Postharvest chitosan application maintains the quality of spinach through suppression of bacterial growth and elicitation. Hortic Environ Biotechnol 63:217–227

Methela NJ, Pande A, Islam MS, Rahim W, Hussain A, Lee DS, Yun BW (2023) Chitosan-GSNO nanoparticles: a positive modulator of drought stress tolerance in soybean. BMC Plant Biol 23:639

Mohamed NG, Abdel-Hakeem MA (2023) Chitosan nanoparticles enhance drought tolerance in tomatoes (*Solanum lycopersicum* L) via gene expression modulation. Plant Gene 34:100406

Mohammadi MA, Cheng Y, Aslam M, Jakada BH, Wai MH, Ye K, Qin Y (2021) ROS and oxidative response systems in plants under biotic and abiotic stresses: revisiting the crucial role of phosphite triggered plants defense response. Front Microbiol 12:631318

Mohan K, Ganesan AR, Ezhilarasi PN, Kondamareddy KK, Rajan DK, Sathishkumar P, Conterno L (2022) Green and eco-friendly approaches for the extraction of chitin and chitosan: a review. Carbohydr Polym 287:119349

Mohan N, Pal A, Saharan V (2024) Nano-armored wheat: enhancing heat stress resilience and yield via zinc–salicylic acid–chitosan bionanoconjugates. J Polym Environ 32(12):6725–6741

Moolphuerk N, Lawson T, Pattanagul W (2022) Chitosan mitigates the adverse effects and improves photosynthetic activity in rice (*Oryza sativa* L.) seedlings under drought condition. J Crop Improv 36:638–655

Mukarram M, Ali J, Dadkhah-Aghdash H, Kurjak D, Kačík F, Ďurkovič J (2023) Chitosan-induced biotic stress tolerance and crosstalk with phytohormones, antioxidants, and other signalling molecules. Front Plant Sci 14:1217822

Mukhopadhyay R, Sarkar B, Jat HS, Sharma PC, Bolan NS (2021) Soil salinity under climate change: challenges for sustainable agriculture and food security. J Environ Manag 280:111736

Mulaudzi T, Nkuna M, Sias G, Doumbia IZ, Njomo N, Iwuoha E (2022) Antioxidant capacity of chitosan on sorghum plants under salinity stress. Agriculture 12:1544

Mura P, Maestrelli F, Cirri M, Mennini N (2022) Multiple roles of chitosan in mucosal drug delivery: an updated review. Mar Drugs 20:335

Mustafa G, Shehzad MA, Tahir MHN, Nawaz F, Akhtar G, Bashir MA, Ghaffar A (2022) Pretreatment with chitosan arbitrates physiological processes and antioxidant defense system to increase drought tolerance in alfalfa (*Medicago sativa* L.). J Soil Sci Plant Nutr 22:2169–2186

Nadarajah KK (2020) ROS homeostasis in abiotic stress tolerance in plants. Int J Mol Sci 21:5208

Narasimhamurthy K, Udayashankar AC, De Britto S, Lavanya SN, Abdelrahman M, Soumya K, Jogaiah S (2022) Chitosan and chitosan-derived nanoparticles modulate enhanced immune response in tomato against bacterial wilt disease. Int J Biol Macromol 220:223–237

Özkurt N, Bektaş Y (2022) Alleviation of salt stress with chitosan foliar application and its effects on growth and development in tomato (*Solanum lycopersicum* L.). Turk J Agric Res 9:342–351

Pande A, Mun BG, Methela NJ, Rahim W, Lee DS, Lee GM, Yun BW (2022) Heavy metal toxicity in plants and the potential NO-releasing novel techniques as the impending mitigation alternatives. Front Plant Sci 13:1019647

Pongprayoon W, Maksup S, Phaonakrop N, Jaresitthikunchai J, Uawisetwathana U, Panya A, Roytrakul S (2022a) Phosphoproteome analysis reveals chitosan-induced resistance to osmotic stress in rice (*Oryza sativa* L.) seedlings. J Plant Interact 17:894–910

Pongprayoon W, Siringam T, Panya A, Roytrakul S (2022b) Application of chitosan in plant defense responses to biotic and abiotic stresses. Appl Sci Eng Prog 15(1):107–158

Priyaadharshini M, Sritharan N, Senthil A, Marimuthu S (2019) Physiological studies on effect of chitosan nanoemulsion in pearl millet under drought condition. J Pharmacogn Phytochem 8(3):3304–3307

Qiao M, Hong C, Jiao Y, Hou S, Gao H (2024) Impacts of drought on photosynthesis in major food crops and the related mechanisms of plant responses to drought. Plants 13(13):1808

Qu DY, Gu WR, Zhang LG, Li CF, Chen XC, Li J, Wei S (2019) Role of chitosan in the regulation of the growth, antioxidant system and photosynthetic characteristics of maize seedlings under cadmium stress. Russ J Plant Physiol 66:140–151

Quitadamo F, De Simone V, Beleggia R, Trono D (2021) Chitosan-induced activation of the antioxidant defense system counteracts the adverse effects of salinity in durum wheat. Plants 10(7):1365

Rasheed R, Ashraf MA, Arshad A, Iqbal M, Hussain I (2020) Interactive effects of chitosan and cadmium on growth, secondary metabolism, oxidative defense, and element uptake in pea (*Pisum sativum* L.). Arab J Geosci 13:1–14

Ray SR, Bhuiyan MJH, Hossain MA, Hasan AK, Sharmin S (2016) Chitosan ameliorates growth and biochemical attributes in mungbean varieties under saline condition. Res Agric Livest Fish 3(1):45–51

Riseh RS, Tamanadar E, Hajabdollahi N, Vatankhah M, Thakur VK, Skorik YA (2022) Chitosan microencapsulation of rhizobacteria for biological control of plant pests and diseases: recent advances and applications. Rhizosphere 23:100565

Riseh RS, Vazvani MG, Kennedy JF (2023) The application of chitosan as a carrier for fertilizer: a review. Int J Biol Macromol 252:126483

Riseh RS, Vatankhah M, Hassanisaadi M, Varma RS (2024a) A review of chitosan nanoparticles: nature's gift for transforming agriculture through smart and effective delivery mechanisms. Int J Biol Macromol 260:129522

Riseh RS, Vazvani MG, Vatankhah M, Kennedy JF (2024b) Chitin-induced disease resistance in plants: a review. Int J Biol Macromol 266:131105

Riseh RS, Vazvani MG, Vatankhah M, Kennedy JF (2024c) Chitosan coating of seeds improves the germination and growth performance of plants: a review. Int J Biol Macromol 278:134750

Sadak MS, Talaat IM (2021) Attenuation of negative effects of saline stress in wheat plant by chitosan and calcium carbonate. Bull Natl Res Cent 45:1–12

Sadak MS, Tawfik MM, Bakhoum GS (2022) Role of chitosan and chitosan-based nanoparticles on drought tolerance in plants: probabilities and prospects. In: Role of chitosan and chitosan-based nanomaterials in plant sciences. Academic Press, London, pp 475–501

Sadeghipour O (2021) Chitosan application improves nickel toxicity tolerance in soybean. J Soil Sci Plant Nutr 21(3):2096–2104

Saha S, Hossen MZ, Mallick S, Alam MS, Quadir QF (2024) Impact of chitosan coatings on shelf-life, nutrient elements and biochemical qualities of country beans (*Phaseolus lunatus* L.) at postharvest storage. J Hortic Postharvest Res 7(Special Issue):1–14

Samarah NH, Al-Quraan NA, Massad RS, Welbaum GE (2020) Treatment of bell pepper (*Capsicum annuum* L.) seeds with chitosan increases chitinase and glucanase activities and enhances emergence in a standard cold test. Sci Hortic 269:109393

Samdurkar AN, Choudhary AD, Varshney L, Badere RS (2021) Foliar spray of gamma ray degraded chitosan enhances the yield in wheat. Plant Physiol Rep 26:561–569

Seleiman MF, Al-Suhaibani N, Ali N, Akmal M, Alotaibi M, Refay Y, Battaglia ML (2021) Drought stress impacts on plants and different approaches to alleviate its adverse effects. Plants 10(2):259

Selvaraj M, Ramasami N, Jagan EG (2022) The effect of chitosan nanoparticles on immune responses in plants. In: Role of chitosan and chitosan-based nanomaterials in plant sciences. Academic Press, London, pp 185–196

Sen SK, Mandal P (2016) Solid matrix priming with chitosan enhances seed germination and seedling invigoration in mung bean under salinity stress. J Cent Eur Agric 17(3). https://doi.org/10.5513/JCEA01/17.3.1773

Sen SK, Chouhan D, Das D, Ghosh R, Mandal P (2020) Improvisation of salinity stress response in mung bean through solid matrix priming with normal and nano-sized chitosan. Int J Biol Macromol 145:108–123

Shahrajabian MH, Chaski C, Polyzos N, Tzortzakis N, Petropoulos SA (2021) Sustainable agriculture systems in vegetable production using chitin and chitosan as plant biostimulants. Biomolecules 11(6):819

Shakoor A, Saleem MF, Sarwar M, Zia Ul Haq M (2023) Exogenous application of chitosan mediated biochemical, phenological, quality, and yield attributes of heat-stressed cotton (*Gossypium hirsutum* L.). Gesunde Pflanzen 75(5):1755–1767

Shams Peykani L (2018) Effect of chitosan on antioxidant enzyme activity, proline, and malondialdehyde content in *Triticum aestivum* L. and *Zea mays* L. under salt stress condition. Iran J Plant Physiol 5(1):2661

Shehzad MA, Nawaz F, Ahmad F, Ahmad N, Masood S (2020) Protective effect of potassium and chitosan supply on growth, physiological processes and antioxidative machinery in sunflower (*Helianthus annuus* L.) under drought stress. Ecotoxicol Environ Saf 187:109841

Shinde NA, Kawar PG, Dalvi SG (2024) Chitosan-based nanoconjugates: a promising solution for enhancing crop drought-stress resilience and sustainable yield in the face of climate change. Plant Nano Biol 7:100059

Singh RK, Martins V, Soares B, Castro I, Falco V (2020) Chitosan application in vineyards (*Vitis vinifera* L. cv. Tinto Cão) induces accumulation of anthocyanins and other phenolics in berries, mediated by modifications in the transcription of secondary metabolism genes. Int J Mol Sci 21(1):306

Sobral MC, Martins IM, Sobral AJ (2022) Role of chitosan and chitosan-based nanoparticles against heavy metal stress in plants. In: Role of chitosan and chitosan-based nanomaterials in plant sciences. Academic Press, London, pp 273–296

Sofi KG, Metzger C, Riemann M, Nick P (2023) Chitosan triggers actin remodelling and activation of defence genes that is repressed by calcium influx in grapevine cells. Plant Sci 326:111527

Soma F, Takahashi F, Yamaguchi-Shinozaki K, Shinozaki K (2021) Cellular phosphorylation signaling and gene expression in drought stress responses: ABA-dependent and ABA-independent regulatory systems. Plants 10(4):756

Sree Rayanoothala P, Dweh TJ, Mahapatra S, Kayastha S (2024) Unveiling the protective role of chitosan in plant defense: a comprehensive review with emphasis on abiotic stress management. Crop Des 3:100076

Stasińska-Jakubas M, Hawrylak-Nowak B (2022) Protective, biostimulating, and eliciting effects of chitosan and its derivatives on crop plants. Molecules 27(9):2801

Sun W, Shahrajabian MH, Petropoulos SA, Shahrajabian N (2023) Developing sustainable agriculture systems in medicinal and aromatic plant production by using chitosan and chitin-based biostimulants. Plants 12(13):2469

Suwanchaikasem P, Nie S, Idnurm A, Selby-Pham J, Walker R, Boughton BA (2023) Effects of chitin and chitosan on root growth, biochemical defense response and exudate proteome of Cannabis sativa. Plant Environ Interact 4(3):115–133

Turk H (2019) Chitosan-induced enhanced expression and activation of alternative oxidase confer tolerance to salt stress in maize seedlings. Plant Physiol Biochem 141:415–422

Van Dijk M, Morley T, Rau ML, Saghai Y (2021) Systematic review and meta-analysis of global food security projections to 2050. https://doi.org/10.21203/rs.3.rs-530565/v1

Vasconcelos MW (2014) Chitosan and chitooligosaccharide utilization in phytoremediation and biofortification programs: current knowledge and future perspectives. Front Plant Sci 5:616

Wang J, Tao M, Xu L, Fan N, Zhao C, Xiao Z, Wang Z (2023) Chitosan nanocarriers loaded with salicylic acid for controlling fall armyworm (*Spodoptera frugiperda*) and alleviating oxidative stress in maize plants. Environ Sci Nano 10(12):3295–3306

Wu H, Du PR, Miao XR, Hou RQ, Li SN, Zeeshan M, Zhang ZX (2024) O-Carboxymethyl chitosan nanoparticles: a novel approach to enhance water stress tolerance in maize seedlings. Int J Biol Macromol 277:134459

Yang Y, Li HG, Wang J, Wang HL, He F, Su Y, Xia X (2020) ABF3 enhances drought tolerance via promoting ABA-induced stomatal closure by directly regulating ADF5 in Populus euphratica. J Exp Bot 71(22):7270–7285

Younas HS, Abid M, Ashraf M, Shaaban M (2022) Seed priming with silicon and chitosan for alleviating water stress effects in maize (*Zea mays* L.) by improving antioxidant enzyme activities, water status and photosynthesis. J Plant Nutr 45(15):2263–2276

Yu Z, Jiang Q, Yu D, Dong J, Xu Y, Xia W (2022) Physical, antioxidant, and preservation properties of chitosan film doped with proanthocyanidins-loaded nanoparticles. Food Hydrocoll 130:107686

Zahra N, Wahid A, Hafeez MB, Ullah A, Siddique KH, Farooq M (2021) Grain development in wheat under combined heat and drought stress: plant responses and management. Environ Exp Bot 188:104517

Zhao Y, Li Y, Zhang B (2020) Induced resistance in peach fruit as treated by Pichia guilliermondii and their possible mechanism. Int J Food Prop 23(1):34–51

Zhu J, Wang J, Zhan X, Li A, White JC, Gardea-Torresdey JL, Xing B (2021a) Role of charge and size in the translocation and distribution of zinc oxide particles in wheat cells. ACS Sustain Chem Eng 9(34):11556–11564

Zhu J, Zhang X, Qin Z, Zhang L, Ye Y, Cao M, Jiao T (2021b) Preparation of PdNPs doped chitosan-based composite hydrogels as highly efficient catalysts for reduction of 4-nitrophenol. Colloids Surf A Physicochem Eng Asp 611:125889

Zong H, Li K, Liu S, Song L, Xing R, Chen X, Li P (2017a) Improvement in cadmium tolerance of edible rape (*Brassica rapa* L.) with exogenous application of chitooligosaccharide. Chemosphere 181:92–100

Zong H, Liu S, Xing R, Chen X, Li P (2017b) Protective effect of chitosan on photosynthesis and antioxidative defense system in edible rape (*Brassica rapa* L.) in the presence of cadmium. Ecotoxicol Environ Saf 138:271–278

Chapter 5
Acetic Acid and Abiotic Stress Tolerance in Plants

Aslihan Temel and Nihal Gören-Saglam

Abstract Plant growth depends on certain quantities of several abiotic factors such as temperature, water, light, CO_2, and nutrients, and excess or scarcity of these factors results in abiotic stress, which adversely affects plant growth. Because climate change aggravates abiotic stress combinations, the enhancement of abiotic stress tolerance mechanisms in plants is a must. Although breeding and genetic engineering have produced stress-tolerant varieties, nongenetic approaches to induce stress tolerance are regarded as rather practical. Exogenous application of numerous compounds has been shown to improve abiotic stress tolerance. One of these compounds is acetic acid (AA), an organic acid with diverse uses. Recently, numerous papers have shown the AA-induced changes in growth and stress response of several plant species. AA, even at low doses and with short treatment periods, improves stress tolerance. In this chapter, the synthesis of organic acids is described first. Then, genetic and nongenetic strategies that induce abiotic stress are explained. The relationship between AA and plant growth and the examples of the effects of AA on abiotic stress are reviewed. Finally, other uses of AA are briefly explored.

Keywords Acetic acid · Abiotic stress · Plant biostimulants · Acetylation · Stress tolerance · Vinegar

5.1 Introduction

Organic acids (OAs) are weak acids and constitute a water-soluble fraction of the molecules found in soil. OAs with low molecular weight (MW) are more soluble than those with higher MW. OAs in soil originate from plants, microorganisms, and the decomposition of organic matter. Depending on the soil type and other factors,

---

A. Temel · N. Gören-Saglam (✉)
Istanbul University, Faculty of Science, Department of Molecular Biology and Genetics, Vezneciler, Istanbul, Turkey
e-mail: atemel@istanbul.edu.tr; gorenn@istanbul.edu.tr

H. I. Mohamed (ed.), *Elicitors for Sustainable Crop Production*, Sustainability Sciences in Asia and Africa, https://doi.org/10.1007/978-981-95-3533-0_5

the concentration of OAs ranges from 0 to 1 mM for monocarboxylic acids, e.g., acetic, formic, propionic, butyric, valeric, and lactic acid (Adeleke et al. 2017). Plant origins of OAs in soil are root exudates, which include ions, as well as primary and secondary metabolites (Badri and Vivanco 2009). Fatty acids (FAs) are OAs characterized by a long series of methylene groups with the terminal carboxylate group. They are the primary building blocks of lipids and can be classified according to the presence of a double bond. FAs with no double bond are saturated and more stable than unsaturated ones, and are further divided into four classes: short-chain fatty acids (SCFAs), medium-chain, long-chain, and very long-chain FAs. SCFAs are saturated, aliphatic OAs and consist of acetic acid (AA, three carbons), propionic acid (three carbons), butyric acid (four carbons), valeric acid (five carbons), and caproic acid (six carbons) (den Besten et al. 2013; Mihaylova and Stratton 2019).

Bacteria and fungi secrete OAs in response to stress. Acidogenic bacteria degrade organic compounds into OAs such as AA and then acetate by acidogenesis and acetogenesis, respectively (Adeleke et al. 2017). Intestinal bacteria provide their hosts with short-chain fatty acids (SCFAs) by processing glucose (Archer et al. 1998; Wong et al. 2006; Louis and Flint 2016).

Previously, AA was considered a toxic agent (Armstrong et al. 1996; Ludovico et al. 2001); however, Kim et al. (2017) demonstrated the function of AA for the amelioration of drought stress in various species such as *Arabidopsis*. Many excellent papers followed the work of Kim et al. (2017), and they led to the accumulation of enormous data on the utilization of AA for the mitigation of stress. This prompted us to write this chapter. After a short introduction about SCFAs, the causes of and responses to stress, and the approaches to developing stress tolerance in plants will be summarized. Then, the use of some priming agents will be reviewed. Because AA is directly linked to epigenetic mechanisms, the relationship between priming and epigenetics will be discussed. Subsequently, the plant metabolism and production of AA will be explained briefly. Then the effects, particularly harmful, of AA on plant growth will be described in detail. After that, how AA influences plant phenotype and response under stress will be thoroughly reviewed in consecutive subtitles. Finally, there will be a brief glimpse into the effects of AA on species other than plants. This chapter aims to provide information on the effects of AA applications, which have been widely used against abiotic stress in plants in recent years, on plant growth and development, as well as various environmental stresses, accompanied by current literature.

5.2 Biostimulants and Seed Priming in Abiotic Stress

Climate change is responsible for the unfavorable variations in temperature, rainfall, and atmospheric conditions. Human activities, e.g., urbanization, rapid industrialization, deforestation, and burning of agricultural wastes, particularly cause changes in climatic conditions by increasing the atmospheric concentrations of greenhouse gases, which cause global warming and disruption of hydrological cycles (Chaudhry and Sidhu 2022). Alterations in the environmental conditions, e.g., high or low temperatures, drought, flood, light, salinity, heavy metals, and scarcity of nutrients, result in abiotic stress and retard plant growth and development. A large portion of the land area is affected by abiotic stress (Llorens et al. 2020). Adverse external conditions compromise plant growth and productivity and, as a result, food security (Chaudhry and Sidhu 2022). Upon encountering osmotic stress, which decreases water availability, abscisic acid (ABA) biosynthesis is initiated. ABA confers stress response through regulation of nearly 10% of the protein-coding genes and modulation of cellular and physiological mechanisms, one of which is stomatal closure due to ABA-induced elevated reactive oxygen species (ROS) levels in the guard cells (Sah et al. 2016). ABA recognition phosphorylates transcription factors, which in turn regulate the expression of genes encoding late embryogenesis abundant (LEA) proteins, protein kinases, phosphatases, transporters, and enzymes involved in osmoprotectant synthesis and secondary metabolism (Fujita et al. 2011). As a key regulator of abiotic stress tolerance, ABA is utilized for increasing stress tolerance by manipulating its synthesis or signaling (Sah et al. 2016; Chaudhry and Sidhu 2022). To prevent water loss, stomatal closure, which results in less uptake of CO_2, occurs in the plants under abiotic stress (Godoy et al. 2021). Under CO_2-limited conditions, photoprotective mechanisms occur in the photosystems (I and II) and photosynthesis is repressed to prevent excess production of reducing power. When CO_2 levels are insufficient, over-reduction of photosystems causes the transfer of electrons to oxygen and results in the generation of reactive oxygen species (ROS) (Chaves and Oliveira 2004). At low concentrations, ROS act as signaling molecules. When the antioxidant system is incapable of scavenging them, ROS accumulate and cause serious harm to the cellular machinery of plants (Chaudhry and Sidhu 2022). Elevated temperatures cause misfolding or aggregation of proteins, and Rubisco, the key enzyme in carbon fixation, is very sensitive to elevated temperatures. As a result, oxygenation rather than carboxylation is favored at high temperatures. Membrane fluidity is adversely affected by high temperature. Plants cope with the stress conditions by changing leaf orientation, stomatal closure, production of a protective waxy layer, rolling of leaves, regulation of ion and water transport, synthesis of osmoprotectants and polyamines, alteration of phytohormone signaling, and the antioxidant system (Dutta et al. 2020).

Generation of stress-tolerant plants by breeding programs is time-consuming. Plants can also be genetically engineered to have stress tolerance; however, this is not only costly but also a lengthy procedure (Chaves and Oliveira 2004; Godoy et al. 2021). One of the strategies to cope with the adverse effects of abiotic stress is the utilization of biostimulants (BSs), which promote plant growth by improving nutrient use efficiency, tolerance to abiotic stress, quality traits, or availability of confined nutrients in the soil or rhizosphere. BS consists of carboxylic FAs, polyamines, humic substances, inorganic salts, amino acids, and microorganisms (Yakhin et al. 2017; García-García et al. 2020).

Several compounds, including water and stress factors, e.g., UV-B, can also be applied at the pre-germination step to induce mild stress, and this application is

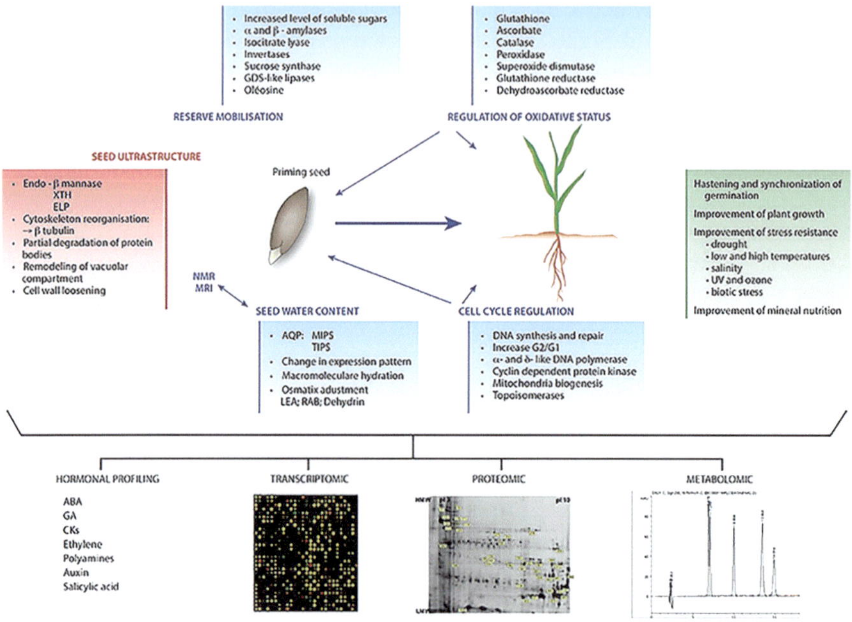

Fig. 5.1 General overview of the biochemical and physiological basis of priming effects. (Lutts et al. 2016)

called seed priming (Aswathi et al. 2022). Primed plants exhibit stronger and faster stimulation of stress responses at morphological, biochemical, physiological, and molecular levels (Zhang and Huang 2020) (Fig. 5.1). Priming upregulates genes involved in germination, the antioxidant system, cell division, and cell wall modification. It activates several enzymes that break macromolecules to sustain embryo growth, facilitates DNA replication and repair, and increases the accumulation of inorganic and organic solutes and reduces the leakage of metabolites. As a result of seed priming, seeds are early- and/or uniformly germinated, and stress-induced oxidative stress in the plants that emerged from primed seeds is diminished (Sen and Puthur 2020; Johnson and Puthur 2021; Aswathi et al. 2022). Plants that develop from primed seeds have higher leaf water potential and improved photosynthesis and nitrogen-use efficiency, as well as enhanced yield (Liu et al. 2017). While severe stress, e.g., water deficit, might be lethal, mild stress can enhance stress tolerance by the induction of stress memory, which is the result of exhaustive adjustment of metabolism upon the initial stress (Wojtyla et al. 2020). Stress priming can also confer cross-tolerance to various abiotic stresses (Zhang and Huang 2020).

5.3 Some Priming Agents

Hydrogen peroxide (H_2O_2), ascorbic acid (AsA), and salicylic acid (SA) are among the most common priming agents (Nair et al. 2022). H_2O_2 is a prominent member of the reactive oxygen species (ROS), which are produced by various metabolic reactions and accumulate as a stress response in plants. Plants are well equipped to cope with the burden of ROS accumulation and are also able to exploit ROS as signaling molecules, particularly H_2O_2, which is diffusible and has a high half-life. Exogenously applied H_2O_2 can confer tolerance to salt, drought, heat, chilling, and heavy metal stress by not only enhancing photosynthesis and the antioxidant system but also activating the expression of nuclear-encoded genes, e.g., the genes encoding heat shock proteins (Hossain et al. 2015). AsA is one of the most abundant nonenzymatic antioxidants, which carry out the detoxification of ROS and are mainly synthesized in plant mitochondria (Akram et al. 2017). AsA, of which the physiologically active form is the ascorbate anion, can react with H_2O_2 and other ROS and regenerate α-tocopherol (Das and Roychoudhury 2014). H_2O_2 can be scavenged by catalase, which functions mainly in peroxisomes, and also by ascorbate peroxidase, which uses AsA as an electron donor to convert H_2O_2 to water and molecular oxygen (Das and Roychoudhury 2014; Akram et al. 2017). APX is widely distributed in the cell, particularly in the cytosol and chloroplasts, and is more effective in the scavenging of H_2O_2 than catalase (Akram et al. 2017). AsA and APX are essential for maintaining chloroplast function, which produces H_2O_2. In addition to detoxification of H_2O_2, AsA is involved in the synthesis of zeaxanthin, which prevents photoinhibition of photosystem II (Das and Roychoudhury 2014; Akram et al. 2017). SA is a plant hormone and, in addition to its essential roles in plant immunity, is involved in abiotic stress tolerance by increasing the synthesis of osmolytes and secondary metabolites, regulating mineral uptake, and enhancing the antioxidant system (Peng et al. 2021; Kaur et al. 2022; Khan et al. 2022). Under optimum conditions, plants have low levels of SA, which increase upon infection. SA antagonizes jasmonic acid (JA), which, in turn, inhibits SA accumulation (Peng et al. 2021). SA regulates JA-responsive genes, abiotic stress response genes, and senescence genes (Morris et al. 2000; Peng et al. 2021). Faster priming effects of SA resemble those induced by drought and might occur through RNA-dependent DNA methylation and chromatin remodeling (Balassa et al. 2022).

5.4 Priming and Epigenetics

Plants adapt to changing environmental conditions by introducing novel phenotypes through epigenetic modifications, as phenotypic variations are not always correlated with genetic diversity (Tirnaz and Batley 2019). Abiotic and biotic factors, particularly viruses and stress, interfere with epigenetic mechanisms. In general, as a stress response, DNA methylation is reduced, transposable elements (TEs) are activated, and a vast array of histone modifications, such as H3K9ac, are added to or removed from chromatin. In addition, a few epigenetic modifications persist after stress. Priming confers acclimation to stress through transcriptional, translational, and epigenetic mechanisms without affecting the DNA sequence (Koc et al. 2020). Primed and non-primed "naive" plants differ in their transcriptional responses to stress (Harris et al. 2023). Epigenetic aspects of priming might also prevail by combining transcription and translation (Koc et al. 2020). The primed state of the plants is defined by memory genes, which are altered by epigenetic modifications, e.g., DNA methylation, histone modifications, and chromatin remodeling. Expression of memory genes can either be induced upon encounter with subsequent stress or maintained after the first exposure to stress (Nair et al. 2022). Epigenetic marks can also be mitotically and meiotically inherited (Nair et al. 2022). Yet, the inheritance of epigenetic marks might favor one parent, e.g., the female germline, and requires repeated stress exposure (Wibowo et al. 2016). Histone modifications, e.g., alterations in H3K4me2/3 levels, prevail over stress-induced epigenetic mechanisms such as DNA methylation and nucleosome occupancy (Harris et al. 2023). Metabolites such as methionine and acetyl-CoA impact epigenetic modification, acting as inhibitors or cofactors. Methionine is converted to S-adenosylmethionine (SAM), which is utilized by DNA and histone methyltransferases as a methyl donor (Ali et al. 2022). Histone modifications usually occur together and accumulate not only in regulated genes but also in silent regions (Temel et al. 2017). Improper priming is prevented by nucleosome assembly (Mozgová et al. 2015). In the absence of stress, the basic region/leucine zipper motif (bZIP) transcription factor ABA INSENSITIVE 5 (ABI5), a positive regulator of ABA response, is restricted by SWI2/SNF2 chromatin remodeling ATPase BRAHMA (Han et al. 2012). Upon sensing ABA, BRAHMA is phosphorylated by sucrose non-fermenting 1-related protein kinases, components of the ABA signaling pathway, and the repressive effect of BRAHMA on ABI5 is eliminated (Peirats-Llobet et al. 2016). In addition to pretranscriptional mechanisms, priming occurs through posttranscriptional mechanisms such as RNA turnover, autophagy, and miRNAs (Nair et al. 2022). Autophagy-deficient mutants have improved thermomemory and fewer insoluble protein aggregates that are induced by heat (Sedaghatmehr et al. 2018). Stress-responsive mRNA is prevented by RNA-binding proteins from degradation (Frei dit Frey et al. 2010). Numerous excellent papers exhaustively explain priming and stress memory (Crisp et al. 2016; Koc et al. 2020; Nair et al. 2022; Harris et al. 2023).

5.5 AA Metabolism in Plants

In eukaryotes, acetate is synthesized from acetaldehyde by aldehyde dehydrogenase and can pass through membranes in its nonionic AA form (Schug et al. 2016; Kudo et al. 2023). Acetaldehyde itself is derived from pyruvate, the end product of glycolysis. Pyruvate is converted by pyruvate decarboxylase into acetaldehyde and CO_2 during the first step of ethanolic fermentation (Tadege et al. 1999; Rahman et al. 2024). Under anaerobic conditions, pyruvate is processed by lactate dehydrogenase to form lactate, and this reaction, which also regenerates NAD+, the oxidized form of nicotinamide adenine dinucleotide (NAD), is referred to as lactic acid fermentation (Tadege et al. 1999). The second step of ethanolic fermentation is the conversion of ethanol to acetaldehyde by alcohol dehydrogenase, along with the oxidation of nicotinamide adenine dinucleotide hydrogen (NADH), the reduced form of NAD (Tadege et al. 1999; Rahman et al. 2024).

Acetate is used by acetyl-CoA synthetase (ACS) enzymes in mitochondria (ACSS1) and cytoplasm (ACSS2), for the generation of acetyl-CoA, which can also be synthesized by pyruvate dehydrogenase complex from pyruvate, in mitochondria and nucleoplasm (Schug et al. 2016), and by ATP citrate lyase from citrate in cytoplasm (Schug et al. 2016; Narita et al. 2018). Acetyl-CoA is the donor of acetyl groups that are transferred by lysine acetyltransferases to proteins to be acetylated (Schug et al. 2016; Narita et al. 2018). In addition to its role in protein acetylation, acetyl-CoA incorporates the tricarboxylic acid cycle for energy generation in mitochondria and is used for lipid synthesis in the cytoplasm (Moffett et al. 2020). In particular tissues, e.g., pollen, acetate can be directly used for lipid biosynthesis (Rahman et al. 2024). Not only lactate but also ethanol accumulates in plants under oxygen stress. Because it can pass through membranes, ethanol is released into the extracellular medium and does not cause a big threat except at high levels (Tadege et al. 1999). Moreover, ethanol, when applied exogenously, enhances drought tolerance by altering gene expression and metabolism and delaying water loss through ABA signaling and acetic acid biosynthesis (Bashir et al. 2022).

5.6 Commercial Production of AA and Its Uses

The worldwide production volume of AA, ethanoic acid, was 17.48 million metric tons in 2022 and is estimated to reach 23.6 million metric tons in 2030 (Global acetic acid market volume 2015–2030). This high demand is fulfilled by two routes, chemical and fermentative (Deshmukh and Manyar 2021). Despite its carbon footprint, chemical production of AA by methanol carbonylation is an efficient solution to meet the market demand. In this reaction, methanol and carbon monoxide are consumed in the presence of catalysts such as rhodium or iridium under high temperatures and pressure (Paulik et al. 1973; Medrano Garcia et al. 2019). Acetaldehyde and hydrocarbons such as butane can be oxidized using various catalysts to obtain AA, usually under higher temperatures and/or pressure (Deshmukh and Manyar 2021). In 2013, BP, the company that had already improved methanol carbonylation to the Cativa® technology, announced the three-step SaaBre process for the conversion of synthesis gas (carbon monoxide and hydrogen derived from natural gas) directly to AA without the need to consume methanol or purify carbon monoxide (BP reveals step-out routes to acetic acid and ethylene with all-new technologies 2013). Chemically obtained AA is used as a substrate for the synthesis of acetate compounds, e.g., vinyl acetate and cellulose acetate (Medrano Garcia et al. 2019).

In addition to its uses in the textile and polymer industries, AA is also used as a food preservative and a flavoring agent. The most commonly used form of AA in the food industry is its vinegar form (Deshmukh and Manyar 2021). The word vinegar is derived from the French words *Vin* (wine) and *Aigre* (sour), and is defined as "a liquid, fit for human consumption, produced from a suitable raw material of agricultural origin, containing starch, sugars, or starch and sugars, by the process of double fermentation, first alcoholic and then acetous" by Codex Alimentarius (Codex Alimentarius 1987; Bhat et al. 2014). AA, whose concentration in vinegar is around 5%, is the main component of vinegar, and several compounds, e.g., salts, fruit acids, and other fermentation products, accompany it (Johnston and Gaas 2006; Bhat et al. 2014). Traditionally, vinegar can be produced from fruit juices such as grapes, apples, and plums by two consecutive fermentation events. The starter culture for the fermentation is also vinegar. In the first step, glucose is anaerobically oxidized by yeasts, e.g., *Saccharomyces cerevisiae*, to ethanol, which is subsequently aerobically oxidized by bacteria belonging to the *Acetobacter* or *Gluconobacter* genera to AA. As described in the Sect. 5.5 above, ethanol is broken down to AA in two reactions catalyzed by the enzymes alcohol dehydrogenase and aldehyde dehydrogenase, in order (Bhat et al. 2014; Budak et al. 2014). Different bacterial species, depending on the plant material, result in not only AA but also other OAs such as citric acid (Budak et al. 2014). Traditionally, AA fermentation occurs slowly in wooden barrels filled with ethanol. This method preserves the taste of the vinegar and is known as the French or Orleans process. Despite the high quality of the product, this method falls short of meeting vinegar demand (Bhat et al. 2014). Another method utilizes a layer of wood shavings to grow bacteria and is known as the trickling/generator method. In this method, air is provided from below the container, temperature is maintained, and the resulting vinegar is recirculated. However, the resulting product may not have high amounts of AA (Bhat et al. 2014; Budak et al. 2014). At the industrial scale, the fermentation medium is stirred to suspend bacteria, aerated, and kept at the desired temperatures. This approach is called submerged fermentation. Bacteria become sensitive in this method; however, they can be continuously cultured in the exponential growth phase to reach high yield (Bhat et al. 2014).

5.7 AA and Plant Growth

AA is present in numerous environments, e.g., atmosphere, soil, rice fields, aquatic environments, and rainwater (Zuo et al. 2012) and is an abundant compound in aquatic sediments (Spencer and Ksander 1995). AA and propionic acid, the smallest members of SCFAs, were defined as highly toxic and exhibit their toxicity in the undissociated form (Van Overbeek and Blondeau 1954). Lynch (1977, 1978) reported the phytotoxicity of AA, which was reflected by reduced root growth in barley. Since then, similar effects of AA were observed in various organisms such as aquatic plants (Spencer and Ksander 1995), reed (Armstrong et al. 1996), *S.*

cerevisiae (Ludovico et al. 2001, 2002; Guaragnella et al. 2008, 2010; Sokolov et al. 2006), *Chlamydomonas reinhardtii* (Zuo et al. 2012), maize (Allen and Allen 2020), and barley (Lynch 1977, 1978; Temel and Kosesakal 2024). In *Hydrilla verticillata*, propagules fail to sprout when exposed to 17 mM AA even for 1 day. The toxic effects of AA depend on the temperature, according to the observations obtained with *Potamogeton pectinatus*, another aquatic plant (Spencer and Ksander 1995). Armstrong et al. (1996) described AA as a phytotoxin originating from decaying reeds and reported that 1.67 mM AA at pH 3.6–4.0 completely inhibited root growth in reed (*Phragmites australis*) and caused lignification in the root epidermis and callus formation in roots. The phytotoxic effects of AA were not caused by its acidic properties (Armstrong et al. 1996). *C. reinhardtii* exhibited increases in respiration and H_2O_2 content and a decrease in photosynthesis when grown in Tris-acetate-phosphate medium, whose pH was adjusted to 5.0 and contained 34.3 mM AA, for 10 min. AA treatment also resulted in cleavage of nuclear DNA after 1 h and demonstrated its programmed cell death (PCD)-inducing effects in *C. reinhardtii*. Even 1 mM AA at pH 5.0 could induce cell death, yet adjustment of the pH of the medium to 4.0 using other acids, such as HCl or $H_2SO_{4,}$ did not cause death. AA treatment did not affect photosynthetic pigments yet induced the synthesis of volatile organic compounds (Zuo et al. 2012). In their interesting paper, Allen and Allen (2020) discussed the biostimulant nature of AA. When AA was applied in an unbuffered manner, i.e., seeds were grown on AA-containing germination papers, AA was mostly in its undissociated, lipid-soluble, and membrane-permeable form and strongly inhibited root growth in maize. They indicated that the undissociated form of AA is responsible for the growth inhibition, and maintenance of higher pH can protect roots from its toxic effects. Based on their findings, pH and/or the concentration of the undissociated form of AA should be considered in treatment regimes. The effects of AA on barley were first studied by Lynch (1977). Stimulation of barley root growth by 0.1–1 mM concentration of AA was reversed when pH was reduced from 6.5 to 3.5. Again, in barley, Temel and Kosesakal (2024) applied 2.5–5 mM AA to 2-day-old barley seedlings for 2 days without pH adjustment and reported dramatic decreases in root and shoot growth, fresh weight (FW), water content (WC), and photosynthetic pigment levels. Shoots of AA-treated seedlings accumulated UV-absorbing compounds, and roots of AA-treated plants lost cell viability.

5.8 Amelioration of Abiotic Stress by AA Treatment

Kim et al. (2017) were the first to demonstrate the stress-ameliorative effects of AA. This study on Arabidopsis plants showed that drought tolerance is proportional to the amount of acetate and that the drought response involves the acetate-mediated jasmonic acid (JA) pathway. They also revealed that the application of AA to the roots enhances drought tolerance in plants such as *Arabidopsis*, rice, and wheat. They suggested that this effect is specific to AA, as other organic acids like butyric acid and formic acid did not affect the phenotype. The effects of AA in plants are mediated through various processes such as root and shoot growth (Lynch 1977; Mirfattahi et al. 2022; Temel and Kosesakal 2024), water potential (Li et al. 2021a), reactive oxygen species (ROS) (Mahmud et al. 2023; Mirfattahi and Eshghi 2023; Temel and Kosesakal 2024), antioxidant systems (Rahman et al. 2021; Li et al. 2022; Mirfattahi and Eshghi 2023), pigment content (Li et al. 2022; Mirfattahi et al. 2022; Mahmud et al. 2023; Temel and Kosesakal 2024), osmolyte synthesis (Rahman et al. 2019; Li et al. 2022; Mahmud et al. 2023), secondary metabolite content (Mirfattahi et al. 2022; Temel and Kosesakal 2024), reduction of water loss (Utsumi et al. 2019; Mahmud et al. 2023), element uptake (Rahman et al. 2019), cell death (Zuo et al. 2012; Temel and Kosesakal 2024), transpiration (Li et al. 2021a, 2022), photosynthesis (Li et al. 2021a, 2022), stomatal conductance (Li et al. 2021a, 2022), hormone content (Li et al. 2021a, 2022), and gene expression (Li et al. 2021a; Mahmud et al. 2023). The observation that AA absorbed by the roots is incorporated into organic acids and amino acids like glutamine has been interpreted as an indication that AA enhances nitrogen-use efficiency (Isaji et al. 2018; Ogawa et al. 2021; Kudo et al. 2023) (Fig. 5.2). The plant species, stress conditions, and their corresponding papers were listed in Table 5.1.

5.8.1 Role of AA Against Drought Stress

The stress tolerance-enhancing effect of AA (acetic acid) has been reported to depend on its concentration (Utsumi et al. 2019) and the pH of the environment (Allen and Allen 2020). Under nonstress conditions, AA has been shown to inhibit plant growth in species such as maize (Allen and Allen 2020), barley (Temel and Kosesakal 2024), and cassava (*Manihot esculenta*) (Utsumi et al. 2019). Based on the results from these studies, the AA-mediated stress response differs from the ABA-dependent and ABA-independent mechanisms typically associated with plant responses to drought. In this mechanism, during the initial stages of drought, when soil moisture begins to decrease but the plant's water reserves are not yet depleted, the histone deacetylase HDA6 alters the plant's metabolism. As a result, AA is synthesized from pyruvate, which then promotes JA synthesis and increases chromatin acetylation. In more advanced or severe drought conditions, where the plant begins to lose water, osmolytes accumulate, and both ABA-dependent and ABA-independent gene pathways are activated. This leads to stomatal closure to prevent further water loss by the plant (Fig. 5.3) (Kudo et al. 2023).

In their pioneering work, Kim et al. (2017) first discovered that histone deacetylase (HDA6) mutants of Arabidopsis exhibited drought tolerance without any changes in the expression of ABA-responsive genes. Instead, the acetate biosynthesis pathway was upregulated both in wild-type and mutant plants, and the enhancement was more apparent in the mutants. They saw that drought caused the release of HDA6 from chromatin and resulted in the accumulation of acetylated histones on the acetate biosynthesis genes. Acetate levels were correlated with drought

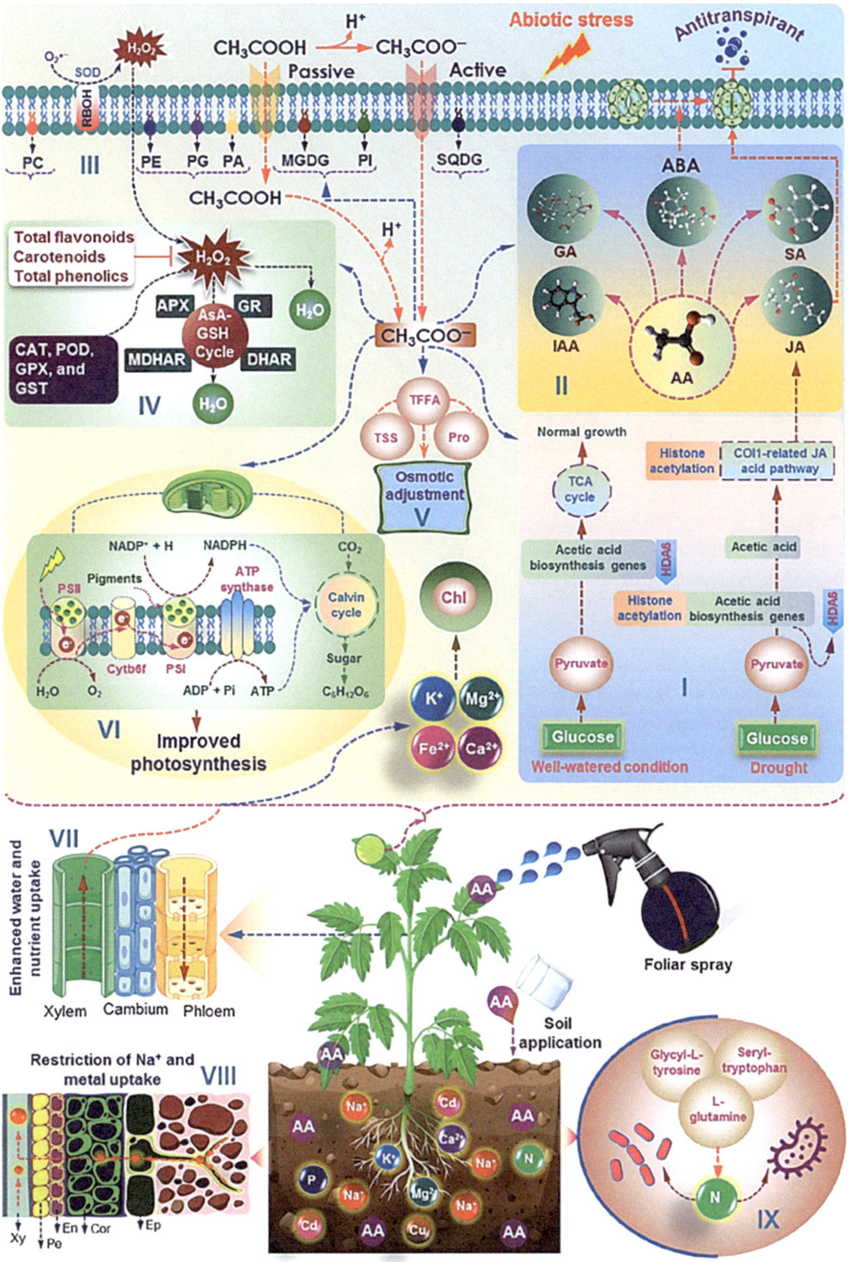

Fig. 5.2 A schematic view showing the role of AA in abiotic stress. (Rahman et al. 2024)

Table 5.1 List of papers that reported the stress-ameliorative effects of AA

Species	Stress	AA	References
Z. mays, O. sativa, T. aestivum, B. napus, A. thaliana	Drought	20–30 mM AA	Kim et al. (2017)
M. esculenta	Drought	10 mM AA	Utsumi et al. (2019)
O. sativa	Drought	30 mM AA	Ogawa et al. (2021)
B. × *hybrida*	Drought	10 mM AA	Allen and Allen (2021)
G. max	Drought	20 mM AA	Rahman et al. (2021)
C. lanceolata	Drought	200 mM AA	Li et al. (2022)
M. domestica	Drought	10 mM AA	Sun et al. (2022)
Z. mays	Drought	0.5–25 mM AA	Mahmud et al. (2023)
V. radiata	Salinity	20 mM AA	Rahman et al. (2019)
L. perenne	Salinity	20 mM AA	Zhang et al. (2021)
F. × *ananassa*	Salinity	2 mM AA	Mirfattahi et al. (2022); Mirfattahi and Eshghi (2023)
A. sativa	Salinity	10 mM sodium acetate	Kappachery et al. (2024)
S. myrtillacea	High and low temperatures	100 mM AA	He et al. (2024)

tolerance, and exogenously applied AA at 10, 20, or 30 mM concentrations indeed conferred drought tolerance. 357 genes, including JA response genes, were upregulated by AA treatment. They interpreted this as plants' uptake of AA from the soil and converting it to acetyl-CoA to be used as a substrate for histone acetylation. Yet, they also indicated that *Arabidopsis* plants could absorb a small amount of AA from the soil. They showed that AA could also confer drought tolerance in maize, rapeseed, rice, and wheat. Moreover, this effect is specific to AA, as other OAs, such as butyric acid and formic acid, did not affect the phenotype. As the authors explained, the inefficiency of other OAs to mitigate drought might be attributed to epigenetic regulation driven by AA, as it connects metabolism and epigenetic regulation. Drought itself affects histone acetylation, and the enzymes involved in histone acetylation/deacetylation work with other proteins in drought response (Li et al. 2021b). Several papers followed the work of Kim et al. (2017). Induction of AA biosynthesis genes prolongs survival under drought stress in *Arabidopsis* (Rasheed et al. 2018). Treatment of cassava plants with 10 mM AA for 7 days ameliorated plant growth as observed by elevated levels of photosynthetic pigments and was proposed to induce drought avoidance due to increased stomatal closure and leaf temperature. Yet, 20–50 mM AA-induced leaf wilting (Utsumi et al. 2019). Improved survival conferred by exogenously applied AA was shown in cotton (*Gossypium hirsutum* L.). Cotton seedlings were irrigated with 8 mM AA for 9 days. Leaves of the seedlings treated with AA had relatively higher water potential, photosynthetic rate, transpiration, and stomatal conductance, and higher levels of ABA and JA (Li et al. 2021a). Ogawa et al. (2021) demonstrated that AA mimics drought and, to exert its

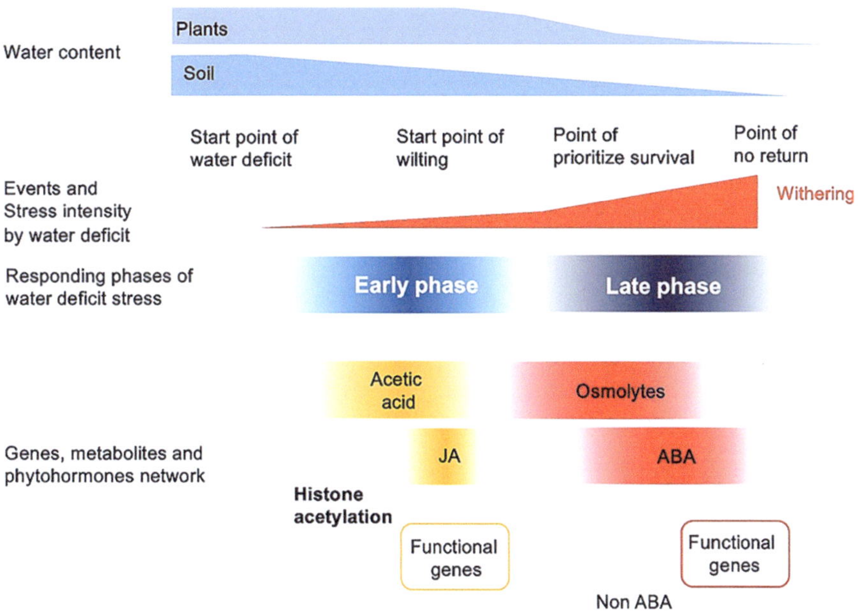

Fig. 5.3 Drought response in plants. (Kudo et al. 2023)

effects, needs to be applied for 3 days without any interruption. AA primes JA signaling in the case of drought stress rather than the induction of JA-mediated drought tolerance. Although the physiological responses to AA and drought overlap, the genes regulated by AA are distinct from those regulated by drought. The effects of AA are independent of its acidity, as previously indicated by Kim et al. (2017).

Allen and Allen (2021) determined that AA, when adjusted to an optimum concentration, can be a cheap and effective transpirant. In their study with *Begonia* × *hybrida*, plant trays were submerged in 10 mM AA solution for 6 days and then subjected to drought by withholding water until a certain percentage, e.g., 10% of plants died. In another study, soybean seedlings were treated with 20 mM AA by foliar spraying before being exposed to drought. AA-treated plants, under drought or well-watered conditions, had improved growth, elevated levels of photosynthetic pigments and water use efficiency, reduced oxidative stress, and accumulation of solutes. AA-treated soybean plants were able to maintain mineral balance (Rahman et al. 2021). Li et al. (2022) reported similar findings in Chinese fir (*Cunninghamia lanceolata* (Lamb.) Hook.), a fast-growing timber species with drought-sensitive traits. 3-year-old seedlings were watered with 200 mM AA solution for 9 days. AA resulted in maintenance of water status, improvement of photosynthesis, and scavenging of ROS. According to their data, AA can inhibit plant growth when applied at an elevated concentration (300 mM) and can be influenced by soil properties such as pH. Perennial woody apple (*Malus domestica*) plants treated with 10 mM AA exhibited less-inflicted phenotypes, i.e., less decrease in FW, reduced oxidative stress, and altered expression of the genes involved in mitogen-activated protein kinase (MAPK) signaling (Sun et al. 2022). Mahmud et al. (2023) tested the stress-counteracting effects of AA in maize by two methods, seed priming and exogenous treatment. In the first approach, maize seeds were imbibed in 0.5 mM AA overnight. In the second treatment, maize seedlings were supplied with 25 mM AA for 48 h through their roots. Both treatment regimes regulated the antioxidant system, increased leaf growth, reduced transpiration and stomatal conductance, induced the gene encoding δ-pyrroline-5-carboxylate synthetase, and resulted in the accumulation of proline.

5.8.2 Role of AA Against Salinity Stress

AA alleviates the harmful effects of salt stress and drought, and has also been shown to mitigate the toxic effects of salinity in mung bean (*Vigna radiata*) (Rahman et al. 2019). Zhang et al. (2021) showed that salinity increased endogenous AA content in perennial ryegrass (*Lolium perenne* L.), a salinity-sensitive plant, by inducing a key gene involved in its biosynthesis. Exogenously applied AA delayed leaf senescence induced by salinity, maintained high K^+/Na^+ ratios in leaves, and led to the accumulation of several hormones, such as JA, in leaves, especially under salinity stress. Mirfattahi et al. (2022) demonstrated that AA can induce salinity tolerance in strawberry (*Fragaria* × *ananassa*), a plant species known for being sensitive to abiotic stress. They also indicated that the timing of the use of AA is important. They applied 1 or 2 mM AA by foliar spraying in three regimens before, during, and after the salt treatment. They reported that 1 mM AA was more effective in causing an avoidance strategy, as demonstrated by the maintenance of leaf and fruit parameters, proline and carbohydrate contents, and the inhibition of sodium and chloride accumulation. The same group (Mirfattahi and Eshghi 2023) later reported improved fruit yield, modulation of the antioxidant system, and accumulation of ABA, cytokinin, gibberellin, and SA in the same species. Recently, Kappachery et al. (2024) reported that 10 mM sodium acetate treatment of oat (*Avena sativa*) plants for 3 weeks resulted in a highly improved phenotype, such as higher amounts of photosynthetic pigments and proline, less oxidative stress and membrane damage, upregulated expression of antioxidant enzymes, and increased synthesis of not only ABA but also Indole Acetic Acid (IAA). Photosynthesis was also improved by sodium acetate treatment.

5.8.3 Role of AA Against Heat Stress

An intriguing paper by et al. (Dewhirst et al. 2021) reported that methyl acetate was released from leaves upon desiccation or heat stress in poplar (*Populus trichocarpa*). This emission coincided with stomatal closure and suppression of photosynthesis and transcription. As the authors suggested, a mass spectrometry–based approach could provide a noninvasive mechanism to trace the acetate pathway. In a recent paper, He et al. (2024) studied the effects of AA in *Salix myrtillacea*, a dioecious wood-producing plant, and demonstrated that AA was effective in alleviating both high- and low-temperature stresses. These promoting effects also varied with the sex of the plant. In addition to improvements in photosynthesis, the antioxidant system, and the accumulation of hormones, AA treatment also favored certain species in the soil microbiota and resulted in better carbon uptake in female plants. In an earlier paper (Kong et al. 2022), the same group reported increased tolerance against drought in the same species. AA-induced tolerance against high temperature contrasts with the observations in *S. cerevisiae,* as AA was previously shown to be more toxic to it than ethanol under high process temperatures (Pinto et al. 1989).

5.8.4 Role of AA Against Cold Stress

Cold stress is one of the most important environmental stress factors that negatively affect plant growth and development, along with salinity and drought. In addition to drought stress, cold stress is one of the most harmful environmental stresses that higher plants face. Cold stress is an important factor that limits the geographical distribution of plants (Ritonga and Chen 2020). Unfortunately, literature reviews indicate that there are not many studies on the effects of acetic acid on cold stress. It is reported that acetic acid alleviates the effects of cold stress in plants. Researchers have shown that acetic acid increases cold tolerance (0 °C) by acting on enzymatic antioxidants, including POD, SOD, and CAT, while nonenzymatic antioxidants such as phenols play an important role in freezing tolerance (−6 °C). In the same study, it was reported that total soluble carbohydrates, photosynthetic pigments, and maximum quantum efficiency of photosystem II (Fv/Fm) increased in plants treated with acetic acid under both cold and freezing stresses (Sanie Khatam et al. 2023). It has also been reported that the application of AA, together with melatonin, has a reducing effect on cold stress damage in Mexican lime plants (Khatam et al. 2024). The molecular mechanisms explaining the biochemical pathways by which acetic acid can increase cold tolerance in crop plants need further investigation.

5.8.5 Role of AA Against Heavy Metal Stress

Heavy metals such as copper, lead, and cadmium are important environmental pollutants. Heavy metal accumulation in soils is a major concern in agricultural production due to its adverse effects on food safety, crop yield, and environmental health of soil organisms. Literature reviews indicate that researchers, particularly in recent years, have conducted studies demonstrating the potential of acetic acid to alleviate heavy metal toxicity in various plants (Zhu et al. 2021; Hossain et al. 2019,

2020, 2022). When studies on the effects of acetic acid on heavy metal stress in plants are examined, it is evident that researchers primarily focus on its effects on cadmium (Cd) stress. Researchers showed that AA enhanced the resistance to Cd (Xie et al. 2013; Ehsan et al. 2014). AA had positive effects on the growth and some physiological parameters of Cd-exposed sunflower plants (Hawrylak-Nowak et al. 2015). Another group demonstrated that AA increased Cd resistance in tomatoes (Zhu et al. 2021). In 2022, Chowardhara et al. conducted a study examining the effects of putrescine, ethanol, and acetic acid on Cd toxicity and accumulation in Indian mustard. The study observed that AA pre-application significantly improved both the belowground and aboveground parts of Cd-applied mustard plants. Additionally, the AA application enhanced the antioxidant system. Based on their data, the researchers concluded that AA could mitigate heavy metal toxicity (Chowardhara et al. 2022). In a study conducted with Cu, it was shown that Na-acetate application caused a decrease in Cu accumulation in the aboveground and belowground parts of lentil seedlings under Cu stress. In the same study, it was reported that there was an increase in proline levels, an enhancement in the antioxidant defense mechanism, and increased resistance to Cu stress (Hossain et al. 2019). It is emphasized that more comprehensive genomic, transcriptomic, proteomic, and metabolic studies should be conducted to elucidate the tolerance mechanisms against heavy metal stress in plants (Rahman et al. 2024).

5.9 The Effects of AA on Other Species

The most remarkable data on the cellular and toxic effects of AA were obtained from yeast (Chaves et al. 2021; Guaragnella and Bettiga 2021). In exponential-phase cells of *S. cerevisiae*, 120 mM AA treatment for 200 min resulted in a mortality rate of 99%. The toxic effects of lower concentrations, e.g., 20–80 mM, can be ameliorated by cycloheximide, an inhibitor of protein synthesis. The authors described their observations as AA, up to 120 mM concentration, can induce a PCD pathway with an apoptotic phenotype, e.g., chromatin condensation along the nuclear envelope, yet elevated concentrations induce necrotic changes (Ludovico et al. 2001). The same group later observed that stationary cells were more resistant to the toxic effects of AA and explained these findings by the differences in mitochondrial function between cells. They indeed described that AA-induced PCD (AA-PCD) involved accumulation of ROS and translocation of cytochrome c (cyt c) from mitochondria to the cytosol and mitochondrial dysfunction (Ludovico et al. 2002). Another group demonstrated that AA-PCD in *S. cerevisiae* depends on the production of H_2O_2 (Guaragnella et al. 2008) and can occur without the release of cyt c (Guaragnella et al. 2010). AA-PCD was attributed to the intracellular acidification by AA but not other acids, such as HCl, as intracellular acidification facilitates the protonation of superoxide anion ($O_2^{\cdot-}$) to hydroperoxyl radical (HO_2) (Sokolov et al. 2006). Single-strand breaks (Ribeiro et al. 2006) and rRNA degradation (Mroczek and Kufel 2008) accompany mitochondrial changes in *S. cerevisiae* treated with AA. Deletion of the genes encoding protein kinases and the proteins of the histone deacetylase complex confers resistance to AA-PCD (Sousa et al. 2013). In contrast, *S. cerevisiae* mutants lacking RTT109, a histone acetyltransferase, exhibited higher levels of glycerol and increased transcription of several stress-responsive genes, such as GPX1, and were resistant to AA (Cheng et al. 2016). 150 mM AA (pH 3.0) treatment for 200 min resulted in 874 differentially expressed rather downregulated proteins. The proteins that were repressed by AA treatment were involved in amino acid metabolism, protein synthesis, and carbohydrate metabolism, while AA-induced genes were involved in ubiquitin-dependent protein catabolic process, cell cycle, and stress response (Dong et al. 2017). Chaves et al. (2021) explained that the cellular effects of AA depend on temperature and defined this process as regulated cell death. They rather focused on the improvement of AA-resistant yeast strains.

AA has more complicated effects on human cells, particularly cancer cells. Consumption of vinegar, which contains low amounts of AA, as a beverage increases insulin sensitivity (Johnston et al. 2004) and can ameliorate lifestyle-related diseases (Samad et al. 2016). Okabe et al. (2012) demonstrated that when applied topically on the gastric wall of mice, AA caused necrosis of the gastric tumor. They later showed that a low concentration (0.3%) of AA was able to induce in vitro cell death, and cancer cells were more sensitive to AA than normal cells. They suggested that AA-induced cell death was independent of pH (Okabe et al. 2014). The same group reported that cancer cell–selective death induced by AA occurs through increased production of ROS and lipid peroxidation, and uptake of AA is favored in cancer cells (Terasaki et al. 2018), and AA increases the efficiency of a cancer treatment based on laser irradiation (Kurokawa et al. 2022). AA-affected pathways in various organisms are related to a degree and require oxidative stress. In addition to its oxidative effects, acetate is essential for immunity (Balmer et al. 2016). However, acetate was shown to promote cancer cell survival under hypoxia due to increased histone acetylation and enhanced lipid synthesis (Gao et al. 2016) and induce endothelial-to-mesenchymal transition, an important pathophysiological process (Zhu et al. 2023). SCFAs such as AA are also important for health, especially in metabolic and inflammatory processes (Campos-Perez and Martinez-Lopez 2021; Challa and Lewandowski 2022).

5.10 Conclusions

The studies conducted so far on AA, as mentioned above, demonstrate the effects of AA on plant growth and development. The data obtained indicate that AA has a particularly ameliorative effect on abiotic stress, especially salinity and drought, in plants. Priming applications in studies involving AA have also garnered attention. In recent years, there has been an increase in studies demonstrating the effects of chemical priming on abiotic stress. In this context, the fact that AA is cheaper and more readily available compared to other chemicals used in priming applications and that its ameliorative effects on abiotic stress have been identified places AA in a significant position. It is obvious that AA influences a wide spectrum of plant growth and characteristics, e.g., photosynthetic performance, pigment composition, solute biosynthesis, and histone acetylation. With the increase in physiological, biochemical, and molecular-level studies on AA, the mechanism by which AA alleviates abiotic stress factors will begin to be elucidated. It is clear that the data obtained will shed light on combating abiotic stress factors, which are among the most significant challenges of our time and severely impact agricultural production, leading to yield losses. In our opinion, the influence of AA on plant growth at the molecular level, particularly proteomics and metabolomics, warrants investigation, as most data on the effects of AA on gene expression have been obtained from non-plant species. Our group has been and is still interested in contributing to this topic. We apologize to the authors, who may think that their work has been unmentioned.

References

Adeleke R, Nwangburuka C, Oboirien B (2017) Origins, roles and fate of organic acids in soils: a review. S Afr J Bot 108:393–406. https://doi.org/10.1016/j.sajb.2016.09.002

Akram NA, Shafiq F, Ashraf M (2017) Ascorbic acid-a potential oxidant scavenger and its role in plant development and abiotic stress tolerance. Front Plant Sci 8:613. https://doi.org/10.3389/fpls.2017.00613

Ali S, Khan N, Tang Y (2022) Epigenetic marks for mitigating abiotic stresses in plants. J Plant Physiol 275:153740. https://doi.org/10.1016/j.jplph.2022.153740

Allen MM, Allen DJ (2020) Biostimulant potential of acetic acid under drought stress is confounded by pH-dependent root growth inhibition. Front Plant Sci 11:647. https://doi.org/10.3389/fpls.2020.00647

Allen MM, Allen DJ (2021) Acetic acid is a low cost antitranspirant that increases begonia survival under drought stress. Sci Hortic 287:110257–110264. https://doi.org/10.1016/j.scienta.2021.110257

Archer SY, Meng S, Shei A, Hodin RA (1998) P21 supWAF1/sup is required for butyrate-mediated growth inhibition of human colon cancer cells. Proc Natl Acad Sci 95:6791–6796. https://doi.org/10.1073/pnas.95.12.6791

Armstrong J, Afreen-Zobayed F, Armstrong W (1996) Phragmites die-back: sulphide- and acetic acid-induced bud and root death, lignifications, and blockages within aeration and vascular systems. New Phytol 134:601–614. https://doi.org/10.1111/j.1469-8137.1996.tb04925.x

Aswathi KPR, Kalaji HM, Puthur JT (2022) Seed priming of plants aiding in drought stress tolerance and faster recovery: a review. Plant Growth Regul 97:235–253. https://doi.org/10.1007/s10725-021-00755-z

Badri DV, Vivanco JM (2009) Regulation and function of root exudates. Plant Cell Environ 32:666–681. https://doi.org/10.1111/j.1365-3040.2008.01926.x

Balassa G, Balassa K, Janda T et al (2022) Expression pattern of RNA interference genes during drought stress and MDMV infection in maize. J Plant Growth Regul 41:2048–2058. https://doi.org/10.1007/s00344-022-10651-z

Balmer ML, Ma EH, Bantug GR, Grahlert J, Pfister S et al (2016) Memory CD8 T cells require increased concentrations of acetate induced by stress for optimal function. Immunity 44:1312–1324. https://doi.org/10.1016/j.immuni.2016.03.016

Bashir K, Todaka D, Rasheed S et al (2022) Ethanol-mediated novel survival strategy against drought stress in plants. Plant Cell Physiol 63:1181–1192. https://doi.org/10.1093/pcp/pcac114

Bhat VS, Akhtar R, Amin T (2014) An overview on the biological production of vinegar. Int J Fermented Foods 3:139–155. https://doi.org/10.5958/2321-712X.2014.01315.5
BP reveals step-out routes to acetic acid and ethylene with all-new technologies (2013). https://www.bp.com/en/global/corporate/news-and-insights/press-releases/bp-reveals-new-technologies.html, access date: 10.11.2024
Budak NH, Aykin E, Seydim AC, Greene AK, Guzel-Seydim ZB (2014) Functional properties of vinegar. J Food Sci 79:R757–R764. https://doi.org/10.1111/1750-3841.12434
Campos-Perez W, Martinez-Lopez E (2021) Effects of short chain fatty acids on metabolic and inflammatory processes in human health. Biochim Biophys Acta Mol Cell Biol Lipids 1866:158900. https://doi.org/10.1016/j.bbalip.2021.158900
Challa AA, Lewandowski ED (2022) Short-chain carbon sources: exploiting pleiotropic effects for heart failure therapy. JACC Basic Transl Sci 7:730–742. https://doi.org/10.1016/j.jacbts.2021.12.010
Chaudhry S, Sidhu GPS (2022) Climate change regulated abiotic stress mechanisms in plants: a comprehensive review. Plant Cell Rep 41:1–31. https://doi.org/10.1007/s00299-021-02759-5
Chaves MM, Oliveira MM (2004) Mechanisms underlying plant resilience to water deficits: prospects for water-saving agriculture. J Exp Bot 55:2365–2384. https://doi.org/10.1093/jxb/erh269
Chaves SR, Rego A, Martins VM, Santos-Pereira C, Sousa MJ, Côrte-Real M (2021) Regulation of cell death induced by acetic acid in yeasts. Front Cell Dev Biol 9:642375. https://doi.org/10.3389/fcell.2021.642375
Cheng C, Zhao X, Zhang MM, Bai F (2016) Absence of Rtt109p, a fungal-specific histone acetyltransferase, results in improved acetic acid tolerance of *Saccharomyces cerevisiae*. FEMS Yeast Res 16:1–9. https://doi.org/10.1093/femsyr/fow010
Chowardhara B, Saha B, Borgohain P, Awasthi JP et al (2022) Effect of ethanol, putresciene and acetic acid on cadmium accumulation and toxicity in Indian mustard. S Afr J Bot 147:42–52. https://doi.org/10.1016/j.sajb.2021.12.019
Codex Alimentarius (1987) Codex regional standard for vinegar. CodexStandard 162. FAO/OMS, Geneva
Crisp PA, Ganguly D, Eichten SR, Borevitz JO, Pogson BJ (2016) Reconsidering plant memory: intersections between stress recovery, RNA turnover, and epigenetics. Sci Adv 2:e1501340. https://doi.org/10.1126/sciadv.1501340
Das K, Roychoudhury A (2014) Reactive oxygen species (ROS) and response of antioxidants as ROS-scavengers during environmental stress in plants. Front Environ Sci 2:53. https://doi.org/10.3389/fenvs.2014.00053
den Besten G, van Eunen K, Groen AK, Venema K, Reijngoud DJ, Bakker BM (2013) The role of short-chain fatty acids in the interplay between diet, gut microbiota, and host energy metabolism. J Lipid Res 54:2325–2340. https://doi.org/10.1194/jlr.R036012
Deshmukh G, Manyar H (2021) Production pathways of acetic acid and its versatile applications in the food industry. IntechOpen. https://doi.org/10.5772/intechopen.92289
Dewhirst RA, Lei J, Afseth CA, Castanha C, Wistrom CM, Mortimer JC, Jardine KJ (2021) Are methanol-derived foliar methyl acetate emissions a tracer of acetate-mediated drought survival in plants? Plants 10:411. https://doi.org/10.3390/plants10020411
Dong B, Zheng X, Liu H, Able JA, Yang H, Zhao H & Liu M (2017) Effects of drought stress on pollen sterility, grain yield, abscisic acid and protective enzymes in two winter wheat cultivars. Front Plant Sci 8:1008
Dutta P, Chakraborti S, Chaudhuri KM, Mondal S (2020) Physiological responses and resilience of plants to climate change. In: Rakshit A, Singh HB, Singh AK, Singh US, Fraceto L (eds) New frontiers in stress management for durable agriculture. Springer, Singapore, pp 3–20
Ehsan S, Ali S, Noureen S, Mahmood K et al (2014) Citric acid assisted phytoremediation of cadmium by Brassica napus L. Ecotoxicol Environ Saf 106:164–172. https://doi.org/10.1016/j.ecoenv.2014.03.007
Frei dit Frey N, Muller P, Jammes F, Kizis D, Leung J, Perrot-Rechenmann C, Bianchi MW (2010) The RNA binding protein Tudor-SN is essential for stress tolerance and stabilizes levels of

stress-responsive mRNAs encoding secreted proteins in Arabidopsis. Plant Cell 22:1575–1591. https://doi.org/10.1105/tpc.109.070680

Fujita Y, Fujita M, Shinozaki K, Yamaguchi-Shinozaki K (2011) ABA-mediated transcriptional regulation in response to osmotic stress in plants. J Plant Res 124:509–525. https://doi.org/10.1007/s10265-011-0412-3

Gao X, Lin SH, Ren F, Li JT, Chen JJ et al (2016) Acetate functions as an epigenetic metabolite to promote lipid synthesis under hypoxia. Nat Commun 7:11960. https://doi.org/10.1038/ncomms11960

García-García AL, García-Machado FJ, Borges AA, Morales-Sierra S, Boto A, Jiménez-Arias D (2020) Pure organic active compounds against abiotic stress: a biostimulant overview. Front Plant Sci 11:575829. https://doi.org/10.3389/fpls.2020.575829

Global acetic acid market volume 2015–2030. https://www.statista.com/statistics/1245203/acetic-acid-market-volume-worldwide/, access date: 10.11.2024

Godoy F, Olivos-Hernández K, Stange C, & Handford M (2021) Abiotic stress in crop species: improving tolerance by applying plant metabolites. Plants 10(2):186

Guaragnella N, Bettiga M (2021) Acetic acid stress in budding yeast: from molecular mechanisms to applications. Yeast 38:391–400. https://doi.org/10.1002/yea.3651

Guaragnella N, Antonacci L, Giannattasio S, Marra E, Passarella S (2008) Catalase T and Cu, Zn-superoxide dismutase in the acetic acid-induced programmed cell death in *Saccharomyces cerevisiae*. FEBS Lett 582:210–214. https://doi.org/10.1016/j.febslet.2007.12.007

Guaragnella N, Bobba A, Passarella S, Marra E, Giannattasio S (2010) Yeast acetic acid-induced programmed cell death can occur without cytochrome c release which requires metacaspase YCA1. FEBS Lett 584:224–228. https://doi.org/10.1016/j.febslet.2009.11.072

Han SK, Sang Y, Rodrigues A, Wu MF et al (2012) The SWI2/SNF2 chromatin remodeling ATPase BRAHMA represses abscisic acid responses in the absence of the stress stimulus in *Arabidopsis*. Plant Cell 24:124892–124906. https://doi.org/10.1105/tpc.112.105114

Harris CJ, Amtmann A, Ton J (2023) Epigenetic processes in plant stress priming: open questions and new approaches. Curr Opin Plant Biol 75:102432. https://doi.org/10.1016/j.pbi.2023.102432

Hawrylak-Nowak B, Dresler S, Matraszek R (2015) Exogenous malic and acetic acids reduce cadmium phytotoxicity and enhance cadmium accumulation in roots of sunflower plants. Plant Physiol Biochem 94:225–234. https://doi.org/10.1016/j.plaphy.2015.06.012

He Y, Guo Z, Liu R, Li Q, He Y, Feng J (2024) Enhancing temperature resilience in dioecious *Salix myrtillacea*: the role of acetic acid in eco-physiological and microbial adaptations. Ind Crop Prod 218:118888. https://doi.org/10.1016/j.indcrop.2024.118888

Hossain MA, Bhattacharjee S, Armin SM et al (2015) Hydrogen peroxide priming modulates abiotic oxidative stress tolerance: insights from ROS detoxification and scavenging. Front Plant Sci 6:420. https://doi.org/10.3389/fpls.2015.00420

Hossain M, Hasanuzzaman M, Sohag M et al (2019) Acetate-induced modulation of ascorbate: glutathione cycle and restriction of sodium accumulation in shoot confer salt tolerance in Lens culinaris Medik. Physiol Mol Biol Plants 25:443–455. https://doi.org/10.1007/s12298-018-00640-6

Hossain MS, Abdelrahman M, Tran CD et al (2020) Insights into acetate mediated copper homeostasis and antioxidant defense in lentil under excessive copper stress. Environ Pollut 258:113544. https://doi.org/10.1016/j.envpol.2019.113544

Hossain MS, Abdelrahman M, Tran CD et al (2022) Modulation of osmoprotection and antioxidant defense by exogenously applied acetate enhances cadmium stress tolerance in lentil seedlings. Environ Pollut 308:119687. https://doi.org/10.1016/j.envpol.2022.119687

Isaji S, Yoshinaga N, Teraishi M et al (2018) Biosynthesis and accumulation of GABA in rice plants treated with acetic acid. J Pest Sci 43:214–219. https://doi.org/10.1584/jpestics.D18-036

Johnson R, Puthur JT (2021) Seed priming as a cost-effective technique for developing plants with cross tolerance to salinity stress. Plant Physiol Biochem 162:247–257. https://doi.org/10.1016/j.plaphy.2021.02.034

Johnston CS, Gaas CA (2006) Vinegar: medicinal uses and antiglycemic effect. MedGenMed 8(2):61
Johnston CS, Kim CM, Buller AJ (2004) Vinegar improves insulin sensitivity to a high-carbohydrate meal in subjects with insulin resistance or type 2 diabetes. Diabetes Care 27:281–282. https://doi.org/10.2337/diacare.27.1.281
Kappachery S, AlHosani M, Khan TA et al (2024) Modulation of antioxidant defense and PSII components by exogenously applied acetate mitigates salinity stress in *Avena sativa*. Sci Rep 14:620. https://doi.org/10.1038/s41598-024-51302-5
Kaur G, Tak Y, Asthir B (2022) Salicylic acid: a key signal molecule ameliorating plant stresses. Cereal Res Commun 50:617–626. https://doi.org/10.1007/s42976-021-00236-z
Khan MIR, Poor P, Janda T (2022) Salicylic acid: a versatile signaling molecule in plants. J Plant Growth Regul 41:1887–1890. https://doi.org/10.1007/s00344-022-10692-4
Khatam AS, Rastegar S, Khankahdani HH (2024) Modulating cold tolerance in Mexican lime (Citrus aurantifolia Swingle): a combined approach using melatonin, acetic acid and mannitol. Sci Hortic 338:113634. https://doi.org/10.1016/j.scienta.2024.113634
Kim JM, To T, Matsui A et al (2017) Acetate-mediated novel survival strategy against drought in plants. Nat Plants 3:17097. https://doi.org/10.1038/nplants.2017.97
Koc A, Markovic D, Ninkovic V, Martinez G (2020) Molecular mechanisms regulating priming and stress memory. In: Priming-mediated stress and cross-stress tolerance in crop plants. Elsevier, Amsterdam, pp 247–265. https://doi.org/10.1016/B978-0-12-817892-8.00016-7
Kong X, Guo Z, Yao Y et al (2022) Acetic acid alters rhizosphere microbes and metabolic composition to improve willows drought resistance. Sci Total Environ 844:157132. https://doi.org/10.1016/j.scitotenv.2022.157132
Kudo T, To TK, Kim JM (2023) Simple and universal function of acetic acid to overcome the drought crisis. Stress Biol 3:15. https://doi.org/10.1007/s44154-023-00094-1
Kurokawa H, Ito H, Matano D et al (2022) Acetic acid enhances the effect of photodynamic therapy in gastric cancer cells via the production of reactive oxygen species. J Clin Biochem Nutr 71:206–211. https://doi.org/10.3164/jcbn.22-34
Li C, Kong X, Luo Z et al (2021a) Exogenous application of acetic acid improves the survival rate of cotton by increasing abscisic acid and jasmonic acid contents under drought stress. Acta Physiol Plant 43:32. https://doi.org/10.1007/s11738-021-03202-8
Li S, He X, Gao Y, Zhou C, Chiang VL, Li W (2021b) Histone acetylation changes in plant response to drought stress. Genes 12:1409. https://doi.org/10.3390/genes12091409
Li M, Gao S, Luo J et al (2022) Mitigation effects of exogenous acetic acid on drought stress in Cunninghamia lanceolata. Plant Soil. https://doi.org/10.1007/s11104-022-05628-6
Liu S, Li X, Larsen DH, Zhu X, Song F, Liu F (2017) Drought priming at vegetative growth stage enhances nitrogen-use efficiency under post-anthesis drought and heat stress in wheat. J Agron Crop Sci 203:29–40. https://doi.org/10.1111/jac.12190
Llorens E, González-Hernández AI, Scalschi L, Fernández-Crespo E, Camañes G, Vicedo B, García-Agustín P (2020) Priming-mediated stress and cross-stress tolerance in crop plants: concepts and opportunities. Elsevier Inc., Amsterdam. ISBN 978-0-12-817893-5
Louis P, Flint HJ (2016) Formation of propionate and butyrate by the human colonic microbiota. Environ Microbiol 19:29–41. https://doi.org/10.1111/1462-2920.13589
Ludovico P, Sousa MJ, Silva MT et al (2001) *Saccharomyces cerevisiae* commits to a programmed cell death process in response to acetic acid. Microbiology 147:2409–2415. https://doi.org/10.1099/00221287-147-9-2409
Ludovico P, Rodrigues F, Almeida A, Silva MT, Barrientos A, Côrte-Real M (2002) Cytochrome c release and mitochondria involvement in programmed cell death induced by acetic acid in *Saccharomyces cerevisiae*. Mol Biol Cell 13:2598–2606. https://doi.org/10.1091/mbc.E01-12-0161
Lutts S, Benincasa P, Wojtyla L, Kubala S, Pace R, Lechowska K et al (2016) Seed priming: new comprehensive approaches for an old empirical technique. In: New challenges in seed biology-basic and translational research driving seed technology, vol 46. https://doi.org/10.5772/64420

Lynch JM (1977) Phytotoxicity of acetic acid produced in the anaerobic decomposition of wheat straw. J Appl Bacteriol 42:81–87. https://doi.org/10.1111/j.1365-2672.1977.tb00672.x
Lynch JM (1978) Production and phytotoxicity of acetic acid in anaerobic soils containing plant residues. Soil Biol Biochem 10:131–135. https://doi.org/10.1016/0038-0717(78)90083-4
Mahmud S, Kamruzzaman M, Bhattacharyya S et al (2023) Acetic acid positively modulates proline metabolism for mitigating PEG-mediated drought stress in maize and arabidopsis. Front Plant Sci 14:1167238. https://doi.org/10.3389/fpls.2023.1167238
Medrano Garcia JD, Ruiz-Femenia R, Caballero JA (2019) Revisiting classic acetic acid synthesis: optimal hydrogen consumption and carbon dioxide utilization. Comput Aided Chem Eng 46:145–150. https://doi.org/10.1016/B978-0-12-818634-3.50025-4
Mihaylova MM, Stratton MS (2019) Short chain fatty acids as epigenetic and metabolic regulators of neurocognitive health and disease. In: Ferguson BS (ed) Nutritional epigenomics. Academic Press, Cambridge, pp 381–397. https://doi.org/10.1016/B978-0-12-816843-1.00023-0
Mirfattahi Z, Eshghi S (2023) Acetic acid alleviates salinity damage and improves fruit yield in strawberry by mediating hormones and antioxidant activity. Erwerbs-obstbau 65:1403–1412. https://doi.org/10.1007/s10341-023-00840-9
Mirfattahi Z, Eshghi S, Gharaghani A et al (2022) Time and rate of acetate foliar spray can ameliorate adverse effect of NaCl stress on strawberry. S Afr J Bot 150:797–805. https://doi.org/10.1016/j.sajb.2022.08.034
Moffett JR, Puthillathu N, Vengilote R et al (2020) Acetate revisited: a key biomolecule at the nexus of metabolism, epigenetics, and oncogenesis, part 2: acetate and ACSS2 in health and disease. Front Physiol 11:580171. https://doi.org/10.3389/fphys.2020.580171
Morris K, Mackerness SAH, Page T et al (2000) Salicylic acid has a role in regulating gene expression during leaf senescence. Plant J 23:677–685. https://doi.org/10.1046/j.1365-313x.2000.00836.x
Mozgová I, Wildhaber T, Liu Q et al (2015) Chromatin assembly factor CAF-1 represses priming of plant defence response genes. Nat Plants 1:15127. https://doi.org/10.1038/nplants.2015.127
Mroczek S, Kufel J (2008) Apoptotic signals induce specific degradation of ribosomal RNA in yeast. Nucleic Acids Res 36:2874–2888. https://doi.org/10.1093/nar/gkm1100
Nair AU, Bhukya DPN, Sunkar R et al (2022) Molecular basis of priming-induced acquired tolerance to multiple abiotic stresses in plants. J Exp Bot 73:3355–3371. https://doi.org/10.1093/jxb/erac089
Narita T, Weinert BT, Choudhary C (2018) Functions and mechanisms of non-histone protein acetylation. Nat Rev Mol Cell Biol 20:156–174. https://doi.org/10.1038/s41580-018-0081-3
Ogawa D, Suzuki Y, Yokoo T et al (2021) Acetic-acid-induced jasmonate signaling in root enhances drought avoidance in rice. Sci Rep 11:6280. https://doi.org/10.1038/s41598-021-85355-7
Okabe S, Kodama Y, Cao H et al (2012) Topical application of acetic acid in cytoreduction of gastric cancer. A technical report using mouse model. J Gastroenterol Hepatol 27:40–48
Okabe S, Okamoto T, Zhao CM et al (2014) Acetic acid induces cell death: an in vitro study using normal rat gastric mucosal cell line and rat and human gastric cancer and mesothelioma cell lines. J Gastroenterol Hepatol 29:65–69. https://doi.org/10.1111/jgh.12775
Paulik FE, Hershman A, Coeur C et al (1973) Patent US 3769329
Peirats-Llobet M, Han SK, Gonzalez-Guzman M et al (2016) A direct link between abscisic acid sensing and the chromatin-remodeling ATPase BRAHMA via core ABA signaling pathway components. Mol Plant 9:136–147. https://doi.org/10.1016/j.molp.2015.10.003
Peng Y, Yang J, Li X, Zhang Y (2021) Salicylic acid: biosynthesis and signaling. Annu Rev Plant Biol 72:761–791. https://doi.org/10.1146/annurev-arplant-081320-092855
Pinto I, Cardoso H, Leão C, van Uden N (1989) High enthalpy and low enthalpy death in *Saccharomyces cerevisiae* induced by acetic acid. Biotechnol Bioeng 33:1350–1352. https://doi.org/10.1002/bit.260331019
Rahman M, Mostofa MG, Rahman A et al (2019) Acetic acid: a cost-effective agent for mitigation of seawater-induced salt toxicity in mung bean. Sci Rep 9:1–15. https://doi.org/10.1038/s41598-019-51178-w

Rasheed S, Bashir K, Kim JM, Ando M, Tanaka M, & Seki M (2018) The modulation of acetic acid pathway genes in Arabidopsis improves survival under drought stress. Sci Rep 8(1):7831
Rahman M, Mostofa MG, Keya SS et al (2021) Acetic acid improves drought acclimation in soybean: an integrative response of photosynthesis, osmoregulation, mineral uptake and antioxidant defense. Physiol Plant 172:334–350. https://doi.org/10.1111/ppl.13191
Rahman MM, Keya SS, Sahu A et al (2024) Acetic acid: a cheap but chief metabolic regulator for abiotic stress tolerance in plants. Stress Biol 4:34. https://doi.org/10.1007/s44154-024-00167-9
Ribeiro GF, Côrte-Real M, and Johansson B (2006) Characterization of DNA damage in yeast apoptosis induced by hydrogen peroxide, acetic acid, and hyperosmotic shock. Mol Biol Cell 17:4584–4591 https://doi.org/10.1091/mbc.e06-05-0475
Ritonga FN, Chen S (2020) Physiological and molecular mechanism involved in cold stress tolerance in plants. Plants 9(5):560. https://doi.org/10.3390/plants9050560
Sah SK, Reddy KR, Li J (2016) Abscisic acid and abiotic stress tolerance in crop plants. Front Plant Sci 7:571. https://doi.org/10.3389/fpls.2016.00571
Samad A, Azlan A, Ismail A (2016) Therapeutic effects of vinegar: a review. Curr Opin Food Sci 8:56–61. https://doi.org/10.1016/j.cofs.2016.03.001
Sanie Khatam A, Rastegar S, Aboutalebi Jahromi A, Hassanzadeh Khankahdani H, Bagherian SAA (2023) Physio-biochemical mechanisms of acetic acid involved in cold and freezing tolerance in Mexican lime plants (Citrus aurantifolia Swingle). Erwerbs-obstbau 65(5):1693–1704. https://doi.org/10.1007/s10341-023-00885-w
Schug ZT, Voorde JV, Gottlieb E (2016) The metabolic fate of acetate in cancer. Nat Rev Cancer 16:708–717. https://doi.org/10.1038/nrc.2016.87
Sedaghatmehr M, Thirumalaikumar VP, Kamranfar I et al (2018) A regulatory role of autophagy for resetting the memory of heat stress in plants. Plant Cell Environ 42:1054–1064. https://doi.org/10.1111/pce.13426
Sen A, Puthur JT (2020) Seed priming-induced physiochemical and molecular events in plants coupled to abiotic stress tolerance: an overview. In: Priming-mediated stress and cross-stress tolerance in crop plants. Elsevier. https://doi.org/10.1016/B978-0-12-817892-8.00018-0
Sokolov S, Knorre D, Smirnova E et al (2006) Ysp2 mediates death of yeast induced by amiodarone or intracellular acidification. Biochim Biophys Acta 1757:1366–1370. https://doi.org/10.1016/j.bbabio.2006.07.005
Sousa M, Duarte AM, Fernandes TR et al (2013) Genome-wide identification of genes involved in the positive and negative regulation of acetic acid-induced programmed cell death in *Saccharomyces cerevisiae*. BMC Genomics 14:838. https://doi.org/10.1186/1471-2164-14-838
Spencer DF, Ksander GG (1995) Differential effects of the microbial metabolite, acetic acid, on sprouting of aquatic plant propagules. Aquat Bot 52:107–119. https://doi.org/10.1016/0304-3770(95)00488-L
Sun T, Zhang J, Zhang Q et al (2022) Exogenous application of acetic acid enhances drought tolerance by influencing the MAPK signaling pathway induced by ABA and JA in apple plants. Tree Physiol 42:1827–1840. https://doi.org/10.1093/treephys/tpac034
Tadege M, Dupuis II, Kuhlemeier C (1999) Ethanolic fermentation: new functions for an old pathway. Trends Plant Sci 4:320–325. https://doi.org/10.1016/s1360-1385(99)01450-8
Temel A, Kosesakal T (2024) Effects of acetic acid treatment on growth and pigment contents in barley. Acta Bot Croat 83:145–152. https://doi.org/10.37427/botcro-2024-015
Terasaki M, Ito H, Kurokawa H, Tamura M, Okabe S, Matsui H, & Hyodo I (2018) Acetic acid is an oxidative stressor in gastric cancer cells. J Clin Biochem Nutr 63(1):36–41
Temel A, Janack B, Humbeck K (2017) Drought stress-related physiological changes and histone modifications in barley primary leaves at HSP17 gene. Agronomy 7:43. https://doi.org/10.3390/agronomy7020043
Tirnaz S, Batley J (2019) Epigenetics: potentials and challenges in crop breeding. Mol Plant 12:1309–1311. https://doi.org/10.1016/j.molp.2019.09.006

Utsumi Y, Utsumi C, Tanaka M et al (2019) Acetic acid treatment enhances drought avoidance in cassava (*Manihot esculenta* Crantz). Front Plant Sci 10:521. https://doi.org/10.3389/fpls.2019.00521

Van Overbeek J, Blondeau R (1954) Mode of action of phytotoxic oils. Weeds 3:55–65. https://doi.org/10.2307/4040131

Wibowo A, Becker C, Marconi G et al (2016) Hyperosmotic stress memory in Arabidopsis is mediated by distinct epigenetically labile sites in the genome and is restricted in the male germline by DNA glycosylase activity. elife 5:e135465. https://doi.org/10.7554/eLife.13546

Wojtyla Ł, Paluch-Lubawa E, Sobieszczuk-Nowicka E, Garnczarska M (2020) Drought stress memory and subsequent drought stress tolerance in plants. In: Hossain MA, Liu F, Burritt DJ, Fujita M, Huang B (eds) Priming-mediated stress and cross-stress tolerance in crop plants. Elsevier, pp 115–131. https://doi.org/10.1016/b978-0-12-817892-8.00007-6

Wong JMW, de Souza R, Kendall CWC et al (2006) Colonic health: fermentation and short chain fatty acids. J Clin Gastroenterol 40:235–243. https://doi.org/10.1097/00004836-200603000-00015

Xie X, Weiss DJ, Weng B et al (2013) The short-term effect of cadmium on low molecular weight organic acid and amino acid exudation from mangrove (Kandelia obovata (S., L.) Yong) roots. Environ Sci Pollut Res 20:997–1008. https://doi.org/10.1007/s11356-012-1031-9

Yakhin OI, Lubyanov AA, Yakhin IA, Brown PH (2017) Biostimulants in plant science: a global perspective. Front Plant Sci 7:2049. https://doi.org/10.3389/fpls.2016.02049

Zhang X, Huang B (2020) Drought priming-induced heat tolerance: metabolic pathways and molecular mechanisms. In: Priming-mediated stress and cross-stress tolerance in crop plants. Elsevier, pp 149–160. https://doi.org/10.1016/B978-0-12-817892-8.00009-X

Zhang J, Zhang Q, Xing J et al (2021) Acetic acid mitigated salt stress by alleviating ionic and oxidative damages and regulating hormone metabolism in perennial ryegrass (*Lolium perenne* L.). Grass Res 1:1–10

Zhu L, Guo J, Sun Y et al (2021) Acetic acid-producing endophyte Lysinibacillus fusiformis orchestrates jasmonic acid signaling and contributes to repression of cadmium uptake in tomato plants. Front Plant Sci 12:670216. https://doi.org/10.3389/fpls.2021.670216

Zhu X, Wang Y, Soaita I et al (2023) Acetate controls endothelial-to-mesenchymal transition. Cell Metab 35:1163–1178. https://doi.org/10.1016/j.cmet.2023.05.010

Zuo Z, Zhu Y, Bai Y, Wang Y (2012) Acetic acid-induced programmed cell death and release of volatile organic compounds in *Chlamydomonas reinhardtii*. Plant Physiol Biochem 51:175–184. https://doi.org/10.1016/j.plaphy.2011.11.003

Chapter 6
Salicylic Acid and Its Effects on Plants and Alleviating Abiotic Stress

Moh Sajid Ansari, Adnan Khan, Abrar A. Khan, and Heba I. Mohamed

Abstract Abiotic stresses are potent growth and yield inhibitors for horticultural crops. Phytohormone treatments will be a potential solution against abiotic stresses. Salicylic acid (SA), a phytohormone, is pivotal in enhancing plants and alleviating various abiotic stresses by modulating key physiological and biochemical processes. As an effective plant growth regulator, SA enhances morpho-physiological parameters, including plant height, root and shoot growth, fresh and dry biomass, photosynthetic pigments, and leaf area. Through its ability to regulate antioxidant enzymes and reactive oxygen species (ROS), SA minimizes oxidative damage, improves water use efficiency, and promotes photosynthetic efficiency. SA also interacts with other signaling molecules and phytohormones, influencing gene expression related to stress responses. For instance, exogenous SA has been shown to enhance drought and salinity tolerance by improving chlorophyll content, $K+/Na+$ ratios, mineral uptake, and antioxidant defense in crops. Additionally, SA application during heat and heavy metal stress regulates antioxidant activity, decreases malondialdehyde content, and enhances growth, as seen in crops. The combination of SA with other regulators, such as melatonin and silicon, further enhances stress resilience by improving nutrient uptake and reducing the accumulation of heavy metals. These findings emphasize SA's potential as a critical tool for improving plant performance, resilience, and productivity under increasingly adverse environmental conditions, offering valuable strategies for sustainable agriculture.

Keywords Salicylic acid · Phytohormone · Abiotic stress alleviation · Antioxidants · Reactive oxygen species · Sustainable agriculture

M. S. Ansari (✉) · A. Khan · A. A. Khan
Department of Botany, Faculty of Life Sciences, Aligarh Muslim University, Aligarh, Uttar Pradesh, India

H. I. Mohamed
Biological and Geological Science Department, Faculty of Education, Ain Shams University, Cairo, Egypt

H. I. Mohamed (ed.), *Elicitors for Sustainable Crop Production*, Sustainability Sciences in Asia and Africa, https://doi.org/10.1007/978-981-95-3533-0_6

6.1 Introduction

Abiotic stress poses a significant threat to global agriculture, severely impacting crop growth and productivity. These stresses, exacerbated by rapid global warming, are expected to increase in intensity, further reducing the availability of fertile soils and leading to substantial yield losses (Subashchandrabose et al. 2013). The primary challenge for current and future agriculture is to enhance food production to meet the demands of a growing population, making mitigation of abiotic stress effects a critical global concern (Boscaiu and Fita 2020). Abiotic stresses, including saline stress, drought stress, heat stress, cold stress, and heavy metals, are becoming increasingly prevalent due to climate change and disrupt nearly all plant functions, from seed germination to maturity, resulting in significant economic losses in horticultural crops (Altaf et al. 2022a; Chaudhry and Sidhu 2022; Liu et al. 2022). To combat these challenges, researchers emphasize understanding the interplay between environmental factors and physiological processes to enhance agricultural techniques and crop resilience (Shahid et al. 2022; Qin et al. 2022).

Global warming, a pressing concern, has significantly increased ambient temperatures, posing further threats to plant growth and the environment. Elevated temperatures disrupt physiological processes in plants, leading to substantial growth and yield losses (Ashraf and Harris 2013). Studies indicate that a 1 °C increase in seasonal temperature can result in over 17% yield losses (Al-Shoaibi 2020). Heat stress adversely affects seed germination, canopy growth, photosynthesis, and chlorophyll fluorescence, exacerbating the challenges faced by agriculture (Yi et al. 2019; Rasheed et al. 2021). The adverse effects of salinity stress pose a significant threat to both crop yield and global food security (Rasheed et al. 2021; Fang et al. 2021). Approximately 7% of the world's soils (1 billion hectares) are affected by salinity, and it is projected that by 2050, half of the world's arable land could be salinized due to improper fertilization, irrigation practices, and industrial activities (Hopmans et al. 2021; Ivushkin et al. 2019).

To address the growing challenges of abiotic stress, improving crop tolerance is essential. However, most studies have focused on plant responses to single stress conditions, despite the reality that multiple stresses often occur simultaneously in the field, leading to greater crop losses (Prasch and Sonnewald 2015; Aycan et al. 2021). Combined stresses are particularly lethal, necessitating strategies to mitigate their harmful impacts on plants (Pandey et al. 2017). Among the various approaches, the application of plant growth regulators, such as salicylic acid (SA), has proven to be effective in enhancing plant stress tolerance (Cohen et al. 2021).

Salicylic acid (SA), a phytohormone, is essential for controlling plant growth, development, and defense reactions against numerous abiotic stresses (Shaukat et al. 2022; Chen et al. 2023). The present chemical exerts a substantial impact on the photosynthetic apparatus in leaves, root growth, antioxidant defense mechanisms, secondary metabolite synthesis, and proline metabolism, therefore augmenting the plant's ability to withstand environmental pressures (Prakash et al. 2021; Santisree et al. 2020). Systematic research indicates that SA improves plant resistance to many abiotic stresses, including salt, drought, temperature fluctuations, and heavy metal toxicity (Liu et al. 2022; Khan et al. 2022). Having a comprehensive understanding of the role of SA in abiotic stress tolerance is crucial for developing sustainable agricultural practices, particularly in the face of climate change, when the severity of stress may intensify. Salicylic acid, represented by the chemical formula C7H6O3, is recognized as the sixth plant hormone and is categorized as a phenolic compound organically produced by plants. The chemical being studied has a hydroxyl group, an aromatic ring, and a carboxylic group, as described by Ding and Ding (2020) and Bagautdinova et al. (2022). The chemical synthesis of saccharide (SA) by Kolbe and his colleagues in 1859 enhanced its accessibility and economic viability (Maruri-Lopez et al. 2019; Ding and Ding 2020). Finally, Felix Hoffmann's discovery that acetylsalicylic acid decreased stomach irritation led to the widespread global use of aspirin (Klessig et al. 2018). Surprisingly, the role of SA as a secondary metabolite in plant biology was not acknowledged until the latter half of the twentieth century (Wani et al. 2017; Sharma et al. 2020).

Salicylic acid is essential for stimulating plant defense systems, enhancing photosynthesis, and controlling growth and development in stressful situations. It also functions as a key signaling molecule in response to various abiotic stresses, including salinity, drought, heat, cold, and heavy metal toxicity. Additionally, it plays a role in developmental pathways that are necessary for typical plant development (Aycan et al. 2021; Zhang and Sonnewald 2017). The phytochemical SA has been demonstrated to improve plant tolerance through the regulation of ion equilibrium, preservation of cell membrane integrity, and modulation of the antioxidant system (Hassan et al. 2021). Given the intricate characteristics of abiotic stresses and their significant impact on agriculture, it is crucial to comprehend the regulatory mechanisms of plant hormones such as SA to devise methods to enhance crop productivity and resilience in response to environmental stresses. Today, SA is recognized as an essential phytohormone and signaling molecule that greatly improves plant stress tolerance, making it a primary area of interest in the advancement of robust crops for future agricultural systems (Fig. 6.1).

This chapter aims to explore how salicylic acid (SA) improves plant tolerance to various abiotic stresses, such as salinity, drought, extreme temperatures, and heavy metal toxicity. It examines the biochemical and physiological roles of SA in enhancing stress resilience, including its impact on antioxidant defense, photosynthesis, and gene regulation. It also discusses the practical application of SA in developing stress-resistant crops, contributing to sustainable agriculture and global food security in the face of climate change.

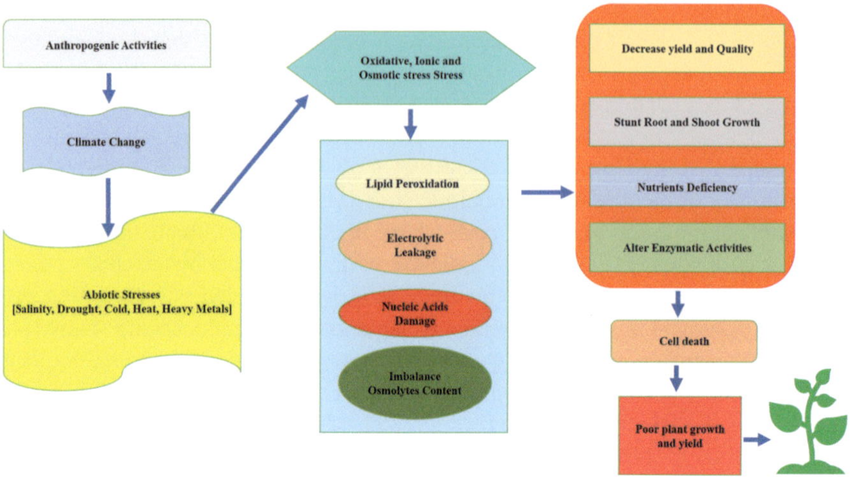

Fig. 6.1 Effects of abiotic stress factors on horticultural crop performance

6.2 Biosynthesis and Signaling of SA Under Environmental Stress Conditions

Salicylic acid (SA) is naturally produced in all plants and serves as an essential biological signaling molecule (Khan et al. 2015). Two processes, namely, the shikimic acid and malonic acid pathways, are responsible for its production (Lefevere et al. 2020). The shikimic acid pathway, alternatively referred to as the isochorismate (IC) pathway, only occurs within the chloroplast (Fig. 6.2). Through the action of the enzyme isochorismate synthase (ICS), chorismate is converted into isochorismate (Ding and Ding 2020; Lefevere et al. 2020). It is believed that around 90% of the sarcoplasmic acid (SA) produced during stress responses is synthesized by this cellular route (Khan et al. 2015).

The phenylalanine ammonia-lyase (PAL) pathway occurs entirely in the cytosol (Ding and Ding 2020; Janda et al. 2020). Through this enzyme pathway, chorismate mutase first converts chorismate into prephenic acid, which is then reduced to L-phenylalanine. According to Lefevere et al. (2020), the enzyme PAL facilitates the transformation of L-phenylalanine into trans-cinnamic acid (t-CA). Biochemical compounds that can be produced from t-CA include benzaldehyde and ortho-coumaric acid. Direct synthesis of salicylic acid from ortho-coumaric acid is possible. The enzyme aldehyde oxidase facilitates the conversion of benzoic acid from benzaldehyde. By the action of benzoic acid 2-hydroxylase, this benzoic acid is transformed into SA (Maruri-Lopez et al. 2019; Ding and Ding 2020).

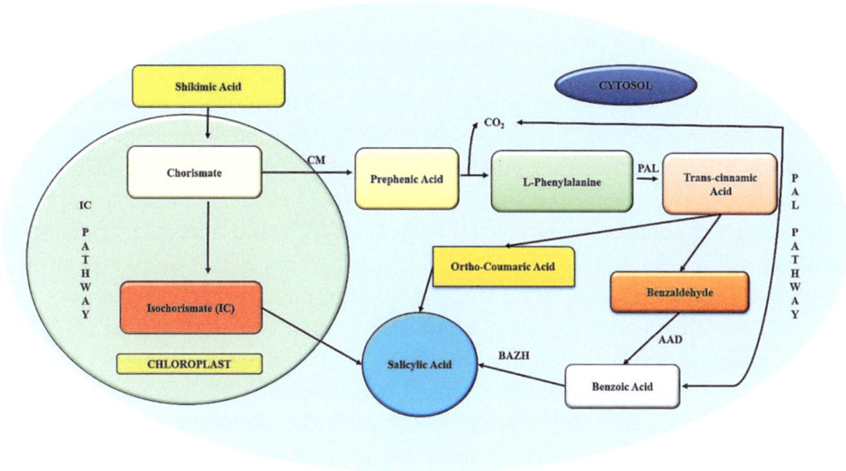

Fig. 6.2 Schematic representation of salicylic acid biosynthesis pathways in plants

6.3 Salicylic Acid: Its Role in Plants and Abiotic Stress Alleviation

Salicylic acid, well acknowledged for its therapeutic uses in human well-being, has become a crucial hormone in plant physiology, namely in controlling growth, development, and reaction to nonliving pollutants. Recent studies have highlighted SA's essential role as a signaling molecule, crucial for enhancing plant resilience under challenging environmental conditions (Lu 2009; Kaur et al. 2022). It has been shown to regulate a wide range of physiological, morphological, and metabolic processes, including the upregulation of antioxidant systems, protection of the photosynthetic apparatus, and promotion of secondary metabolite production, all of which contribute to improved stress tolerance in horticultural crops (Hernández et al. 2017; Fu et al. 2021).

6.3.1 Salinity Stress Alleviation Through Salicylic Acid

Soil salinity is a significant environmental challenge, particularly in horticulture, severely limiting crop productivity. The growing concern over food security has brought soil salinity into sharp focus, with research indicating that it affects approximately 6% of irrigated land and over 20% of all cultivated fields worldwide (Mir and Somasundaram 2021; Pirasteh-Anosheh et al. 2016). The impact of salinity on horticultural crops is profound, leading to impaired cell division, genotoxicity, oxidative damage, nutrient deficiencies, and membrane disruption, ultimately restricting plant growth, yield, and quality (Saraf et al. 2018).

Plants have evolved various strategies to cope with salt stress, and SA has been identified as a key molecule in modulating these defense mechanisms. Researchers have confirmed SA's protective effects in various horticultural crops, including tomatoes, spinach, strawberries, jujube, eggplant, and cucumbers, where it has been shown to improve growth characteristics, enhance chlorophyll content, and reduce electrolyte leakage under saline conditions (Kim et al. 2017; Wen et al. 2023).

In cucumber, for instance, the application of SA (1 mM) significantly improved leaf chlorophyll content, relative water content (RWC), and mineral nutrient accumulation, while also reducing electrolyte leakage (Sahu 2013). Similarly, in olive plants, SA supplementation was found to enhance photosynthetic rates, root volume, and overall vegetative growth under salinity stress (Aliniaeifard et al. 2016). Further studies on garlic, calendula, and tomato plants have shown that SA application not only mitigates the adverse effects of salinity but also promotes growth, increases bulb size, and improves physiological traits, including reduced sodium accumulation and enhanced potassium levels (Shama et al. 2016; Mimouni et al. 2016). The accumulation, growth, and biomass production of osmolytes in spinach are greatly enhanced by SA. Additionally, SA has been shown to upregulate antioxidant enzyme activity, hence mitigating the adverse impacts of salinity (Nigam et al. 2022). The role of SA in enhancing stress tolerance extends to other crops such as potatoes and strawberries, where it has been shown to improve growth characteristics, protein content, and antioxidant enzyme activity under saline conditions (Sajid and Aftab 2012; Roshdy et al. 2021).

The results of this study highlight the significance of salicylic acid (SA) in reducing oxidative damage and boosting abiotic stress tolerance in short-season crops. This suggests potential pathways for enhancing plant resilience against growing environmental stressors (Radhakrishnan and Lee 2013; Altaf et al. 2022a) (Table 6.1).

6.3.2 Drought Stress Alleviation Through Salicylic Acid

Elevated temperatures from global warming have caused substantial water evaporation, exacerbating the prevailing drought conditions. This environmental issue presents a significant risk to the development and yield of horticulture crops (Toscano et al. 2019; Hanaka et al. 2021). The adverse effects of drought stress on plant performance are attributed to its influence on soil conditions, causing disturbances in the regular metabolic processes of plants (Kour et al. 2022). The repercussions encompass lowered seed germination, inhibited seedling growth, reduced branch length, nutritional imbalances, and reduced biomass production, all of which result in decreased agricultural output (Rahdari and Hoseini 2012). Furthermore, the environmental conditions of drought can impede the normal functioning of photosynthesis and pigmentation in leaves, leading to the decline of nucleic acid metabolism and protein synthesis, finally resulting in plant dysplasia (Tiwari et al. 2020).

Table 6.1 Impact of salicylic acid on various plant species subjected to abiotic stress

Plant species	Abiotic stresses	SA concentrations	Impacts of SA on various parameters	References
Salinity stress				
Pistacia vera L. and *P. atlantica* L.	Salinity stress	0.5–1.0 mM	The concentration of H_2O_2 was elevated, while the index of lipid peroxidation showed a reduction. Additionally, growth and various physiological and biochemical activities improved	Gharibiyan et al. (2023)
Lactuca sativa L.	Salinity stress	0.25, 0.50, or 1.00 mM	Decreased levels of malondialdehyde (MDA) were noted in conjunction with an elevation in flavonoid and phenolic concentrations, nutrient absorption, and antioxidant enzyme activity	Kusvuran and Yilmaz (2023)
Hypericum perforatum L.	Salinity stress	0.25 mM	Photosynthetic performance improved while endogenous Abscisic Acid (ABA) levels declined, with upregulated expression of genes involved in melatonin biosynthesis	Kwon et al. (2023)
Phaseolus vulgaris L.	Salinity stress	0.75–1.50 mM	Augmented growth regulators (auxins, cytokinins, and gibberellins), characteristics of growth, membrane stability, macronutrients, proline concentrations, and biochemical activity	Youssef et al. (2023)
Vigna unguiculata L.	Salinity stress	100 ppm	Improved growth parameters, along with increased nitrogen, phosphorus, potassium, and enhanced anatomical characteristics	El-Taher et al. (2021)
Rosa damascene L.	Salinity stress	0.5, 1, and 2 mM	Growth, as well as physiological and biochemical traits, showed significant improvement	Omidi et al. (2022)
Mentha pulegium L.	Salinity stress	0.5 mM	Increases in both growth and biochemical characteristics were noted	Azad et al. (2021)
Salvia offcinalis L.	Salinity stress	0.5 mM	Significant increases were observed in the levels of calcium, potassium, phosphorus, essential oil, chlorophyll content, and peltate gland density	Es-sbihi et al. (2021)
Pisum sativum L.	Salinity stress	1.0 mM	Growth and photosynthesis showed enhancements, with upregulation of Na+/H+ anti-porters (SOS1, NHX1)	Ahmad et al. (2020)

(continued)

Table 6.1 (continued)

Plant species	Abiotic stresses	SA concentrations	Impacts of SA on various parameters	References
Triticum aestivum L.	Salinity stress	0.5 mM	Growth parameters, biochemical traits, and levels of inorganic solutes improved	Loutfy et al. (2020)
Lycopersicon esculentum L.	Salinity stress	0.5–1.0 mM	Fruit size (length and diameter), the number of fruits per plant, and dry matter content all increased	Naeem et al. (2020)
Triticum aestivum L.	Salinity stress	0.5–0.75 mM	Antioxidant enzyme activity, osmoprotectant levels, and potassium uptake were upregulated	Alsahli et al. (2019)
Carthamus tinctorius L.	Salinity stress	50–100 mg L^{-1}	Salicylic acid (SA) significantly boosted the total phenolics, flavonoids, and flavanols content, along with antioxidant activity under in vitro conditions	Golkar et al. (2019)
Solanum lycopersicum L.	Salinity stress	10 mM	Enhanced growth, sugar concentration, net photosynthetic rate, and stomatal conductance	Gharbi et al. (2018)
Oryza sativa	Salinity stress	0.5–1.0 mM	Increased expression of *GSNOR1* and Oryza sativa Nitric Oxide Associated (gene) *(OsNOA)* genes was accompanied by heightened antioxidant enzyme activity	Kim et al. (2018)
Tanacetum parthenium L.	Salinity stress	200–300 ppm	Enhanced essential oil content and antioxidant enzyme activity	Mallahi et al. (2018)
Oryza sativa L.	Salinity stress	1.0 mmol L^{-1}	Both endogenous salicylic acid (SA) levels and antioxidant enzyme activity showed an upward trend	Jini and Joseph (2017)
Helianthus annuus L.	Salinity stress	200 mg L^{-1}	K+ uptake increased along with a higher Soil Plant Analysis Development (chlorophyll content meter reading) (SPAD) value (chlorophyll content)	Noreen et al. (2020)
Beta vulgaris L.	Salinity stress	0, 0.5, and 1 mM	All parameters, including chlorophyll content, leaf area, stomatal conductance, photosynthetic efficiency (Quantum Yield of Photosystem II (ΦPSII) (FPSII), Fv/Fm) proline, and total soluble sugars, increased, while membrane permeability decreased	Cheraghabadi et al. (2016)

(continued)

Table 6.1 (continued)

Plant species	Abiotic stresses	SA concentrations	Impacts of SA on various parameters	References
Lens esculenta L.	Salinity stress	0.5 mM	Pyrroline-5-carboxylate reductase and γ-glutamyl kinase increased, but proline oxidase activity dropped. Also increased were glycine betaine synthase enzymes	Misra and Saxena (2009)
Matricaria chamomilla L.	Salinity stress	50 mM	Total phenolic content increased, with no change in root lignin levels. PAL and shikimate dehydrogenase activities were elevated	Kováčik et al. (2009)
Cucumis sativus L.	Salinity stress	0.25, 0.50, and 1.00 mM	Growth characteristics and nutrient content increased, while membrane permeability decreased	Yildirim et al. (2008)
Hordeum vulgare L.	Salinity stress	0, 0.5–1 mg L^{-1}	Increasing plant height, root and shoot biomass (fresh and dried weights), length, photosynthetic pigments, leaf area, and yield improves morpho-physiological parameters	Hanif et al. (2024)
Drought stress				
Vicia faba L.	Drought stress	0.5 mM	Modulation of antioxidant enzyme activity led to a reduction in oxidative damage	Dawood et al. (2022)
Eriocephalus africanus L.	Drought stress	1-, 2- and 3-mM SA; foliar spray	The generation of essential oils, phenolic compounds, and radical scavenging activity has been increased	Khalil et al. (2022)
Mentha piperita L.	Drought stress	0, 1, 2, and 2.5 mM; foliar spray	Elevated levels of essential oil, total phenolic content, and dry leaf weight were seen in the plant	Abdi and Karami (2020)
Portulaca oleracea L.	Drought stress	0.5–1 mM	Improved gas exchange, enhanced secondary metabolite accumulation, and better antioxidant defense, alongside regulation of membrane lipid fatty acids	Saheri et al. (2020)
Carthamus tinctorius L.	Drought stress	250 mM; foliar application	Enhancements in photosynthetic rates, anthocyanin concentration, and phenylalanine ammonia-lyase activity are all distinctive features of this plant	Chavoushi et al. (2020)

(continued)

Table 6.1 (continued)

Plant species	Abiotic stresses	SA concentrations	Impacts of SA on various parameters	References
Ammi visnaga L.	Drought stress	1–2 mM	Increasing fruit yield per plant, polyphenolic content, and radical scavenging activity	Osama et al. (2019)
Brassica rapa L.	Drought stress	1.5 mM SA; foliar spray	Elevated proline levels contribute to the maintenance of redox homeostasis	La et al. (2019)
Cicer arietinum L.	Drought stress	150 mg/L; foliar application	All parameters, including relative water content, phenolic compounds, chlorophyll content, fluorescence, shoot and root biomass, sugars, and proteins, demonstrated increases	Khan et al. (2019)
Triticum aestivum L.	Drought stress	10- and 100-mM SA; seed priming	Antioxidant enzyme activities improved, contributing to better growth under stress conditions	Kolupaev et al. (2018)
Olea europaea L.	Drought stress	10, 100, and 1000 mM	Enhanced reactive oxygen species (ROS) detoxification, regulation of macro and micronutrient uptake, and improved cellular homeostasis	Brito et al. (2018)
Fragaria × *ananassa* L.	Drought stress	0.01–0.05 mM	Elevated amounts of total soluble carbs, total proteins, and free proline, at the same time, decreased lipid peroxidation	Mozafari et al. (2018)
Sesamum indicum L.	Drought stress	0.6 mM SA; foliar spray	Increased capsule number per plant, seed density per square meter, proline concentration, and relative water content	Najafabadi and Ehsanzadeh (2017)
Triticum aestivum L.	Drought stress	1 mM SA; foliar spray.	Decreased electrolyte leakage, improved lipid peroxidation, increased catalase (CAT), peroxidase (POD), and ascorbate peroxidase (APX) activities, and reduced MDA content	Sedaghat et al. (2017)
Hibiscus syriacus L.	Drought stress	500 μmol L^{-1}	The external application of melatonin (MT) or salicylic acid (SA) improves drought tolerance in hibiscus plants by modulating the photosynthetic system, antioxidant enzyme activities, and activating drought-response genes. Between the two, SA appears more effective in providing drought resilience	Yan et al. (2024)

(continued)

Table 6.1 (continued)

Plant species	Abiotic stresses	SA concentrations	Impacts of SA on various parameters	References
Lactuca sativa L.	Drought stress	SA (1.0 mM)	Increased growth, yield, chlorophyll a, b, carotenoids, K+/Na+ and Ca2+/Na+ ratios, nutrient absorption, and antioxidant capability are all indicators	Kiremit et al. (2024)
Solanum lycopersicum L.	Drought stress	200 mg L^{-1} or 250 mg L^{-1}	This concentration (250 mg/L) increases antioxidant enzyme activity and gene expression, as well as plant development-related morpho-physiological parameters	Rai et al. (2024)
Sesamum indicum L.	Drought stress	10–5 M SA; seed priming	Increased biomass and a higher percentage of seeds that germinate effectively	Silva et al. (2017)
Heat stress				
Capsicum frutescens L.	Heat stress	1–2 mM	The biomass, growth metrics, relative water content, photosynthetic pigments, and enzyme activity of the plant were all improved	Saeed et al. (2023)
Pisum sativum L.	Heat stress	50 mg L^{-1}	Enhancements were made to both the germination rate and physiological properties	Tamindžić et al. (2023)
Capsicum annuum L.	Heat stress	0.1, 0.20, and 0.50 mM	Enhanced biochemical parameters, increased antioxidant enzyme activity, and enhanced overall fruit yield were observed	Preet et al. (2023)
Pleurotus ostreatus L.	Heat stress	0.05 mmol L^{-1}	The buildup of secondary metabolites and antibiotics has increased during the process	Hu et al. (2022)
Brassica napus L.	Heat stress	0.13 mM	Enhanced all physiological parameters and antioxidant enzyme activities	Ghani et al. (2021)
Medicago sativa L.	Heat stress	0.25–0.5 mM	Improvements were made to growth factors as well as the effectiveness of photosynthesis	Wassie et al. (2020)
Lablab purpureus L.	Heat stress	0.5–1 mM	Photosynthetic efficiency, physiological characteristics, leaf anatomy, pollen viability, yield, antioxidant activity, enzyme function, and stress-responsive gene expression improved	Rai et al. (2020)

(continued)

Table 6.1 (continued)

Plant species	Abiotic stresses	SA concentrations	Impacts of SA on various parameters	References
Capsicum annuum L.	Heat stress	0.002, 0.01, 0.05, and 0.25 mM	The activation of the antioxidant defense system and osmotic regulation helped maintain the effectiveness of photosynthesis	Zhang et al. (2020)
Euphorbia pulcherrima L.	Heat stress	200–400 mM	Improvements were made to the morphological characteristics of plants as well as the antioxidant enzyme activity	Lin et al. (2019)
Solanum lycopersicum L.	Heat stress	1 mM	Enhanced biochemical parameters, increased photosynthetic efficiency, regulated antioxidant defense, decreased electrolyte leakage, and increased efficiency in water use are all benefits of this plant	Jahan et al. (2019)
Oryza sativa L.	Heat stress	100 mg L^{-1}	Improvements in seedling characteristics, total soluble sugars and proteins, nitrogen, phosphorus, and potassium content, as well as enhanced thermo-tolerance, were seen in genotypes	Akasha et al. (2019)
Oryza sativa L.	Heat stress	0.01, 0.1, 1, 10, and 50 mM	Pollen viability and antioxidant enzyme activity increased, with elevated H_2O_2 content and reduced ROS	Feng et al. (2018)
Triticum aestivum L.	Heat stress	100 mM	Salicylic acid (SA) modulated heat-stable genes and proteins to protect the starch production pathway from heat stress	Kumar et al. (2015)
Cold stress				
Vitis vinifera L.	Cold stress	0.1–0.5 mM	Reduced chlorophyll and carotenoid levels while raising the concentrations of H_2O_2, proline, membrane stability index (MSI), MDA, and antioxidants	Jalili et al. (2023)
Zea mays L.	Cold stress	50 mg L^{-1}	Reduced relative electrolyte conductivity (EC), ROS, and MDA, and increased antioxidants, proline, RWC, and osmotic adjustment capability	Zhang et al. (2021)

(continued)

Table 6.1 (continued)

Plant species	Abiotic stresses	SA concentrations	Impacts of SA on various parameters	References
Triticum aestivum L. cv.	Cold stress	100 mM containing 0.1% ethanol and 0.05% tween 20	Enhanced plant fresh and dry weight, plant height, plant yield, gas exchange, chlorophyll fluorescence, malondialdehyde (MDA) levels, and antioxidant activity	Wang et al. (2021)
Cucumis sativus L.	Cold stress	1.0 mM SA, 0.1 mM	Cold stress genes were expressed more frequently, and antioxidant activity was increased	Pan et al. (2020)
Triticum aestivum L. cv.	Cold stress	100 mmol SA	Improved cold tolerance, increased H_2O_2 and MDA content, enhanced antioxidant activity, proline content, soluble protein, and cold stress gene transcripts	Ignatenko et al. (2019)
Spinacia oleracea L.	Cold stress	0.5–1 mM	Freezing tolerance improved with reduced ion leakage and oxidative stress, along with elevated proline, antioxidants, and Ascorbic Acid (AsA), indicating the involvement of NO and H_2O_2 signaling	Shin et al. (2018)
Phaseolus vulgaris L.	Cold stress	0–3 mM SA	Enhanced yield, growth, proline, soluble sugars, protein, ascorbic acid content, MDA, antioxidant activity, and expression of cold-responsive genes	Soliman et al. (2018)
Catharanthus roseus L.	Cold stress	0.5, 1- and 2-mM	Increased nodes per plant, plant height, flowering time, antioxidant activity, and gene expression	Khandan-Mirkohi et al. (2017)
Anthurium andraeanum L.	Cold stress	1, 2, and 4 mM	Reduced chilling injury, electrolyte leakage (EL), increased PAL activity, reduced Polyphenol Oxidase (PPO) activity, enhanced (2,2-Diphenyl-1-picrylhydrazyl (used for free radical scavenging activity)) (DPPH) scavenging, glycine betaine content, proline, and antioxidant activity	Aghdam et al. (2016)
Dendrobium officinale L.	Cold stress	0.5, 1.0, 1.5, and 2.0 mmol L^{-1}	Increased levels of MDA, antioxidant activity, net photosynthetic rate, fluorescence of chlorophyll, and D1 protein as measured by SDS-PAGE and western blotting in thylakoids	Huang et al. (2016)

(continued)

Table 6.1 (continued)

Plant species	Abiotic stresses	SA concentrations	Impacts of SA on various parameters	References
Brassica napus L.	Cold stress	50, 500, and 5000 mM	Decreased photosynthetic pigments, MDA, proline, protein, and CAT content, and increased antioxidant activity	Keshavarz et al. (2016)
Zea mays L.,	Cold stress	0.3–0.5 mM	Enhanced levels of free o-hydroxy cinnamic acid in plants, elevated putrescine content, reduced spermidine levels in seeds, and increased spermine and spermidine levels in roots	Szalai et al. (2016)
Citrullus lanatus L.	Cold stress	5, 10, 50, 100, 500, 1000, and 2000 mM	Enhanced levels of antioxidant activity, photosynthetic activity, and gene expression were observed	Cheng et al. (2016)
Triticum aestivum L.	Cold stress	1 mM SA and 0.1 mM SNP	Decreased MDA, H_2O_2, and O_2^- content, while antioxidant activity increased	Esim and Atici (2015)
Heavy metal stress				
Allium cepa L.	Heavy stress	Pb^{2+} (50 µM and 500 µM)	Reduced DNA damage by increasing the activity of superoxide dismutases, catalases, ascorbate, and guaiacol peroxidases	Kaur et al. (2021)
Carthamus tinctorius L.	Heavy stress	500 µM $ZnSO_4$	The activity of oxidative stress was decreased. Ascorbate, PC Phytochelatin, and Glutathione levels all increased throughout this period	Eisalou et al. (2021)
Lactuca sativa L.	Heavy stress	0, 0.5, 1, 1.5, or 2 mM	Increases in the levels of ascorbic acid, peroxidase enzyme activity, and proline content can be observed	Talabany and Albarzinji (2023)
Capsicum annum L.	Heavy stress	SA 0.5 mM	Elevated activity of nitrate reductases, glutamine synthetase, nitrogen metabolism, and glutamate synthase	Kaya et al. (2022)
Portulaca oleraceae L.	Heavy stress	0–100 µM	SA resulted in reduced overall growth and physiological parameters. However, SA treatments led to an increased accumulation of lead (Pb) and nickel (Ni) in the shoots and roots	Rasouli et al. (2024)
Triticum aestivum L.	Heavy stress	4.0 mM Si, 150 µM SA, 4.0 mM Si + 150 µM SA	Effectively reduced arsenic concentration in wheat seedlings while simultaneously boosting silicon content in the roots and shoots	El-Mogy et al. (2024)

Salicylic acid (SA) has been recognized as a crucial factor in mitigating the adverse effects of drought stress on horticultural crops. It plays a protective role by minimizing photosynthetic damage, enhancing antioxidant defense mechanisms, and controlling oxidative stress, among other protective functions (Hara et al. 2012). It has been demonstrated that, in tomato plants under water stress, the exogenous application of SA dramatically increases photosynthetic efficiency, chlorophyll content, nitrogen reductase (NR) levels, relative water content (RWC), proline concentration, antioxidant enzyme activity, and membrane stability index (MSI) (Hayat et al. 2008).

Additionally, studies have shown that SA foliar spray enhances proline levels, photosynthetic properties, and leaf morphological features in drought-stressed peas (El-Saadony et al. 2017). Under drought conditions, SA treatment significantly enhanced the growth and yield attributes, photosynthetic pigment, and ornamental features of petunias (Zarghami Moghaddam et al. 2014). A range of horticultural crops has shown comparable protective effects: tomatoes (Chakma et al. 2021; Fan et al. 2022), muskmelons (Korkmaz et al. 2007), eggplants (Wakchaure et al. 2020), sweet potatoes (Huang et al. 2022), cucumbers (Baninasab 2010), watermelons (Silva et al. 2023), peppers (Kaya 2021), and strawberries (Mozafari et al. 2018).

In strawberries, for example, total soluble solid content, proline concentration, and the activity of antioxidant enzymes like superoxide dismutase (SOD), peroxidase (POD), and catalase (CAT) were all raised by SA supplementation during drought stress. Furthermore, strawberry leaf malondialdehyde (MDA) levels were dramatically lowered by SA application, which also reduced oxidative damage (Sun et al. 2013). Application of salicylic acid (SA) improved water stress tolerance and reduced fatty acid levels in pumpkin, as reported by Biareh et al. (2022). In olive trees, SA has been shown to regulate growth, improve root architecture, increase water use efficiency (WUE), and enhance the photosynthetic system (Brito et al. 2018). Exogenous SA application in olive trees under drought conditions led to improved growth, yield, plant height, plant leaf area, proline concentration, soluble carbohydrate levels, and chlorophyll content (Shafiei et al. 2019). Similarly, SA-treated muskmelons exhibited significant improvements in plant root dry weight (PRDW) and seedling health index (Korkmaz et al. 2007). In African eggplants, SA treatment boosted antioxidant enzyme activity, phenolic content, proline, and overall growth status while dramatically reducing electrolyte leakage under drought stress (Opabode et al. 2019).

These findings collectively underscore the vital role of SA in enhancing drought stress tolerance in horticultural crops, providing a promising strategy for mitigating the impact of water scarcity on agricultural productivity (Table 6.1).

6.3.3 *Cold Stress Alleviation Through Salicylic Acid*

Cold stress is a critical environmental factor that poses a substantial threat to crop production, significantly affecting the sustainability of agriculture. Extended periods of exposure to low temperatures can cause substantial physiological and metabolic disturbances in plants, perhaps resulting in plant death (Atayee and Noori

2020). The adverse effects of cold stress include hindered seed germination, leaf chlorosis and necrosis, inhibition of seedling growth, altered enzyme activity, and reduced leaf expansion. These factors collectively lead to tissue damage and decreased plant survival (Yadav 2010). Additionally, cold stress can severely impair reproductive development, further hindering crop productivity.

Salicylic acid has been recognized as a critical mechanism in enhancing cold stress resistance in many horticultural crops. Ge et al. (2020) have shown that the application of SA effectively regulates growth and increases the fatty acid content in bell peppers under cold stress conditions. The studies conducted by Koç (2013) and Korkmaz (2005) revealed that the external application of SA greatly enhanced the ability of pepper seedlings to withstand cold stress. In these investigations, the application of SA to the leaves resulted in higher plant height, fresh weight, proline content, glucose concentration, and leaf phosphorus levels. These parameters are important indicators of improved plant resistance.

In strawberries, SA treatment significantly improved cold stress tolerance by reducing electrolyte leakage (EL) and mitigating oxidative damage, as evidenced by decreased hydrogen peroxide (H_2O_2) and superoxide (O_2^-) production in leaves. Additionally, SA increased proline accumulation and boosted antioxidant enzyme activities in spinach leaves, both of which are critical for combating oxidative stress (Shin et al. 2018; Min et al. 2018). In bananas, the foliar application of SA was found to reduce H_2O_2 levels and enhance the activity of key antioxidant enzymes such as superoxide dismutase (SOD), peroxidase (POD), and catalase (CAT) under cold stress (Guo-Zhang et al. 2003). Similarly, in eggplant seedlings, SA pretreatment led to significant improvements in antioxidant enzyme activity and upregulated the expression of antioxidant enzyme genes, thereby decreasing oxidative damage in leaves (Chen et al. 2011). Moreover, a study carried out on olive plants revealed that the application of SA had a positive impact on growth and resulted in a significant rise in proline accumulation, antioxidant enzyme activity, and phenolic content in leaves following cold stress exposure (Hashempour et al. 2014). Previous studies have demonstrated that the implementation of SA greatly improved the growth and resilience to stress in various horticultural crops, including peppers (Park et al. 2014), spinach (Aydin and Nalbantoğlu 2011), *Catharanthus roseus* (Khandan-Mirkohi et al. 2017), lemon fruits (Siboza et al. 2014), radishes (Devi et al. 2018), watermelons (Jing-Hua et al. 2008), muskmelons (Korkmaz et al. 2007), and melons (Kim et al. 2016).

Collectively, these findings highlight the pivotal role of SA in enhancing cold stress tolerance across a wide range of horticultural crops, offering a promising strategy for improving crop resilience in cold environments (Table 6.1).

6.3.4 Heat Stress Alleviation Through Salicylic Acid

Global warming has emerged as a critical concern due to its profound impact on plant development, particularly through the increased occurrence of heat stress. This environmental stressor has significantly hindered horticultural crop production worldwide, with rising temperatures leading to severe disruptions in cellular homeostasis. The resulting imbalance can drastically slow down growth and, in extreme cases, lead to plant death (Kotak et al. 2007). The detrimental impact of heat stress on certain horticultural crops has been documented in studies conducted by Han et al. (2013), Zhang et al. (2013), Nijabat et al. (2020), and Guo et al. (2020). When exposed to high temperatures, these plants exhibit reduced growth, decreased biomass yield, and damaged photosynthetic systems.

Salicylic acid (SA) is acknowledged as an essential element in enhancing the ability of horticultural crops to withstand damage caused by heat stress. Application of this treatment has resulted in significant improvements in the heat stress tolerance of strawberries, tomatoes, potatoes, grape berries, and watermelons (Wen et al. 2008; Coltro et al. 2014; Ram et al. 2014). Applications of SA to potatoes have demonstrated a substantial improvement in root and shoot growth, especially under conditions of high-temperature stress (Rudić et al. 2022).

Heat stress typically leads to increased reactive oxygen species (ROS) accumulation, altered antioxidant enzyme activity, and oxidative damage in plants. However, SA treatment has been shown to protect the photosynthetic apparatus, balance the antioxidant enzyme system, and reduce the generation of malondialdehyde (MDA) and hydrogen peroxide (H_2O_2). This preventive mechanism preserves pigment molecules and reduces cellular membrane damage in heat-stressed tomato seedlings (Jahan et al. 2019). According to Khan et al. (2014), adding SA to heat-stressed tomatoes improves root length, volume, area, and diameter. When applied foliarly in a heated greenhouse, it increases mineral nutrients, chlorophyll, TSS, TA, and fruit quality (Karlidag et al. 2009).

The benefits of SA extend beyond tomatoes. Recent studies have shown that using SA in melons can improve their ability to withstand heat stress (Widiastuti et al. 2013). Exposure of young grape plants to salicylic acid (SA) enhances the function of antioxidant enzymes, including superoxide dismutase (SOD), peroxidase (POD), and gosipylase (GR), controls calcium equilibrium, and improves tolerance to heat stress (Wang and Li 2006). Several experimental studies have yielded further verification of the role of SA in enhancing plant development under heat stress conditions. For example, the use of SA has been demonstrated to improve the development of tobacco plants (Liu et al. 2017), revive the photosynthetic system in grapevine leaves (Wang et al. 2010), maintain the carbohydrate concentration in peppers (Otálora et al. 2020), attain a greater maximum quantum efficiency of PSII (Fv/Fm), and increase the relative chlorophyll content (SPAD value) in sage plants (Lin et al. 2021). Studies have shown that SA can reduce oxidative damage and boost antioxidant enzyme activity in peppers exposed to heat stress (Saeed et al. 2023).

In summary, salicylic acid plays a pivotal role in mitigating the adverse effects of heat stress on horticultural crops by enhancing their physiological and biochemical resilience, thus offering a promising approach to sustaining crop productivity in the face of global warming (Table 6.1).

6.3.5 Heavy Metal Stress Alleviation Through Salicylic Acid

Heavy metals (HMs) are significant soil pollutants that pose a major challenge to global crop production. Exposure to HM-contaminated soils often results in stunted plant growth, reduced yield, and impaired development (Noor et al. 2022). Elevated levels of heavy metals (HMs) interfere with several metabolic processes in horticulture crops, causing negative impacts on the growth of roots and shoots and modifying their physical composition and structure.

Interestingly, low levels of HMs can sometimes stimulate plant growth through a phenomenon known as hormesis (Małkowski et al. 2019). The toxicity of HMs has been shown to notably reduce growth in several horticultural species, including lettuce (Haghighi et al. 2010), pepper (Altaf et al. 2022b), watermelon (Shirani Bidabadi et al. 2018), tomato (Xie et al. 2020), cucumber (Munawar et al. 2022), and petunia (Milusheva et al. 2019). To deal with heavy metal (HM) toxicity, plants have developed intricate physiological and cellular defense mechanisms. These systems include the capacity to store and transport metals and to activate HM transporters through specialized signaling pathways (Noor et al. 2022). Under heavy metal (HM) stress, plants respond by either inhibiting HM absorption or initiating detoxification mechanisms, such as the synthesis of antioxidants to neutralize reactive oxygen species (Kalaivanan and Ganeshamurthy 2016).

Salicylic acid (SA) has emerged as a vital regulator in enhancing plant tolerance to HM stress. For instance, Cd (cadmium) exposure in pea seedlings significantly hampers growth and causes oxidative damage by increasing electrolyte leakage (EL), malondialdehyde (MDA), and proline levels. However, SA application has been shown to counteract these effects, improving growth traits, boosting the antioxidant enzyme system, enhancing photosynthetic pigments, and reducing EL levels, proline production, and MDA concentration. Moreover, SA treatment significantly decreases Cd accumulation in the shoots of pea plants (Popova et al. 2009).

Furthermore, a study conducted by Gupta and Seth (2021) demonstrated that the growth of tomato seedlings is adversely affected by chromium (Cr) toxicity. Conversely, the application of SA has a beneficial effect on their growth by improving leaf photosynthesis, pigment concentration, and gas exchange properties, while simultaneously decreasing oxidative damage and Cr buildup. Furthermore, the application of SA therapy has been shown to alleviate the negative consequences of high boron toxicity in watermelon seedlings. The effects encompass a reduction in the rate of photosynthetic absorption, an elevation in pigment concentrations, and suppression of seedling vegetative development. The systematic use of SA helps to restore the leaf photosynthetic process, maintain carotenoid concentration, regulate reactive oxygen species (ROS) homeostasis, and restrict boron absorption from the root to the shoot (Moustafa-Farag et al. 2020). By supplementing potatoes with SA, the levels of MDA, H_2O_2, O_2^-, and EL are significantly reduced. Additionally, the translocation of Cd from the root to the shoot is decreased in the presence of Cd toxicity. Moreover, SA enhances the accumulation of proline, increases gas exchange properties, and boosts antioxidant enzyme activity (Li et al. 2019). Furthermore, empirical investigations have shown that the application of SA treatment increases the activity of antioxidant enzymes and promotes the ability of two peanut cultivars to withstand Cd stress (Xu et al. 2015) (Table 6.1).

Numerous studies have reported that SA application significantly improves growth, photosynthesis, and antioxidant mechanisms in horticultural plants, helping maintain ROS and mineral homeostasis while reducing heavy metal accumulation from plant roots to shoots (Wang et al. 2022; Chen et al. 2023). In summary, SA plays a crucial role in regulating the growth of horticultural plants under HM stress by enhancing their defense mechanisms. Exogenous SA application effectively improves photosynthesis, strengthens the antioxidant enzyme system, reduces oxidative damage, and modulates growth, making it a promising strategy to support plant survival in HM-contaminated environments. The use of SA could prove to be a key approach in ensuring the global sustainability of agricultural crop production, particularly in areas affected by heavy metal pollution (Fig. 6.3).

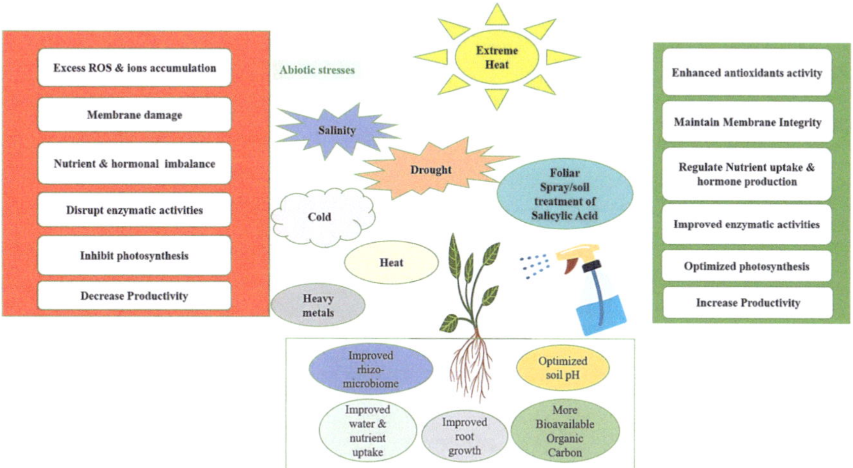

Fig. 6.3 Diagrammatic overview of abiotic stress-induced effects in plants and the contribution of salicylic acid to plant tolerance mechanisms against diverse abiotic stresses

6.4 Conclusion

In conclusion, the signaling molecule salicylic acid (SA) is essential for regulating important physiological activities in plants, such as photosynthesis, metabolism, and growth. It plays a crucial role in their response to different abiotic stimuli. Multiple studies have empirically shown its efficacy in alleviating the adverse impacts of abiotic stressors such as salinity, drought, severe temperatures, and heavy metals. SA enhances stress tolerance by modulating antioxidant enzyme activities, which reduce reactive oxygen species (ROS) levels, improve physiological traits, and promote overall growth. Furthermore, SA's role in enhancing plant resistance to cold stress and alleviating heat-induced damage has been widely acknowledged. The application of SA, whether through foliar sprays or incorporation into growing media, has proven beneficial across different plant species and under diverse environmental stress conditions.

Future research into SA's dual role in promoting growth and enhancing stress resilience will provide further insights into optimizing plant responses. This understanding will be crucial in developing strategies that allow crops to achieve higher yields and maintain a balance between growth and defense mechanisms. As a plant growth regulator with immense potential, SA holds promise for improving agricultural productivity, especially in challenging environmental conditions.

References

Abdi G, Karami L (2020) Salicylic acid effects on some physiochemical properties and secondary metabolite accumulation in Mentha piperita L. under water deficit stress. Adv Hortic Sci 34:81–91

Aghdam MS, Jannatizadeh A, Sheikh-Assadi M, Malekzadeh P (2016) Alleviation of postharvest chilling injury in anthurium cut flowers by salicylic acid treatment. Sci Hortic 202:70–76

Ahmad F, Kamal A, Singh A, Ashfaque F, Alamri S, Siddiqui MH (2020) Salicylic acid modulates antioxidant system, defense metabolites, and expression of salt transporter genes in Pisum sativum under salinity stress. J Plant Growth Regul 41(5):1905–1918

Akasha A, Ashraf M, Shereen A, Mahboob W, Faisal S (2019) Heat tolerance screening studies and evaluating salicylic acid efficacy against high temperature in rice (Oryza sativa L.) genotypes. J Plant Biochem Physiol 7:235

Aliniaeifard S, Hajilou J, Tabatabaei SJ (2016) Photosynthetic and growth responses of olive to proline and salicylic acid under salinity condition. Not Bot Horti Agrobot Cluj-Napoca 44(2):579–585. https://doi.org/10.15835/nbha44210419

Alsahli A, Mohamed AK, Alaraidh I, Al-Ghamdi A, Al-Watban A, El-Zaidy M, Alzahrani SM (2019) Salicylic acid alleviates salinity stress through the modulation of biochemical attributes and some key antioxidants in wheat seedlings. Pak J Bot 51:1551–1559

Al-Shoaibi AA (2020) Combined effects of salinity and temperature on germination, growth, and gas exchange in two cultivars of *Sorghum bicolor*. J Taibah Univ Sci 14:812–822

Altaf MA, Hao Y, He C, Mumtaz MA, Shu H, Fu H, Wang Z (2022a) Physiological and biochemical responses of pepper (Capsicum annuum L.) seedlings to nickel toxicity. Front Plant Sci 13:950392

Altaf MA, Shahid R, Altaf MM, Kumar R, Naz S, Kumar A, Bano S (2022b) Melatonin: first-line soldier in tomato under abiotic stress, current and future perspective. Plant Physiol Biochem 185:188–197

Ashraf M, Harris PJ (2013) Photosynthesis under stressful environments: an overview. Photosynthetica 51:163–190

Atayee AR, Noori MS (2020) Alleviation of cold stress in vegetable crops. J Sci Agric 4:38–44

Aycan M, Baslam M, Asiloglu R, Mitsui T, Yildiz M (2021) Development of new high-salt tolerant bread wheat (Triticum aestivum L.) genotypes and insight into the tolerance mechanisms. Plant Physiol Biochem 166:314–327

Aydin B, Nalbantoğlu B (2011) Effects of cold and salicylic acid treatments on nitrate reductase activity in spinach leaves. Turk J Biol 35(4):443–448

Azad N, Rezayian M, Hassanpour H, Niknam V, Ebrahimzadeh H (2021) Physiological mechanism of salicylic acid in Mentha pulegium L. under salinity and drought stress. Braz J Bot 44:359–369

Bagautdinova ZZ, Omelyanchuk N, Tyapkin AV, Kovrizhnykh VV, Lavrekha VV, Zemlyanskaya EV (2022) Salicylic acid in root growth and development. Int J Mol Sci 23(4):2228

Baninasab B (2010) Induction of drought tolerance by salicylic acid in seedlings of cucumber (Cucumis sativus L.). J Hortic Sci Biotechnol 85(3):191–196

Biareh V, Shekari F, Sayfzadeh S, Zakerin H, Hadidi E, Beltrão JGT, Mastinu A (2022) Physiological and qualitative response of Cucurbita pepo L. to salicylic acid under controlled water stress conditions. Horticulturae 8(1):79

Boscaiu M, Fita A (2020) Physiological and molecular characterization of crop resistance to abiotic stresses. Agronomy 10:1308

Brito C, Dinis LT, Meijón M, Ferreira H, Pinto G, Moutinho-Pereira J, Correia C (2018) Salicylic acid modulates olive tree physiological and growth responses to drought and rewatering events in a dose-dependent manner. J Plant Physiol 230:21–32

Chakma R, Biswas A, Saekong P, Ullah H, Datta A (2021) Foliar application and seed priming of salicylic acid affect growth, fruit yield, and quality of grape tomato under drought stress. Sci Hortic 280:109904. https://doi.org/10.1016/j.scienta.2021.109904

Chaudhry S, Sidhu GPS (2022) Climate change regulated abiotic stress mechanisms in plants: a comprehensive review. Plant Cell Rep 41:1–31

Chavoushi M, Najafi F, Salimi A, Angaji SA (2020) Effect of salicylic acid and sodium nitroprusside on growth parameters, photosynthetic pigments and secondary metabolites of safflower under drought stress. Sci Hortic 259:108823

Chen S, Zimei L, Cui J, Jiangang D, Xia X, Liu D, Yu J (2011) Alleviation of chilling-induced oxidative damage by salicylic acid pretreatment and related gene expression in eggplant seedlings. Plant Growth Regul 65:101–108

Chen S, Zhao CB, Ren RM, Jiang JH (2023) Salicylic acid had the potential to enhance tolerance in horticultural crops against abiotic stress. Front Plant Sci 14:1141918

Cheng F, Lu J, Gao M, Shi K, Kong Q, Huang Y, Bie Z (2016) Redox signaling and CBF-responsive pathway are involved in salicylic acid-improved photosynthesis and growth under chilling stress in watermelon. Front Plant Sci 7:1519

Cheraghabadi MM, Roshanfekr H, Hasibi P, Meskarbashi M (2016) Effect of foliar application of salicylic acid on some physiological traits of sugar beet in salt stress conditions Iran. J Field Crop Sci 46(4):591–604

Cohen I, Zandalinas SI, Huck C, Fritschi FB, Mittler R (2021) Meta-analysis of drought and heat stress combination impact on crop yield and yield components. Physiol Plant 171(1):66–76

Coltro S, Broetto L, Rotilli MCC, Moraes AJD, Barp FK, Braga GC (2014) Heat shock and salicylic acid on postharvest preservation of organic strawberries. Rev Ceres 61:306–312

Dawood MF, Zaid A, Latef AAH (2022) Salicylic acid spraying-induced resilience strategies against the damaging impacts of drought and/or salinity stress in two varieties of Vicia faba L. seedlings. J Plant Growth Regul 41:1919–1942

Devi J, Bhatia S, Alam MS, Dhillon TS (2018) Effect of calcium and salicylic acid on quality retention in relation to antioxidative enzymes in radish stored under refrigerated conditions. J Food Sci Technol 55:1116–1126

Ding P, Ding Y (2020) Stories of salicylic acid: a plant defense hormone. Trends Plant Sci 25(6):549–565

Eisalou AV, Namdjoyan S, Soorki AA (2021) The combined effect of melatonin and salicylic acid improved the tolerance of safflower seedlings to zinc toxicity. Acta Physiol Plant 43:1–14

El-Mogy MM, El-Beltagi HS, Almutairi HH, Hamed LM, Sattar A, Sher A, Ali Q (2024) Application of silicon with salicylic acid up-regulate physio-biochemical mechanisms of wheat in conferring tolerance to arsenic induced oxidative stress. Silicon:1–14

El-Saadony FM, Nawar DA, Zyada HG (2017) Effect of foliar application with salicylic acid, garlic extract, and proline on growth, yield, and leaf anatomy of pea (Pisum sativum L.) grown under drought stress. Middle East J Appl Sci 7(3):633–650

El-Taher AM, Abd El-Raouf HS, Osman NA, Azoz SN, Omar MA, Elkelish A, Abd El-Hady MA (2021) Effect of salt stress and foliar application of salicylic acid on morphological, biochemical, anatomical, and productivity characteristics of cowpea (Vigna unguiculata L.) plants. Plants 11:115

Esim N, Atici Ö (2015) Effects of exogenous nitric oxide and salicylic acid on chilling-induced oxidative stress in wheat (Triticum aestivum). Front Life Sci 8:124–130

Es-Sbihi FZ, Hazzoumi Z, Aasfar A, Amrani Joutei K (2021) Improving salinity tolerance in Salvia officinalis L. by foliar application of salicylic acid. Chem Biol Technol Agric 8:1–12

Fan S, Wu H, Gong H, Guo J (2022) Salicylic acid mediates selenium-induced tolerance to drought stress in tomato plants. Sci Hortic 300:111092. https://doi.org/10.1016/j.scienta.2022.111092

Fang S, Hou X, Liang X (2021) Response mechanisms of plants under saline-alkali stress. Front Plant Sci 12:667458

Feng B, Zhang C, Chen T, Zhang X, Tao L, Fu G (2018) Salicylic acid reverses pollen abortion of rice caused by heat stress. BMC Plant Biol 18:1–16

Fu L, Li Q, Chen C, Zhang Y, Liu Y, Xu L, Zhou Y, Li C, Zhou D, Rittmann BE (2021) Benzoic and salicylic acid are the signaling molecules of Chlorella cells for improving cell growth. Chemosphere 265:129084

Ge W, Zhao Y, Kong X, Sun H, Luo M, Yao M, Ji S (2020) Combining salicylic acid and trisodium phosphate alleviates chilling injury in bell pepper (Capsicum annuum L.) through enhancing fatty-acid desaturation efficiency and water retention. Food Chem 327:127057

Ghani MA, Abbas MM, Ali B, Ziaf K, Azam M, Anjum R, Jillani U (2021) Role of salicylic acid in heat stress tolerance in tri-genomic Brassica napus L. Bioagro 33:13–20

Gharbi E, Lutts S, Dailly H, Quinet M (2018) Comparison between the impacts of two different modes of salicylic acid application on tomato (Solanum lycopersicum) responses to salinity. Plant Signal Behav 13:e1469361

Gharibiyan P, Roozban MR, Rahemi M, Vahdati K (2023) Exogenous salicylic acid improves growth and physiological status of two pistacia species under salinity stress. Erwerbs-obstbau 65:1441–1452

Golkar P, Taghizadeh M, Yousefian Z (2019) The effects of chitosan and salicylic acid on elicitation of secondary metabolites and antioxidant activity of safflower under in vitro salinity stress. Plant Cell Tissue Organ Cult 137:575–585

Guo R, Wang X, Han X, Chen X, Wang-Pruski G (2020) Physiological and transcriptomic responses of water spinach (Ipomoea aquatica) to prolonged heat stress. BMC Genomics 21:1–15

Guo-Zhang K, Zheng-Xun W, Gu-Chou SUN (2003) Participation of H2O2 in enhancement of cold chilling by salicylic acid in banana seedlings. J Integr Plant Biol 45(5):567

Gupta S, Seth CS (2021) Salicylic acid alleviates chromium (VI) toxicity by restricting its uptake, improving photosynthesis, and augmenting antioxidant defense in Solanum lycopersicum L. Physiol Mol Biol Plants 27:2651–2664

Haghighi M, Kafi M, Fang P, Gui-Xiao L (2010) Humic acid decreased hazardous of cadmium toxicity on lettuce (L.). J Fruit Ornam Plant Res 72(1):49–61

Han Y, Fan S, Zhang Q, Wang Y (2013) Effect of heat stress on the MDA, proline and soluble sugar content in leaf lettuce seedlings. Agric Sci 4(5B):112

Hanaka A, Ozimek E, Reszczyńska E, Jaroszuk-Ściseł J, Stolarz M (2021) Plant tolerance to drought stress in the presence of supporting bacteria and fungi: an efficient strategy in horticulture. Horticulturae 7(10):390. https://doi.org/10.3390/horticulturae7100390

Hanif S, Mahmood A, Javed T, Bibi S, Zia MA, Asghar S, Ali B (2024) Exogenous application of salicylic acid ameliorates salinity stress in barley (Hordeum vulgare L.). BMC Plant Biol 24:1–16

Hara M, Furukawa J, Sato A, Mizoguchi T, Miura K (2012) Abiotic stress and the role of salicylic acid in plants. In: Ahmad P, Prasad MNV (eds) Abiotic stress responses in plants: metabolism, productivity, and sustainability. Springer, pp 235–251. https://doi.org/10.1007/978-1-4614-0634-1_12

Hashempour A, Ghasemnezhad M, Fotouhi Ghazvini R et al (2014) The physiological and biochemical responses to freezing stress of olive plants treated with salicylic acid. Russ J Plant Physiol 61:443–450

Hassan MU, Chattha MU, Khan I, Chattha MB, Barbanti L, Aamer M, Iqbal MM, Nawaz M, Mahmood A, Ali A, Aslam MT (2021) Heat stress in cultivated plants: nature, impact, mechanisms, and mitigation strategies—a review. Plant Biosyst 155(2):211–234

Hayat S, Hasan SA, Fariduddin Q, Ahmad A (2008) Growth of tomato (Lycopersicon esculentum) in response to salicylic acid under water stress. J Plant Interact 3(4):297–304. https://doi.org/10.1080/17429140801933799

Hernández JA, Diaz-Vivancos P, Barba-Espín G, Clemente-Moreno MJ (2017) On the role of salicylic acid in plant responses to environmental stresses. In: Hayat S, Ahmad A (eds) Salicylic acid: a multifaceted hormone. Springer, pp 17–34. https://doi.org/10.1007/978-81-322-3862-1_2

Hopmans JW, Qureshi A, Kisekka I, Munns R, Grattan S, Rengasamy P, Ben-Gal A, Assouline S, Javaux M, Minhas P, Taleisnik E (2021) Critical knowledge gaps and research priorities in global soil salinity. Adv Agron 169:1–191

Hu Y, Chai Q, Wang Y, Chen Y, Dong H, Shen J, Wen Q (2022) Effects of heat stress and exogenous salicylic acid on secondary metabolites biosynthesis in Pleurotus ostreatus (Jacq.) P. Kumm. Life 12:915

Huang C, Wang D, Sun L, Wei L (2016) Effects of exogenous salicylic acid on the physiological characteristics of Dendrobium officinale under chilling stress. Plant Growth Regul 79:199–208

Huang C, Liao J, Huang W, Qin N (2022) Salicylic acid protects sweet potato seedlings from drought stress by mediating abscisic acid-related gene expression and enhancing the antioxidant defense system. Int J Mol Sci 23(23):14819. https://doi.org/10.3390/ijms232314819

Ignatenko A, Talanova V, Repkina N, Titov A (2019) Exogenous salicylic acid treatment induces cold tolerance in wheat through promotion of antioxidant enzyme activity and proline accumulation. Acta Physiol Plant 41:1–10

Ivushkin K, Bartholomeus H, Bregt AK, Pulatov A, Kempen B, De Sousa L (2019) Global mapping of soil salinity change. Remote Sens Environ 231:111260

Jahan MS, Wang Y, Shu S, Zhong M, Chen Z, Wu J, Guo S (2019) Exogenous salicylic acid increases the heat tolerance in tomato (Solanum lycopersicum L) by enhancing photosynthesis efficiency and improving antioxidant defense system through scavenging of reactive oxygen species. Sci Hortic 247:421–429

Jalili I, Ebadi A, Askari MA, KalatehJari S, Aazami MA (2023) Foliar application of putrescine, salicylic acid, and ascorbic acid mitigates frost stress damage in Vitis vinifera cv. 'Giziluzum'. BMC Plant Biol 23:135

Janda T, Szalai G, Pál M (2020) Salicylic acid signalling in plants. Int J Mol Sci 21(7):2655

Jing-Hua Y, Yuan G, Yan-Man L, Xiao-Hua Q, Zhang MF (2008) Salicylic acid-induced enhancement of cold tolerance through activation of antioxidative capacity in watermelon. Sci Hortic 118(3):200–205

Jini D, Joseph B (2017) Physiological mechanism of salicylic acid for alleviation of salt stress in rice. Rice Sci 24:97–108

Kalaivanan D, Ganeshamurthy AN (2016) Mechanisms of heavy metal toxicity in plants. In: Abiotic stress physiology of horticultural crops. Springer, pp 85–102. https://doi.org/10.1007/978-81-322-2725-0
Karlidag H, Yildirim E, Turan M (2009) Exogenous applications of salicylic acid affect quality and yield of strawberry grown under antifrost heated greenhouse conditions. J Plant Nutr Soil Sci 172(2):270–276
Kaur G, Sharma P, Rathee S, Singh HP, Batish DR, Kohli RK (2021) Salicylic acid pre-treatment modulates Pb2+-induced DNA damage vis-à-vis oxidative stress in Allium cepa roots. Environ Sci Pollut Res 28:51989–52000
Kaur G, Tak Y, Asthir B (2022) Salicylic acid: a key signal molecule ameliorating plant stresses. Cereal Res Commun 50(4):617–626. https://doi.org/10.1007/s42976-022-00183-4
Kaya C (2021) Nitrate reductase is required for salicylic acid-induced water stress tolerance of pepper by upregulating the AsA-GSH pathway and glyoxalase system. Physiol Plant 172(2):351–370. https://doi.org/10.1111/ppl.13199
Kaya C, Sarıoglu A, Ashraf M, AlYemeni MN, Ahmad P (2022) The combined supplementation of melatonin and salicylic acid effectively detoxifies arsenic toxicity by modulating phytochelatins and nitrogen metabolism in pepper plants. Environ Pollut 297:118727
Keshavarz H, Sanavy SAMM, Moghadam RSG (2016) Impact of foliar application with salicylic acid on biochemical characters of canola plants under cold stress condition. Not Sci Biol 8:98–105
Khalil N, Elhady SS, Diri RM, Fekry MI, Bishr M, Salama O, El-Zalabani SM (2022) Salicylic acid spraying affects secondary metabolites and radical scavenging capacity of drought-stressed Eriocephalus africanus L. Agronomy 12:2278
Khan AR, Cheng Z, Ghazanfar B, Khan MA, Yongxing Z (2014) Acetyl salicylic acid and 24-epibrassinolide enhance root activity and improve root morphological features in tomato plants under heat stress. Acta Agric Scand B Soil Plant Sci 64(4):304–311
Khan MIR, Fatma M, Per TS, Anjum NA, Khan NA (2015) Salicylic acid-induced abiotic stress tolerance and underlying mechanisms in plants. Front Plant Sci 6:462
Khan N, Bano A, Babar MA (2019) Metabolic and physiological changes induced by plant growth regulators and plant growth-promoting rhizobacteria and their impact on drought tolerance in Cicer arietinum L. PLoS One 14:e0213040
Khan MIR, Poor P, Janda T (2022) Salicylic acid: a versatile signaling molecule in plants. J Plant Growth Regul 41:1887–1890
Khandan-Mirkohi A, Khalili Halbi M, Salami SA, Lesani H (2017) Improving effects of mild cold stress and salicylic acid on growth and physiology of periwinkle (Catharanthus roseus Don.). Int J Hortic Sci Technol 4(1):67–78
Kim YH, Choi KI, Khan AL, Waqas M, Lee IJ (2016) Exogenous application of abscisic acid regulates endogenous gibberellins homeostasis and enhances resistance of oriental melon (Cucumis melo var. L.) against low temperature. Sci Hortic 207:41–47
Kim Y, Kim S, Shim IS (2017) Exogenous salicylic acid alleviates salt-stress damage in cucumber under moderate nitrogen conditions by controlling endogenous salicylic acid levels. Hortic Environ Biotechnol 58:247–253. https://doi.org/10.1007/s13580-017-0187-3
Kim Y, Mun BG, Khan AL, Waqas M, Kim HH, Shahzad R, Lee IJ (2018) Regulation of reactive oxygen and nitrogen species by salicylic acid in rice plants under salinity stress conditions. PLoS One 13:e0192650
Kiremit MS, Akınoğlu G, Mitrovica B, Rakıcıoğlu S (2024) Enhancing drought-salinity stress tolerance in lettuce: synergistic effects of salicylic acid and melatonin. S Afr J Bot 172:212–226
Klessig DF, Choi HW, Dempsey DMA (2018) Systemic acquired resistance and salicylic acid: past, present, and future. Mol Plant 31(9):871–888
Koç E (2013) The effect of exogenous proline and salicylic acid application on proline and apoplastic protein in cold tolerance of pepper callus cultures. J Integr Plant Biol 15(2):382–385

Kolupaev YE, Karpets YV, Yastreb TO, Lugovaya AA (2018) Combined effect of salicylic acid and nitrogen oxide donor on stress-protective system of wheat plants under drought conditions. Appl Biochem Microbiol 54:418–424
Korkmaz A (2005) Inclusion of acetyl salicylic acid and methyl jasmonate into the priming solution improves low-temperature germination and emergence of sweet pepper. J Integr Plant Biol
Korkmaz A, Uzunlu M, Demirkiran AR (2007) Treatment with acetyl salicylic acid protects muskmelon seedlings against drought stress. Acta Physiol Plant 29:503–508
Kotak S, Larkindale J, Lee U, von Koskull-Döring P, Vierling E, Scharf KD (2007) Complexity of the heat stress response in plants. Curr Opin Plant Biol 10(3):310–316
Kour D, Khan SS, Kaur T, Kour H, Singh G, Yadav A, Yadav AN (2022) Drought-adaptive microbes as bioinoculants for horticultural crops. Heliyon 8(5). https://doi.org/10.1016/j.heliyon.2022.e09336
Kováčik J, Klejdus B, Hedbavny J, Bačkor M (2009) Salicylic acid alleviates NaCl-induced changes in the metabolism of Matricaria chamomilla plants. Ecotoxicology 18:544–554
Kumar RR, Sharma SK, Goswami S, Verma P, Singh K, Dixit N, Rai RD (2015) Salicylic acid alleviates the heat stress-induced oxidative damage of starch biosynthesis pathway by modulating the expression of heat-stable genes and proteins in wheat (Triticum aestivum). Acta Physiol Plant 37:1–12
Kusvuran S, Yilmaz UD (2023) Ameliorative role of salicylic acid in the growth, nutrient content, and antioxidative responses of salt-stressed lettuce. Acta Sci Pol Hortorum Cultus 22:75–85
Kwon EH, Adhikari A, Imran M, Lee DS, Lee CY, Kang SM, Lee IJ (2023) Exogenous SA applications alleviate salinity stress via physiological and biochemical changes in St John's Wort plants. Plants 12:310
La VH, Lee BR, Zhang Q, Park SH, Islam MT, Kim TH (2019) Salicylic acid improves drought-stress tolerance by regulating the redox status and proline metabolism in Brassica rapa. Hortic Environ Biotechnol 60:31–40
Lefevere H, Bauters L, Gheysen G (2020) Salicylic acid biosynthesis in plants. Front Plant Sci 11:338
Li Q, Wang G, Wang Y, Yang D, Guan C, Ji J (2019) Foliar application of salicylic acid alleviates cadmium toxicity by modulating reactive oxygen species in potato. Ecotoxicol Environ Saf 172:317–325
Lin KH, Huang SB, Wu CW, Chang YS (2019) Effects of salicylic acid and calcium chloride on heat tolerance of poinsettia. HortScience 54:499–504
Lin KH, Lin TY, Wu CW, Chang YS (2021) Protective effects of salicylic acid and calcium chloride on sage plants (Salvia officinalis L. and Salvia elegans Vahl) under high-temperature stress. Plants 10(10):2110
Liu ZM, Yue MM, Yang DY, Zhu SB, Ma NN, Meng QW (2017) Over-expression of SlJA2 decreased heat tolerance of transgenic tobacco plants via salicylic acid pathway. Plant Cell Rep 36:529–542
Liu J, Qiu G, Liu C, Li H, Chen X, Fu Q, Lin Y, Guo B (2022) Salicylic acid, a multifaceted hormone, combats abiotic stresses in plants. Life 12(7):886
Loutfy N, Sakuma Y, Gupta DK, Inouhe M (2020) Modifications of water status, growth rate, and antioxidant system in two wheat cultivars as affected by salinity stress and salicylic acid. J Plant Res 133:549–570
Lu H (2009) Dissection of salicylic acid-mediated defense signaling networks. Plant Signal Behav 4(8):713–717
Małkowski E, Sitko K, Zieleźnik-Rusinowska P, Gieroń Ż, Szopiński M (2019) Heavy metal toxicity: physiological implications of metal toxicity in plants. In: Plant metallomics and functional omics: a system-wide perspective. Springer, pp 253–301. https://doi.org/10.1007/978-3-030-19103-0_10
Mallahi T, Saharkhiz MJ, Javanmardi J (2018) Salicylic acid changes morpho-physiological attributes of feverfew (Tanacetum parthenium L.) under salinity stress. Acta Ecol Sin 38:351–355

Maruri-Lopez I, Aviles-Baltazar NY, Buchala A, Serrano M (2019) Intra and extracellular journey of the phytohormone salicylic acid. Front Plant Sci 10:423
Milusheva DI, Atanassova BY, Iakimova ET (2019) Application of in vitro test system for evaluating the tolerance of Ageratum houstonianum and Petunia x hybrida to cadmium toxicity. Bulg J Agric Sci 25(2):300–309
Mimouni H, Wasti S, Manaa A, Gharbi E, Chalh A, Vandoorne B, Ahmed HB (2016) Does salicylic acid (SA) improve tolerance to salt stress in plants? A study of SA effects on tomato plant growth, water dynamics, photosynthesis, and biochemical parameters. Omics 20(3):180–190
Min K, Showman L, Perera A, Arora R (2018) Salicylic acid-induced freezing tolerance in spinach (Spinacia oleracea L.) leaves explored through metabolite profiling. Environ Exp Bot 156:214–227
Mir RA, Somasundaram R (2021) Salicylic acid and salt stress tolerance in plants: a review. J Stress Physiol Biochem 17(3):32–50
Misra N, Saxena P (2009) Effect of salicylic acid on proline metabolism in lentil grown under salinity stress. Plant Sci 177:181–189
Moustafa-Farag M, Mohamed HI, Mahmoud A, Elkelish A, Misra AN, Guy KM, Zhang M (2020) Salicylic acid stimulates antioxidant defense and osmolyte metabolism to alleviate oxidative stress in watermelons under excess boron. Plants 9(6):724
Mozafari AA, Havas F, Ghaderi N (2018) Application of iron nanoparticles and salicylic acid in in vitro culture of strawberries (Fragaria × ananassa Duch.) to cope with drought stress. Plant Cell Tissue Organ Cult 132:511–523
Munawar S, Ghani MA, Ali B, Azam M, Rashid MZ, Anjum R, Abbas MM (2022) Attenuation of nickel toxicity in cucumber (Cucumis sativus L.) through exogenous application of salicylic acid and humic acid. Saudi. J Biol Sci 29(5):3320–3326
Naeem M, Basit A, Ahmad I, Mohamed HI, Wasila H (2020) Effect of salicylic acid and salinity stress on the performance of tomato plants. Gesunde Pflanzen 72:4
Najafabadi MY, Ehsanzadeh P (2017) Salicylic acid effects on osmoregulation and seed yield in drought-stressed sesame. Agron J 109:1414–1422
Nigam B, Dubey RS, Rathore D (2022) Protective role of exogenously supplied salicylic acid and PGPB (Stenotrophomonas sp.) on spinach and soybean cultivars grown under salt stress. Sci Hortic 293:110654. https://doi.org/10.1016/j.scienta.2022.110654
Nijabat A, Bolton A, Mahmood-ur-Rehman M, Shah AI, Hussain R, Naveed NH et al (2020) Cell membrane stability and relative cell injury in response to heat stress during early and late seedling stages of diverse carrot (Daucus carota L.) germplasm. HortScience 55(9):1446–1452
Noor I, Sohail H, Sun J, Nawaz MA, Li G, Hasanuzzaman M, Liu J (2022) Heavy metal and metalloid toxicity in horticultural plants: tolerance mechanism and remediation strategies. Chemosphere 303:135196
Noreen S, Mahmood S, Faiz S, Akhter S (2020) Plant growth regulators for cotton production in changing environment. In: Cotton production and uses. Agronomy, crop protection, and postharvest technologies, pp 119–144. https://doi.org/10.1007/978-981-15-1472-2_8
Omidi M, Khandan-Mirkohi A, Kafi M, Zamani Z, Ajdanian L, Babaei M (2022) Biochemical and molecular responses of Rosa damascena mill. cv. Kashan to salicylic acid under salinity stress. BMC Plant Biol 22:373
Opabode JT, Okewale MO, Ibrahim OR, Francisca A (2019) Growth, biochemical, and physiological responses of water-stressed African eggplant seedlings to exogenous salicylic acid. Asian J Biol Sci 12(4):724–732
Osama S, El Sherei M, Al-Mahdy DA, Bishr M, Salama O (2019) Effect of salicylic acid foliar spraying on growth parameters, γ-pyrones, phenolic content and radical scavenging activity of drought-stressed Ammi visnaga L. plant. Ind Crop Prod 134:1–10
Otálora G, Piñero MC, Collado-González J, López-Marín J, Del Amor FM (2020) Exogenous salicylic acid modulates the response to combined salinity-temperature stress in pepper plants (Capsicum annuum L. var. Tamarin). Plants 9(12):1790

Pan DY, Fu X, Zhang XW, Liu FJ, Bi HG, Ai XZ (2020) Hydrogen sulfide is required for salicylic acid–induced chilling tolerance of cucumber seedlings. Protoplasma 257:1543–1557
Pandey P, Irulappan V, Bagavathiannan MV, Senthil-Kumar M (2017) Impact of combined abiotic and biotic stresses on plant growth and avenues for crop improvement by exploiting physio-morphological traits. Front Plant Sci 8:537
Park SY, Kim HT, Oh MM (2014) Effect of exogenous application of salicylic acid or nitric oxide on chilling tolerance and disease resistance in pepper seedlings. J Bio-Environ Control 23(4):329–336
Pirasteh-Anosheh H, Ranjbar G, Pakniyat H, Emam Y (2016) Physiological mechanisms of salt stress tolerance in plants: an overview. In: Hakeem KR (ed) Plant-environment interaction: responses and approaches to mitigate stress. Wiley, pp 141–160. https://doi.org/10.1002/9781119080992.ch8
Popova LP, Maslenkova LT, Yordanova RY, Ivanova AP, Krantev AP, Szalai G, Janda T (2009) Exogenous treatment with salicylic acid attenuates cadmium toxicity in pea seedlings. Plant Physiol Biochem 47(3):224–231
Prakash V, Singh VP, Tripathi DK, Sharma S, Corpas FJ (2021) Nitric oxide (NO) and salicylic acid (SA): a framework for their relationship in plant development under abiotic stress. Plant Biol 23(1):39–49
Prasch CM, Sonnewald U (2015) Signaling events in plants: stress factors in combination change the picture. Environ Exp Bot 114:4–14
Preet T, Ghai N, Jindal SK, Sangha MK (2023) Salicylic acid and 24-epibrassinolide induced thermotolerance in bell pepper through enhanced antioxidant enzyme system and heat shock proteins. J Agric Sci Technol 25:171–183
Qin X, Zhang K, Fan Y, Fang H, Nie Y, Wu XL (2022) The bacterial MtrAB two-component system regulates the cell wall homeostasis responding to environmental alkaline stress. Microbiol Spectr 10:e02311–e02322
Radhakrishnan R, Lee IJ (2013) Regulation of salicylic acid, jasmonic acid, and fatty acids in cucumber (Cucumis sativus L.) by spermidine promotes plant growth against salt stress. Acta Physiol Plant 35:3315–3322
Rahdari P, Hoseini SM (2012) Drought stress: a review. In: International conference on biotechnology and environment management, Int J Agron Plant Prod, pp 443–446
Rai KK, Rai N, Aamir M, Tripathi D, Rai SP (2020) Interactive role of salicylic acid and nitric oxide on transcriptional reprogramming for high temperature tolerance in Lablab purpureus L.: structural and functional insights using computational approaches. J Biotechnol 309:113–130
Rai GK, Magotra I, Khanday DM, Choudhary SM, Bhatt A, Gupta V, Kumar P (2024) Boosting drought tolerance in tomatoes through stimulatory action of salicylic acid imparted antioxidant defense mechanisms. Agronomy 14:1227
Ram A, Verma P, Gadi BR (2014) Effect of fluoride and salicylic acid on seedling growth and biochemical parameters of watermelon (Citrullus lanatus). Fluoride 47(1):49–55
Rasheed A, Seleiman MF, Nawaz M, Mahmood A, Anwar MR, Ayub MA, Aamer M, El-Esawi MA, El-Harty EH, Batool M, Hassan MU, Wu Z, Li H (2021) Agronomic and genetic approaches for enhancing tolerance to heat stress in rice: a review. Not Bot Horti Agrobo Cluj-Napoca 49:12501
Rasouli F, Jalalian S, Hayati F, Hassanpouraghdam MB, Asadi M, Ebrahimzadeh A, Baglieri A (2024) Salicylic acid foliar application meliorates Portulaca oleraceae L. growth responses under Pb and Ni over-availability while keeping reliable phytoremediation potential. Int J Phytoremediation:1–15
Roshdy AED, Alebidi A, Almutairi K, Al-Obeed R, Elsabagh A (2021) The effect of salicylic acid on the performances of salt-stressed strawberry plants, enzymes activity, and salt tolerance index. Agronomy 11(4):775. https://doi.org/10.3390/agronomy11040775
Rudić J, Pantelić D, Oljača J, Momčilović I (2022) Effects of elevated temperature and salicylic acid on heat shock response and growth of potato microplants. Horticulturae 8(5):372

Saeed F, Rasul S, Batool S, Zafar ZU, Manzoor H (2023) Exogenous applications of salicylic acid alleviate the damaging effects of heat stress in chili (Capsicum frutescens L.) through improved antioxidant defense system. Int J Appl Exp Biol 2(1):59–68

Saheri F, Barzin G, Pishkar L, Boojar MMA, Babaeekhou L (2020) Foliar spray of salicylic acid induces physiological and biochemical changes in purslane (Portulaca oleracea L.) under drought stress. Biologia 75:2189–2200

Sahu GK (2013) Salicylic acid: role in plant physiology and stress tolerance. In: Maheshwari DK (ed) Molecular stress physiology of plants. Springer, pp 217–239. https://doi.org/10.1007/978-81-322-0807-5_9

Sajid ZA, Aftab F (2012) Role of salicylic acid in amelioration of salt tolerance in potato (Solanum tuberosum L.) under in vitro conditions. Pak J Bot 44:37–42

Santisree P, Jalli LCL, Bhatnagar-Mathur P, Sharma KK (2020) Emerging roles of salicylic acid and jasmonates in plant abiotic stress responses. In: Protective chemical agents ameliorating plant abiotic stress: biochemical and molecular perspectives, pp 342–373. https://doi.org/10.1002/9781119552154.ch17

Saraf R, Saingar S, Chaudhary S, Chakraborty D (2018) Response of plants to salinity stress and the role of salicylic acid in modulating tolerance mechanisms: physiological and proteomic insights. In: Azooz M, Ahmad P (eds) Plant-abiotic stress tolerance. Springer, pp 75–92. https://doi.org/10.1007/978-3-319-75670-6_4

Sedaghat M, Tahmasebi-Sarvestani Z, Emam Y, Mokhtassi-Bidgoli A (2017) Physiological and antioxidant responses of winter wheat cultivars to strigolactone and salicylic acid in drought. Plant Physiol Biochem 119:59–69

Shafiei N, Khaleghi E, Moallemi N (2019) Effect of salicylic acid on some morphological and biochemical characteristics of Olive (Olea europaea cv.'Konservalia') under water stress. Plant Produc 42(1):15–30

Shahid R, Li S, Gao J, Altaf MA, Jahanger A, Shakoor A (2022) The carbon emission trading policy of China: does it really boost the environmental upgrading? Energies 15:6065

Shama MA, Moussa SA, Abo El Fadel NI (2016) Salicylic acid efficacy on resistance of garlic plants (Allium sativum L.) to water salinity stress on growth, yield, and its quality. Alex Sci Exch J 37(3):483–490

Sharma A, Sidhu GPS, Araniti F, Bali AS, Shahzad B, Tripathi DK, Breestic M, Skalicky M, Landi M (2020) The role of salicylic acid in plants exposed to heavy metals. Molecules 25(3):540. https://doi.org/10.3390/molecules25030540

Shaukat K, Zahra N, Hafeez MB, Naseer R, Batool A, Batool H, Raza A, Wahid A (2022) Role of salicylic acid-induced abiotic stress tolerance and underlying mechanisms in plants. In: Emerging plant growth regulators in agriculture. Academic Press, pp 73–98. https://doi.org/10.1016/B978-0-323-91005-7.00008-4

Shin H, Min K, Arora R (2018) Exogenous salicylic acid improves freezing tolerance of spinach (Spinacia oleracea L.) leaves. Cryobiology 81:192–200

Shirani Bidabadi S, Abolghasemi R, Zheng SJ (2018) Grafting of watermelon (Citrullus lanatus cv. Mahbubi) onto different squash rootstocks as a means to minimize cadmium toxicity. Int J Phytoremediation 20(7):730–738

Siboza XI, Bertling I, Odindo AO (2014) Salicylic acid and methyl jasmonate improve chilling tolerance in cold-stored lemon fruit (Citrus limon). J Plant Physiol 171(18):1722–1731

Silva ACD, Suassuna JF, Melo ASD, Costa RR, Andrade WL, Silva DCD (2017) Salicylic acid as an attenuator of drought stress on germination and initial development of sesame. Rev Bras Eng Agríc Ambient 21:156–162

Silva JM, da Silva Júnior GB, Bonifácio A, Dutra AF, de Mello Prado R, de Alcântara NF, de Sousa RS (2023) Exogenous salicylic acid alleviates water stress in watermelon plants. Ann Appl Biol 182(1):121–130

Soliman MH, Alayafi AA, El Kelish AA, Abu-Elsaoud AM (2018) Acetylsalicylic acid enhances tolerance of Phaseolus vulgaris L. to chilling stress, improving photosynthesis, antioxidants and expression of cold stress responsive genes. Bot Stud 59:1–17

Subashchandrabose S, Ramakrishnan B, Megharaj M, Venkateswarlu K, Naidu R (2013) Mixotrophic cyanobacteria and microalgae as distinctive biological agents for organic pollutant degradation. Environ Int 51:59–72

Sun CH, Wang D, Hu YL, Li XH, Zhang WD, Sun J, Gao XL (2013) Effects of salicylic acid on physiological characteristics of strawberry leaves under drought stress. Eur J Hortic Sci 1:106–111

Szalai G, Pál M, Árendás T, Janda T (2016) Priming seed with salicylic acid increases grain yield and modifies polyamine levels in maize. Cereal Res Commun 44:537–548

Talabany NK, Albarzinji IM (2023) Effects of salicylic acid on some growth and physiological characteristics of lettuce (Lactuca sativa L.) under cadmium stress conditions. Sci J Univ Zakho 11:37–44

Tamindžić G, Ignjatov M, Miljaković D, Červenski J, Milošević D, Nikolić Z, Vasiljević S (2023) Seed priming treatments to improve heat stress tolerance of garden pea (Pisum sativum L.). Agriculture 13:439

Tiwari RK, Lal MK, Naga KC, Kumar R, Chourasia KN, Subhash S, Sharma S (2020) Emerging roles of melatonin in mitigating abiotic and biotic stresses of horticultural crops. Sci Hortic 272:109592. https://doi.org/10.1016/j.scienta.2020.109592

Toscano S, Ferrante A, Romano D (2019) Response of Mediterranean ornamental plants to drought stress. Horticulturae 5(1):6. https://doi.org/10.3390/horticulturae5010006

Wakchaure GC, Minhas PS, Meena KK, Kumar S, Rane J (2020) Effect of plant growth regulators and deficit irrigation on canopy traits, yield, water productivity, and fruit quality of eggplant (Solanum melongena L.) grown in the water-scarce environment. J Environ Manag 262:110320. https://doi.org/10.1016/j.jenvman.2020.110320

Wang LJ, Li SH (2006) Salicylic acid-induced heat or cold tolerance in relation to Ca2+ homeostasis and antioxidant systems in young grape plants. Plant Sci 170(4):685–694

Wang L-J, Fan L, Loescher W, Duan W, Liu G-J, Cheng J-S, Luo H-B, Li S-H (2010) Salicylic acid alleviates decreases in photosynthesis under heat stress and accelerates recovery in grapevine leaves. BMC Plant Biol 10:1–10

Wang W, Wang X, Huang M, Cai J, Zhou Q, Dai T, Jiang D (2021) Alleviation of field low-temperature stress in winter wheat by exogenous application of salicylic acid. J Plant Growth Regul 40:811–823

Wang J, Allan AC, Wang WQ, Yin XR (2022) The effects of salicylic acid on quality control of horticultural commodities. NZ J Crop Hortic Sci 50(2–3):99–117

Wani AB, Chadar H, Wani AH, Singh S, Upadhyay N (2017) Salicylic acid to decrease plant stress. Environ Chem Lett 15(1):101–123

Wassie M, Zhang W, Zhang Q, Ji K, Cao L, Chen L (2020) Exogenous salicylic acid ameliorates heat stress-induced damages and improves growth and photosynthetic efficiency in alfalfa (Medicago sativa L.). Ecotoxicol Environ Saf 191:110206

Wen PF, Chen JY, Wan SB, Kong WF, Zhang P, Wang W et al (2008) Salicylic acid activates phenylalanine ammonia-lyase in grape berry in response to high temperature stress. Plant Growth Regul 55:1–10

Wen C, Zhang Z, Shi Q, Duan X, Du J, Wu C, Li X (2023) Methyl jasmonate- and salicylic acid-induced transcription factor ZjWRKY18 regulates triterpenoid accumulation and salt stress tolerance in jujube. Int J Mol Sci 24(4):3899. https://doi.org/10.3390/ijms24043899

Widiastuti A, Yoshino M, Hasegawa M, Nitta Y, Sato T (2013) Heat shock-induced resistance increases chitinase-1 gene expression and stimulates salicylic acid production in melon (Cucumis melo L.). Physiol Mol Plant Pathol 82:51–55

Xie Y, Tan H, Sun G, Li H, Liang D, Xia H, Tang Y (2020) Grafting alleviates cadmium toxicity and reduces its absorption by tomato. J Soil Sci Plant Nutr 20:2222–2229

Xu LL, Fan ZY, Dong YJ, Kong J, Bai XY (2015) Effects of exogenous salicylic acid and nitric oxide on physiological characteristics of two peanut cultivars under cadmium stress. Biol Plantarum 59:171–182

Yadav SK (2010) Cold stress tolerance mechanisms in plants: a review. Agron Sustain Dev 30(3):515–527

Yan R, Liu J, Zhang S, Guo J (2024) Exogenous melatonin and salicylic acid enhance the drought tolerance of hibiscus (Hibiscus syriacus L.) by regulating photosynthesis and antioxidant system. J Soil Sci Plant Nutr 24:497–511

Yi F, Wang Z, Baskin CC, Baskin JM, Ye R, Sun H, Zhang Y, Ye X, Liu G, Yang X, Huang Z (2019) Seed germination responses to seasonal temperature and drought stress are species-specific but not related to seed size in a desert steppe: implications for effect of climate change on community structure. Ecol Evol 9:2149–2159

Yildirim E, Turan M, Guvenc I (2008) Effect of foliar salicylic acid applications on growth, chlorophyll, and mineral content of cucumber grown under salt stress. J Plant Nutr 31:593–612

Youssef SM, López-Orenes A, Ferrer MA, Calderón AA (2023) Foliar application of salicylic acid enhances the endogenous antioxidant and hormone systems and attenuates the adverse effects of salt stress on growth and yield of French bean plants. Horticulturae 9:75

Zarghami Moghaddam M, Shoor M, Ganjeali A, Moshtaghi N, Tehranifar A (2014) Effect of salicylic acid on morphological and ornamental characteristics of Petunia hybrida at drought stress. Indian J Fundam Appl Life Sci 4(3):61–66

Zhang H, Sonnewald U (2017) Differences and commonalities of plant responses to single and combined stresses. Plant J 90(5):839–855

Zhang Y, Xu L, Zhu X, Gong Y, Xiang F, Sun X, Liu L (2013) Proteomic analysis of heat stress response in leaves of radish (Raphanus sativus L.). Plant Mol Biol Report 31:195–203

Zhang Z, Lan M, Han X, Wu J, Wang-Pruski G (2020) Response of ornamental pepper to high-temperature stress and role of exogenous salicylic acid in mitigating high temperature. J Plant Growth Regul 39:133–146

Zhang Q, Li D, Wang Q, Song X, Wang Y, Yang X, Yang D (2021) Exogenous salicylic acid improves chilling tolerance in maize seedlings by improving plant growth and physiological characteristics. Agronomy 11:1341

Chapter 7
Optimizing Plant Growth and Resilience under Abiotic Stresses through Arbuscular Mycorrhizal Symbiosis

Debabrata Nath, Vandana Kumari, and Mattaparthi Lakshmi Durga

Abstract Abiotic stresses, intensified by climate change and unsustainable agricultural practices, severely constrain plant productivity and ecosystem stability. Arbuscular mycorrhizal fungi (AMF), as key biofertilizers, offer a promising nature-based solution to enhancejb plant resilience and promote sustainable agriculture. AMF form obligate symbiotic associations with most terrestrial plant roots, fundamentally improving host hydromineral acquisition (particularly phosphorus and nitrogen) and growth, especially under suboptimal environmental conditions. There is a pivotal role of AMF symbiosis in mitigating diverse abiotic stresses, including drought, salinity, heat, and heavy metal toxicity. The underlying mechanisms conferring enhanced tolerance involve multifaceted physiological and molecular adaptations: improved nutrient status and ion homeostasis, modulation of phytohormone networks, transcriptional regulation of stress-responsive genes, accumulation of compatible osmolytes, and enhanced antioxidant defense systems. By stabilizing essential metabolic processes, AMF fortify the plant's intrinsic adaptive capacity and contribute to yield stability under abiotic stress conditions. This chapter highlights current advances in understanding AMF-mediated stress mitigation, emphasizing their significant contribution to plant growth regulation and yield stability under abiotic stress conditions. It further underscores their strategic importance in developing climate-resilient and sustainable agricultural systems.

D. Nath
International Rice Research Institute (IRRI)-South Asia Regional Centre (ISARC), Varanasi, Uttar Pradesh, India

V. Kumari (✉)
Department of Soil Science, Advance Centre on Sericulture, Dr. Kalam Agricultural College (Bihar Agricultural University), Kishanganj, Bihar, India

M. L. Durga
Division of Agricultural Engineering, ICAR-Indian Institute of Agricultural Research, New Delhi, India

H. I. Mohamed (ed.), *Elicitors for Sustainable Crop Production*, Sustainability Sciences in Asia and Africa, https://doi.org/10.1007/978-981-95-3533-0_7

Keywords Abiotic stress · Arbuscular mycorrhizal fungi · Salinity · Drought · Heavy metal

7.1 Introduction

Global population growth has diminished arable land and agricultural output through industrialization, urbanization, and globalization. Concurrently, climate change intensifies extreme weather, including droughts, heatwaves, and floods (Nath et al. 2025b; Kastner et al. 2021). While export and biofuel crops offer economic and energy advantages, they can compromise food security and soil condition (Byerlee et al. 2017). Sustainable agriculture, integrating natural resource use with minimal environmental harm, is vital for enduring economic and ecological stability (Finucane et al. 2020; Ghosh et al. 2019). Soil, fundamental to food systems, demands sustainable management. Arbuscular mycorrhizal fungi (AMF) significantly improve crop performance under both favorable and stressful conditions by enhancing nutrient acquisition, notably phosphorus, and strengthening plant tolerance (Sahoo et al. 2022; Brevik et al. 2020). This mutualism entails resource exchange: AMF release enzymes and hormones aiding nutrient solubilization and root growth, while host plants provide carbon for energy (Riaz et al. 2021; Chanclud and Morel 2016). These processes collectively boost plant water and nutrient absorption, elevate pathogen resistance, and aid nutrient release from complex organic matter (Chanclud and Morel 2016; Wang et al. 2022). In farming systems, AMF use enhances soil nutrient availability and plant photosynthetic capacity, diminishing reliance on synthetic fertilizers and pesticides (Luty et al. 2021). AMF colonization also fortifies plants against abiotic stress via intricate molecular and physiological interactions, resulting in better photosynthesis, gas exchange, and water use efficiency (Birhane et al. 2012). The symbiosis increases tolerance to drought, salinity, temperature extremes, and heavy metals (Wahab et al. 2023; Salam et al. 2017). Conversely, disturbances to microbial communities can degrade soil health, impair nutrient cycling and water retention, and reduce crop output (Nath et al. 2017, 2025a; O'Callaghan et al. 2022). Comprehending AMF-influenced microbial dynamics is essential for managing soil fertility sustainably, conserving ecosystems, and enhancing yield efficiency under stress (Wild et al. 2022; Bhattacharjee et al. 2022; Tiwari et al. 2022; Dong et al. 2022). AMF-colonized plants exhibit greater resilience to abiotic stresses like drought, salinity, and temperature, enabling higher productivity and survival (Sangwan and Prasanna 2022). Utilizing these symbiotic associations can advance agricultural efficiency and foster sustainable farming for future needs (Xiao et al. 2022). This chapter examines the ecosystem services delivered by AMF and their prospective contribution to sustainable soil fertility management and crop production amidst various abiotic stresses.

7.2 Background of Arbuscular Mycorrhizal Fungi

Arbuscular mycorrhizal fungi (AMF) are fundamental to plant adaptation and stress tolerance, establishing mutualistic partnerships across diverse plant taxa. These symbioses enable sophisticated signaling mechanisms between fungal and plant partners, resulting in augmented physiological performance, including elevated photosynthetic rates, refined gas exchange, and superior water absorption efficiency (Birhane et al. 2012). Substantial evidence confirms AMF's critical function in reinforcing plant protection against diverse abiotic and biotic challenges, such as drought, salinity, herbivore damage, thermal variability, metal contamination, and pathogenic invasions. Such associations promote heightened resilience, increased vitality, and improved yield under environmentally constrained settings (Salam et al. 2017). Roughly 90% of terrestrial flora, spanning angiosperms, bryophytes, and pteridophytes, engage in symbiotic alliances with AMF. Key fungal structures (vesicles, arbuscules, and extraradical hyphae) are indispensable for root system expansion and soil ecosystem establishment, enabling effective resource interchange between plants and their rhizosphere environment. AMF-generated hyphal networks substantially amplify host capacity to exploit soil resources, driving developmental improvements and enhanced physiological condition (Bowles et al. 2016). These organisms markedly advance soil fertility through increased nutrient accessibility and efficient assimilation/distribution within plant organs (Rouphael et al. 2015). Furthermore, AMF actively modifies soil architecture and physical characteristics, supporting enduring pedological sustainability.

By stimulating the mobilization of carbon from aerial plant parts to root-associated fungal structures, AMF also accelerates the decomposition of organic matter and nutrient cycling within the soil. Given their fundamental role in enhancing plant nutrition and soil health, AMF are increasingly recognized as valuable biofertilizers in sustainable agriculture. This chapter highlights their significance in improving crop performance and nutrient efficiency under stressful environmental conditions. Harnessing the potential of AMF across various agroecosystems presents a promising avenue for enhancing plant growth, optimizing yield, and supporting resilient agricultural systems (Fig. 7.1).

7.3 Characteristics of AMF Symbiosis

Arbuscular mycorrhizal fungi (AMF) have coexisted symbiotically with terrestrial plants for ~400 million years (Selosse et al. 2015). This mutualism originates from complex coevolutionary processes, delivering significant ecological and agronomic advantages (Begum et al. 2019; Van der Heijden et al. 2015). AMF enhance host physiology by mediating bidirectional resource transfer through extensive extraradical mycelia that surpass root depletion zones. Concurrent colonization of multiple plant forms common mycorrhizal networks (CMNs), recognized as critical

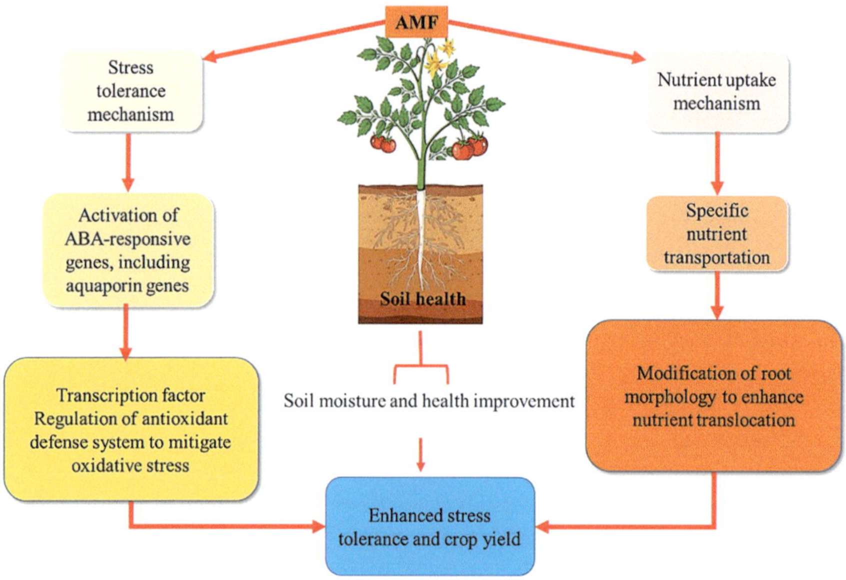

Fig. 7.1 A schematic representation of mycorrhizal functions in regulating ecosystem processes and enhancing plant growth under abiotic stress conditions

belowground infrastructure in terrestrial ecosystems. CMNs regulate plant community dynamics, including invasive species establishment (Pringle et al. 2009) and facilitate interplant flux of nitrogen, phosphorus, and other nutrients. AMF further enable nutrient translocation from fungal tissues to hosts, a mechanism essential for augmenting plant tolerance to abiotic stressors (Nath and Meena 2018).

7.3.1 Crop Defense Mechanism of AMF

Plant immunity is enhanced by arbuscular mycorrhizal fungi (AMF), making plants more resistant to pests and pathogens (Begum et al. 2019). Plant growth and development are also enhanced by AMF, which produce antibiotic compounds that protect plants from specific pathogens (Edlinger et al. 2022). The fungi modulate plants' defense pathways, allowing them to colonize without activating the immune system. Furthermore, AMF enhances plant resilience to environmental stress by gaining access to nutrients that are otherwise inaccessible (Nie et al. 2015). The mycorrhizal colonization of crops boosts crop defenses by reallocating defense molecules (Babikova et al. 2014). As a result of colonization, plants acquire more nutrients and grow faster, as well as resist phloem-feeding insects better than those that are not colonized. In response to AMF-mediated defenses, aphid-repellent volatile compounds are produced, preventing pest infestations and disease transmission. Additionally, the symbiotic exchange of nutrients, water, and carbon between AMF and plant roots enhances disease resistance by producing antibacterial and antifungal secondary metabolites (War et al. 2012). When pathogens are present, AMF trigger resistance mechanisms, including the release of phenolic compounds and the formation of physical barriers that inhibit pathogen growth (Nanjundappa et al. 2019; George and Ray 2023). Using Glomus intraradices, for example, significantly reduced *Fusarium solani* DNA in mycorrhizospheres, rhizospheres, and bulk soils (Chen et al. 2018). It is important to understand how AMF interacts with plants, soil, and other microorganisms to develop agronomic practices that increase crop yields, reduce soil erosion, and minimize pesticide use (Khaliq et al. 2022). Using AMF can reduce the use of chemical treatments and promote biocides that are safer for the environment (Loo et al. 2022; Liang et al. 2022).

7.3.2 Enhancing Crop Productivity and Soil Health Through Arbuscular Mycorrhizal Fungi

AMF improve soil structure and fertility, which are crucial to nutrient cycling and erosion resistance (Dhiman et al. 2022; Watts-Williams 2022). As a result of excessive chemical fertilizer and pesticide use, pollution and health risks are associated with agriculture, emphasizing the importance of integrating AMF. It remains challenging to produce large quantities of AMF, and crop rotation, reduced tillage, monoculture, and careful use of biocides significantly affect AMF populations (Watts-Williams 2022; Gao et al. 2020). AMF are indispensable in tropical agriculture where phosphorus limitation prevails. Given that ~75% of soil phosphorus exists in plant-unavailable forms, AMF-mediated acquisition becomes critical (Adisa et al. 2019). Crop dependence on AMF for nutrient assimilation varies with root architecture, growth kinetics, and edaphic factors. In *Zea mays* and *Linum usitatissimum*, AMF symbiosis constitutes the primary phosphorus source (Reicosky et al. 2021; Kaur 2020). This mutualism enhances phosphate and nitrogen assimilation efficiency while concurrently elevating abiotic stress tolerance. Through coordinated morphological, physiological, and molecular adaptations, AMF colonization augments:

- Hydraulic and mineral nutrient uptake
- Stress resilience mechanisms
- Biomass accumulation under environmental constraints

Furthermore, AMF stimulate phytochemical biosynthesis (e.g., essential oils) and upregulate antioxidant enzymes, including superoxide dismutase (SOD), peroxidase (POD), and ascorbate peroxidase (APX), thereby enhancing medicinal species' metabolic profiles (Babikova et al. 2014). As illustrated in Fig. 7.2, this symbiosis is fundamental to plant vigor and development.

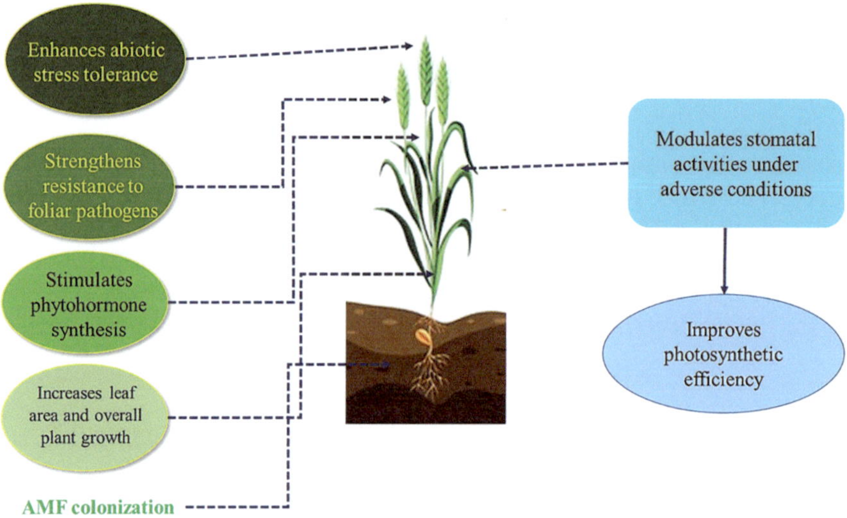

Fig. 7.2 Role of arbuscular mycorrhizal fungi (AMF) in plant growth and stress tolerance

7.4 Role of AMF in Mitigating Abiotic Stress

The balance of nitrogen dynamics in soil is enhanced by crop rotation, such as alternating corn and soybeans. While corn depletes nitrogen, soybeans replenish it through nitrogen fixation (Kaur 2020). However, the benefits of rotation extend beyond nutrient management. Arbuscular mycorrhizal fungi (AMF) play a vital role in improving yield potential (Salomon et al. 2021). The non-mycorrhizal fungi preserve beneficial microbes by acting as reservoirs for soil inoculum. In addition to increasing soil productivity, mycorrhizal plants promote the spread of these microbes (Li et al. 2022). AMF colonization was higher in maize following sunflower (*Helianthus annuus*) than in maize following mustard (*Brassica juncea*) (Diagne et al. 2020)). Corn planted after canola (*Brassica napus* L.) showed significantly lower AMF colonization rates than corn planted after mycorrhizal crops. The colonization of AMF is enhanced when mycorrhizal hosts like alfalfa (*Medicago sativa* L.) or brome grass (*Bromus* spp.) are planted before corn. By incorporating mycorrhizal crops after non-mycorrhizal crops, the inhibitory effects of non-mycorrhizal vegetation can be mitigated (Konvalinková et al. 2017).

7.4.1 *Mitigation of Drought Stress*

Drought stress significantly impairs plant development through diminished transpiration, oxidative injury (Hasanuzzaman et al. 2013), and disruption of enzymatic activity, ion homeostasis, and nutrient uptake (Ahanger and Agarwal 2017).

Arbuscular mycorrhizal symbioses confer drought resilience across multiple crops (*Triticum aestivum*, *Zea mays*, *Fragaria ananassa*) (Begum et al. 2019; Yooyongwech et al. 2016; Moradtalab et al. 2019) via interconnected mechanisms:

- *Hyphal Network Expansion*: Extraradical hyphae enhance soil exploration and resource acquisition (Gutjahr and Paszkowski 2013; Zhang et al. 2016)
- *Biochemical Modulation*: Osmotic adjustment through osmolyte (e.g., proline) and antioxidant (e.g., glutathione) accumulation (Begum et al. 2019; Ruiz-Sánchez et al. 2010; Yooyongwech et al. 2013; Rani 2016)
- *Hormonal Regulation*: ABA-mediated stomatal response optimization (Ludwig-Müller 2010)
- *Morphological Adaptation*: Augmented root proliferation, foliar expansion, and biomass production (Al-Karaki et al. 2004; Gholamhoseini et al. 2013)
- *Physiological Enhancement*: Improved plant water status, stomatal conductance, and gas exchange parameters (Morte et al. 2000; Mena-Violante et al. 2006)

Additionally, Arbuscular mycorrhizal (AM) symbiosis activates antioxidant defense systems, boosting photosynthetic efficiency in both C_3 and C_4 plants (Li et al. 2019).

7.4.2 Mitigation of Salinity Stress

Soil salinization threatens global food security by impairing plant growth, reducing carbon assimilation and yield (Ahanger et al. 2017a), and inducing ROS-mediated cellular damage (Ahanger et al. 2017b, 2018). Arbuscular mycorrhizal (AM) symbiosis mitigates salinity stress, enhancing crop growth/yield (Talaat and Shawky 2014; Abdel Latef and Chaoxing 2014) through:

- *Physiological Optimization*: Improved photosynthesis, stomatal conductance (Ait-El-Mokhtar et al. 2019), leaf water potential, water use efficiency (WUE) (El-Nashar 2017; Elhindi et al. 2017), leaf area, biomass (Borde et al. 2010; Elhindi et al. 2017), and chlorophyll (Cekic et al. 2012; Elhindi et al. 2017)
- *Nutrient & Ion Regulation*: Enhanced acquisition of P, N, K^+, Ca^{2+}, and Mg^{2+} (Hashem et al. 2018) with reduced Na^+ accumulation (Cekic et al. 2012) and ionic homeostasis
- *Biochemical Responses*: Proline-Osmoprotectant synthesis (Santander et al. 2019), cytokinins, polyamines—hormone modulation (Hameed et al. 2014; Kapoor et al. 2013), organic acid accumulation (Sheng et al. 2011), and oxidative stress alleviation via reduced lipid peroxidation (Abdel Latef and Chaoxing 2014)

Synergistically, these mechanisms underscore the integral role of AMF in bolstering plant resilience to salinity stress, supporting sustainable agricultural productivity (Table 7.1).

Table 7.1 Role of arbuscular mycorrhizal fungi in enhancing plant resilience to abiotic stresses

Host plant	AMF strains	Abiotic stresses	Responses	References
Euonymus maackii Rupr	*Rhizophagus intraradices*	Salinity	Demonstrated improved photosynthetic efficiency, enhanced nutrient uptake, and boosted antioxidant enzyme activity	Klinsukon et al. (2021)
Citrullus lanatus L.	*Glomus mosseae Gigaspora gigantean*	Salinity	Exhibited expanded leaf area, larger fruit size, stronger root growth, increased nutrient accumulation, and heightened antioxidant enzyme activity	Mathur et al. (2018)
Eucalyptus camaldulensis	*Glomus* spp., *Gigaspora albida*	Salinity	Showed increased levels of photosynthetic pigments, decreased leaf proline content, and improved physiological and biochemical traits	Cabral et al. (2016)
Zea mays L.	*Rhizophagus intraradices, Funneliformis mosseae*	High temperature	Displayed enhanced growth due to better photosynthetic transpiration rate and pigment accumulation	Liu et al. (2016)
Triticum aestivum	*Rhizophagus irregularis, Funneliformis mosseae*	High temperature	Demonstrated improved nutrient absorption and increased grain yield	Zhu et al. (2016)
Zea mays L.	*Glomus tortuosum*	Temperature stress	Recorded higher shoot nitrogen, phosphorus, potassium, and copper levels, along with increased nitrate reductase activity	Chu et al. (2016)
Elymus nutans Griseb.	*Funneliformis mosseae*	Cold stress	Showed improved antioxidant enzyme activity, increased photosynthetic pigments, and enhanced plant growth	Nogales et al. (2020)
Hordeum vulgare L	*Glomus versiforme, Rhizophagus irregularis*	Cold stress	Demonstrated higher antioxidant and osmoprotectant levels, improved potassium uptake, and increased phenolic metabolism	Reva et al. (2021)
Cucumis sativus L.	*Rhizophagus irregularis*	Cold stress	Showed increased efficiency in photosynthesis and improved carbon sink dynamics	Mirshad and Puthur (2016)

(continued)

Table 7.1 (continued)

Host plant	AMF strains	Abiotic stresses	Responses	References
Saccharum arundinaceum	*Glomus* spp.	Drought	Plants exhibited higher levels of antioxidant enzymes, phenolics, chlorophyll content, and overall biomass accumulation	Tallapragada et al. (2016)
Triticum aestivum	*Glomus mosseae*	Drought	Showed enhanced chlorophyll content, greater antioxidant enzyme activity, and increased N, P, and K concentrations	Chitarra et al. (2016)
Ipomoea batatas	*Glomus* spp.	Drought	Osmoprotectants played a role in modulating the plant's osmotic potential	Pal and Pandey (2016)
Lonicera japonica Thunb.	*Rhizophagus intraradices*, *Glomus versiforme*	Cadmium (Cd) stress	Reduced Cd accumulation in shoots and roots, with a greater retention in roots, indicating improved Cd tolerance	Garg and Singh (2018)
Cajanus cajan L.	*Rhizophagus irregularis*	Heavy metal stress (cadmium and zinc)	Observed increased root biomass, higher macro- and micronutrient levels, and elevated proline production	Shahvali et al. (2020)
Zea mays L.	*Glomus intraradices*	Heavy metal stress (cadmium)	Noted effects on soil alkalinization, cadmium immobilization, and reduced Cd bioavailability	Cui et al. (2019)

7.4.3 Mitigation of Heavy Metal Stress

Arbuscular mycorrhizal fungi (AMF) bolster plant defense systems, stimulate growth, and alleviate heavy metal toxicity in contaminated soils. Excessive accumulation of metals like Cd severely inhibits plant development, causing symptoms such as stunted roots/shoots, chlorosis, and mortality (Begum et al. 2019; Liu et al. 2013).

AMF employ diverse mechanisms for heavy metal detoxification:

- *Metal Sequestration*: Metals (e.g., Al, Cd) are immobilized within fungal structures (hyphae, cell walls, vacuoles) or via chelation, reducing phytotoxicity and enhancing nutrient uptake in hosts like *Triticum aestivum* (Ouziad et al. 2005; Punamiya et al. 2010).
- *Growth Dilution & Nutrient Enhancement*: Colonization increases plant biomass and essential nutrient acquisition (P, Cu, Zn), diluting metal concentrations in tissues (Begum et al. 2019; Kanwal et al. 2015; Miransari 2017).
- *Soil Immobilization*: AMF contribute to metal fixation within the soil matrix (Andrade and Silveira 2008; Garg and Chandel 2012).

- *Restricted Translocation*: Uptake regulation and reduced root-to-shoot transfer occur via AMF's high cation-exchange capacity (Dong et al. 2008; Li et al. 2015). This mechanism stabilizes polluted soils and limits accumulation, as seen in *Lolium* spp. under Cd/Ni/Zn stress (Takács and Vörös 2003).
- *Nutrient-Mediated Detoxification*: Enhanced Si transfer to hosts (e.g., *Glycine max*, *Zea mays*) occurs via fungi like *Rhizophagus irregularis* (Yost and Fox 1982; Clark and Zeto 2000).
- *Soil Chemistry Modification*: Detoxification involves pH elevation, Cd storage in extraradical mycelium, glomalin binding, and polyphosphate precipitation (Janouškova and Pavlíková 2010; Li et al. 2016).
- *Intraplant Metal Partitioning*: AMF alter metal speciation and compartmentalization, exemplified by Cd sequestration in *Oryza sativa* vacuoles/cell walls (Li et al. 2016) and modified Cd chemical forms enhancing tolerance in *Medicago sativa* (Wang et al. 2012).

7.4.4 Mitigation of Temperature (High and Low) Stress

Arbuscular mycorrhizal fungi (AMF) play a significant role in shaping plant community responses to rising soil temperatures (Bunn et al. 2009). Elevated temperatures not only reduce plant vigor but also negatively impact key physiological processes, including seed germination, growth, biomass accumulation, and reproductive development. Heat stress can lead to leaf damage, premature senescence, fruit discoloration, lower yields, and heightened oxidative stress (Hasanuzzaman et al. 2013). However, plants colonized by AMF generally exhibit improved growth and stress tolerance under high-temperature conditions compared to uninoculated plants (Gavito et al. 2005). For instance, the *Glomus fasciculatum* association has been reported to support plant growth under heat stress (Maya and Matsubara 2013). Beyond promoting growth, AMF contribute to enhanced water retention, synthesis of secondary metabolites, protein accumulation, and overall plant immunity (Zhu et al. 2010b). Similarly, under cold stress, AMF facilitates efficient water use, osmotic adjustments, and chlorophyll biosynthesis (Abdel Latef and Chaoxing 2011). These symbiotic interactions lead to increased metabolite accumulation and modifications in protein composition in cold-stressed plants, including tomatoes and various vegetables (Begum et al. 2019; Abdel Latef and Chaoxing 2011).

7.4.5 AMF and Combined Abiotic Stresses

In the presence of increasing soil temperatures, AMF interactions can influence the responses of plant communities (Bunn et al. 2009). The effects of heat stress include reduced plant vigor, stunted growth, decreased biomass, damage to leaves and reproductive organs, premature senescence, discoloration of fruit, reduced yield, and increased oxidative stress (Hasanuzzaman et al. 2013). Plants inoculated with AMF grow more rapidly under heat stress compared to plants not inoculated (Gavito et al. 2005). For instance, the *Glomus fasciculatum* association enhances plant growth at high temperatures (Maya and Matsubara 2013). Plants treated with AMF grow faster, retain more water, produce secondary metabolites, and become less susceptible to cold stress (Zhu et al. 2010b). Furthermore, chlorophyll synthesis also helps plants survive during cold weather by enhancing water use efficiency and adjusting osmotic pressure (Zhu et al. 2010a). Tomatoes and other vegetables, which are cold-stressed, benefit from symbiotic interactions that increase metabolite concentrations and change protein content (Abdel Latef and Chaoxing 2011).

7.5 Conclusion

Arbuscular mycorrhizal fungi (AMF) play a pivotal role in enhancing nutrient uptake, particularly phosphorus, nitrogen, zinc, and other essential micronutrients. Beyond nutrient acquisition, AMF helps in mitigating environmental stresses, such as salinity, drought and temperatures (high or low), thereby improving crop yields across various crops and vegetables. Beyond nutrient uptake, AMF improve soil structure, stimulate beneficial microbial activity, and strengthen plant defense mechanisms collectively contributing to improved crop productivity and resilience. Future research should prioritize identifying the genes, proteins, and signaling pathways involved in AMF-mediated stress tolerance to better understand their regulatory networks and functional mechanisms. Breeding or selecting cultivars with enhanced AMF compatibility and climate resilience will further optimize these benefits. Overall, harnessing AMF as a biofertilizer offers a promising and environmentally friendly approach to strengthen climate-smart agriculture, ensure food security, and build resilient agroecosystems.

References

Abdel Latef AA, Chaoxing H (2011) Arbuscular mycorrhizal influence on growth, photosynthetic pigments, osmotic adjustment and oxidative stress in tomato plants subjected to low temperature stress. Acta Physiol Plant 33:1217–1225

Abdel Latef AA, Chaoxing HJ (2014) Does the inoculation with Glomus mosseae improve salt tolerance in pepper plants? J Plant Growth Regul 33:644–653

Adisa IO, Pullagurala VLR, Peralta-Videa JR, Dimkpa CO, Elmer WH, Gardea-Torresdey JL, White JC (2019) Recent advances in nano-enabled fertilizers and pesticides: a critical review of mechanisms of action. Environ Sci Nano 6:2002–2030

Ahanger MA, Agarwal RM (2017) Potassium up-regulates antioxidant metabolism and alleviates growth inhibition under water and osmotic stress in wheat (Triticum aestivum L.). Protoplasma 254(4):1471–1486

Ahanger MA, Tittal M, Mir RA, Agarwal RM (2017a) Alleviation of water and osmotic stress-induced changes in nitrogen metabolizing enzymes in Triticum aestivum L. cultivars by potassium. Protoplasma 254(5):1953–1963

Ahanger MA, Tomar NS, Tittal M, Argal S, Agarwal RM (2017b) Plant growth under water/salt stress: ROS production; antioxidants and significance of added potassium under such conditions. Physiol Mol Biol Plants 23(4):731–744

Ahanger MA, Alyemeni MN, Wijaya L, Alamri SA, Alam P, Ashraf M et al (2018) Potential of exogenously sourced kinetin in protecting Solanum lycopersicum from NaCl-induced oxidative stress through up-regulation of the antioxidant system, ascorbate–glutathione cycle and glyoxalase system. PLoS One 13(9):e0202–e0175

Ait-El-Mokhtar M, Laouane RB, Anli M, Boutasknit A, Wahbi S, Meddich A (2019) Use of mycorrhizal fungi in improving tolerance of the date palm (Phoenix dactylifera L.) seedlings to salt stress. Sci Hortic 253:429–438

Al-Karaki G, Mcmichael B, Zak J (2004) Field response of wheat to arbuscular mycorrhizal fungi and drought stress. Mychorrhiza 14:263–269

Andrade SAL, Silveira APD (2008) Mycorrhiza influence on maize development under cd stress and P supply. Braz J Plant Physiol 20(1):39–50

Babikova Z, Gilbert L, Bruce T, Dewhirst SY, Pickett JA, Johnson D (2014) Arbuscular mycorrhizal fungi and aphids interact by changing host plant quality and volatile emission. Funct Ecol 28:375–385

Begum N, Qin C, Ahanger MA, Raza S, Khan MI, Ashraf M, Ahmed N, Zhang L (2019) Role of arbuscular mycorrhizal fungi in plant growth regulation: implications in abiotic stress tolerance. Front Plant Sci 10:1068

Bhattacharjee A, Qafoku O, Richardson JA, Anderson LN, Schwarz K, Bramer LM, Lomas GX, Orton DJ, Zhu Z, Engelhard MH (2022) A mineral-doped micromodel platform demonstrates fungal bridging of carbon hot spots and hyphal transport of mineral-derived nutrients. Msystems 7:e0091322

Birhane E, Sterck F, Fetene M, Bongers F, Kuyper T (2012) Arbuscular mycorrhizal fungi enhance photosynthesis, water use efficiency, and growth of frankincense seedlings under pulsed water availability conditions. Oecologia 169:895–904

Borde M, Dudhane M, Jite PK (2010) AM fungi influences the photosynthetic activity, growth and antioxidant enzymes in Allium sativum L. under salinity condition. Not Sci Biol 2:64–71

Bowles TM, Barrios-Masias FH, Carlisle EA, Cavagnaro TR, Jackson LE (2016) Effects of arbuscular mycorrhizae on tomato yield, nutrient uptake, water relations, and soil carbon dynamics under deficit irrigation in field conditions. Sci Total Environ 566:1223–1234

Brevik EC, Slaughter L, Singh BR, Steffan JJ, Collier D, Barnhart P, Pereira P (2020) Soil and human health: current status and future needs. Air Soil Water Res 13:1178622120934441

Bunn R, Lekberg Y, Zabinski C (2009) Arbuscular mycorrhizal fungi ameliorate temperature stress in thermophilic plants. Ecology 90(5):1378–1388

Byerlee D, Falcon WP, Naylor R (2017) The tropical oil crop revolution: food, feed, fuel, and forests. Oxford University Press, Oxford

Cabral C, Ravnskov S, Tringovska I, Wollenweber B (2016) Arbuscular mycorrhizal fungi modify nutrient allocation and composition in wheat (Triticum aestivum L.) subjected to heat-stress. Plant Soil 408:385–399

Cekic FO, Unyayar S, Ortas I (2012) Effects of arbuscular mycorrhizal inoculation on biochemical parameters in capsicum annuum grown under long term salt stress. Turk J Bot 36:63–72

Chanclud E, Morel JB (2016) Plant hormones: a fungal point of view. Mol Plant Pathol 17:1289–1297

Chen M, Arato M, Borghi L, Nouri E, Reinhardt D (2018) Beneficial services of arbuscular mycorrhizal fungi–from ecology to application. Front Plant Sci 9:1270

Chitarra W, Pagliarani C, Maserti B, Lumini E, Siciliano I, Cascone P, Schubert A, Gambino G, Balestrini R, Guerrieri E (2016) Insights on the impact of arbuscular mycorrhizal symbiosis on tomato tolerance to water stress. Plant Physiol 171:1009–1023

Chu X, Fu J, Sun Y, Xu Y, Miao Y, Xu Y, Hu T (2016) Effect of arbuscular mycorrhizal fungi inoculation on cold stress-induced oxidative damage in leaves of Elymus nutans Griseb. S Afr J Bot 104:21–29

Clark RB, Zeto SK (2000) Mineral acquisition by arbuscular mycorrhizal plants. J Plant Nutr 23:867–902

Cui G, Ai S, Chen K, Wang X (2019) Arbuscular mycorrhiza augments cadmium tolerance in soybean by altering accumulation and partitioning of nutrient elements, and related gene expression. Ecotoxicol Environ Saf 171:231–239

Dhiman M, Sharma L, Kaushik P, Singh A, Sharma MM (2022) Mycorrhiza: an ecofriendly biotool for better survival of plants in nature. Sustainability 14:10220

Diagne N, Ngom M, Djighaly PI, Fall D, Hocher V, Svistoonoff S (2020) Roles of arbuscular mycorrhizal fungi on plant growth and performance: Importance in biotic and abiotic stressed regulation. Diversity 12(10):370

Dong Y, Zhu YG, Smith FA, Wang Y, Chen B (2008) Arbuscular mycorrhiza enhanced arsenic resistance of both white clover Trifolium repens L. and ryegrass Lolium perenne L. plants in an arsenic-contaminated soil. Environ Pollut 155:174–181

Dong H, Huang L, Zhao L, Zeng Q, Liu X, Sheng Y, Shi L, Wu G, Jiang H, Li F (2022) A critical review of mineral–microbe interaction and co-evolution: mechanisms and applications. Natl Sci Rev 9:nwac128

Edlinger A, Garland G, Hartman K, Banerjee S, Degrune F, García-Palacios P, Hallin S, Valzano-Held A, Herzog C, Jansa J (2022) Agricultural management and pesticide use reduce the functioning of beneficial plant symbionts. Nat Ecol Evol 6:1145–1154

Elhindi KM, El-Din SA, Elgorban AM (2017) The impact of arbuscular mycorrhizal fungi in mitigating salt-induced adverse effects in sweet basil (Ocimum basilicum L.). Saudi J Biol Sci 24:170–179

El-Nashar YI (2017) Response of snapdragon Antirrhinum majus L. to blended water irrigation and arbuscular mycorrhizal fungi inoculation: uptake of minerals and leaf water relations. Photosynthetica 55(2):201–209

Finucane ML, Acosta J, Wicker A, Whipkey K (2020) Short-term solutions to a long-term challenge: rethinking disaster recovery planning to reduce vulnerabilities and inequities. Int J Environ Res Public Health 17:482

Gao X, Guo H, Zhang Q, Guo H, Zhang L, Zhang C, Gou Z, Liu Y, Wei J, Chen A (2020) Arbuscular mycorrhizal fungi (AMF) enhanced the growth, yield, fiber quality and phosphorus regulation in upland cotton (Gossypium hirsutum L.). Sci Rep 10:2084

Garg N, Chandel S (2012) Role of arbuscular mycorrhizal (AM) fungi on growth, cadmium uptake, osmolyte, and phytochelatin synthesis in Cajanus cajan (L.) Millsp. under NaCl and Cd stresses. J Plant Growth Regul 31(3):292–308

Garg N, Singh S (2018) Arbuscular mycorrhiza Rhizophagus irregularis and silicon modulate growth, proline biosynthesis and yield in Cajanus cajan L. Millsp.(pigeonpea) genotypes under cadmium and zinc stress. J Plant Growth Regul 37:46–63

Gavito ME, Olsson PA, Rouhier H, Medinapeñafiel A, Jakobsen I, Bago A (2005) Temperature constraints on the growth and functioning of root organ cultures with arbuscular mycorrhizal fungi. New Phytol 168:179–188

George NP, Ray JG (2023) The inevitability of arbuscular mycorrhiza for sustainability in organic agriculture—a critical review. Front Sustain Food Syst 7:1124688

Gholamhoseini M, Ghalavand A, Dolatabadian A, Jamshidi E, Khodaei Joghan A (2013) Effects of arbuscular mycorrhizal inoculation on growth, yield, nutrient uptake and irrigation water productivity of sunflowers grown under drought stress. Agric Water Manag 117:106–114

Ghosh A, Khanra S, Haldar G, Bhowmick TK, Gayen K (2019) Diverse cyanobacteria resource from north east region of India for valuable biomolecules: phycobiliprotein, carotenoid, carbohydrate and lipid. Curr Biochem Eng 5:21–33

Gutjahr C, Paszkowski U (2013) Multiple control levels of root system remodeling in arbuscular mycorrhizal symbiosis. Front Plant Sci 4:204

Hameed A, Dilfuza E, Abd-Allah EF, Hashem A, Kumar A, Ahmad P (2014) Salinity stress and arbuscular mycorrhizal symbiosis in plants. In: Miransari M (ed) Use of microbes for the alleviation of soil stresses, vol 1. Springer Science+Business Media, New York, pp 139–159

Hasanuzzaman M, Gill SS, Fujita M (2013) Physiological role of nitric oxide in plants grown under adverse environmental conditions. In: Tuteja N, Gill SS (eds) Plant acclimation to environmental stress. Springer Science+Business Media, New York, pp 269–322

Hashem A, Alqarawi AA, Radhakrishnan R, Al-Arjani AF, Aldehaish HA, Egamberdieva D et al (2018) Arbuscular mycorrhizal fungi regulate the oxidative system, hormones and ionic equilibrium to trigger salt stress tolerance in Cucumis sativus L. Saudi J Biol Sci 25(6):1102–1114

Janouškova M, Pavlíková D (2010) Cadmium immobilization in the rhizosphere of arbuscular mycorrhizal plants by the fungal extraradical mycelium. Plant Soil 332:511–520

Kanwal S, Bano A, Malik RN (2015) Effects of arbuscular mycorrhizal fungi on metals uptake, physiological and biochemical response of Medicago sativa L. with increasing Zn and Cd concentrations in soil. Am J Plant Sci 6:2906–2923

Kapoor R, Evelin H, Mathur P, Giri B (2013) Arbuscular mycorrhiza: approaches for abiotic stress tolerance in crop plants for sustainable agriculture. In: Tuteja N, Gill SS (eds) Plant acclimation to environmental stress. LLC: Springer Science+Business Media, pp 359–401

Kastner T, Chaudhary A, Gingrich S, Marques A, Persson UM, Bidoglio G, Le Provost G, Schwarzmüller F (2021) Global agricultural trade and land system sustainability: implications for ecosystem carbon storage, biodiversity, and human nutrition. One Earth 4:1425–1443

Kaur J (2020) Effects of arbuscular mycorrhizal fungi (AMF) on growth and herbivore defenses in sorghum sudangrass (Sorghum X drummondii). Master's Thesis, The University of Texas Rio Grande Valley, Rio Grande Valley, TX, USA

Khaliq A, Perveen S, Alamer KH, Zia Ul Haq M, Rafique Z, Alsudays IM, Althobaiti AT, Saleh MA, Hussain S, Attia H (2022) Arbuscular mycorrhizal fungi symbiosis to enhance plant–soil interaction. Sustainability 14:7840

Klinsukon C, Lumyong S, Kuyper TW, Boonlue S (2021) Colonization by arbuscular mycorrhizal fungi improves salinity tolerance of eucalyptus (Eucalyptus camaldulensis) seedlings. Sci Rep 11:4362

Konvalinková T, Püschel D, Řezáčová V, Gryndlerová H, Jansa J (2017) Carbon flow from plant to arbuscular mycorrhizal fungi is reduced under phosphorus fertilization. Plant Soil 419:319–333

Li H, Chen XW, Wong MH (2015) Arbuscular mycorrhizal fungi reduced the ratios of inorganic/organic arsenic in rice grains. Chemosphere 145:224–230

Li H, Luo N, Zhang LJ, Zhao HM, Li YW, Cai QY et al (2016) Do arbuscular mycorrhizal fungi affect cadmium uptake kinetics, subcellular distribution and chemical forms in rice? Sci Total Environ 571:1183–1190

Li J, Meng B, Chai H, Yang X, Song W, Li S et al (2019) Arbuscular mycorrhizal fungi alleviate drought stress in C3 (Leymus chinensis) and C4 (Hemarthria altissima) grasses via altering antioxidant enzyme activities and photosynthesis. Front Plant Sci 10:499

Li W, Li W-B, Xing L-J, Guo S-X (2022) Effect of arbuscular mycorrhizal fungi (AMF) and plant growth-promoting rhizobacteria (PGPR) on microorganism of phenanthrene and pyrene contaminated soils. Int J Phytoremediation 25:240–251

Liang T, Chen G, Zhao J, Sinha S, Zhang W (2022) AMF-Placer 2.0: open source timing-driven analytical mixed-size placer for large-scale heterogeneous FPGA. arXiv 2022,arXiv:2210.08682

Liu X, Song Q, Tang Y, Li W, Xu J, Wu J et al (2013) Human health risk assessment of heavy metals in soil–vegetable system: a multi-medium analysis. Sci Total Environ 463–464:530–540

Liu N, Chen X, Song F, Liu F, Liu S, Zhu X (2016) Effects of arbuscular mycorrhiza on growth and nutrition of maize plants under low temperature stress. Philipp Agric Sci 99:246–252

Loo WT, Chua K-O, Mazumdar P, Cheng A, Osman N, Harikrishna JA (2022) Arbuscular mycorrhizal symbiosis: a strategy for mitigating the impacts of climate change on tropical legume crops. Plants 11:2875

Ludwig-Müller J (2010) Hormonal responses in host plants triggered by arbuscular mycorrhizal fungi. In: Koltai H, Kapulnik Y (eds) Arbuscular mycorrhizas: physiology and function. Springer, Dordrecht, pp 169–190
Luty L, Musiał K, Zioło M (2021) The role of selected ecosystem services in different farming systems in Poland regarding the differentiation of agricultural land structure. Sustainability 13:6673
Mathur S, Sharma MP, Jajoo A (2018) Improved photosynthetic efficacy of maize (Zea mays) plants with arbuscular mycorrhizal fungi (AMF) under high temperature stress. J Photochem Photobiol B Biol 180:149–154
Maya MA, Matsubara Y (2013) Influence of arbuscular mycorrhiza on the growth and antioxidative activity in Cyclamen under heat stress. Mycorrhiza 23(5):381–390
Mena-Violante HG, Ocampo-Jimenez O, Dendooven L, Martinez-Soto G, Gonzalez-Castafeda J, Davies FT et al (2006) Arbuscular mycorrhizal fungi enhance fruit growth and quality of Chile ancho Capsicum annuum L. cv San Luis plants exposed to drought. Mycorrhiza 16:261–267
Miransari M (2017) Arbuscular mycorrhizal fungi and heavy metal tolerance in plants. In: Wu QS (ed) Arbuscular mycorrhizas and stress tolerance of plants. Springer Nature, Singapore, pp 174–161
Mirshad P, Puthur JT (2016) Arbuscular mycorrhizal association enhances drought tolerance potential of promising bioenergy grass (Saccharum arundinaceum retz.). Environ Monit Assess 188:425
Moradtalab N, Roghieh H, Nasser A, Tobias EH, Günter N (2019) Silicon and the association with an arbuscular-mycorrhizal fungus (Rhizophagus clarus) mitigate the adverse effects of drought stress on strawberry. Agronomy 9:41
Morte A, Lovisolo C, Schubert A (2000) Effect of drought stress on growth and water relations of the mycorrhizal association Helianthemum almeriense–Terfezia claveryi. Mycorrhiza 10:115–119
Nanjundappa A, Bagyaraj DJ, Saxena AK, Kumar M, Chakdar H (2019) Interaction between arbuscular mycorrhizal fungi and Bacillus spp. in soil enhancing growth of crop plants. Fungal Biol Biotechnol 6:23
Nath D, Maurya BR, Meena VS (2017) Documentation of five potassium-and phosphorus-solubilizing bacteria for their K and P-solubilization ability from various minerals. Biocatalysis and agricultural biotechnology 10:174–181
Nath D, Meena VS (2018) Mycorrhizae: a potential microorganism and its implication in agriculture. In: Role of rhizospheric microbes in soil: volume 1: stress management and agricultural sustainability, pp 251–276. https://doi.org/10.1007/978-981-10-8402-7_10
Nath D, Durga, ML, Kumari V, Singh S (2025a) Restoring the Soil Functions Through Biochar-Microbiome Interactions. In Biochar: A Precious Resource from Biological Waste: Applications for Soil, Plant and Environmental Health (pp. 305–319). Singapore: Springer Nature Singapore
Nath D, Bhowmick MK, Kumar S, Peramaiyan P, Singh S, ... Laik R, Kumari V (2025b) Climate-Smart Rice Production Technologies for Improving Productivity and Resilience. In Climate-Smart Agricultural Technologies: Approaches for Field Crops Production Systems (pp. 161–181). Singapore: Springer Nature Singapore
Nie M, Bell C, Wallenstein MD, Pendall E (2015) Increased plant productivity and decreased microbial respiratory C loss by plant growth-promoting rhizobacteria under elevated CO2. Sci Rep 5:9212
Nogales A, Ribeiro H, Nogales-Bueno J, Hansen LD, Gonçalves EF, Coito JL, Rato AE, Peixe A, Viegas W, Cardoso H (2020) Response of mycorrhizal 'Touriga Nacional' variety grapevines to high temperatures measured by calorespirometry and near-infrared spectroscopy. Plants 9:1499
O'Callaghan M, Ballard RA, Wright D (2022) Soil microbial inoculants for sustainable agriculture: limitations and opportunities. Soil Use Manag 38:1340–1369
Ouziad F, Hildebrandt U, Schmelzer E, Bothe H (2005) Differential gene expressions in arbuscular mycorrhizal-colonized tomato grown under heavy metal stress. J Plant Physiol 162:634–649
Pal A, Pandey S (2016) Role of arbuscular mycorrhizal fungi on plant growth and reclamation of barren soil with wheat (Triticum aestivum L.) crop. Int J Soil Sci 12:25–31

Pringle A, Bever JD, Gardes M, Parrent JL, Rillig MC, Klironomos JN (2009) Mycorrhizal symbioses and plant invasions. Ann Rev Ecol Evol Syst 40:699–715
Punamiya P, Datta R, Sarkar D, Barber S, Patel M, Da P (2010) Symbiotic role of Glomus mosseae in phytoextraction of lead in vetiver grass Chrysopogon zizanioides L. J Hazard Mater 177:465–474
Rani B (2016) Effect of arbuscular mycorrhiza fungi on biochemical parameters in wheat Triticum aestivum L. under drought conditions. Doctoral dissertation, CCSHAU, Hisar
Reicosky D, Calegari A, dos Santos DR, Tiecher T (2021) Cover crop mixes for diversity, carbon and conservation agriculture. In: Cover crops and sustainable agriculture. CRC Press, Boca Raton, pp 169–208
Reva M, Cano C, Herrera M-A, Bago A (2021) Arbuscular mycorrhizal inoculation enhances endurance to severe heat stress in three horticultural crops. HortScience 56:396–406
Riaz M, Kamran M, Fang Y, Wang Q, Cao H, Yang G, Deng L, Wang Y, Zhou Y, Anastopoulos I (2021) Arbuscular mycorrhizal fungi-induced mitigation of heavy metal phytotoxicity in metal contaminated soils: a critical review. J Hazard Mater 402:123919
Rouphael Y, Franken P, Schneider C, Schwarz D, Giovannetti M, Agnolucci M (2015) Arbuscular mycorrhizal fungi act as bio-stimulants in horticultural crops. Sci Hortic 196:91–108
Ruiz-Sánchez M, Aroca R, Muñoz Y, Polón R, Ruiz-Lozano JM (2010) The arbuscular mycorrhizal symbiosis enhances the photosynthetic efficiency and the antioxidative response of rice plants subjected to drought stress. J Plant Physiol 167:862–869
Sahoo G, Wani AM, Swamy SL, Roul PK, Dash AC, Sharma A (2022) Livelihood strategy and sustainability aspects in industrialization as a source of employment in rural areas. In: Social morphology, human welfare, and sustainability. Springer, Berlin/Heidelberg, pp 643–670
Salam EA, Alatar A, El-Sheikh MA (2017) Inoculation with arbuscular mycorrhizal fungi alleviates harmful effects of drought stress on damask rose. Saudi J Biol Sci 25(8):1772–1780
Salomon M, Watts-Williams S, McLaughlin M, Brien C, Jewell N, Berger B, Cavagnaro T (2021) Evaluation of commercial composts and potting mixes and their ability to support arbuscular mycorrhizal fungi with maize (Zea mays) as host plant. Waste Manag 134:187–196
Sangwan S, Prasanna R (2022) Mycorrhizae helper bacteria: unlocking their potential as bioenhancers of plant–arbuscular mycorrhizal fungal associations. Microb Ecol 84:1–10
Santander C, Sanhueza M, Olave J, Borie F, Valentine C, Cornejo P (2019) Arbuscular mycorrhizal colonization promotes the tolerance to salt stress in lettuce plants through an efficient modification of ionic balance. J Soil Sci Plant Nutr 19(2):321–331
Selosse MA, Strullu-Derrien C, Martin FM, Kamoun S, Kenrick P (2015) Plants, fungi and oomycetes: a 400-million years affair that shapes the biosphere. New Phytol 206:501–506
Shahvali R, Shiran B, Ravash R, Fallahi H, Đeri BB (2020) Effect of symbiosis with arbuscular mycorrhizal fungi on salt stress tolerance in GF677 (peach almond) rootstock. Sci Hortic 272:109535
Sheng M, Tang M, Zhang F, Huang Y (2011) Influence of arbuscular mycorrhiza on organic solutes in maize leaves under salt stress. Mycorrhiza 21:423–430
Takács T, Vörös I (2003) Effect of metal non-adapted arbuscular mycorrhizal fungi on Cd, Ni and Zn uptake by ryegrass. Acta Agron Hung 51:347–354
Talaat NB, Shawky BT (2014) Protective effects of arbuscular mycorrhizal fungi on wheat (Triticum aestivum L.) plants exposed to salinity. Environ Exp Bot 98:20–31
Tallapragada P, Dikshit R, Seshagiri S (2016) Influence of Rhizophagus spp. and Burkholderia seminalis on the growth of tomato (Lycopersicon esculatum) and bell pepper (Capsicum annuum) under drought stress. Commun Soil Sci Plant Anal 47:1975–1984
Tiwari J, Ma Y, Bauddh K (2022) Arbuscular mycorrhizal fungi: an ecological accelerator of phytoremediation of metal contaminated soils. Arch Agron Soil Sci 68:283–296
Van der Heijden MG, Martin FM, Selosse MA, Sanders IR (2015) Mycorrhizal ecology and evolution: the past, the present, and the future. New Phytol 205:1406–1423

Wahab A, Muhammad M, Munir A, Abdi G, Zaman W, Ayaz A et al (2023) Role of arbuscular mycorrhizal fungi in regulating growth, enhancing productivity, and potentially influencing ecosystems under abiotic and biotic stresses. Plants 12(17):3102

Wang Y, Jing H, Gao Y (2012) Arbuscular mycorrhizal colonization alters subcellular distribution and chemical forms of cadmium in Medicago sativa L. and resists cadmium toxicity. PLoS One 7:3161–3164

Wang F, Zhang L, Zhou J, Rengel Z, George TS, Feng G (2022) Exploring the secrets of hyphosphere of arbuscular mycorrhizal fungi: processes and ecological functions. Plant Soil 418:1–22

War AR, Paulraj MG, Ahmad T, Buhroo AA, Hussain B, Ignacimuthu S, Sharma HC (2012) Mechanisms of plant defense against insect herbivores. Plant Signal Behav 7:1306–1320

Watts-Williams SJ (2022) Track and trace: how soil labelling techniques have revealed the secrets of resource transport in the arbuscular mycorrhizal symbiosis. Mycorrhiza 32:257–267

Wild B, Gerrits R, Bonneville S (2022) The contribution of living organisms to rock weathering in the critical zone. npj Mater Degrad 6:98

Xiao L, Lai S, Chen M, Long X, Fu X, Yang H (2022) Effects of grass cultivation on soil arbuscular mycorrhizal fungi community in a tangerine orchard. Rhizosphere 24:100583

Yooyongwech S, Phaukinsang N, Cha-Um S, Supaibulwatana K (2013) Arbuscular mycorrhiza improved growth performance in Macadamia tetraphylla L. grown under water deficit stress involves soluble sugar and proline accumulation. Plant Growth Regul 69:285–293

Yooyongwech S, Samphumphuang T, Tisarum R, Theerawitaya C, Chaum S (2016) Arbuscular mycorrhizal fungi (AMF) improved water deficit tolerance in two different sweet potato genotypes involves osmotic adjustments via soluble sugar and free proline. Sci Hort 198:107–117

Yost RS, Fox RL (1982) Influence of mycorrhizae on the mineral contents of cowpea and soybean grown in an oxisol. Agron J 74(3):475–481

Zhang X, Li W, Fang M, Jixian Y, Meng S (2016) Effects of arbuscular mycorrhizal fungi inoculation on carbon and nitrogen distribution and grain yield and nutritional quality in rice (Oryza sativa L.). J Sci Food Agric 97:2919–2925

Zhu XC, Song FB, Xu HW (2010a) Arbuscular mycorrhizae improve low temperature stress in maize via alterations in host water status and photosynthesis. Plant Soil 331:129–137

Zhu XC, Song FB, Xu HW (2010b) Effects of arbuscular mycorrhizal fungi on photosynthetic characteristics of maize under low temperature stress. Acta Ecol Sin 21:470–475

Zhu X, Song F, Liu F (2016) Altered amino acid profile of arbuscular mycorrhizal maize plants under low temperature stress. J Plant Nutr Soil Sci 179:186–189

Chapter 8
Amino Acids in the Regulation of Abiotic Stress Tolerance

Rania Francis, Keyanna Radi, Sabine Hilal, and Tony Kevork Sajyan

Abstract Amino acids play a crucial role in regulating plant responses to abiotic stress, ensuring cellular stability, metabolic homeostasis, and stress adaptation. This chapter explores the chemical nature, biosynthesis, and classification of amino acids, emphasizing their involvement in primary and secondary metabolism. The mechanisms by which amino acids mediate stress tolerance are discussed, including their roles in stress perception, signal transduction, gene regulation, and hormonal interactions, particularly with abscisic acid, ethylene, cytokinins, and jasmonic acid.

This chapter highlights several key aspects of amino acid-mediated abiotic stress tolerance. First, it examines how amino acids contribute to osmoprotection, antioxidant defense, and metabolic balance under stress conditions. Second, it discusses the specific functions of proline, glycine betaine (GB), glutamate, glutamine, arginine, serine, and threonine in enhancing plant resilience. Proline plays a major role in osmoprotection and reactive oxygen species (ROS) scavenging, GB stabilizes cellular structures, glutamate, and glutamine participate in nitrogen assimilation and detoxification. Arginine and polyamine precursors contribute to nitric oxide signaling and antioxidant enzyme regulation, while serine and threonine are involved in metabolic cross talk and signal transduction.

Furthermore, the chapter explores the interplay between amino acids and hormonal pathways in modulating stress responses. Finally, it emphasizes the potential applications of amino acid-based strategies in improving crop tolerance through

R. Francis (✉)
Department of Agriculture and Food Engineering, School of Engineering, Holy Spirit University of Kaslik, Jounieh, Lebanon
e-mail: rania.francis@usek.edu.lb

K. Radi · S. Hilal
Faculty of Agricultural and Food Sciences, American University of Beirut, Beirut, Lebanon

T. K. Sajyan
Department of Agriculture and Food Engineering, School of Engineering, Holy Spirit University of Kaslik, Jounieh, Lebanon

Faculty of Agricultural and Food Sciences, American University of Beirut, Beirut, Lebanon

H. I. Mohamed (ed.), *Elicitors for Sustainable Crop Production*, Sustainability Sciences in Asia and Africa, https://doi.org/10.1007/978-981-95-3533-0_8

biotechnological and agronomic approaches. By elucidating these biochemical and physiological mechanisms, this chapter provides insights into how amino acids enhance plant resilience, paving the way for the development of stress-resistant crops.

Keywords Amino acids · Abiotic stress tolerance · Osmoprotection · Signal transduction · Reactive oxygen species (ROS)

8.1 Introduction

Plants are constantly exposed to various abiotic stresses, such as drought, salinity, temperature extremes, and nutrient deficiencies, which significantly affect their growth, development, and productivity. Adaptation to these stresses involves complex physiological and biochemical adjustments, among which amino acid metabolism plays a pivotal role. Amino acids not only serve as fundamental building blocks for proteins but also act as key signaling molecules and metabolic intermediates that modulate stress perception, signal transduction, and the activation of stress-responsive genes (Heinemann and Hildebrandt 2021). Different amino acids fulfill distinct functions during stress conditions; for example, proline and γ-aminobutyric acid (GABA) accumulate to mitigate osmotic imbalance and oxidative damage, while branched-chain amino acids (BCAAs) such as leucine, isoleucine, and valine contribute as compatible osmolytes or alternative energy substrates (Shim et al. 2023).

The intricate cross talk between amino acid metabolism and phytohormonal signaling pathways further enhances plant resilience. Abscisic acid (ABA) orchestrates various drought-responsive processes by regulating amino acid biosynthesis and catabolism, thereby influencing stomatal behavior and antioxidant defenses (Hasan et al. 2022). Moreover, amino acids like cysteine and GABA interact with multiple phytohormones—including auxins, cytokinins, salicylic acid (SA), and ethylene—to fine-tune stress responses through diverse molecular networks (Islam et al. 2024a, b).

Recent advances in genetic engineering have opened promising avenues to enhance amino acid synthesis in crops, aiming to improve stress tolerance and nutritional quality simultaneously (Verma et al. 2020).

This chapter aims to elucidate the mechanisms by which amino acids regulate plant responses to abiotic stresses, emphasizing their roles in signaling and hormonal interactions. Additionally, it explores specific amino acids contributing to stress tolerance and reviews genetic engineering approaches designed to enhance amino acid-mediated resilience, providing insights into future directions for sustainable crop improvement.

8.2 Mechanisms of Amino Acid-Mediated Stress Regulation

8.2.1 Stress Perception, Signal Transduction, and Regulation of Stress-Responsive Genes

Amino acid metabolism plays a critical role in plant responses to abiotic stress. Different amino acids contribute to distinct functions during stress perception. Some amino acids act as signaling molecules, enabling plants to detect and respond to stress. Others serve as alternative substrates during energy deficiency and as precursors for stress-induced proteins and secondary metabolites, which are essential for plant adaptive mechanisms in response to various environmental stresses (Heinemann and Hildebrandt 2021).

Plants perceive abiotic stress factors such as drought, salinity, and extreme temperatures through various receptors and signaling pathways (Fig. 8.1). These stress

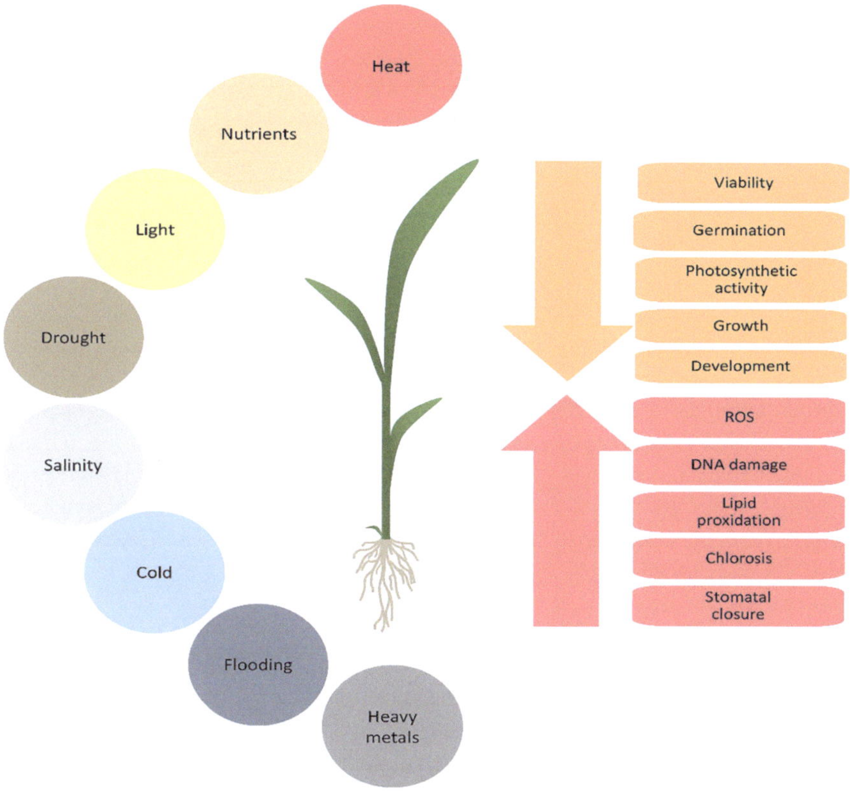

Fig. 8.1 Major abiotic stresses in plants and their effects on physiological and biochemical processes affecting plant growth and development. (Reprinted from Nawaz et al. (2022). Open Access journal)

signals lead to the activation of specific genes involved in stress responses. Upon stress perception, amino acid metabolism is adjusted to optimize energy production and protect against oxidative damage. The metabolism of different amino acid families is regulated differentially during stress. For example, the glutamate family pathway is activated to produce proline and GABA, which are involved in metabolic responses to stress. Specifically, the amino acid GABA is synthesized upon sensing changes in cellular Ca^{2+} levels, which activate the glutamate decarboxylase enzyme involved in GABA synthesis. Increased cytosolic Ca^{2+} concentrations strengthen the Ca^{2+}/CaM stress response signal, initiating the expression of genes and enzymes involved in stress-related metabolic pathways. Additionally, GABA activates genes linked to phytohormone biosynthesis, triggering physiological stress responses. These responses include the enhancement of antioxidant defenses, modulation of H_2O_2 levels, and facilitation of interactions between phytohormones and other signaling molecules (Islam et al. 2024a, b). Calcium perception also mediates proline biosynthesis during osmotic stress. ABA mediates proline biosynthesis by activating transcription factors that will activate proline biosynthesis through several signal transduction pathways, like the SnRK2 pathway, the CDPKs pathway, or the mitogen-activated protein kinase (MAPK) pathway. ABA can trigger an increase in Ca^{2+} levels, which can be sensed by the CDPKs, which are calcium-sensing proteins. These proteins will phosphorylate ABA-responsive transcription factors, leading to the activation of the P5CS1 gene involved in proline biosynthesis (Yan et al. 2025). Proline has also been shown to enhance the expression of certain salt stress-responsive genes and boost the salt tolerance of *Pancratium maritimum* by increasing the levels of stress-protective proteins and safeguarding the protein turnover system from stress-induced damage (He et al. 2017).

BCAAs like isoleucine, leucine, and valine also function as compatible osmolytes like proline. The three BCAAs share four common enzymes involved in their synthesis: acetohydroxyacid synthase, ketolacid reductoisomerase, dihydroxyacid dehydratase, and branched-chain aminotransferase. It is established that BCAA accumulation can result from either de novo synthesis or from protein degradation. Under nitrogen deficiency conditions, the OsbZIP18 transcription factor is induced and was found to upregulate the expression of *OsBCAT1* and *OsBCAT2* genes by binding to their promoters. These two genes are involved in BCAA biosynthesis. Under drought conditions, the DROUGHT-INDUCED BRANCHED-CHAIN AMINO ACID AMINOTRANSFERASE (OsDIAT) gene in rice was found to be induced and mediated the accumulation of BCAAs (Shim et al. 2023). Another study reported that *Arabidopsis* under osmotic stress accumulated BCAAs through an ABA-mediated protein degradation through proteases, rather than de novo synthesis of the amino acids (Huang and Jander 2017).

Glutamate is an amino acid that functions as a signaling molecule, mediating stress tolerance in plants through glutamate-like receptors (GLRs)-mediated calcium signaling. This signaling pathway activates various mechanisms, helping plants adapt to a broad spectrum of abiotic stresses. For instance, in *Arabidopsis* subjected to cold stress, two GLRs, AtGLR1.2 and 1.3, were found to mediate stress tolerance through the activation of the jasmonate signaling pathways as well as the

CBF/DREB1 binding factors that activate cold-responsive genes (Zheng et al. 2018). The CBF/DREB binding factors were found to play a role in abiotic stress responses in many crops, like soybean (Yamasaki and Randall 2016), lettuce (Park et al. 2020), and rice (Debnath et al. 2023).

8.2.2 Role in Hormonal Cross Talk

8.2.2.1 Interaction with Abscisic Acid

In response to environmental stresses, specific amino acids, such as serine, proline, and leucine, function as signaling molecules, whereas others are implicated as precursors in the biosynthesis of phytohormones and secondary metabolites that contribute to stress-related signaling networks (Hildebrandt et al. 2015). In particular, the phytohormone ABA has been shown to mediate several drought-responsive pathways, conferring plant resistance to water stress. The activation of the ABA signaling cascade leads to an increase in the expression of genes involved in the synthesis of metabolites, as well as physiological adjustments that confer plants drought resistance. In response to stress, ABA perception via the ABA receptor PYR/PYL/RCAR suppresses the activity of group A PP2Cs phosphatases that activate SnRK2s. This activation subsequently triggers the phosphorylation or activation of downstream targets, such as bZIP transcription factors and ion channels (You et al. 2019).

The production of ABA is tightly linked to the sulfur-containing amino acid cysteine. An elevation in the levels of cysteine within guard cell chloroplasts, triggered by the perception of sulfate signals, leads to the production of ABA. This function of cysteine as a signaling molecule has been observed in *Arabidopsis* under water stress conditions (Batool et al. 2018). As soil dries, sulfate is transported to the guard cells through the xylem, where it is incorporated into cysteine. The resulting increase in cysteine levels activates two key enzymes in the ABA biosynthesis pathway, NCED3 and AAO3, thereby promoting ABA production in the leaves. In turn, ABA triggers the degradation of cysteine into hydrogen sulfide, which acts as a signaling molecule activating the SnRK2.6 protein kinase, leading to various physiological responses such as stomatal closure to limit water losses (Hasan et al. 2022). SnRKs are protein kinases that mediate signaling pathways involved in amino acid degradation during unfavorable conditions. Conversely, under favorable conditions, they are inhibited and the TOR signaling pathway is activated, regulating cell growth and metabolism, as well as promoting anabolic processes such as amino acid synthesis, while inhibiting catabolic pathways like autophagy (Wang et al. 2018).

Under energy deprivation, ABA can also activate SnRK1 protein kinase, which promotes amino acid catabolism that can act as alternative substrates for mitochondrial respiration (Hildebrandt et al. 2015). Additionally, several studies have pointed out that ABA-mediated protein turnover leads to enhanced free amino acid content during abiotic stress conditions. ABA was found to mediate the accumulation of proline under hypoxic stress in rice and osmotic stress in wheat (Pál et al. 2018). ABA also mediates the accumulation of branched-chain amino acids (BCAAs) like leucine and valine in drought-stressed *Arabidopsis* through its regulation of protein degradation (Huang and Jander 2017).

8.2.2.2 Synergy with Other Phytohormones

Phytohormones are key metabolites involved in plant responses to abiotic stress. Recent studies have shown that phytohormones and amino acids often work synergistically to enhance abiotic stress resistance in plants (Fig. 8.2). For example, GABA, a nonprotein amino acid, has been shown to interact with various phytohormones, including auxin, cytokinins, ethylene, ABA, gibberellins, and SA, to modulate abiotic stress tolerance (Islam et al. 2024a, b). Under heavy metal stress, IAA and ABA play a key role in regulating aluminum stress by enhancing the gene expression of aluminum-activated malate transporters (ALMTs) (Podlesakova et al. 2019). ALMTs act as transporters of GABA, which can further regulate malate efflux, thereby inhibiting malate transport in plants under aluminum stress (Ramesh et al. 2018).

Furthermore, phytohormones can activate signaling cascades that regulate gene expression related to amino acid biosynthesis or catabolism, affecting the availability of amino acids. For example, under salt stress, SA regulates the activity of the

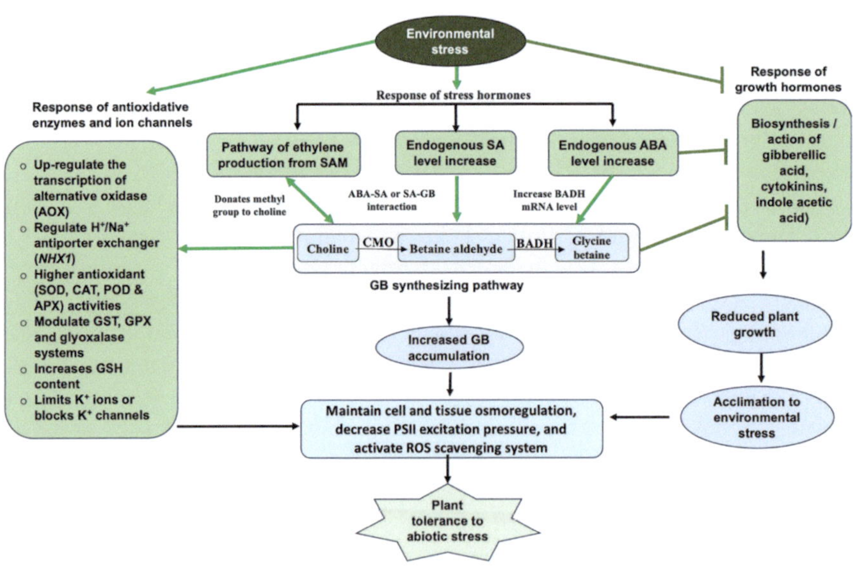

Fig. 8.2 Schematic model of glycine betaine interaction with other molecules and different molecular pathways under abiotic stress. (Reprinted from Jarin et al. (2024). Open Access journal)

GDH enzyme, which is responsible for the production of GABA's precursor, glutamate. The conversion of glutamate into GABA is carried out by the GAD, whose activity was shown to be influenced by the hormone ABA (Yang et al. 2016). Additionally, phytohormones may enhance stress tolerance by regulating the metabolism of amino acid derivatives like proline and glycine betaine (GB), which function as osmoprotectants (Sharma et al. 2019). In wheat exposed to heat stress, SA alleviated the stress by enhancing the activity of γ-glutamyl kinase, an enzyme involved in proline synthesis, while reducing the activity of proline oxidase, which is responsible for proline degradation. This led to increased proline accumulation, helping restore osmotic balance. Cross talk between ABA and cytokinins was also found to be responsible for the decrease in proline degradation during osmotic stress (Yan et al. 2025). Additionally, SA treatment inhibited enzymes involved in ethylene synthesis, which is crucial for nitrogen assimilation. Since nitrogen assimilation is essential for proline synthesis, this indicates that SA and ethylene work together with proline to mitigate the effects of heat stress (Khan et al. 2013). Similarly, Khan et al. (2014) further explored the relationship between GB, phytohormones, and abiotic stress resistance in mungbean (*Vigna radiata L.*). Under salinity stress, treatment with SA increased methionine and GB, while suppressing ethylene formation. This was linked to higher glutathione (GSH) content and lower oxidative stress, improving photosynthesis and growth. The results suggest that SA enhances GB accumulation through increased Met and suppresses ethylene production, boosting the antioxidant system and mitigating salt stress effects (Khan et al. 2014).

Exogenous applications of several phytohormones further reinforce their interconnection with amino acids during abiotic stress conditions. For example, in faba bean seeds exposed to salinity stress, the exogenous application of cytokinin and SA increased the levels of amino acids such as leucine, lysine, valine, glutamic acid, and several other nonessential amino acids (Ghassemi-Golezani and Samea-Andabjadid 2022). Similarly, in bitter melon subjected to salinity stress, the application of jasmonic acid enhanced amino acid accumulation by modulating metabolic processes, leading to improved stress resistance (Alisofi et al. 2019).

8.3 Amino Acids and their Specific Roles in Abiotic Stress Tolerance

8.3.1 Proline

Proline is a proteinogenic multifunctional amino acid that plays an important role in enhancing plant tolerance to various abiotic stressors, such as drought, salinity, extreme temperatures, and heavy metal toxicity. Proline accumulation is a conserved response in plants; however, the extent of accumulation depends on the plant species and the level of stress (Siddique et al. 2018). The abundance of proline under environmental constraints results from an increase in the de novo synthesis of the amino acid, reduced degradation, and/or protein hydrolysis (Saibi and Brini 2020). Proline is synthesized in the cytosol and then transported to the chloroplasts, where it accumulates. This accumulation is most prominent in photosynthetic tissues (Forlani et al. 2019), where, under stress conditions, proline levels can rise up to 100-fold compared to normal concentrations, which are typically much lower. This significant accumulation serves as a hallmark of stress adaptation, linked to various protective mechanisms that help mitigate cellular damage and maintain cellular integrity under adverse environmental conditions (Table 8.1).

8.3.1.1 Role of Proline in Osmotic Regulation and Ion Homeostasis

Proline is an important osmotic regulator that accumulates in the cytosol in response to drought and salt stress, two primary causes of osmotic stress in plants. Drought stress reduces water availability, leading to decreased turgor pressure in cells and stunted growth (Takahashi et al. 2020). Stomatal closure, while conserving water, limits carbon dioxide uptake, impairing photosynthesis and reducing chlorophyll content, which further compromises energy production. While drought and salinity share many similarities in their effects on plants, salinity primarily disrupts ion homeostasis. Excessive accumulation of sodium and chloride ions in plant cells replaces essential nutrients like potassium, causing metabolic disturbances. Furthermore, high salt concentrations increase the osmotic potential in the soil, hindering water absorption and creating a state of physiological drought, even when soil moisture is present (Kamran et al. 2020). In response, plants accumulate osmolytes such as proline, which help protect against water deficit by increasing cellular osmotic pressure and water retention, thereby preserving turgor pressure and preventing dehydration (Forlani et al. 2019). Proline also functions as an osmoprotectant, mitigating the denaturing effects of high ionic stress, primarily induced by salinity, thereby preserving the structural integrity of proteins and cellular components (Nahar et al. 2016).

Several studies have highlighted the importance of proline in improving osmoregulation and resilience of plants under hyperosmotic stress. For example, proline accumulation is crucial for peanuts' response to pre-flowering drought stress. Both drought-tolerant and drought-sensitive cultivars showed increased proline accumulation under stress. However, the drought-tolerant cultivar exhibited higher proline levels, better osmotic regulation, and lower oxidative damage (Zhang et al. 2017). Similarly, a comparative study between eggplant and its wild ancestor, *Solanum insanum*, found that *S. insanum* exhibited greater salt tolerance, primarily due to its higher ability to accumulate proline (Brenes et al. 2020). In tomatoes, salt tolerance was correlated with proline accumulation, resulting in decreased sodium concentration (Gharsallah et al. 2016). In transgenic *Sorghum bicolor* with an overexpressed gene encoding proline biosynthesis, proline overaccumulation resulted in higher stress tolerance compared to the control (Reddy et al. 2015). Nevertheless, the literature presents conflicting evidence regarding the role of proline in stress tolerance. While some studies found that higher proline concentrations are found in

Table 8.1 Effect of proline on crops under different abiotic stress conditions

Crop	Abiotic stress	Endogenous accumulation/ exogenous application	Effects	Reference
Peanut	Drought	Endogenous	Improved osmotic regulation and decreased oxidative damage	Zhang et al. (2017)
Tomato	Salinity	Endogenous	Reduced Na+ accumulation and increased K+ and Ca2+ accumulation	Gharsallah et al. (2016)
Fennel	Drought	Exogenous	Improved plant water relations and increased grain and essential oil yield	Zali and Ehsanzadeh (2018)
Maize	Drought	Exogenous	Promoted growth of shoots and roots, reduced cell damage, maintained water balance, preserved chlorophyll, boosted antioxidant enzyme activities	Khan et al. (2025)
Onions	Drought	Exogenous	Improved leaf water content, photosynthetic efficiency, cell membrane integrity, and upregulated osmoprotectants contents	Semida et al. (2020)
Melon	Salinity	Exogenous	Improved growth parameters and chlorophyll content	Castañares et al. (2023)
Eggplant	Salinity	Exogenous	Improved osmoregulation and antioxidant enzyme activity	Kusvuran et al. (2020)
Pea	Heavy metal and salinity	Exogenous	Reduced lipid peroxidation and electrolyte leakage, enhanced membrane stability index	Shahid et al. (2014)
Rice	Heavy metal	Exogenous	Reduced lipid peroxidation	Yu et al. (2017)
Sugar beet	Drought	Exogenous	Improved antioxidant enzyme activity, phenolic compounds, relative water content, and yield parameters	AlKahtani et al. (2021)
Citrus	Low temperature	Exogenous	Improved activity of antioxidant enzymes ascorbate peroxidase and catalase	Mohammadrezakhani et al. (2019)

(continued)

Table 8.1 (continued)

Crop	Abiotic stress	Endogenous accumulation/ exogenous application	Effects	Reference
Date palm	Heavy metal	Exogenous	Improved photosynthetic activity, water status, and activity of antioxidant enzymes	Zouari et al. (2016)
Rice	Heat and drought	Exogenous	Enhanced activities of superoxide dismutase, peroxidase, and catalase, increased levels of total soluble proteins, leaf proline, and glycine betaine, reduced lipid peroxidation	Hanif et al. (2021)

more drought-tolerant cultivars, other reports found that proline accumulates in more sensitive cultivars. For instance, a study on common bean genotypes found a negative correlation between proline accumulation and growth parameters. This indicates that higher levels of proline are associated with greater growth inhibition, suggesting that its accumulation may not directly correlate with enhanced drought tolerance but rather with stress sensitivity (Arteaga et al. 2020).

The exogenous application of proline has emerged as a promising strategy to enhance plant tolerance to drought and salt stress. The effects of foliar application of proline on fennel genotypes under both non-stress and drought stress conditions were studied. Drought stress led to significant decreases in leaf water potential, relative water content, plant above-ground dry mass, and water use efficiency, alongside reductions in carotenoids and chlorophyll concentrations. However, the application of exogenous proline improved plant water relations, primarily through enhanced osmoregulation. This suggests that proline supplementation helped mitigate some of the negative effects of drought stress, improving the plant's ability to retain water and maintain cellular function under challenging conditions (Zali and Ehsanzadeh 2018). Several studies have documented the effectiveness of exogenous proline application in enhancing plant resilience to drought and salt stress. By promoting osmoregulation and osmoprotection, proline improves the stress tolerance of crops such as maize (Khan et al. 2025), quinoa (Elewa et al. 2017), and onions (Semida et al. 2020) under drought conditions, as well as melon (Castañares et al. 2023), maize (Alam et al. 2016), and eggplant (Kusvuran et al. 2020) under salt stress.

8.3.1.2 Role of Proline in Antioxidative Defense

Reactive oxygen species (ROS) are molecules that contain oxygen, such as hydrogen peroxide (H2O2), superoxide anion, and hydroxyl radicals. These molecules are naturally produced as by-products of cellular metabolism, particularly during processes like photosynthesis and respiration. Under normal conditions, ROS are regulated by antioxidant systems within the plant, which help balance their levels. However, when plants experience stress, ROS production can increase significantly, exceeding the plants' antioxidant capacity to neutralize them. The accumulation of ROS leads to oxidative damage, including lipid peroxidation, protein denaturation, and DNA damage, impairing plant growth and function (Rejeb et al. 2014). Proline serves as an effective scavenger of ROS. For example, in chickpea plants subjected to heavy metal stress, elevated proline levels were correlated with enhanced antioxidant activity (Bhagyawant et al. 2019).

Proline is believed to function through both direct and indirect mechanisms, although there is ongoing debate regarding its direct scavenging capabilities (Renzetti et al. 2024). Some studies have highlighted the capacity of proline to directly interact with ROS, thereby mitigating oxidative stress. For example, in vitro experiments have shown that proline can effectively quench singlet oxygen and superoxide radicals, neutralizing these harmful molecules and reducing their detrimental effects on cellular structures (Rehman et al. 2021).

The role of proline in oxidative defense extends beyond direct scavenging, as it also enhances the activity of key antioxidant enzymes, including catalase, peroxidase, polyphenol oxidase, and ascorbate peroxidase, thereby strengthening the plant's capacity to mitigate oxidative stress. Research has highlighted the relationship between proline accumulation, increased antioxidant enzyme activity, and stress tolerance. In the roots and leaves of date palm plants, proline application resulted in increased activity of antioxidant enzymes, superoxide dismutase, catalase, and glutathione peroxidase (Zouari et al. 2016). Similarly, these three enzymes were upregulated in rice subjected to heat and drought stress following proline application (Hanif et al. 2021). Low-temperature stress, as well as exogenous proline treatment, resulted in the upregulation of peroxidase and catalase enzymes in citrus (Mohammadrezakhani et al. 2019). In sugar beet plants under drought stress, proline application resulted in increased tolerance and increased antioxidant activity (AlKahtani et al. 2021).

8.3.1.3 Role of Proline in Protein and Membrane Stabilization

Besides being an osmoprotectant and ROS scavenger, proline acts as a stabilizer for both proteins and cell membranes under stress. It promotes the orderly arrangement of water molecules around proteins, aiding in proper folding and preventing unfolding. By enhancing the water-binding capacity of proteins and protecting them from dehydration, proline helps maintain their stability and function. Additionally, proline supports protein renaturation and prevents aggregation. It also stabilizes cellular membranes by interacting with phospholipid headgroups, regulating membrane fluidity, and forming a protective hydration layer that preserves membrane integrity during stress (Renzetti et al. 2024). In pea plants exposed to heavy metal and salt stress, exogenous application of proline helped detoxify the stress caused by NiCl2 and NaCl by reducing lipid peroxidation and electrolyte leakage and enhancing membrane stability index (MSI) (Shahid et al. 2014). In onions, proline contributes to drought tolerance through increased cell membrane integrity, photosynthetic efficiency, and leaf water content (Semida et al. 2020). Foliar application of proline and silicon improved the membrane stability index in cowpea subjected to water deficit stress (Merwad et al. 2018). In a study involving rice seedlings, proline application has been shown to effectively reduce chromium-induced lipid peroxidation, highlighting its protective role in maintaining membrane integrity and protein function under metal stress (Yu et al. 2017).

8.3.2 Glycine Betaine

GB is a secondary metabolite derived from the amino acid choline. While GB does not fit the traditional definition of an amino acid since it lacks a free amino group, it serves important physiological roles similar to those of amino acids. In plants, GB is predominantly synthesized within the chloroplasts. Although numerous studies have demonstrated that GB accumulates as a response to abiotic stress, certain plant species, such as tomato, tobacco, and *Arabidopsis*, do not exhibit significant GB accumulation under stress conditions. Nevertheless, GB is a widely recognized and extensively studied compound that has been shown to enhance plant tolerance to a broad spectrum of abiotic stresses, despite its biosynthesis being species- and cultivar-specific (Jarin et al. 2024). GB is one of the many compatible osmolytes produced by plants, which confer stress tolerance to plants by scavenging ROS, promoting ion homeostasis, and maintaining osmotic regulation. Table 8.2 shows the effect of GB on crops under different abiotic stress conditions.

8.3.2.1 Role of GB in Oxidative Defense

GB serves as a crucial ROS scavenger in plants under various abiotic stress conditions. While GB does not directly remove ROS, it enhances the plant's antioxidative capacity by regulating the activity of ROS-scavenging enzymes (Kumar et al. 2017). Increased GB accumulation may also influence the transport of ROS by either directly blocking K^+ channels or reducing the RO-driven efflux of K+ ions. Exogenous application of GB has been demonstrated to enhance the activity of antioxidant enzymes and improve abiotic stress resistance in various crops, including mustard (Islam et al. 2021), Indian mustard (Islam et al. 2024a, b), peach (Wang et al. 2019), flax (Gupta et al. 2021), tomato (Zhang et al. 2022), maize (Shafiq et al. 2021), and pepper (Li et al. 2023). Additionally, GB has been shown to regulate

Table 8.2 Effect of glycine betaine on crops under different abiotic stress conditions

Crop	Abiotic stress	Effect	Reference
Mustard	–	Enhanced ROS-scavenging system, improved yield and photosynthetic efficiency	Islam et al. (2021)
Indian mustard	Salinity	Enhanced water content, photosynthetic efficiency, osmolyte accumulation, and ROS scavenging system	Islam et al. (2024a, b)
Peach	Low temperature (cold stress)	Inhibited ROS accumulation, activated antioxidant enzymes	Wang et al. (2019)
Flax	Drought	Enhanced ROS-scavenging system and endogenous glycine betaine accumulation	Gupta et al. (2021)
Tomato	Low temperature (cold stress)	Upregulated ROS-scavenging genes, improved seed germination and viability, reduced malondialdehyde (MDA) and hydrogen peroxide levels, increased proline accumulation	Zhang et al. (2022)
Maize	Drought	Improved growth, yield, and chlorophyll content, increased osmolytes, phenolics, ascorbic acid, and ROS-scavenging enzyme activity	Shafiq et al. (2021)
Pepper	Low temperature/ low light	Improved growth, root activity, and chlorophyll content, enhanced PSII efficiency, photochemical reactions, and light energy distribution, reduced oxidative damage	Li et al. (2023)
Wheat	Salinity	Enhanced ROS scavenging system through upregulation of enzymatic and nonenzymatic antioxidants, reduced Na+ levels	Badran et al. (2015)
Cotton	Heavy metal	Improved biomass, enhanced photosynthetic pigments, proteins and antioxidant enzymes, minimized oxidative damage	Farooq et al. (2016)
Sugarcane	Drought	Improved gas exchange and leaf water status, reduced photoinhibition, and enhanced energy use in PSII	de Oliveira Maia Júnior et al. (2020)
Maize	Salinity	Regulated Na+ homeostasis, upregulated PM H+-ATPase genes	Zhu et al. (2022)
Glycyrrhiza uralensis	Salinity	Increased osmoregulatory substances and antioxidant activity, enhanced Na+ secretion	Dong et al. (2024)
Safflower	Salinity	Upregulated antioxidant defense system, reduced Na+/K+ ratio	Alasvandyari et al. (2017)

ROS homeostasis in wheat plants under salinity stress. This regulation occurs through the upregulation of key genes, including alternative oxidase, H+/Na+ antiporter, and salt overly sensitive1 (Badran et al. 2015). In cotton under cadmium stress, GB treatment improved the antioxidant system, as demonstrated by a reduction in H_2O_2 and electrolyte leakage in leaves and roots (Farooq et al. 2016).

8.3.2.2 Role of GB in Photosystem Protection

GB plays a pivotal role in protecting photosystems under stress conditions, particularly during heat stress coupled with excessive light. In such circumstances, plants undergo photoinhibition, a process characterized by a decline in photosynthetic efficiency. This stress-induced damage is most pronounced at the photosystem II (PSII) reaction center, specifically impacting the D1 protein, which is essential for PSII functionality. Photoinhibition is often accompanied by the generation of ROS, which exacerbate damage by impairing the repair mechanisms of PSII (Jarin et al. 2024). Studies have shown that GB can protect and repair PSII by preserving its structural integrity, enabling it to maintain functionality under stress conditions. Furthermore, GB mitigates lipid peroxidation and reduces oxidative stress by enhancing the activity of antioxidant enzymes, thereby lowering ROS levels within the chloroplasts. A study showed that the increased accumulation of GB in a transgenic tomato line was linked to improved tolerance to heat-induced photoinhibition. The enhanced tolerance was linked to an increase in D1 protein levels, which facilitated the recovery of PSII after heat-induced damage (Li et al. 2014). Low temperature and low-light stress significantly limit the productivity, yield, and quality of peppers. Pre-spraying with GB has been shown to enhance chlorophyll content, reduce photoinhibition, and improve photochemical efficiency, leading to better photosynthetic capacity. The enhanced tolerance to stress in pepper plants is attributed to the redistribution of energy within the PSII center mediated by GB and facilitated by the encoded protein CapsB. This process reduces the dissipation of excess energy, allowing for more efficient light capture by the PSII center. As a result, photosynthetic efficiency is increased, and PSII recovers faster from its photoactivated state, accelerating the plant's recovery from photoinhibition (Li et al. 2023). Similarly, the foliar application of GB in sugarcane effectively alleviated the effects of drought stress on the photochemical efficiency of PSII. GB promoted a more efficient energy utilization in PSII, thereby reducing photoinhibition (de Oliveira Maia Júnior et al. 2020).

8.3.2.3 Role of GB in Ion Homeostasis

GB is a crucial osmolyte that plays a significant role in ion homeostasis, especially under salinity stress conditions. The ratio of potassium to sodium is a key determinant of salinity tolerance in plants. GB plays a role in ion homeostasis, particularly by regulating the K+/Na+ ratio. GB helps plants withstand salinity stress caused by Na+ retention by increasing K+ accumulation, thereby increasing K^+/Na^+ or Ca^+/Na^+ ratios (Jarin et al. 2024). In maize, GB treatment improves salt tolerance by regulating sodium homeostasis under saline stress. GB reduced the Na+/K+ ratio by decreasing Na+ accumulation and promoting Na+ efflux from roots and leaf protoplasts. GB increased the activity of vacuolar Na+/H+ antiporters and upregulated genes involved in Na+ transport, such as ZmMHA2, ZmMHA4, and ZmNHX1 (Zhu et al. 2022). Similarly, the exogenous application of GB in maize alleviated salt stress by reducing the accumulation of Na+ and Cl− ions, while enhancing the uptake of Ca^{2+} and K^{+} ions (Dustgeer et al. 2021).

In the medicinal plant *Glycyrrhiza uralensis*, GB application led to an increase in the rate of salt secretion in the leaves in particular the secretion of sodium ions (Dong et al. 2024). In safflower seedlings under salt stress, GB application improved ion homeostasis by decreasing the Na+/K+ ratio and increasing potassium uptake (Alasvandyari et al. 2017).

8.3.3 Glutamate and Glutamine

Glutamate and glutamine are essential amino acids in plant metabolism, serving as key nitrogen carriers and signaling molecules. Glutamate is central to nitrogen assimilation, linking carbon and energy metabolism while acting as a precursor for γ-aminobutyric acid (GABA) and plant defense compounds such as SA (Liao et al. 2022). Glutamine, a major nitrogen donor, regulates gene expression involved in metabolism, transport, and stress responses, highlighting its role in nutrient signaling and adaptation (Kan et al. 2015). Together, they are fundamental to nitrogen use efficiency and plant resilience.

8.3.3.1 Nitrogen Assimilation

Glutamate plays a central role in nitrogen metabolism, facilitating both nitrogen assimilation and redistribution within plant cells. The primary enzymatic pathway involved in nitrogen assimilation is the glutamine synthetase (GS)–glutamine:2-oxoglutarate aminotransferase (GOGAT) cycle, which is essential for incorporating inorganic nitrogen into organic molecules. Within this cycle, GS catalyzes the ATP-dependent conversion of glutamate and ammonium into glutamine, marking a crucial step in nitrogen assimilation (Campolo et al. 2023). This reaction is fundamental for nitrogen homeostasis, enabling plants to assimilate ammonium derived from root uptake and internal recycling processes (Bernard and Habash 2009).

The glutamine synthesized by GS is subsequently converted into glutamate by GOGAT, utilizing 2-oxoglutarate as a co-substrate. This reaction not only completes ammonium assimilation but also replenishes the glutamate pool, thereby sustaining amino acid biosynthesis and other essential metabolic processes. The GS–GOGAT cycle is tightly regulated, ensuring a dynamic balance between nitrogen uptake and metabolic demand. This regulation links nitrogen assimilation to carbon metabolism, as 2-oxoglutarate, a key intermediate in the tricarboxylic acid (TCA) cycle, is required for glutamate synthesis (Zayed et al. 2023). The integration of glutamate within this system highlights its fundamental role in plant growth, nitrogen use efficiency, and adaptation to fluctuating nitrogen availability (Gunka and Commichau 2012).

Glutamate also plays a key role in nitrogen redistribution through transamination reactions, particularly under stress conditions. These reactions facilitate the transfer of amino groups between amino acids and keto acids, allowing for nitrogen mobilization within plant tissues. Enzymes such as aspartate aminotransferase and alanine aminotransferase play central roles in these processes, maintaining nitrogen homeostasis and mitigating ammonium toxicity (Dutta Banik et al. 2020). Under environmental stressors, such as heavy metal exposure, transaminase activity increases, promoting nitrogen redistribution to sustain metabolic functions and mitigate stress-induced damage.

GS and glutamate synthase further contribute to nitrogen redistribution by converting ammonium ions into glutamine and glutamate, both of which are crucial for nitrogen transport within the plant. These metabolites facilitate nitrogen translocation across different tissues, particularly under conditions of nitrogen limitation or abiotic stress. The regulation of transamination and amino acid metabolism under these conditions highlights their critical role in nitrogen balance, allowing plants to allocate nitrogen efficiently to metabolic pathways essential for survival and stress adaptation (Zayed et al. 2023).

Ammonium assimilation plays a vital role in plant stress responses, ensuring nitrogen homeostasis and resilience under fluctuating environmental conditions. The regulation of ammonium uptake and assimilation enables plants to optimize nitrogen balance, mitigating stress-induced damage. Under osmotic stress, plants adjust the expression of key nitrogen assimilation genes, including ammonium transporters and GS, to enhance nitrogen uptake and integration into metabolic pathways. In wheat, nitrogen priming has been shown to promote root development and delay leaf senescence by improving ammonium absorption and assimilation, thereby alleviating growth limitations under nitrogen-deficient conditions (Lu et al. 2024).

8.3.3.2 ROS Detoxification

Glutamate plays a central role in antioxidant defense by serving as a precursor in the synthesis of GSH, one of the most important regulators of redox homeostasis in plant cells. GSH is essential for neutralizing ROS and preventing oxidative damage to cellular components, including proteins, lipids, and nucleic acids. The biosynthesis of GSH requires glutamate, cysteine, and glycine, with glutamate acting as a key determinant of GSH production. The glutamate-cystine antiporter system facilitates cystine uptake, ensuring a sufficient supply of cysteine, the rate-limiting precursor in GSH formation. The conversion of these substrates into GSH is catalyzed by γ-glutamylcysteine synthetase (GCL) and glutamine synthetase (GS), which sustain redox balance by maintaining adequate GSH levels. As a key antioxidant, GSH directly scavenges ROS and protects cellular components from oxidative damage (Aboushousha et al. 2023).

Glutamine also contributes to nitrogen-based defense mechanisms that enhance resistance to oxidative stress. One of its key roles is in the biosynthesis of proline, an amino acid that functions as both an osmoprotectant and a ROS scavenger in plants. Under oxidative stress conditions, the activation of genes involved in proline biosynthesis leads to increased proline accumulation, which stabilizes proteins, maintains osmotic balance, and mitigates oxidative damage. Additionally, glutamine participates in nitrogen remobilization, ensuring that nitrogen resources are efficiently allocated to biosynthetic pathways crucial for stress responses and redox regulation (Gulyás et al. 2017).

8.3.3.3 Role in Abiotic Stressors

Glutamate and glutamine play a crucial role in enhancing plant tolerance to drought and salinity stress by modulating key metabolic and signaling pathways. These amino acids contribute to osmoprotection by regulating proline biosynthesis, ion homeostasis, and antioxidant defense mechanisms, thereby mitigating the adverse effects of water deficit and osmotic stress (Heinemann and Hildebrandt 2021). In *Brassica napus*, glutamate-mediated stress responses involve a complex interplay between SA and calcium-dependent protein kinase (CPK5) signaling, which enhances proline accumulation and activates redox-regulating genes to restore cellular reducing potential. Similarly, exogenous application of glutamate in *Populus alba* and *P. glandulosa* has been shown to increase endogenous glutamate levels, induce proline synthesis, and reduce electrolyte permeability, thereby improving plant adaptability under drought conditions. By facilitating the accumulation of proline and stabilizing cellular ion balance, glutamate and glutamine contribute to the structural and metabolic integrity of plants exposed to prolonged drought stress.

The role of glutamate and glutamine in proline biosynthesis is particularly significant in drought tolerance mechanisms. As precursors in the proline metabolic pathway, they promote the upregulation of proline synthesis-related genes, leading to increased proline accumulation and improved stress adaptation. Proline functions as an osmoprotectant, stabilizing proteins and membranes while acting as a molecular chaperone to reduce oxidative damage. In addition to proline induction, glutamate also plays a role in reducing electrolyte permeability, thereby mitigating oxidative stress and osmotic imbalances in drought-affected plants. Through these mechanisms, glutamate and glutamine reinforce cellular resilience, allowing plants to maintain metabolic function and growth despite limited water availability.

The maintenance of ion homeostasis is another critical function of glutamate and glutamine under drought and salinity stress. By enhancing the expression of redox-regulating genes, these amino acids contribute to restoring cellular reducing potential, thereby counteracting oxidative stress-induced damage. The ability of glutamate and glutamine to modulate ion balance ensures the preservation of membrane stability and metabolic efficiency, which are essential for plant survival under osmotic stress conditions (La et al. 2020; Geng et al. 2023).

Beyond their role in drought and salinity adaptation, glutamate and glutamine are also involved in thermotolerance mechanisms, particularly through the stabilization of chaperone proteins and antioxidant systems. In maize seedlings, glutamate-mediated thermotolerance is linked to the activation of osmoregulatory pathways, glyoxalase, and non-glyoxalase systems, which collectively reinforce plant defense responses against heat stress. Additionally, glutamate and glutamine enhance the activity of antioxidant enzymes, reducing membrane damage and improving cellular stability under high-temperature conditions (Qiu et al. 2022). These findings underscore the broader role of glutamate and glutamine in stress adaptation, highlighting their ability to integrate osmotic regulation, antioxidant defense, and chaperone stabilization in response to diverse abiotic stressors.

8.3.4 Arginine and Polyamine Precursors

Arginine is a key amino acid in plants, functioning as both a nitrogen reservoir and a central player in nitrogen metabolism. It supports protein synthesis, nitrogen transport, and stress adaptation, making it essential for plant growth and metabolic regulation (Frémont et al. 2013).

Polyamines, including putrescine, spermidine, and spermine, are small, positively charged molecules that regulate growth, stress tolerance, and developmental processes. They play a particularly significant role in cell division, flowering, and fruit development, while also contributing to membrane stability and antioxidant defense under abiotic stress conditions (Kusano et al. 2007).

8.3.4.1 Nitric Oxide Signaling

Arginine serves as a precursor for nitric oxide (NO) synthesis, with its biosynthesis regulated through multiple metabolic pathways. NO production primarily involves two competing enzymatic reactions: The NO synthase (NOS) pathway and the arginase pathway, both of which utilize L-arginine as a common substrate. The reciprocal regulation of these pathways ensures that the balance between NO synthesis and arginase activity is modulated by cytokines and environmental signals. However, dysregulation of these enzymatic processes, including the overexpression of NOS and arginase, can contribute to pathophysiological conditions, emphasizing the need for precise metabolic control (Jacobs 2011).

NO plays a dual role in plant stress responses, acting as both a protective and potentially damaging signaling molecule. Under stress conditions, NO contributes to membrane stabilization and ROS suppression, thereby enhancing stress tolerance in plants. However, despite its critical role in stress adaptation, the specific mechanisms underlying NO-polyamine cross talk remain largely unexplored, including the effects of NO on polyamine biosynthesis. The interplay between NO and polyamines suggests a complex regulatory network, but further research is required to fully elucidate their interactions in abiotic stress responses.

Beyond its direct involvement in NO production, arginine also interacts with key hormonal pathways, particularly ABA and ethylene, which play essential roles in plant stress adaptation. The crosstalk between arginine metabolism and hormonal signaling has been linked to enhanced tolerance under stress conditions, enabling plants to modulate their physiological responses to environmental fluctuations (Tripathi et al. 2023). This interplay underscores the broader functional significance of arginine beyond NO biosynthesis, reinforcing its role in hormone-mediated stress mitigation.

Polyamine metabolism further contributes to abiotic stress tolerance, with compounds such as putrescine, spermidine, and spermine playing crucial roles in plant defense mechanisms. The manipulation of polyamine biosynthesis pathways has been shown to enhance stress resilience, confirming the protective roles of these compounds. Additionally, polyamines interact with ABA and NO in plant responses to abiotic stress, integrating multiple regulatory pathways to optimize stress adaptation (Shi and Chan 2014). These findings highlight the importance of polyamines as key modulators in stress tolerance, contributing to hormonal and redox balance in plants.

8.3.4.2 Regulation of Antioxidant Enzymes

Polyamines play a critical role in maintaining redox balance by modulating antioxidant enzyme activity and facilitating ROS scavenging. Their involvement in oxidative stress mitigation is closely associated with GSH metabolism, a key system for cellular redox homeostasis. The biosynthesis of GSH, driven by GCL and GS, is upregulated in response to oxidative stress, reinforcing enzymatic defense mechanisms that protect cellular components from damage (Dickinson et al. 2004). Moreover, polyamines contribute to ROS detoxification by stabilizing cellular structures, thereby enhancing plant resilience to environmental stressors (Gulyás et al. 2017).

8.3.4.3 Role in Abiotic Stressors

Arginine and its polyamine precursors are vital in plant responses to abiotic stress, contributing to osmolyte accumulation, cellular stability, and heavy metal detoxification. Under salinity and drought stress, these compounds regulate ion transporters and osmotic balance, improving water retention and ion homeostasis. Additionally, arginine-derived polyamines modulate stress-responsive gene expression, enhancing antioxidant defenses and metabolic stability, which supports plant survival under osmotic stress conditions (Shi and Chan 2014).

In response to extreme temperatures, polyamines protect cellular structures, including membranes and proteins, from stress-induced damage. Their capacity to enhance antioxidant enzyme activity, such as superoxide dismutase (SOD) and catalase (CAT), mitigates oxidative stress, improving plant thermotolerance (Ebeed and Ahmed 2023).

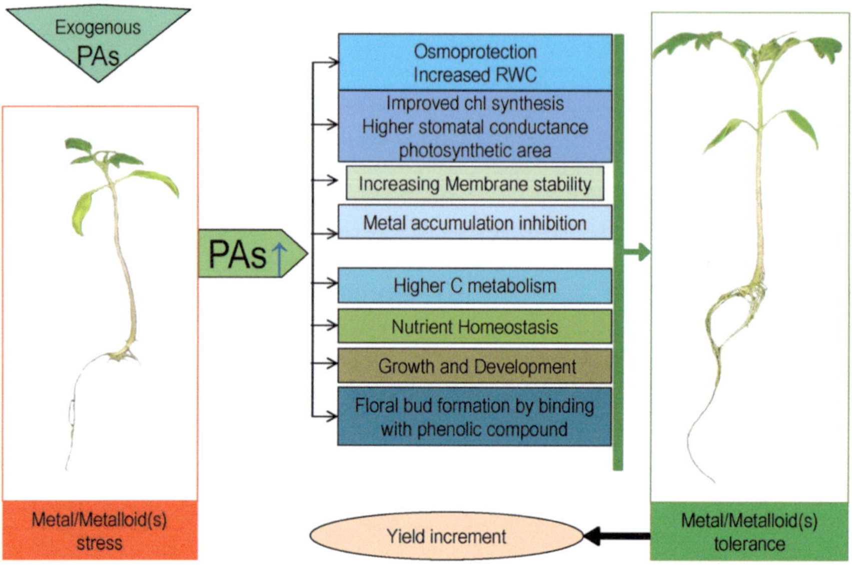

Fig. 8.3 Metal/metalloid(s) stress tolerance by polyamines (PAs). The blue arrow represents the upregulation of PA accumulation, the black arrow represents enhancement, and the green arrow represents tolerance. (Reprinted from Hasanuzzaman et al. (2019). Open Access journal)

Polyamines also aid in heavy metal detoxification by chelating toxic ions, reducing their bioavailability, and preventing cellular toxicity. Their interaction with metal transporters enables the sequestration and compartmentalization of heavy metals, minimizing oxidative damage and increasing stress resilience (Fig. 8.3) (Hasanuzzaman et al. 2019). These mechanisms highlight the significant role of arginine-derived polyamines in stress adaptation across diverse environmental conditions.

8.3.5 Serine and Threonine

8.3.5.1 Cross Talk in Metabolic Pathways

Amino acids are fundamental to plant responses to abiotic stress, functioning not only as metabolic intermediates but also as key regulators of energy homeostasis and stress-associated metabolite production. Among these, serine and threonine are deeply embedded in central metabolic pathways, influencing glycolysis, photorespiration, and amino acid biosynthesis.

Serine is vital for glycolysis and photorespiration, playing a central role in maintaining metabolic balance under stress (He et al. 2023). Additionally, it contributes to amino acid biosynthesis and acts as a precursor for purine and pyrimidine formation, both of which are essential for maintaining plant metabolism. Threonine, on the other hand, is integral to energy metabolism and regulates metabolic responses, particularly through its links to methionine and isoleucine biosynthesis, which are tightly controlled under stress conditions (Joshi et al. 2010).

During abiotic stress, plants undergo metabolic adjustments to sustain energy production. This often involves amino acid oxidation, where specific amino acids are diverted into respiratory pathways to offset energy deficiencies (Heinemann and Hildebrandt 2021). A key adaptive response is proline accumulation, which functions as an osmoprotectant, safeguarding cellular structures while facilitating redox balance through the transfer of reducing equivalents. Additionally, protein turnover plays a significant role in metabolic plasticity, as proteins are degraded to provide amino acids for energy generation (Hildebrandt 2018).

Beyond primary metabolism, amino acids also contribute to secondary metabolite synthesis, which enhances stress tolerance. For example, GB, a well-established osmoprotectant, plays a crucial role in stabilizing cellular components under stress conditions. Similarly, branched-chain amino acids (BCAAs) such as leucine, valine, and isoleucine accumulate during osmotic stress, supporting metabolic flexibility and adaptation (Joshi et al. 2010).

8.3.5.2 Role in Signal Transduction

Serine/threonine kinases are integral to stress signal transduction, allowing plants to process environmental cues and regulate cellular responses. Among these, mitogen-activated protein kinase (MAPK) cascades, including extracellular signal-regulated kinases (ERKs), c-Jun N-terminal kinases (JNKs), and p38 MAP kinases, serve as crucial regulators of cell survival, differentiation, and programmed cell death under stress conditions. Apoptosis signal-regulating kinase 1 (ASK1), a serine/threonine kinase, plays a key role in activating both JNK and p38 MAPK pathways in response to oxidative and endoplasmic reticulum stress, linking external stress perception to intracellular phosphorylation cascades. JNKs also function as tissue-specific regulators of synaptic activity, indicating their role in stress-induced neurodegenerative mechanisms.

Beyond individual kinase activity, stress signaling pathways are highly integrated with hormonal networks, particularly ABA and cytokinin. This interplay enables stress-responsive kinases to regulate gene expression, influencing both hormone biosynthesis and signal transduction pathways. Under abiotic stress conditions, these signaling mechanisms refine plant adaptation strategies, ensuring precise regulation of hormone-driven responses (Harrison 2012).

Plants perceive stress through the activation of transduction cascades, where signaling networks interact with metabolic pathways to modulate key transporters, transcription factors, and enzymatic functions. In response to salt stress, for instance, plants detect ion imbalances and initiate signaling pathways that regulate defense mechanisms, enhancing overall salt tolerance and resilience (Shah et al. 2021).

8.3.5.3 Role in Abiotic Stressors

Serine and threonine are essential regulators of plant responses to abiotic stress, playing key roles in signal transduction pathways that govern adaptation to drought, osmotic fluctuations, extreme temperatures, and heavy metal exposure. These amino acids participate in kinase-mediated phosphorylation cascades, which influence protein stability, enzymatic activity, osmolyte accumulation, and detoxification mechanisms during stress conditions (He et al. 2023).

Within these pathways, serine/threonine protein kinases, particularly those in the MAPK cascade, are critical in drought and osmotic stress tolerance. They regulate osmoprotectant production, enhance membrane stability, and contribute to stress signaling networks (Joshi et al. 2010; Heinemann and Hildebrandt 2021). In tomato, the overexpression of SlMAPK3 has been shown to improve drought resilience by increasing the synthesis of proline and sugars, which enhance osmotic adjustment (Chakraborty et al. 2013). These physiological changes correlate with higher chlorophyll retention and enhanced root biomass accumulation, suggesting that MAPK-dependent signaling mechanisms offer protection against water-deficient conditions (Joshi et al. 2010). Additionally, the calcineurin B-like (CBL–CBL-interacting protein kinase (CIPK)) pathway, which involves serine/threonine kinases, mediates calcium-dependent signal transduction to facilitate osmotic stress adaptation.

Similarly, temperature stress, whether extreme heat or cold, activates serine/threonine kinase pathways, allowing plants to stabilize proteins, regulate enzymatic activity, and maintain metabolic homeostasis. MAPK cascades are implicated in the phosphorylation of downstream targets, which reinforces structural integrity and functional efficiency in stress-responsive proteins. Additionally, the CBL–CIPK pathway plays a role in cold tolerance, integrating calcium signaling mechanisms to fine-tune stress perception and metabolic regulation (Kaya et al. 2024; He et al. 2023).

Beyond osmotic and temperature stress, serine/threonine kinases are vital in heavy metal tolerance, particularly in detoxification pathways for toxic elements such as cadmium (Cd2+). The overexpression of SlMAPK3 in tomato has been linked to enhanced cadmium resistance, primarily through increased antioxidant enzyme activity, which reduces oxidative stress markers like malondialdehyde and H_2O_2 (Hasanuzzaman et al. 2019). These findings suggest that MAPK-dependent phosphorylation events help regulate metal detoxification pathways and stress-protective proteins, enabling plants to counteract the harmful effects of heavy metal accumulation.

Collectively, these studies underscore the pivotal role of serine and threonine kinases in orchestrating stress adaptation strategies, where phosphorylation cascades contribute to osmotic regulation, protein stability, calcium homeostasis, and metal detoxification. By integrating these kinase-driven mechanisms, plants enhance their resilience against diverse environmental stressors, ensuring survival and sustained growth under fluctuating conditions (Heinemann and Hildebrandt 2021).

8.4 Genetic Engineering for Enhanced Amino Acid Synthesis

Genetic engineering offers promising avenues to enhance amino acid synthesis in plants, contributing to improved nutritional quality and agricultural sustainability. Continued research and technological innovations are essential to overcome existing challenges and fully realize the potential of genetically engineered crops in addressing global food security and nutritional deficiencies (Verma et al. 2020).

One primary approach involves the overexpression of genes encoding key enzymes in amino acid biosynthesis. For instance, overexpressing dihydrodipicolinate synthase, a crucial enzyme in lysine biosynthesis, has led to increased lysine content in crops like maize and canola (Hildebrandt et al. 2015). Similarly, enhancing methionine content has been achieved by manipulating the expression of enzymes involved in its biosynthetic pathway (Nawaz et al. 2022). These modifications not only boost the targeted amino acid levels but also contribute to overall plant growth and stress tolerance (Sharma et al. 2019).

Another strategy focuses on modifying seed storage proteins to enhance their essential amino acid composition. By altering the amino acid sequences of these proteins, researchers have developed crop varieties with improved nutritional profiles (Hayat et al. 2012). For example, transgenic rice expressing a modified storage protein exhibited increased lysine content, offering a potential solution to lysine deficiency in diets relying heavily on rice (Szabados and Savouré 2010).

Inhibiting the catabolic pathways of essential amino acids presents an alternative method to enhance their accumulation. By suppressing enzymes responsible for amino acid degradation, plants can retain higher levels of these nutrients (Chen and Murata 2011). For instance, downregulating threonine deaminase, which degrades threonine, resulted in increased threonine content in potato tubers. This approach ensures that the synthesized amino acids are preserved, enhancing the nutritional value of the crop (Sharma et al. 2019).

The advent of CRISPR-Cas9 technology has revolutionized genetic engineering by enabling precise genome editing (Verma et al. 2020). This system allows for targeted modifications in genes involved in amino acid metabolism without introducing foreign DNA, addressing public concerns regarding genetically modified organisms (Nawaz et al. 2022). For example, using CRISPR-Cas9 to knock out a gene encoding an amino acid transporter led to increased accumulation of certain amino acids in rice grains, demonstrating the potential of this technology in crop improvement (Zhu 2016).

Despite these advancements, several challenges persist. Metabolic pathway engineering can lead to unintended effects, such as the accumulation of toxic intermediates or growth retardation (Verma et al. 2020). Therefore, a comprehensive understanding of the interconnected metabolic networks is essential to mitigate potential negative outcomes. Future research should focus on integrating multi-omics approaches, including genomics, transcriptomics, and metabolomics, to elucidate the complex regulatory mechanisms governing amino acid metabolism (Nawaz et al. 2022). Additionally, field trials are necessary to assess the performance of genetically engineered crops under diverse environmental conditions, ensuring their stability and efficacy (Hildebrandt et al. 2015).

8.5 Conclusions and Perspectives

Amino acids are vital components in plants' defense mechanisms against abiotic stresses, contributing to key processes such as osmotic regulation, antioxidant defense, and overall metabolic balance. This chapter highlights the significant role of amino acids in maintaining plant homeostasis under environmental stress and their potential as a focal point for improving crop resilience. Advances in metabolic engineering, including the manipulation of biosynthetic pathways, have enabled the enhancement of amino acid levels, offering new ways to strengthen plants against stress. Techniques like CRISPR-Cas9 allow for precise edits in plant genomes, facilitating targeted improvements in amino acid composition without introducing foreign genes.

Despite these promising developments, challenges persist in fully optimizing amino acid biosynthesis in plants. Fine-tuning metabolic pathways to enhance amino acid accumulation while avoiding negative effects on plant growth requires a more comprehensive understanding of plant metabolism. In addition, integrating systems biology and multi-omics strategies will provide deeper insights into amino acid signaling networks and their interactions with other metabolic pathways. To translate these molecular advances into practical applications, extensive field trials are necessary to assess the performance and stability of genetically modified plants in diverse agricultural environments. Ultimately, the ability to enhance amino acid biosynthesis through genetic manipulation holds great potential for addressing global food security and combating the impact of climate change on crop yields.

References

Aboushousha R, van der Velden J, Hamilton N, Peng Z, MacPherson M, Erickson C, White S, Wouters EFM, Reynaert NL, Seward DJ, Li J, Janssen-Heininger YMW (2023) Glutaredoxin attenuates glutathione levels via deglutathionylation of Otub1 and subsequent destabilization of system x_C. Sci Adv 9:eadi5192

Alam R, Das DK, Islam MR, Murata Y, Hoque MA (2016) Exogenous proline enhances nutrient uptake and confers tolerance to salt stress in maize (*Zea mays L.*). Progress Agric 27:409–417

Alasvandyari F, Mahdavi B, Madah Hosseini S (2017) Glycine betaine affects the antioxidant system and ion accumulation and reduces salinity-induced damage in safflower seedlings. Arch Biol Sci 69:139–147

Alisofi S, Einali A, Sangtarash MH (2019) Jasmonic acid-induced metabolic responses in bitter melon (*Momordica charantia*) seedlings under salt stress. J Hortic Sci Biotechnol 95:247–259

AlKahtani MDF, Hafez YM, Attia K, Rashwan E, Husnain LA, AlGwaiz HIM, Abdelaal KAA (2021) Evaluation of silicon and proline application on the oxidative machinery in drought-stressed sugar beet. Antioxidants 10:398

Arteaga S, Yabor L, Díez MJ, Prohens J, Boscaiu M, Vicente O (2020) The use of proline in screening for tolerance to drought and salinity in common bean (*Phaseolus vulgaris* L.) genotypes. Agronomy 10:817

Badran EG, Abogadallah GM, Nada RM, Nemat Alla MM (2015) Role of glycine in improving the ionic and ROS homeostasis during NaCl stress in wheat. Protoplasma 252:835–844

Batool S, Uslu VV, Rajab H, Ahmad N, Waadt R, Geiger D, Malagoli M, Xiang CB, Hedrich R, Rennenberg H, Herschbach C, Hell R, Wirtz M (2018) Sulfate is incorporated into cysteine to trigger ABA production and stomatal closure. Plant Cell 30:2973–2987

Bernard SM, Habash DZ (2009) The importance of cytosolic glutamine synthetase in nitrogen assimilation and recycling. New Phytol 182:608–620

Bhagyawant SS, Narvekar DT, Gupta N, Bhadkaria A, Koul KK, Srivastava N (2019) Variations in the antioxidant and free radical scavenging under induced heavy metal stress expressed as proline content in chickpea. Physiol Mol Biol Plants 25:683–696

Brenes M, Solana A, Boscaiu M, Fita A, Vicente O, Calatayud Á, Prohens J, Plazas M (2020) Physiological and biochemical responses to salt stress in cultivated eggplant (*Solanum melongena L.*) and in *S. insanum L.*, a close wild relative. Agronomy 10:651

Campolo N, Mastrogiovanni M, Mariotti M, Radi R (2023) Multiple oxidative post-translational modifications of human glutamine synthetase mediate peroxynitrite-dependent enzyme inactivation and aggregation. J Biol Chem 299:102889

Castañares JL, Corvalán B, Larraburu EE (2023) Effect of exogenous proline on physiological and growth traits of melon seedlings under salt stress. Rev Investig Agropecu 49:50–55

Chakraborty U, Pradhan B, Lama R (2013) Chapter 27 – Metabolome analyses for understanding abiotic stress responses in plants to evolve management strategies. In: Tuteja N, Gill SS (eds) Climate change and plant abiotic stress tolerance, 1st edn. Wiley-Blackwell Publishing Ltd, pp 215–240

Chen THH, Murata N (2011) Glycine betaine: an effective protectant against abiotic stress in plants. Trends Plant Sci 16:489–495

de Oliveira Maia Júnior S, de Andrade JR, dos Santos CM, Santos JV, dos Santo Silva LK, Clemente PRA, Ferreira VM, Silba JV, Endres L (2020) Foliar-applied glycine betaine minimizes drought stress-related impact to gas exchange and the photochemical efficiency of PSII in sugarcane. Theor Exp Plant Physiol 32:315–329

Debnath S, Aisha S, Malakar A, Perveen K, Alfagham AT, Khanam MN, Ahmad RA, Pramanik B, Mohammed YA (2023) Understanding the cross-talk of major abiotic-stress-responsive genes in rice: a computational biology approach. J King Saud Univ Sci 35:102786

Dickinson DA, Levonen AL, Moellering DR, Forman HJ (2004) Human glutamate cysteine ligase gene regulation through the electrophile response element. Free Radic Biol Med 37:1159–1169

Dong X, Ma X, Zhao Z, Ma M (2024) Exogenous betaine enhances salt tolerance of *Glycyrrhiza uralensis* through multiple pathways. BMC Plant Biol 24:165

Dustgeer Z, Seleiman MF, Khan I, Chattha MU, Ali EF, Alhammad BA, Jalal RS, Refay Y, Hassan MU (2021) Glycine-betaine induced salinity tolerance in maize by regulating the physiological attributes, antioxidant defense system and ionic homeostasis. Not Bot Horti Agrobot Cluj-Napoca 49:12248

Dutta Banik D, Zamboni N, Sauer U (2020) Transaminases and amino acid metabolism in nitrogen homeostasis and stress response. Metab Eng 57:143–155

Ebeed HT, Ahmed HS (2023) Chapter 12 – Role of polyamines in regulating physiological and molecular responses of plants under abiotic stress. In: Sharma A, Pandey S, Bhardwaj R, Zheng B, Tripathi DK (eds) The role of growth regulators and phytohormones in overcoming environmental stress, 1st edn. Elsevier, pp 263–287

Elewa TA, Sadak MS, Saad AM (2017) Proline treatment improves physiological responses in quinoa plants under drought stress. Biosci Res 14:21–33

Farooq M, Ali S, Hameed A, Bharwana S, Rizwan M, Ishaque W, Farid M, Mahmood K, Iqbal Z (2016) Cadmium stress in cotton seedlings: physiological, photosynthesis and oxidative damages alleviated by glycinebetaine. S Afr J Bot 104:61–68

Forlani G, Trovato M, Funck D, Signorelli S (2019) Chapter 3 – Regulation of proline accumulation and its molecular and physiological functions in stress defence. In: Hossain M, Kumar V, Burritt D, Fujita M, Mäkelä P (eds) Osmoprotectant-mediated abiotic stress tolerance in plants, recent advances and future perspectives, 1st edn. Springer, Cham, pp 73–97

Frémont N, Riefler M, Stolz A, Schmülling T (2013) The *Arabidopsis* tumor prone5 gene encodes an acetylornithine aminotransferase required for arginine biosynthesis and root meristem maintenance in blue light. Plant Physiol 161:1127–1140

Geng Y, Zeng W, Huang LC, Zheng RJ, Yu JJ, An Y, Qu MQ, Liu Y, Han X (2023) Effect of exogenous application glutamate on drought tolerance of *Populus alba* × *P. glandulosa* "84K". For Res 36:59–71

Gharsallah C, Fakhfakh H, Grubb D, Gorsane F (2016) Effect of salt stress on ion concentration, proline content, antioxidant enzyme activities and gene expression in tomato cultivars. AoB Plants 8:plw055

Ghassemi-Golezani K, Samea-Andabjadid S (2022) Exogenous cytokinin and salicylic acid improve amino acid content and composition of faba bean seeds under salt stress. J Crop Health 74:935–945

Gulyás Z, Simon-Sarkadi L, Badics E, Kocsy G (2017) Redox regulation of free amino acid levels in *Arabidopsis thaliana*. Physiol Plant 159:264–276

Gunka K, Commichau FM (2012) Control of glutamate homeostasis in *Bacillus subtilis*: interplay between glutamate synthase, glutamine synthetase, and ammonium transporters. J Bacteriol 194:4145–4156

Gupta P, Rai R, Vasudev S, Yadava DK, Dash PK (2021) Ex-foliar application of glycine betaine and its impact on protein, carbohydrates and induction of ROS scavenging system during drought stress in flax (*Linum usitatissimum*). J Biotechnol 337:80–89

Hanif S, Saleem MF, Sarwar M, Irshad M, Shakoor A, Wahid MA, Khan HZ (2021) Biochemically triggered heat and drought stress tolerance in rice by proline application. J Plant Growth Regul 40:305–312

Harrison MA (2012) Chapter 2 – Cross-talk between phytohormone signaling pathways under both optimal and stressful environmental conditions. In: Khan N, Nazar R, Iqbal N, Anjum N (eds) Phytohormones and abiotic stress tolerance in plants, 1st edn. Springer, Berlin, Heidelberg, pp 49–76

Hasan MM, Liu XD, Waseem M, Guang-Qian Y, Alabdallah NM, Jahan MS, Fang XW (2022) ABA activated SnRK2 kinases: an emerging role in plant growth and physiology. Plant Signal Behav 17:2071024

Hasanuzzaman M, Alhaithloul HAS, Parvin K, Fujita M (2019) Polyamine action under metal/metalloid stress: regulation of biosynthesis, metabolism, and molecular interactions. Int J Mol Sci 20:3217

Hayat S, Hayat Q, Alyemeni MN, Wani AS, Pichtel J, Ahmad A (2012) Role of proline under changing environments: a review. Plant Signal Behav 7:1456–1466

He G, Wu C, Hunag J, Zhou R (2017) Effect of exogenous proline on metabolic response of *Tetragenococcus halophilus* under salt stress. J Microbiol Biotechnol 27:1681–1691

He L, Ding Y, Zhou X, Yin Y (2023) Serine signaling governs metabolic homeostasis and health. Trends Endocrinol Metab 34:245–260

Heinemann B, Hildebrandt TM (2021) The role of amino acid metabolism in signaling and metabolic adaptation to stress-induced energy deficiency in plants. J Exp Bot 72:4634–4645

Hildebrandt TM (2018) Synthesis versus degradation: directions of amino acid metabolism during *Arabidopsis* abiotic stress response. Plant Mol Biol 98:121–135

Hildebrandt TM, Nunes Nesi A, Araújo WL, Braun HP (2015) Amino acid catabolism in plants. Mol Plant 8:1563–1579

Huang T, Jander G (2017) Abscisic acid-regulated protein degradation causes osmotic stress-induced accumulation of branched-chain amino acids in *Arabidopsis thaliana*. Planta 246:737–747

Islam S, Parrey ZA, Shah SH, Mohammad F (2021) Glycine betaine mediated changes in growth, photosynthetic efficiency, antioxidant system, yield and quality of mustard. Sci Hortic 285:110170

Islam S, Mohammad F, Shakeel A, Corpas FJ (2024a) Glycine betaine: a multifaceted protectant against salt stress in Indian mustard through ionic homeostasis, ROS scavenging and osmotic regulation. Physiol Plant 176:e14530

Islam SN, Kouser S, Hassan P, Asgher M, Shah AA, Khan NA (2024b) Gamma-aminobutyric acid interactions with phytohormones and its role in modulating abiotic and biotic stress in plants. Stress Biol 4:36

Jacobs NL (2011) Chapter 2 – Alternative metabolic pathways of arginine and their pathophysiological roles. In: Jacobs NL (ed) Arginine amino acid, 1st edn. Nova Science Publishers

Jarin A, Ghosh UK, Hossain MS, Mahmud A, Khan MAR (2024) Glycine betaine in plant responses and tolerance to abiotic stresses. Discov Agric 2:127

Joshi V, Joung JG, Fei Z, Jander G (2010) Interdependence of threonine, methionine, and isoleucine metabolism in plants: accumulation and transcriptional regulation under abiotic stress. Amino Acids 39:933–947

Kamran M, Parveen A, Ahmar S, Malik Z, Hussain S, Chattha MS, Saleem MH, Adil M, Heidari P, Chen JT (2020) An overview of hazardous impacts of soil salinity in crops, tolerance mechanisms, and amelioration through selenium supplementation. Int J Mol Sci 21:148

Kan CC, Chung TY, Juo YA, Hsieh MH (2015) Glutamine rapidly induces the expression of key transcription factor genes involved in nitrogen and stress responses in rice roots. BMC Genomics 16:731

Kaya C, Uğurlar F, Adamakis IDS (2024) Molecular mechanisms of CBL-CIPK signaling pathway in plant abiotic stress tolerance and hormone crosstalk. Int J Mol Sci 25:1098

Khan MIR, Iqbal N, Masood A, Per TS, Khan NA (2013) Salicylic acid alleviates adverse effects of heat stress on photosynthesis through changes in proline production and ethylene formation. Plant Signal Behav 8:e26374

Khan MI, Asgher M, Khan NA (2014) Alleviation of salt-induced photosynthesis and growth inhibition by salicylic acid involves glycinebetaine and ethylene in mungbean (*Vigna radiata* L.). Plant Physiol Biochem 80:67–74

Khan P, Abdelbacki AMM, Albaqami M, Jan R, Kim KM (2025) Proline promotes drought tolerance in maize. Biology 14:41

Kumar V, Shriram V, Hoque TS, Hasan MM, Burritt DJ, Hossain MA (2017) Chapter 5 – Glycinebetaine-mediated abiotic oxidative-stress tolerance in plants: physiological and biochemical mechanisms. In: Sarwat M, Ahmad A, Abdin M, Ibrahim M (eds) Stress signaling in plants: genomics and proteomics perspective, vol 2, 1st edn. Springer, Cham, pp 111–133

Kusano T, Yamaguchi K, Berberich T, Takahashi Y (2007) Advances in polyamine research in 2007. J Plant Res 120:345–350

Kusvuran S, Bayat R, Ustun AS, Elliatioglur SS (2020) Exogenous proline improves osmoregulation, physiological and biochemical responses of eggplant under salt stress. Fresenius Environ Bull 29:152–161

La HV, Lee BR, Islam T, Mamun A, Park SH, Bae DW, Kim TH (2020) Characterization of glutamate-mediated hormonal regulatory pathway of the drought responses in relation to proline metabolism in *Brassica napus* L. Plants 9:512

Li M, Li Z, Li S, Guo S, Meng Q, Li G, Yang X (2014) Genetic engineering of glycine betaine biosynthesis reduces heat-enhanced photoinhibition by enhancing antioxidative defense and alleviating lipid peroxidation in tomato. Plant Mol Biol Report 32:42–45

Li N, Pu K, Ding D, Yang Y, Niu T, Li J, Xie J (2023) Foliar spraying of glycine betaine alleviated growth inhibition, photoinhibition, and oxidative stress in pepper (*Capsicum annuum* L.) seedlings under low temperatures combined with low light. Plants 12:2563

Liao HS, Chung YH, Hsieh MH (2022) Glutamate: a multifunctional amino acid in plants. Plant Sci 318:111238

Lu Y, Gao L, Hu J, Liu X, Jiang D, Cao W, Dai T, Tian Z (2024) Low nitrogen priming improves nitrogen uptake and assimilation adaptation to nitrogen deficit stress in wheat seedling. Planta 259:107

Merwad AM, Desoky EM, Rady MM (2018) Response of water deficit-stressed *Vigna unguiculata* performances to silicon, proline or methionine foliar application. Sci Hortic 228:132–144

Mohammadrezakhani S, Hajilou J, Rezanejad F, Zaare-Nahandi F (2019) Assessment of exogenous application of proline on antioxidant compounds in three Citrus species under low temperature stress. J Plant Interact 14:347–358

Nahar K, Hasanuzzaman M, Fujita M (2016) Chapter 5 – Roles of osmolytes in plant adaptation to drought and salinity. In: Iqbal N, Nazar R, Khan N (eds) Osmolytes and plants acclimation to changing environment: emerging omics technologies, 1st edn. Springer, New Delhi, pp 37–68

Nawaz MA, Huang Y, Bie Z, Ahmed W, Rehman AU, Barozai MYK, Kar F (2022) Nutrient regulation and amino acid involvement in plant responses to abiotic stresses. J Plant Growth Regul 41:305–323

Pál M, Tajti J, Szalai G, Peeva V, Vegh B, Janda T (2018) Interaction of polyamines, abscisic acid and proline under osmotic stress in the leaves of wheat plants. Sci Rep 8:1283

Park S, Shi A, Mou B (2020) Genome-wide identification and expression analysis of the CBF/DREB1 gene family in lettuce. Sci Rep 10:5733

Podlesakova K, Ugena L, Spichal L, Dolezal K, De Diego N (2019) Phytohormones and polyamines regulate plant stress responses by altering GABA pathway. Nat Biotechnol 48:53–65

Qiu XM, Sun YY, Wang JQ, Xiang RH, Li ZG (2022) Involvement of osmoregulation, glyoxalase, and non-glyoxalase systems in signaling molecule glutamic acid-boosted thermotolerance in maize seedlings. Protoplasma 259:1507–1520

Ramesh SA, Kamran M, Sullivan W, Chirkova L, Okamoto M, Degryse F, McLaughlin M, Gilliham M, Tyerman SD (2018) Aluminum activated malate transporters can facilitate GABA transport. Plant Cell 30:1147–1164

Reddy PS, Jogeswar G, Rasineni GK, Maheswari M, Reddy AR, Varshney RK, Kishor PK (2015) Proline over-accumulation alleviates salt stress and protects photosynthetic and antioxidant enzyme activities in transgenic sorghum (*Sorghum bicolor* (*L.*) *Moench*). Plant Physiol Biochem 94:104–113

Rehman AU, Bashir F, Ayaydin F, Kóta Z, Páli T, Vass I (2021) Proline is a quencher of singlet oxygen and superoxide both in in vitro systems and isolated thylakoids. Physiol Plant 172:7–18

Rejeb KB, Abdelly C, Savouré A (2014) How reactive oxygen species and proline face stress together. Plant Physiol Biochem 80:278–284

Renzetti M, Funck D, Trovato M (2024) Proline and ROS: a unified mechanism in plant development and stress response? Plants 14:2

Saibi W, Brini F (2020) Proline, a peculiar amino acid with astucious functions in development and salt tolerance process in plants. J Food Nutr Metab 3:1–8

Semida WM, Abdelkhalik A, Rady MO, Marey RA, El-Mageed TAA (2020) Exogenously applied proline enhances growth and productivity of drought stressed onion by improving photosynthetic efficiency, water use efficiency and up-regulating osmoprotectants. Sci Hortic 272:109580

Shafiq S, Akram NA, Ashraf M, García-Caparrós P, Ali OM, Latef AAHA (2021) Influence of glycine betaine (natural and synthetic) on growth, metabolism and yield production of drought-stressed maize (*Zea mays* L.) plants. Plants 10:2540

Shah WH, Rasool A, Saleem S, Mushtaq NU, Tahir I, Hakeem KR, Rehman RU (2021) Understanding the integrated pathways and mechanisms of transporters, protein kinases, and transcription factors in plants under salt stress. Int J Genomics 2021:5578727

Shahid MA, Balal RM, Pervez MA, Abbas T, Aqeel MA, Javaid MM, Garcia-Sanchez F (2014) Exogenous proline and proline-enriched *Lolium perenne* leaf extract protects against phytotoxic effects of nickel and salinity in *Pisum sativum* by altering polyamine metabolism in leaves. Turk J Bot 38:924–926

Sharma A, Shahzad B, Kumar V, Kohli SK, Sidhu GPS, Bali AS, Handa N, Kapoor D, Bhardwaj R, Zheng B (2019) Phytohormones regulate accumulation of osmolytes under abiotic stress. Biomolecules 9:285

Shi H, Chan Z (2014) Improvement of plant abiotic stress tolerance through modulation of the polyamine pathway. J Integr Plant Biol 56:114–121

Shim JS, Jeong HI, Bang SW, Jung SE, Kim G, Kim YS, Redillas MCF, Oh SJ, Seo JS, Kim JK (2023) Drought-induced branched-chain amino acid aminotransferase enhances drought tolerance in rice. Plant Physiol 191:1435–1447

Siddique A, Kandpal G, Kumar P (2018) Proline accumulation and its defensive role under diverse stress condition in plants: an overview. J Pure Appl Microbiol 12:1655–1659

Szabados L, Savouré A (2010) Proline: a multifunctional amino acid. Trends Plant Sci 15:89–97

Takahashi F, Kuromori T, Urano K, Yamaguchi-Shinozaki K, Shinozaki K (2020) Drought stress responses and resistance in plants: from cellular responses to long-distance intercellular communication. Front Plant Sci 11:556972

Tripathi DK, Bhat JA, Antoniou C, Kandhol N, Singh VP, Fernie AR, Fotopoulos V (2023) Redox regulation by priming agents toward a sustainable agriculture. Plant Cell Physiol 65:1087–1102

Verma V, Ravindran P, Kumar PP (2020) Plant hormone-mediated regulation of stress responses. BMC Plant Biol 20:1–12

Wang P, Zhao Y, Li Z, Hsu CC, Liu X, Fu L, Hou YJ, Du Y, Xie S, Zhang C, Gao J, Cao M, Huang X, Zhu Y, Tang K, Wang X, Tao WA, Xiong Y, Zhu JK (2018) Reciprocal regulation of the TOR kinase and ABA receptor balances plant growth and stress response. Mol Cell 69:100–112

Wang L, Shan T, Xie B, Ling C, Shao S, Jin P, Zheng Y (2019) Glycine betaine reduces chilling injury in peach fruit by enhancing phenolic and sugar metabolisms. Food Chem 272:530–538

Yamasaki Y, Randall SK (2016) Functionality of soybean CBF/DREB1 transcription factors. Plant Sci 246:80–90

Yan S, Zhan M, Liu Z, Zhang X (2025) Insight into the transcriptional regulation of key genes involved in proline metabolism in plants under osmotic stress. Biochimie 228:8–14

Yang R, Hui Q, Gu Z (2016) Effects of ABA and $CaCl_2$ on GABA accumulation in fava bean germinating under hypoxia-NaCl stress. Biosci Biotechnol Biochem 80:540–546

You J, Zhang Y, Liu A, Li D, Wang X, Dossa K, Zhou R, Yu J, Zhang Y, Wang L, Zhang X (2019) Transcriptomic and metabolomic profiling of drought-tolerant and susceptible sesame genotypes in response to drought stress. BMC Plant Biol 19:267

Yu X, Lin Y, Fan W, Lu M (2017) The role of exogenous proline in amelioration of lipid peroxidation in rice seedlings exposed to Cr (VI). Int Biodeterior Biodegradation 123:106–112

Zali AG, Ehsanzadeh P (2018) Exogenous proline improves osmoregulation, physiological functions, essential oil, and seed yield of fennel. Ind Crop Prod 111:133–140

Zayed O, Hewedy OA, Abdelmoteleb A, Ali M, Youssef M, Roumia A, Seymour D, Yuan ZC (2023) Nitrogen journey in plants: from uptake to metabolism, stress response, and microbe interaction. Biomolecules 13:1443

Zhang M, Wang L, Zhang K, Liu F, Wan Y (2017) Drought-induced responses of organic osmolytes and proline metabolism during pre-flowering stage in leaves of peanut (*Arachis hypogaea L.*). J Integr Agric 16:2197–2205

Zhang Y, Dai T, Liu Y, Wang J, Wang Q, Zhu W (2022) Effect of exogenous glycine betaine on the germination of tomato seeds under cold stress. Int J Mol Sci 23:10474

Zheng Y, Luo L, Wei J, Chen Q, Yang Y, Hu X, Kong X (2018) The glutamate receptors AtGLR1.2 and AtGLR1.3 increase cold tolerance by regulating jasmonate signaling in *Arabidopsis thaliana*. Biochem Biophys Res Commun 506:895–900

Zhu JK (2016) Abiotic stress signaling and responses in plants. Cell 167:313–324

Zhu M, Li Q, Zhang Y, Zhang M, Li Z (2022) Glycine betaine increases salt tolerance in maize (*Zea mays L.*) by regulating Na+ homeostasis. Front Plant Sci 13:978304

Zouari M, Ahmed CB, Zorrig W, Elloumi N, Rabhi M, Delmail D, Rouina BB, Labrousse P, Abdallah FB (2016) Exogenous proline mediates alleviation of cadmium stress by promoting photosynthetic activity, water status and antioxidative enzymes activities of young date palm (*Phoenix dactylifera L.*). Ecotoxicol Environ Saf 128:100–108

Chapter 9
Oligosaccharides Stimulate Defense Against Abiotic Stresses

Tony Kevork Sajyan, Sana AlHajj Ahmad, Israa El Ghandour, Lourd Abd Al Ahad, and Rania Francis

Abstract Oligosaccharides play a crucial role in enhancing plant defense mechanisms against various abiotic stresses, making them valuable in sustainable agriculture. These carbohydrate polymers, including fructooligosaccharides (FOSs), galactooligosaccharides (GOSs), and xylooligosaccharides (XOSs), possess unique structural and physicochemical properties that contribute to stress tolerance. Their biosynthesis and functional attributes enable plants to withstand adverse environmental conditions such as drought, salinity, temperature extremes, and heavy metal toxicity. The mechanisms by which oligosaccharides enhance stress tolerance include stress sensing, signal transduction, and the regulation of gene expression. They assist in scavenging reactive oxygen species (ROS), modulating hormonal pathways, and upregulating defense-related genes. Additionally, epigenetic modifications contribute to the plant's adaptive response. Oligosaccharides help mitigate drought stress by improving osmotic balance and maintaining membrane integrity. Under salinity stress, they support ion homeostasis and activate stress tolerance pathways. Similarly, they contribute to temperature resilience by enhancing cold and heat stress responses, while also aiding in heavy metal detoxification through chelation and antioxidant system enhancement.

The integration of oligosaccharides with modern breeding and genetic engineering offers promising advancements for developing stress-resilient crops. However, challenges remain, including gaps in research, scalability issues, and the need for further biotechnological innovations. Future research should focus on optimizing

T. K. Sajyan (✉)
Faculty of Agricultural and Food Sciences, American University of Beirut, Beirut, Lebanon

Department of Agriculture and Food Engineering, School of Engineering, Holy Spirit University of Kaslik, Jounieh, Lebanon
e-mail: ts71@aub.edu.lb

S. A. Ahmad · I. El Ghandour · L. A. Al Ahad
Faculty of Agricultural and Food Sciences, American University of Beirut, Beirut, Lebanon

R. Francis
Department of Agriculture and Food Engineering, School of Engineering, Holy Spirit University of Kaslik, Jounieh, Lebanon

H. I. Mohamed (ed.), *Elicitors for Sustainable Crop Production*, Sustainability Sciences in Asia and Africa, https://doi.org/10.1007/978-981-95-3533-0_9

their application to enhance crop productivity and environmental sustainability. As oligosaccharides continue to gain attention, their potential to contribute to resilient agricultural systems highlights their importance in addressing global food security challenges.

Keywords Oligosaccharides · Abiotic stress · Plant defense · Gene regulation · Sustainable agriculture

9.1 Introduction

Oligosaccharides play a vital role in plant physiology, particularly in stress tolerance, immune defense, and metabolic regulation. These carbohydrates, composed of short chains of monosaccharide units, function in various biological processes, making them essential for plant adaptation and agricultural applications.

One of the critical functions of oligosaccharides in plants is their role in cold acclimation and freezing tolerance. Certain oligosaccharides, such as fructans, stachyose, and raffinose, help plants withstand freezing conditions by acting as osmoprotectants and stabilizing cellular structures (Hernández and Plou 2023). For example, the moss *Physcomitrium patens*, despite lacking epidermal tissues, exhibits freezing tolerance due to its accumulation of the trisaccharide theanderose (Chavan et al. 2023). However, many major cultivated crops, including wheat, rice, and maize, do not naturally accumulate significant concentrations of these protective osmolytes. Genetic engineering approaches have demonstrated that the overexpression of fructan biosynthetic genes can enhance abiotic stress tolerance, resulting in increased biomass and yield. Additionally, oligosaccharides contribute to oxidative stress management by mitigating reactive oxygen species (ROS) accumulation and protecting the photosynthetic system. The ability of sugars to act as ROS scavengers depends on their localization within cellular compartments or near membranes (Huang et al. 2019).

Beyond their protective role in stress tolerance, oligosaccharides also function as immune inducers in plants. During plant–pathogen interactions, enzymatic degradation of pathogen and plant cell walls releases oligosaccharide fragments that serve as signals to activate plant defense mechanisms (Zhao et al. 2021). For instance, chitooligosaccharides, derived from chitin deacetylation, and β-1,3-glucan oligosaccharides, extracted from brown algae, have been shown to induce ROS accumulation, salicylic acid production, and key enzyme activities involved in plant immunity (González-Pérez et al. 2018). Oligogalacturonic acid, another plant-derived oligosaccharide, enhances resistance by increasing the formation of papillae in wheat cells, which reduces fungal infection success. Recent studies have explored oligosaccharide-based compounds as environmentally friendly alternatives to chemical fungicides, demonstrating their effectiveness in reducing disease severity through the activation of systemic acquired resistance (SAR) pathways (Huang et al. 2019).

The classification of oligosaccharides is based on their molecular structure and physiological functions. According to the Joint Committee on Biochemical Nomenclature, oligosaccharides are naturally occurring carbohydrates composed of three to ten monosaccharide units, linked by α- or β-glycosidic bonds. However, some definitions extend the classification to include disaccharides, such as maltose and sucrose (Mussatto and Mancilha 2007). Oligosaccharides play diverse physiological roles, including energy provision, gut microbiota modulation, and immune enhancement. Functional oligosaccharides, such as marine-derived and glycosaminoglycan oligosaccharides, have attracted significant research interest due to their bioactive properties, including antiaging and cancer-preventive effects (Krishna Perumal et al. 2023). The increasing demand for functional foods and sustainable agricultural practices highlights the significance of oligosaccharides in both plant biology and human health. Despite the complexity of glycan structures, continued research on oligosaccharide-based plant protection and nutrition strategies presents promising opportunities for enhancing crop resilience and human well-being.

The aim of this chapter is to provide an in-depth overview of the roles, mechanisms, and applications of oligosaccharides in plant stress tolerance and immune defense, highlighting their potential in improving agricultural sustainability and food security.

9.2 Chemical Nature and Classification of Oligosaccharides

9.2.1 Structural Types of Oligosaccharides

Carbohydrates with two or more monosaccharides are called oligosaccharides (two to ten monosaccharide units). Simply joining various monosaccharides with glycosidic linkages to form oligosaccharides can result in diverse structures (Lebrilla et al. 2022). They can be obtained by chemical synthesis or enzymatic reaction, or they can be extracted directly from natural sources using physicochemical approaches. Oligosaccharides are essential for many biological functions, such as cellular communication, energy storage, and structural support. They are crucial topics of research in biochemistry, plant science, food science, and medicine because of their intricate chemical structure and wide range of applications (Chen et al. 2024). In terms of chemistry, oligosaccharides are distinguished by the kinds of monosaccharide residues they contain, their order, the kind of glycosidic bonds they have, and the extent of polymerization (Yue et al. 2022). Different classifications of oligosaccharides have been described, such as classification based on the number of monosaccharide units (e.g., trisaccharide, which includes three monosaccharides, such as raffinose) or the function and source (e.g., prebiotics like fructooligosaccharides, galactooligosaccharides, and xylooligosaccharides) (Chavan et al. 2023). The main types of oligosaccharides having functional characteristics are thought to be those made up of the monosaccharide units of glucose, fructose, xylose, and galactose. The most common oligosaccharides being produced on a commercial level are fructooligosaccharides (FOSs), galactooligosaccharides (GOSs), and xylooligosaccharides (XOSs) (Phukon et al. 2022).

9.2.1.1 Fructooligosaccharides (FOSs)

Fruits and vegetables contain naturally occurring organic substances called fructooligosaccharides (FOSs) (Rahim et al. 2021). They are short-chain carbohydrates (short-chain fructans) that have a terminal glucose unit and are mostly made up of fructose units connected by β-(2 → 1) glycosidic linkages. Various structures, including 1-kestose, 1-nystose, and 1-β-fructofuranosylnystose, are generated enzymatically from sucrose or inulin. FOS has a different chemical structure compared to inulin because its degree of polymerization (DP) is lower (Ibrahim 2021; Sánchez-Martínez et al. 2020). To clarify the precise linking locations and sequences inside these oligosaccharides, sophisticated methods such as mass spectrometry have been used (Liew et al. 2024). When a concentrated solution of sucrose is treated with invertase or a particular fungal transferase, some of the glycosidic linkages are broken down and some D-fructofuranosyl units are transferred to O1 of the D-fructofuranosyl unit of another sucrose molecule in a β-D linkage. The resulting trisaccharide is a ketose, whose structure is commonly shortened to GF2, which denotes a disaccharide consisting of two fructofuranosyl units joined to a glucosyl unit (Fig. 9.1) (BeMiller 2018a, b).

FOSs are also sometimes referred to as oligofructose or oligofructans, and they are oligosaccharide fructans (Hernández and Plou 2023). FOSs are commonly utilized as sweeteners, are low in calories, and are not harmful for diabetics. As previously mentioned, they can be found naturally in fruits and vegetables. Onions,

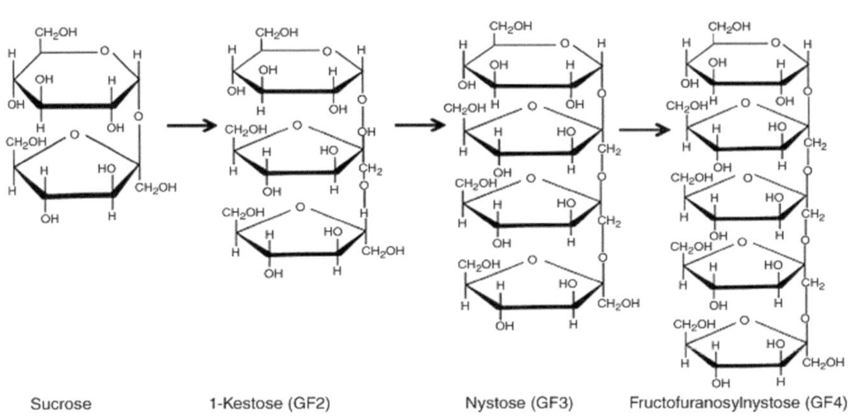

Fig. 9.1 FOS fructose oligomers. (Reprinted from Rahim et al. (2021). Open access journal)

bananas, garlic, chicory, rye, leeks, wheat, Jerusalem artichokes, burdock, and yacón are among the fruits, vegetables, and plants from which FOSs are derived (Table 9.1). The ones with the highest quantities of FOSs have been identified as yacón, chicory, and Jerusalem artichokes. FOSs can be extracted from plants using either methanol or ethanol extraction, which is based on water or aqueous alcohol (Guo et al. 2016). The hydrolysis of inulin is the primary commercial source of FOSs (BeMiller 2018a, b). Chicory roots and Jerusalem artichokes contain high levels of inulin (Ibrahim 2021). A sucrosyl unit terminates the reducing end of a large number of native inulin molecules. As a result, a sucrosyl unit will also terminate the reducing end of some of the FOS molecules produced by the hydrolysis of inulin. Another name for the FOSs produced by partially hydrolyzing inulin is inulodextrins. Whether they are derived from inulin or through the action of an enzyme on sucrose, the primary characteristic of fructooligosaccharide combinations is their prebiotic properties (BeMiller 2018a, b).

FOSs come in a range of chemical forms, with varying levels of chemical bonding states and degree of polymerization (DP) (Alatorre-Santamaría et al. 2022). As a result, it is challenging to pinpoint the exact amount and composition of each form

Table 9.1 Different natural sources of FOSs

Sources	Scientific name	Total FOS (g/100 g)	Reference
Nectarine	*Prunus persica var. nucipersica*	0.89 ± 0.030	Jovanovic-Malinovska et al. (2014)
Banana	*Musa acuminata*	0.5 ± 0.011	Sridevi et al. (2014)
Onion	*Allium cepa*	22.96 ± 5.39	Mabrok et al. (2018)
Garlic	*Allium sativum*	3.34 ± 0.01	Sunu et al. (2019)
Wheat grain	*Triticum*	2.21 ± 0.07	Pepe et al. (2013)
Rye	*Secale cereale*	0.6 ± 0.22	Karppinen et al. (2003)
Oat	*Avena sativa*	0.92 ± 0.08	Koruri et al. (2014)
Chicory root	*Cichorium intybus*	42.39 ± 0.15	Chikkerur et al. (2020)
Jerusalem artichoke	*Helianthus tuberosus*	4.74 ± 0.16	Khuenpet et al. (2017)
Seaweed	*Gracilaria verrucosa*	37.49 ± 0.09	Kim et al. (2015)
Sugarcane bagasse	*Gramineae Saccharum officinarum L.*	35.95 ± 1.64	Ganaie et al. (2017)
Cassava waste	*Manihot esculenta*	34 ± 0.64	Lateef and Gueguim-Kana (2012)
Rice bran	*Oryza sativa*	50 ± 0.04	Sangeetha et al. (2012)
Rice straw	*Oryza sativa L.*	1.61 ± 1.04	Sophonputtanaphoca et al. (2018)
Apple pomace	*Malus domestica Borkh.*	4.72 ± 0.80	Ganaie et al. (2017)
Papaya	*Carica papaya*	15.1 ± 0.16	Rohin et al. (2014)
Beetroot peels	*Beta vulgaris*	2.68 ± 0.61	Ganaie et al. (2017)
Burdock root	*Arctium*	2.4 ± 1.01	Li et al. (2013)

Reprinted from Rahim et al. (2021). Open access journal

of FOS. A study was conducted to examine the variations in FOS composition and relative amounts between onion cultivars that are cultivated in Hirosaki City, Japan. The tested varieties were split into three categories because of different patterns of sugar and FOS accumulation. Enzyme activity measurements revealed that the variations in enzyme activities linked to the FOS biosynthesis and biodegradation pathways are the causes of varietal variations in FOS composition (Maeda et al. 2017).

Osmotic balance is negatively impacted by drought stress, so plants store various organic and inorganic materials that aid in osmotic adjustment. These include sugars like glucose, sucrose, fructose, fructooligosaccharides, and trehalose; sugar alcohols like mannitol; and amino acids like proline and glycine to lower the osmotic potential and enhance cell water retention (Liu et al. 2022). Additionally, fructooligosaccharides contribute to stress tolerance. Transgenic research demonstrates that exotic genes from bacteria such as *Bacillus subtilis* and several plant species, such as wheat, Jerusalem artichokes (*Helianthus tuberosus*), and onions (*Allium cepa*), that are involved in FOS biosynthesis, lead to higher FOS concentrations and better drought resistance in cotton, sugar beet, and tobacco (Dempsey and Thavarajah 2024). A plant reserve carbohydrate known as burdock fructooligosaccharide (BFO) was initially identified in *Arctium lappa* (burdock) roots. According to studies, BFO is a natural elicitor that can cause significant physiological changes and activate plants' defense mechanisms (Guo et al. 2016). A study was conducted by Wang et al. to investigate the mechanism of BFO on tobacco mosaic virus (TMV) in tobacco seedlings. The findings indicated that BFO may boost TMV resistance, and the method may involve activating the expression of several defense genes (Wang et al. 2009).

9.2.1.2 Galactooligosaccharides (GOSs)

Usually made from lactose, galactose units joined by glycosidic linkages make up the majority of galactooligosaccharides (GOSs), which are indigestible carbohydrates. Their functional qualities as prebiotics are influenced by their wide range of glycosidic connections and varying DP in their chemical structure. Lactose, a sugar present in milk, is the primary source of GOSs, which can be produced by a variety of enzymatic and chemical processes. These oligosaccharides are known for their prebiotic qualities, which support gut health by inducing beneficial bacteria. Typically, a GOS is composed of glucose and galactose units. Common links between galactose units include α-(1 $\rightarrow$ 6) and β-(1 $\rightarrow$ 6), whereas between galactose and glucose, there are α-(1 $\rightarrow$ 4) and β-(1 $\rightarrow$ 4) linkages (Zeng et al. 2023).

GOS comes from sources other than lactose, such as from plant-derived galactose or from acid whey. For plant-derived galactose, GOS can be made from gum arabic galactose by using specific yeast strains that catalyze the synthesis without any by-products. The galactose is sourced from plants and comes from the acid hydrolysate of gum arabic because it supplies the substrate required for the enzymatic production of GOS (Wang et al. 2023).

Plant defense against abiotic stressors, including drought and salinity, is greatly aided by galactooligosaccharides, especially raffinose family oligosaccharides (RFOs) (De Koning et al. 2023). It has been proposed that RFOs, which accumulate in many plants under stress, serve as stress protectants (Gu et al. 2018). Raffinose synthase and galactinol synthase, which are essential for the buildup of RFOs in response to environmental stresses, manufacture these chemicals (De Koning et al. 2023). De Koning et al. conducted a study to check whether RFOs play a role in crop tolerance toward abiotic stress in common bean (*Phaseolus vulgaris*). It was found that the amount of raffinose in the leaves noticeably increased under salt stress. The RFO biosynthesis genes' mRNA levels were typically low in the roots, and no galactinol, raffinose, or stachyose was found. These findings imply that both galactinol and raffinose may contribute to the common bean's defense against abiotic stressors in its leaves (De Koning et al. 2023).

9.2.1.3 Xylooligosaccharides (XOSs)

Xylose units make up the majority of xylooligosaccharides (XOSs), which are short-chain polymers that normally have a length of two to seven units. They can have a variety of side chains in their chemical structure, including 4-O-methylglucose and arabinose, which enhance their functional qualities. Because it can affect their biological activities, such as their antioxidant and prebiotic benefits, XOSs' structural diversity is important. XOSs' chemical structure is characterized by β-1,4-linked xylose residues, with different degrees of polymerization (DP) and possible side chains. XOSs are health-promoting substances that are used in a variety of industries, such as food, medicine, and cosmetics (Ali et al. 2025). XOSs are made from plants, especially lignocellulosic biomass. These substances are produced by hydrolyzing xylans, which are an important part of the hemicellulose present in plant cell walls (Thakuria 2025).

Xylooligosaccharides (XOSs) are emerging as a promising biostimulant for enhancing plant tolerance to abiotic stresses, such as drought and salinity. Under saline circumstances, exogenous XOSs decreased protein oxidation and total polyphenols in *Arabidopsis thaliana* by increasing catalase (CAT) gene expression and enzymatic activity, so they might be able to mitigate the detrimental consequences of oxidative stress in saline environments (González-Pérez et al. 2018). Similarly, XOS treatment decreased protein oxidation and lipid peroxidation in *Nicotiana tabacum*, suggesting an effective antioxidant defense system (Páez-Watson et al. 2020). A study was done to evaluate how XOSs would enhance Chinese cabbage's ability to withstand salinity, and the results showed that increased osmolyte accumulation, decreased lipid peroxidation, improved ionic balance, and an improved antioxidative defense system were all linked to the higher salinity tolerance and improved growth in these treatments (Chen et al. 2016).

9.3 Mechanisms of Oligosaccharide-Induced Abiotic Stress Tolerance

Plants are exposed to various abiotic stresses, such as drought, salinity, extreme temperatures, and heavy metal toxicity, which pose significant challenges to plant growth and productivity. Oligosaccharides have emerged as crucial elicitors of plant defense mechanisms against these stresses. This section explores the mechanisms through which oligosaccharides induce abiotic stress tolerance, focusing on stress sensing, signal transduction, and regulation of gene expression.

9.3.1 Stress Sensing and Signal Transduction

Plants perceive abiotic stress through complex sensing mechanisms that trigger a series of signaling pathways. Oligosaccharides act as signaling molecules that improve the plant's ability to sense stress and activate appropriate defense responses. This process involves the modulation of reactive oxygen species (ROS) scavenging and the induction of hormonal pathways.

9.3.1.1 Role in Reactive Oxygen Species (ROS) Scavenging

Reactive oxygen species (ROS) are by-products of several metabolic processes in plants. ROS play dual roles as both signaling molecules and potential toxic agents under stress conditions. Under abiotic stress, ROS accumulation can lead to oxidative damage. However, they also act as important signals that trigger adaptive responses. Oligosaccharides, especially the raffinose family oligosaccharides (RFOs), have been shown to enhance the scavenging capacity of ROS in plants. For example, transgenic *Arabidopsis* plants with higher amounts of raffinose and galactinol had better ROS-scavenging abilities, indicating that RFOs contribute to maintaining redox homeostasis during abiotic stress (ElSayed et al. 2014).

Several enzymatic pathways, including superoxide dismutase (SOD) and peroxidases (POXs) that mitigate oxidative stress, tightly regulate the production of ROS (Li et al. 2022). When combined with oligosaccharides, these enzymes provide a defense against ROS-induced damage. Moreover, under different abiotic stresses, plants produce specific signatures of ROS, leading to acclimation responses that are appropriate to the stress (Choudhury et al. 2017). This indicates that oligosaccharides and ROS signaling pathways interact in a complex manner to mediate stress responses. This interaction is multifaceted, involving direct mechanisms, indirect enhancement of antioxidant systems, and modulation of gene expression.

Research indicates that oligosaccharides can act directly as ROS scavengers. A study found that fructans, a type of RFO, have an effective scavenging capacity against hydroxyl radicals and other reactive species. This direct interaction alleviates oxidative injury by excess ROS under stress conditions (Van den Ende and Valluru 2009). At higher concentrations, sucrose and other oligosaccharides have also been shown to provide protective effects by establishing stable sugar radicals that can more efficiently quench free radicals (Keunen et al. 2013). These oligosaccharides can interact with ROS produced during metabolic processes in vacuoles and apoplastic spaces, thus reducing oxidative stress (Van den Ende and Valluru 2009).

In addition to their direct scavenging abilities, oligosaccharides enhance the plant's overall antioxidant capacity by promoting the development of different antioxidant enzymes. These enzymes include superoxide dismutase (SOD), catalase (CAT), and ascorbate peroxidase (APX), which synergistically work on neutralizing ROS (Huang et al. 2019). The application of RFOs could lead to an increased activity of such enzymes to protect the plant against oxidative stress (You and Chan 2015).

Furthermore, oligosaccharides contribute to metabolic processes involved in NADPH synthesis, a critical cofactor in the antioxidant defense mechanism. By contributing to the oxidative pentose, soluble sugars like RFOs can increase the availability of NADPH, which is essential for the regeneration of reduced forms of antioxidants like glutathione (You and Chan 2015). Oligosaccharides serve this dual role, acting as both substrates for energy production and antioxidant regulators, which reinforces the importance of oligosaccharides in maintaining redox homeostasis under abiotic stress conditions.

Genes encoding antioxidant enzymes have been found to be upregulated when we have increased levels of RFOs. This suggests that oligosaccharides also influence gene expression related to ROS scavenging (Sanyal et al. 2023) (Table 9.2). This is essential for establishing a strong defense mechanism against oxidative damage. These genes are expressed in various ways, depending on the type and intensity of the stress experienced by the plant (Huang et al. 2019).

9.3.1.2 Induction of Hormonal Pathways

Oligosaccharides play a crucial role in inducing hormonal pathways that facilitate stress tolerance. Abscisic acid (ABA) is a key hormone that is involved in plant responses to abiotic stresses. Studies have shown that RFOs can moderate ABA signaling pathways by increasing the expression of genes involved in the ascorbate–glutathione cycle, which is essential for ROS detoxification (Li et al. 2022).

Moreover, additional hormones that are known to be involved in stress responses, such as ethylene and jasmonic acid, can be influenced by oligosaccharides in terms of their synthesis and signaling (Dempsey and Thavarajah 2024). This interaction emphasizes how crucial oligosaccharides are for both direct ROS scavenging and the coordination of hormonal reactions that improve stress resilience, where integrating these hormonal pathways with oligosaccharides forms a comprehensive network that allows plants to effectively detect and respond to abiotic stressors.

Table 9.2 ROS are involved in plant stress responses

Stress response	Relative ROS	Gene or phytohormone	Source	Reference
Waterlogging	$O_2^{\cdot-}$, hydrogen peroxide (H_2O_2)	Ethylene, *OsRBOHH*	Rice	Yamauchi et al. (2017)
Water stress	H_2O_2	*ABA*, *OsCATB*	Rice	Ye et al. (2011)
High temperature	$O_2^{\cdot-}$, H_2O_2	*OsCATB*	Rice	Zhao et al. (2018)
Disease resistance	H_2O_2	Ethylene, *OsEIN2*	Rice	Yang et al. (2017)
Cold temperature	$O_2^{\cdot-}$, H_2O_2	*AtSRC*	Arabidopsis	Kawarazaki et al. (2013)
Plant immunity		*AtRBOHD*	Arabidopsis	Kadota et al. (2014)
Drought	$O_2^{\cdot-}$, H_2O_2	*ABA*, *AtNTL4*	Arabidopsis	Lee et al. (2012)
Aluminum stress	H_2O_2	*AtPRX64*	Tobacco	Wu et al. (2017)
Organic pollutant treatment	H_2O_2, NO	24-Epibrassinolide	Cucumber	Ahammed et al. (2017)
Microbial pathogens	H_2O_2	*SA*, *CaPAL1*	Pepper	Kim and Hwang (2014)

Reprinted from Huang et al. (2019). Open access journal
Some key genes involved in ROS homeostasis and their functions during plant stress response are listed in the table. "–" indicates that the specific types of ROS are not specified in the literature
ABA abscisic acid, *SA* salicylic acid

9.3.2 Regulation of Gene Expression

Beyond their role in signaling, oligosaccharides influence abiotic stress tolerance by regulating gene expression at the transcriptional and epigenetic levels. This regulation involves the upregulation of defense-related genes and the induction of epigenetic modifications that tackle stress resilience.

When plants detect stress signals, they initiate a series of gene expression modifications that activate defensive mechanisms. Oligosaccharides activate several defense-related genes that encode proteins involved in protective mechanisms, such as antioxidant enzymes, osmoprotectants, and stress-responsive transcription factors (Ghadirnezhad Shiade et al. 2024).

The accumulation of ROS under abiotic stress can lead to oxidative damage. Oligosaccharides, especially RFOs, increase the expression of gene encoding for antioxidant enzymes such as superoxide dismutase (SOD), catalase (CAT), and peroxidases. Studies have shown that RFOs can increase the activity of these enzymes, thereby enhancing the plant's ability to scavenge ROS and protect cellular components from oxidative damage (Zandi and Schnug 2022).

Oligosaccharides induce the synthesis of osmoprotectants like proline and glycine betaine, which maintain cellular turgor pressure under stress conditions. Plants treated with oligosaccharides during drought or salinity stress were found to have upregulation in the genes involved in the production of these osmoprotectant enzymes (Nishizawa-Yokoi et al. 2008).

Oligosaccharides affect the expression of several transcription factors that control stress-responsive networks. For example, transcription factors such as dehydration-responsive element-binding (DREB) proteins, myeloblastosis (MYB), and WRKY are often increased in response to oligosaccharide treatment. These transcription factors bind to specific cis-acting regions of target gene promoters, initiating their transcription under stress conditions. The orchestrated action of these transcription factors forms a complex regulatory system that allows the plants to build effective defensive responses against environmental stresses (ElSayed et al. 2014).

The effect of oligosaccharides in mediating the epigenetic modifications that regulate gene expression during abiotic stress responses has been highlighted in various studies. Epigenetics includes all the processes that can alter gene activity without changing the DNA sequence, which leads to modifications that can be transmitted to daughter cells (Weinhold 2006). Epigenetic modifications can alter the transcriptional range of stress-responsive genes without any modifications in the underlying DNA sequence. This epigenetic change can occur as DNA methylation or histone modification (Mittler et al. 2022).

9.4 Types of Abiotic Stresses Mitigated by Oligosaccharides

9.4.1 Drought Stress

Drought stress is considered one of the most damaging abiotic stresses affecting plants. This type of stress is found in the regions exposed to changes in climatic factors such as low precipitation levels and temperature dynamics, leading to water scarcity and, therefore, resulting in a reduction in plant turgor, stunted growth, decreased photosynthesis, and inhibition of root elongation (Liu et al. 2013). In response, the plant utilizes several mechanisms to maintain its osmotic homeostasis through the accumulation of solutes, such as oligosaccharides. These carbohydrates include raffinose, stachyose, and verbascose, which are vital in plant responses to drought stress.

9.4.1.1 Osmotic Adjustment and Water Retention

Plants maintain their osmotic adjustment through the accumulation of solutes, osmoprotectants, and osmolytes (Pintó-Marijuan and Munné-Bosch 2013). Raffinose and stachyose, from the oligosaccharide family, are compounds synthesized to withstand reduced water levels by maintaining cell turgor and water retention. Additionally, they help the cells preserve their hydrated state for a longer time and inhibit the accumulation of reactive oxygen species (ROS) (Soualiou et al. 2022).

By accumulating in leaves, raffinose and stachyose help plants deal with drought by acting as a reserve that may be utilized during water deficit (Egert et al. 2015). These sugars maintain osmotic balance and protect cellular structures as they are a part of the raffinose family oligosaccharide (RFO) biosynthesis pathway (Elango et al. 2022). For example, a study conducted by De Koning et al. shows that raffinose and stachyose accumulate in the leaves of common bean (*Phaseolus vulgaris*) under drought stress. The raffinose level increased significantly by 150% during early drought stress and remained high as conditions were more severe, while stachyose synthase (PvSS) was upregulated, highlighting its role in drought stress adaptation (De Koning et al. 2023).

Moreover, alginate oligosaccharides (AOSs), another member of the oligosaccharide family, regulate stomatal closure, activate antioxidant enzyme genes, stimulate photosynthetic efficiency, and induce abscisic acid (ABA) production to help plants preserve their water. To demonstrate this, a hydroponic study was done to emphasize the role of alginate oligosaccharides (AOSs) in wheat (*Triticum aestivum* L.), which is subjected to drought stress and simulated by polyethylene glycol-6000 (PEG-6000). Results showed that the application of AOS positively affected and improved the seedling, root length, and relative water content. Besides, the antioxidant enzyme activity was enhanced, and AOS upregulated the drought-resistance genes involved in ABA signaling. Therefore, AOS enhances wheat drought resistance via the ABA-dependent pathway (Liu et al. 2013). Moreover, another study conducted by Li et al. (2018) highlighted the role of alginate-derived oligosaccharides (ADOs) presented by increasing stomatal conductance, transpiration rate, and chlorophyll retention (Li et al. 2018).

9.4.1.2 Stabilization of Membrane Integrity

The basic structure of cell membranes includes the lipid bilayer, proteins, fats, and carbohydrates. Membrane proteins have oligosaccharides such as raffinose, stachyose, and verbascose attached to them via the hydroxyl group. Drought stress disrupts cellular membranes through an increase in reactive oxidative species (ROS), superoxide radicals, hydrogen peroxide, and hydroxyl radicals, leading to increased permeability and ion leakage. These high levels indicate oxidative damage that disrupts membrane integrity (Assaha et al. 2016).

Additionally, this causes plants to lose their electrolytes, water, and soluble materials in the intercellular spaces. Hence, when components of the cell pack together, cellular dehydration alters membrane fluidity. At the same time, an increase in ROS causes lipid peroxidation and protein denaturation and can eventually lead to cell death (Hernández-Galván and Mercado-Uribe 2023). However, plants tend to withstand stress by the accumulation of osmolytes and compatible solutes such as sugars (e.g., raffinose and sucrose), sugar alcohols (e.g., galactinol), and amino acids (e.g., proline) (Takahashi et al. 2020). Consequently, in times of stress, this helps stabilize membrane integrity, prevents cellular damage, and maintains physiological balance.

A study highlights the crucial role of sucrosyl oligosaccharides (SOSs), including raffinose family oligosaccharides (RFOs) and fructans, in plant responses to oxidative stress. These compounds protect cells from oxidative damage caused by environmental stressors like drought by stabilizing membranes and scavenging ROS (Van den Ende and Valluru 2009). Another study underscores the role of alginate-derived oligosaccharides (ADOs) in resisting drought stress in cucumber plants (*Cucumis sativus* L.). The research showed that the ADOs stimulated antioxidant levels and enhanced the activity of enzymes that neutralize harmful reactive oxygen species (ROS). In addition, ADOs triggered the expression of genes related to drought resistance and abscisic acid (ABA) signaling. As a result, ADOs help cucumbers withstand drought by increasing ABA levels and activating genes that protect the plant from stress (Li et al. 2018). Consequently, protective proteins such as late embryogenesis abundant (LEA), which are essential in enhancing plant stress, may be produced and, therefore, help stabilize membrane-associated proteins, preventing their denaturation and maintaining overall membrane integrity (Amara et al. 2014).

9.4.2 *Salinity Stress*

Another major abiotic stress affecting plant productivity, development, and crop yield is salinity stress. It causes severe damage to plants by disturbing their growth at the molecular, biochemical, and morphological levels. The production of ROS and peroxidation of significant biological molecules causes oxidative harm to the plant and, therefore, reduces photosynthesis and crop yield (Acharya et al. 2024). High salt concentrations lead to ion toxicity, osmotic stress, oxidative damage, and metabolic imbalances. However, plants tend to deal with salt stress through the accumulation of compatible solutes or osmolytes, ion balance, activation of signaling pathways, and several other mechanisms (Fig. 9.2). Recently, it has been investigated that oligosaccharides play a key role in stimulating plant resilience against such stress (Hasanuzzaman and Fujita 2022).

9.4.2.1 Ion Homeostasis

Plants exposed to salty environments experience stress because the excess salt affects their ability to take up water. This causes two main types of stress, which are osmotic stress and ionic stress. In other words, salinity results in increased toxic levels of sodium (Na^+) ions in the plant's tissues, disrupting the function of enzymes and membranes and triggering ionic imbalances and osmotic stress (Munns 2011; Hasanuzzaman and Fujita 2022). This means that the excess absorption of Na^+ and

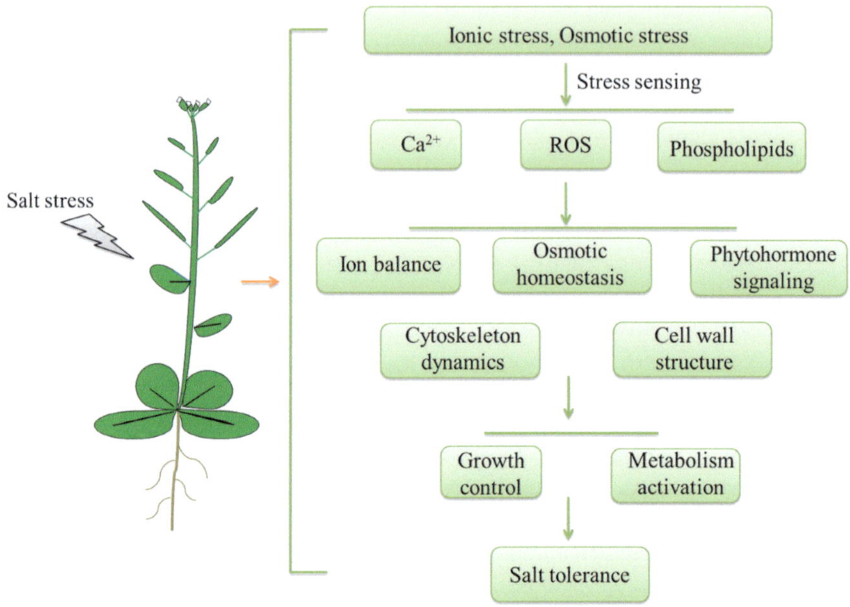

Fig. 9.2 A simplified model of the plant salt stress response. Salt stress primarily causes ionic stress and osmotic stress. After sensing Na+ and hyperosmolality, plants accumulate Ca2+, activate ROS signaling, and alter their phospholipid composition. These signals activate adaptive processes to alleviate salt stress, including maintaining an ion balance and osmotic homeostasis, inducing phytohormone signaling, and regulating cytoskeleton dynamics and the cell wall structure. Subsequently, through an array of signal transduction pathways, plant growth is slowed, and metabolism is activated to increase salt tolerance. (Reprinted from Zhao et al. (2021). Open access journal)

Cl^- induces oxidative stress by damaging biological macromolecules and stimulates ionic stress by inhibiting enzyme activity (Zhou et al. 2024) (Fig. 9.3). On the other hand, potassium (K^+) ions, which are essential for maintaining cellular osmotic balance, are often replaced by Na^+ ions in the cytosol under saline conditions, leading to physiological imbalances, thus affecting plant growth and development (Hualpa-Ramirez et al. 2024).

Moreover, as Na^+ and Cl^- increase in cells, they increase in the cytosol and exceed the capacity of the vacuole to hold them. This results in a larger vacuolar-to-extravacuolar volume ratio, affecting ion toxicity susceptibility (Lu and Fricke 2023). Plants usually tend to counteract this by pumping excess Na^+ ions out of the cytosol using antiporters and specific transporters.

Plants under salinity stress also undergo osmotic adjustments by accumulating compatible solutes or osmoprotectants such as proline, glycine betaine, and sugars that help maintain cell turgor. After that, a reduction in the osmotic potential, specifically in the cytosolic compartment, occurs when osmolytes contribute to regulating osmotic pressure. Additionally, they function as signaling molecules to trigger

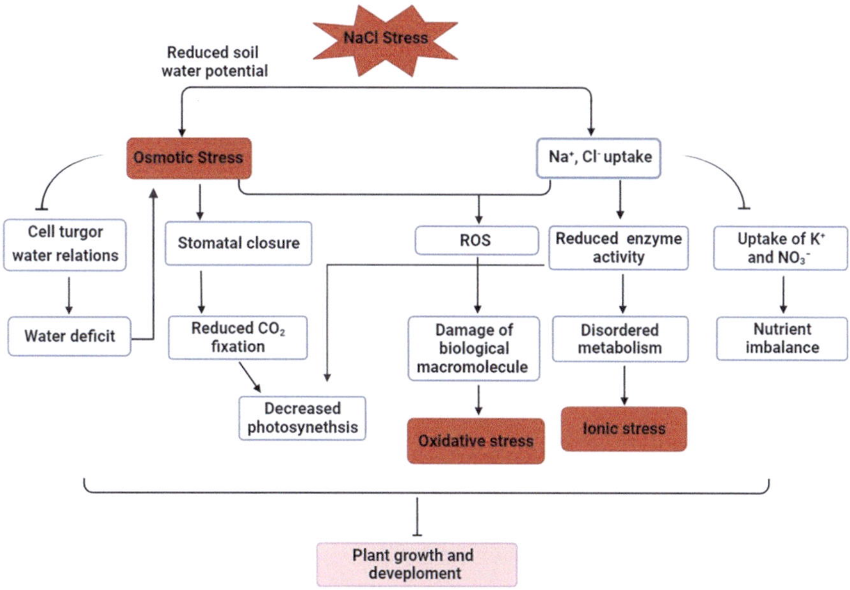

Fig. 9.3 The adverse effects of salt stress imposed on plants. The damages to plants are mainly caused by water deficit and increased intracellular Naþ and Cl contents under saline conditions. Osmotic stress changes cell morphology and leads to stomatal closure, interfering with photosynthesis in plants. The large amount of Naþ and Cl entering the cells causes ionic stress in plants, which gives rise to the reduction of Kþ-dependent enzyme activity, obstruction of nitrogen sources, and finally, metabolic disorders and nutrient imbalance. Osmotic and ionic stress also lead to ROS production, and oxidative stress is finally caused in plants by affecting the organelle-operating components or causing damage to their structure. These adverse effects together repress normal plant growth and development. ROS: reactive oxygen species. (Reprinted from Zhou et al. (2024). Open access journal)

ABA accumulation. The role of oligosaccharides has gained attention recently as they play a key role in increasing plant tolerance and, in this case, against salinity. The accumulation of these molecules, such as raffinose, fructooligosaccharides, and others, helps maintain water balance, thereby preventing cellular dehydration and stabilizing cellular structures. Oligosaccharides enhance the expression of ion transporters like Na^+/H^+ antiporter (Zhao et al. 2021).

The research done by Chen et al. highlights the role of xylooligosaccharides (XOSs) in enhancing the salinity tolerance of Chinese cabbage. It demonstrates that soil salinity negatively impacts several physiological parameters, such as biomass, photosynthesis, and water use efficiency. However, the application of XOSs moderates these effects by inducing antioxidant activity, reducing malondialdehyde content, and increasing the levels of soluble sugars, proline, and soluble proteins. In addition, it helps maintain ionic balance by reducing harmful Na^+/K^+, Na^+/Ca^{2+}, and Na^+/Mg^{2+} ratios. Notably, lower concentrations of XOSs (40–80 mg/L) were most effective, improving growth and salinity stress tolerance through enhanced antioxidative defense and reduced lipid peroxidation (Chen et al. 2016).

Another study investigated whether mixed-oligosaccharide priming, where seeds are soaked in a KP solution before planting, could promote germination under salt and alkaline stress. The results showed that by regulating root elongation and enhancing seedling establishment, KP priming reduces these stresses. This highlights the importance of oligosaccharides in stress tolerance, as KP priming improves energy metabolism and enhances the plant's ability to deal with environmental challenges (Yigang et al. 2024).

The impact of xyloglucan oligosaccharides (XGOs) was demonstrated in a study that assesses their role in salt stress in *Arabidopsis thaliana.* The efficiency of these exogenous XGOs, which were derived from *Tamarindus indica* cell walls, was emphasized in this research, where they alleviate oxidative stress and improve plant performance under saline conditions. The study found that XGOs significantly enhanced antioxidant enzyme activities, including those of catalase and peroxidase, while also improving chlorophyll content and reducing protein oxidation, suggesting a potential mechanism through which XGOs could help maintain ion homeostasis under salt stress (González-Pérez et al. 2018).

9.4.2.2 Activation of Stress Tolerance Pathways

Plants activate stress response pathways to manage salt and maintain water balance. To help the plant deal with this type of stress, these pathways involve multiple signals. It involves activating antioxidant mechanisms to prevent cellular damage, excluding excess harmful substances like Na^+ to reduce toxicity, and maintaining cell water balance through osmotic regulation.

Oligosaccharides are involved in promoting signaling pathways for plants under abiotic stress; thus, they act as "messengers" to help the plant respond effectively. According to Du et al. (2023), alginate oligosaccharides (AOSs) induce plant stress resistance. This type of oligosaccharide is associated with the activation of the ABA-dependent pathway, which appears to engage antioxidant and osmoprotectant mechanisms (Du et al. 2023). Therefore, under high salinity conditions, the rise in endogenous ABA levels leads to stomatal closure, helping regulate water balance and osmotic homeostasis (Zhao et al. 2021).

Moreover, when plants experience stress, they start detecting signals from outside. This happens through protein kinases found on the cell surface, which are involved in signal transduction. The mitogen-activated protein kinase (MAPK) cascade is a crucial pathway that is triggered in response to stressors like salinity. Thus, by raising the levels of Ca^{2+} inside the plant cells, oligosaccharides like mannan oligosaccharides (MOSs) and oligogalacturonic acid (OGA) contribute to activating the MAPK cascade; therefore, the early elevation in Ca^{2+} acts as a signal to activate the MAPK cascade (Peng et al. 2025).

9.4.3 *Temperature Extremes*

Plants have a specific temperature range for efficient growth and development. This is referred to as their optimum growth temperature. When the temperature rises beyond this range, it can negatively affect essential physiological processes such as photosynthesis, respiration, and transpiration (Parthasarathi et al. 2022) and inhibit plant productivity (Hatfield and Prueger 2015). On the other hand, low-temperature (LT) stress can significantly damage plants, especially in tropical and subtropical regions. Consequently, it slows down the process of cell division and growth, reduces the ability of plants to photosynthesize, and inhibits plant metabolism and yield (Devi et al. 2023). Since plants are sessile, their survival relies on effectively activating resistance mechanisms in response to thermal stress (Nievola et al. 2017). This highlights the key role of oligosaccharides in helping plants cope with heat and cold stress, which are temperature extremes.

9.4.3.1 Cold Stress Tolerance

Low-temperature (LT) stress, also known as cold stress, negatively impacts plant growth by affecting cell survival, division, photosynthesis, and water transport, which, in turn, reduces crop productivity. It is important to note that tropical and subtropical plants are more sensitive to this type of abiotic stress when compared to other temperate plants that show some tolerance. LT stress damages various plant processes, leading to reduced yield. To deal with this stress, plants usually activate gene expression changes and ROS-scavenging systems, triggering biochemical and physiological adaptations to enhance tolerance, and accumulate soluble sugars that serve as osmoprotectants to protect cells from damage. These soluble sugars include sucrose, hexoses, and oligosaccharides such as raffinose and stachyose, which are found at elevated levels when plants are in the cold season (Devi et al. 2023).

In extremely cold regions, alfalfa is vulnerable to cold damage during seed germination. A study by Jiang et al. investigated how alginate oligosaccharides (AOSs), along with low-variable-temperature (LVT) stress, affect alfalfa growth in cold regions. The results showed that although LVT stress increased the mean germination rate (MGT), its effect on germination percentage was minimal. At the same time, AOS played a key role in improving root crown diameter and root dry weight, helping alfalfa seedlings adapt to LVT stress. As the temperature rose, the AOS also increased the underground biomass, which, in turn, led to faster growth and higher total biomass. It is evident that the research highlights that AOS enhances alfalfa's ability to cope with cold stress during germination, contributing to better growth in cold regions (Jiang et al. 2023).

Considering the accumulation of compatible solutes, earlier investigations have highlighted their importance. Notably, a study conducted by Noronha et al. found that VviRafS5 is essential for the grapevine's cold stress response, as it promotes raffinose synthesis. Cold exposure and ABA increased *VviRafS5* expression, leading to higher raffinose levels. The enzyme *VviRafS5*, identified as a raffinose synthase, is upregulated during cold acclimation, helping the grapevine better tolerate low temperatures (Noronha et al. 2022). Another study found that RFOs like stachyose, galactinol, sucrose, and raffinose accumulated in different subcellular compartments in cucumber leaves when they were exposed to cold stress. After cold stress removal, these sugars gradually decreased. The expression of galactinol synthase genes increased, except for *CsGS1*. The upregulation of some enzymes, such as GAL3 and AGA, and their localization in distinct compartments indicate that different galactosidases metabolize RFOs following the elimination of stress (Gu et al. 2018).

9.4.3.2 Heat Stress Mitigation

Heat stress is one of the major abiotic stressors that inhibit the growth and development of plants (Zhao et al. 2020). It also causes dehydration and irreversible damage to plant function and reduces the overall plant performance (Mishra et al. 2023). Additionally, elevated temperatures can adversely affect flower organs and tissues, causing changes in their development, structure, and function (Resentini et al. 2023) (Fig. 9.4). However, oligosaccharides, which include fructooligosaccharides and the raffinose family, are one class of chemicals that plants generate to help them cope with stress. Notably, TPS1 was found to be a crucial gene involved in raffinose accumulation under heat stress in *Arabidopsis thaliana.* By decreasing sucrose levels, higher TPS1 activity improved thermotolerance. In contrast, by increasing starch and sucrose accumulation, the disruption of trehalose 6-phosphate signaling improved heat resistance (Reichelt et al. 2023).

Nevertheless, another study investigated the role of oligosaccharins in tomato yield and fruit composition under greenhouse conditions, with varying high temperatures in a greenhouse. Oligosaccharins improved photosynthesis and yields, especially under lower-temperature stress. While leaf treatment had no impact, root application decreased several fruit components. These findings suggest that oligosaccharins enhance tomato yields under stress without harming the fruit quality, offering a natural biostimulant option. Moreover, priming helps plants develop a stress memory for further heat exposure, thus increasing their resistance to heat stress (Hernández et al. 2022). Primed plants perform better under heat stress than non-primed ones, partly due to increased raffinose family oligosaccharide (RFO) content, indicating that RFOs are crucial for heat stress tolerance (Serrano et al. 2019; Yan et al. 2022).

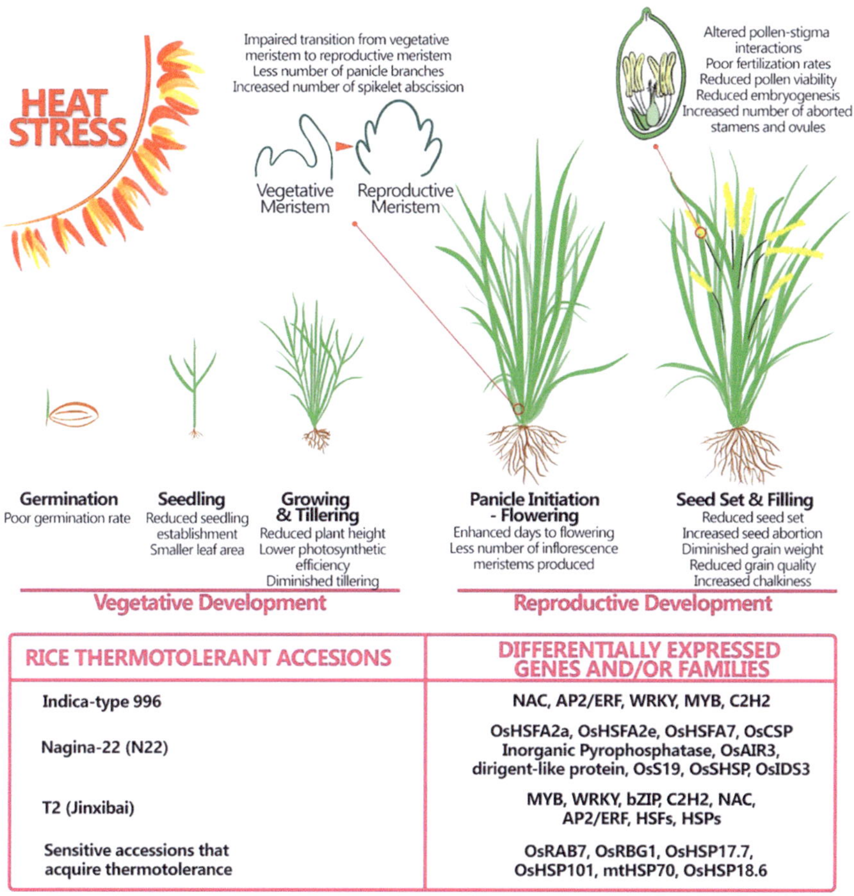

RICE THERMOTOLERANT ACCESIONS	DIFFERENTIALLY EXPRESSED GENES AND/OR FAMILIES
Indica-type 996	NAC, AP2/ERF, WRKY, MYB, C2H2
Nagina-22 (N22)	OsHSFA2a, OsHSFA2e, OsHSFA7, OsCSP Inorganic Pyrophosphatase, OsAIR3, dirigent-like protein, OsS19, OsSHSP, OsIDS3
T2 (Jinxibai)	MYB, WRKY, bZIP, C2H2, NAC, AP2/ERF, HSFs, HSPs
Sensitive accessions that acquire thermotolerance	OsRAB7, OsRBG1, OsHSP17.7, OsHSP101, mtHSP70, OsHSP18.6

Fig. 9.4 Adverse impact of heat stress on the morphological and physiological aspects of rice at different stages of development and possible genes and/or families involved in thermotolerance. Heat stress affects rice productivity by affecting both vegetative and reproductive stages; still, major effects of heat stress are seen during the reproductive stage and can provoke alterations in the number of panicles produced in the latest stages, modifying the starch content and, therefore, the quality of the grain. Some thermotolerant accessions were studied, and specific genes or families involved in thermotolerance were discovered. (Reprinted from Resentini et al. (2023). Open access journal)

9.4.4 Heavy Metal Stress

Heavy metals and metalloids are essential for plants when found in normal amounts, but excess concentrations can be toxic, disrupting growth and causing cellular damage (Angulo-Bejarano et al. 2021). These heavy metals could be either essential metals such as iron (Fe), magnesium (Mg), and manganese (Mn), which are important in plants, using which they enhance the capture and transfer of light energy within chloroplasts (Carrillo and Borthakur 2021), or toxic metals such as cadmium (Cd), lead (Pb), and chromium (Cr) (Angulo-Bejarano et al. 2021). Exposure to heavy metals in plants causes the overproduction of reactive oxygen species (ROS), leading to oxidative stress. This disrupts cellular functions by damaging membranes, enzymes, and signaling pathways, resulting in ATP depletion, mitochondrial damage, and cell death. However, oligosaccharides that accumulate under stress conditions play a significant role in plants' defense mechanisms against heavy metal stress (Mansoor et al. 2023).

The effects of galactoglucomannan oligosaccharides (GGMOs) on reducing cadmium (Cd) toxicity in maize (*Zea mays* L.) protoplasts were studied. The Cd absorption under stress, cell wall regeneration, and protoplast survival were examined. In optimal conditions, protoplasts cultivated with GGMOs showed increased cell wall regeneration, which reduced Cd uptake and improved viability. These findings suggest that GGMOs promote cell wall regeneration, providing a potential solution for mitigating Cd toxicity in plants, marking the first evidence of this effect in Cd-stressed conditions (Hačkuličová et al. 2023). Similarly, the role of GGMOs in plants under cadmium (Cd) stress was examined. The results showed that GGMOs reduced Cd concentration in the leaves, leading to improved plant growth (24%), better photosynthetic efficiency, and a reduction in stress markers like malondialdehyde (MDA) and hydrogen peroxide (H_2O_2). Additionally, it increased auxin levels; therefore, GGMOs help maize plants cope with Cd stress through nutrient uptake (Vivodová et al. 2023).

However, another study conducted by Kučerová et al. examined the protective role of GGMOs against cadmium stress in *Arabidopsis thaliana*. Under cadmium stress, GGMOs increased photosynthetic pigment content and root development, but they had no effect on cadmium accumulation or distribution within the plant. The development of suberin lamellae, which aid in water and solute transport, was also observed, with GGMOs stimulating faster root growth without stress conditions. Instead of changing cadmium absorption, the results show that GGMOs protect plants from cadmium stress by affecting metabolic responses (Kučerová et al. 2014). Moreover, alginate oligosaccharides, another type of oligosaccharide, have a significant role in enhancing crop photosynthesis and metabolite production by reducing environmental stress and activating the antioxidant system to help crops better withstand challenging conditions for survival (Li et al. 2023). Besides, chitooligosaccharide (COS) treatment on edible rape (*Brassica rapa* L.) enhanced the activity of critical antioxidant enzymes, including superoxide dismutase (SOD), catalase (CAT), and peroxidase (POX), which play vital roles in neutralizing the reactive oxygen species (ROS) generated by Cd stress (Zong et al. 2017).

9.5 Conclusions and Future Prospects

Oligosaccharides have emerged as vital components in plant stress physiology, offering a natural and effective means of enhancing tolerance to abiotic stresses such as drought, salinity, extreme temperatures, and heavy metal toxicity. Their ability to modulate signaling pathways, regulate gene expression, and maintain cellular homeostasis underscores their significance in crop resilience. However, while extensive research has demonstrated their beneficial effects, there is still much to be explored regarding their precise mechanisms of action and practical applications in large-scale agriculture.

Future research should focus on elucidating the molecular networks governing oligosaccharide-mediated stress responses through high-throughput sequencing and functional genomic approaches. Integrating oligosaccharides with biostimulants, microbial consortia, and precision agriculture techniques could further enhance their effectiveness in real-world farming conditions. Additionally, the ethical and regulatory aspects of using oligosaccharide-enriched genetically modified crops must be carefully considered to ensure public acceptance and ecological safety.

In the coming years, interdisciplinary collaborations between plant biologists, geneticists, biotechnologists, and agronomists will be essential for translating laboratory findings into field applications. By leveraging cutting-edge technologies such as clustered regularly interspaced short palindromic repeats (CRISPR), metabolomics, and AI-driven breeding, oligosaccharides could become key players in developing climate-resilient crops. Their integration into agricultural systems holds immense potential for achieving food security and sustainability, particularly in regions vulnerable to environmental stressors.

Overall, oligosaccharides represent a promising frontier in plant biotechnology, with the potential to revolutionize stress management strategies in crops. Continued research and innovation will be critical to harnessing their full potential and ensuring a more resilient and sustainable agricultural future.

References

Acharya BR, Gill SP, Kaundal A, Sandhu D (2024) Strategies for combating plant salinity stress: the potential of plant growth-promoting microorganisms. Front Plant Sci 15:1406913

Ahammed GJ, He BB, Qian XJ, Zhou YH, Shi K, Zhou J, Yu JQ, Xia XJ (2017) 24-Epibrassinolide alleviates organic pollutants-retarded root elongation by promoting redox homeostasis and secondary metabolism in *Cucumis sativus* L. Environ Pollut 229:922–931

Alatorre-Santamaría S, Cruz-Guerrero A, Guzmán-Rodríguez F (2022) Chapter 30 – Fructooligosaccharides (FOS). In: Jafari SM, Rashidinejad A, Simal-Gandara J (eds) Handbook of food bioactive ingredients: properties and applications, 1st edn. Springer, Cham, pp 1–30

Ali K, Niaz N, Waseem M, Ashraf W, Hussain M, Khalid MU, Tahir AB, Raza A, Khan IM (2025) Xylooligosaccharides: a comprehensive review of production, purification, characterization, and quantification. Food Res Int 201:115631

Amara I, Zaidi I, Masmoudi K, Ludevid MD, Pagès M, Goday A, Brini F (2014) Insights into late embryogenesis abundant (LEA) proteins in plants: from structure to the functions. Am J Plant Sci 5:3440–3455

Angulo-Bejarano PI, Puente-Rivera J, Cruz-Ortega R (2021) Metal and metalloid toxicity in plants: an overview on molecular aspects. Plants 10:635

Assaha DVM, Liu L, Ueda A, Nagaoka T, Saneoka H (2016) Effect of drought stress on growth, solute accumulation and membrane stability of leafy vegetable, huckleberry (*Solanum scabrum* Mill.). J Environ Biol 37:107–114

BeMiller JN (2018a) Chapter 3 – Oligosaccharides. In: BeMiller JN (ed) Carbohydrate chemistry for food scientists, 3rd edn. Elsevier, pp 49–74

BeMiller JN (2018b) Chapter 10 – Inulin and Konjac glucomannan. In: BeMiller JN (ed) Carbohydrate chemistry for food scientists, 3rd edn. Elsevier, pp 253–259

Carrillo JT, Borthakur D (2021) Methods for metal chelation in plant homeostasis. Plant Physiol Biochem 163:95–107

Chavan AR, Singh AK, Gupta RK, Nakhate SP, Poddar BJ, Gujar VV, Purohit HJ, Khardenavis AA (2023) Recent trends in the biotechnology of functional non-digestible oligosaccharides with prebiotic potential. Biotechnol Genet Eng Rev 39:465–510

Chen W, Guo C, Hussain S, Zhu B, Deng F, Xue Y, Geng M, Wu L (2016) Role of xylo-oligosaccharides in protection against salinity-induced adversities in Chinese cabbage. Environ Sci Pollut Res 23:1254–1264

Chen YJ, Sui X, Wang Y, Zhao ZH, Han TH, Liu YJ, Zhang JN, Zhou P, Yang K, Ye ZH (2024) Preparation, structural characterization, biological activity, and nutritional applications of oligosaccharides. Food Chem X 22:101289

Chikkerur J, Samanta AK, Kolte AP, Dhali A, Roy S (2020) Production of short chain fructo-oligosaccharides from inulin of chicory root using fungal endoinulinase. Appl Biochem Biotechnol 191:695–715

Choudhury FK, Rivero RM, Blumwald E, Mittler R (2017) Reactive oxygen species, abiotic stress and stress combination. Plant J 90:856–867

De Koning R, Wils GE, Kiekens R, De Vuyst L, Angenon G (2023) Impact of drought and salt stress on galactinol and raffinose family oligosaccharides in common bean (*Phaseolus vulgaris*). AoB Plants 15:plad038

Dempsey M, Thavarajah D (2024) Low molecular weight carbohydrates and abiotic stress tolerance in lentil (*Lens culinaris* Medikus): a review. Front Plant Sci 15:1408252

Devi V, Kaur A, Sethi M, Avinash G (2023) Chapter 5 – Effect of low-temperature stress on plant performance and adaptation to temperature change. In: Hussain S, Awan TH, Waraich EA, Awan MI (eds) Plant abiotic stress responses and tolerance mechanisms. IntechOpen, pp 1–28

Du YW, Liu L, Feng NJ, Zheng DF, Liu ML, Zhou H, Deng P, Wang YX, Zhao HM (2023) Combined transcriptomic and metabolomic analysis of alginate oligosaccharides alleviating salt stress in rice seedlings. BMC Plant Biol 23:455

Egert A, Eicher B, Keller F, Peters S (2015) Evidence for water deficit-induced mass increases of raffinose family oligosaccharides (RFOs) in the leaves of three *Craterostigma* resurrection plant species. Front Physiol 6:206

Elango D, Rajendran K, Van der Laan L, Sebastiar S, Raigne J, Thaiparambil NA, El Haddad N, Raja B, Wang W, Ferela A, Chiteri KO, Thudi M, Varshney RK, Chopra S, Singh A, Singh AK (2022) Raffinose family oligosaccharides: friend or foe for human and plant health? Front Plant Sci 13:829118

ElSayed AI, Rafudeen MS, Golldack D (2014) Physiological aspects of raffinose family oligosaccharides in plants: protection against abiotic stress. Plant Biol 16:1–8

Ganaie MA, Soni H, Naikoo GA, Oliveira LTS, Rawat HK, Mehta PK, Narain N (2017) Screening of low-cost agricultural wastes to maximize the fructosyltransferase production and its applicability in generation of fructooligosaccharides by solid state fermentation. Int Biodeterior Biodegrad 118:19–26

Ghadirnezhad Shiade SR, Zand-Silakhoor A, Fathi A, Rahimi R, Minkina T, Rajput VD, Zulfiqar U, Chaudhary T (2024) Plant metabolites and signaling pathways in response to biotic and abiotic stresses: exploring bio stimulant applications. Plant Stress 12:100454

González-Pérez L, Páez-Watson T, Álvarez-Suarez JM, Obando-Rojas MC, Bonifaz-Arcos E, Viteri G, Rivas-Romero F, Tejera E, Rogers HJ, Cabrera JC (2018) Application of exogenous xyloglucan oligosaccharides affects molecular responses to salt stress in *Arabidopsis thaliana* seedlings. J Soil Sci Plant Nutr 18:1187–1205

Gu H, Lu M, Zhang Z, Xu J, Cao W, Miao M (2018) Metabolic process of raffinose family oligosaccharides during cold stress and recovery in cucumber leaves. J Plant Physiol 224:112–120

Guo M, Chen G, Chen K (2016) Chapter 5 – Fructooligosaccharides: effects, mechanisms, and applications. In: Yin H, Du Y (eds) Research progress in oligosaccharins, 1st edn. Springer Nature, pp 51–64

Hačkuličová D, Labancová E, Šípošová K, Bajus M, Vivodová Z, Kollárová K (2023) Galactoglucomannan oligosaccharides mitigate cadmium toxicity in maize protoplasts by improving viability and cell wall regeneration. Plant Physiol Biochem 201:107907

Hasanuzzaman M, Fujita M (2022) Plant responses and tolerance to salt stress: physiological and molecular interventions. Int J Mol Sci 23:4810

Hatfield JL, Prueger JH (2015) Temperature extremes: effect on plant growth and development. Weather Clim Extrem 10:4–10

Hernández L, Plou FJ (2023) Chapter 1 – Fructans: the terminology. In: Van den Ende W, Öner ET (eds) The book of fructans, 1st edn. Academic Press, pp 3–10

Hernández V, Hellín P, Botella MÁ, Vicente E, Fenoll J, Flores P (2022) Oligosaccharins alleviate heat stress in greenhouse-grown tomatoes during the spring-summer season in a semi-arid climate. Agronomy 12:802

Hernández-Galván G, Mercado-Uribe H (2023) Dehydration of biological membranes in a non-condensing environment. Soft Matter 19:9173–9178

Hualpa-Ramirez E, Carrasco-Lozano EC, Madrid-Espinoza J, Tejos R, Ruiz-Lara S, Stange C, Norambuena L (2024) Stress salinity in plants: new strategies to cope with in the foreseeable scenario. Plant Physiol Biochem 208:108507

Huang H, Ullah F, Zhou DX, Yi M, Zhao Y (2019) Mechanisms of ROS regulation of plant development and stress responses. Front Plant Sci 10:800

Ibrahim OO (2021) Technological aspects of fructo-oligosaccharides (FOS), production processes, physiological properties, applications and health benefits. J Food Chem Nanotechnol 7:41–46

Jiang G, Wang S, Xie J, Tan P, Han L (2023) Discontinuous low temperature stress and plant growth regulators during the germination period promote roots growth in alfalfa (*Medicago sativa* L.). Plant Physiol Biochem 197:107624

Jovanovic-Malinovska R, Kuzmanova S, Winkelhausen E (2014) Oligosaccharide profile in fruits and vegetables as sources of prebiotics and functional foods. Int J Food Prop 17:949–965

Kadota Y, Sklenar J, Derbyshire P, Stransfeld L, Asai S, Ntoukakis V, Jones JD, Shirasu K, Menke F, Jones A, Zipfel C (2014) Direct regulation of the NADPH oxidase RBOHD by the PRR-associated kinase BIK1 during plant immunity. Mol Cell 54:43–55

Karppinen S, Myllymäki O, Forssell P, Poutanen K (2003) Fructan content of rye and rye products. Cereal Chem 80:168–171

Kawarazaki T, Kimura S, Iizuka A, Hanamata S, Nibori H, Michikawa M, Imai A, Abe M, Kaya H, Kuchitsu K (2013) A low temperature-inducible protein AtSRC2 enhances the ROSproducing activity of NADPH oxidase AtRbohF. Biochim Biophys Acta 1833:2775–2780

Keunen E, Peshev D, Vangronsveld J, Van Den Ende W, Cuypers A (2013) Plant sugars are crucial players in the oxidative challenge during abiotic stress: extending the traditional concept. Plant Cell Environ 36:1242–1255

Khuenpet K, Jittanit W, Sirisansaneeyakul S, Srichamnong W (2017) Inulin powder production from Jerusalem artichoke (*Helianthus tuberosus* L.) tuber powder and its application to commercial food products. J Food Process Preserv 41:e13097

Kim DS, Hwang BK (2014) An important role of the pepper phenylalanine ammonia-lyase gene (PAL1) in salicylic acid-dependent signaling of the defense response to microbial pathogens. J Exp Bot 65:2295–2306

Kim SW, Hong CH, Jeon SW, Shin HJ (2015) High-yield production of biosugars from *Gracilaria verrucosa* by acid and enzymatic hydrolysis processes. Bioresour Technol 196:634–641

Koruri SS, Banerjee D, Chowdhury R, Bhattacharya P (2014) Studies on prebiotic food additive (inulin) in Indian dietary fibre sources-garlic (*Allium sativum*), wheat (*Triticum* spp.), oat (*Avena sativa*) and dalia (*Bulgur*). Int J Pharm Pharm Sci 6:278–282

Krishna Perumal P, Dong CD, Chauhan AS, Anisha GS, Kadri MS, Chen SW, Singhania RR, Patel AK (2023) Advances in oligosaccharides production from algal sources and potential applications. Biotechnol Adv 67:108195

Kučerová D, Kollárová K, Zelko I, Vatehová Z, Lišková D (2014) Galactoglucomannan oligosaccharides alleviate cadmium stress in *Arabidopsis*. J Plant Physiol 171:518–524

Lateef A, Gueguim-Kana EB (2012) Utilization of cassava wastes in the production of fructosyltransferase by *Rhizopus stolonifer* LAU 07. Rom Biotechnol Lett 17:7309–7316

Lebrilla CB, Liu J, Widmalm G, Prestegard JH (2022) Chapter 3 – Oligosaccharides and polysaccharides. In: Varki A, Cummings RD, Esko JD, Stanley P, Hart GW, Aebi M, Mohnen D, Kinoshita T, Packer NH, Prestegard JH, Schnaar RL, Seeberger PH (eds) Essentials of glycobiology, 4th edn. Cold Spring Harbor Laboratory Press

Lee S, Seo PJ, Lee HJ, Park CM (2012) A NAC transcription factor NTL4 promotes reactive oxygen species production during drought-induced leaf senescence in *Arabidopsis*. Plant J 70:831–844

Li J, Liu X, Zhou B, Zhao J, Li S (2013) Determination of fructooligosaccharides in burdock using HPLC and microwave-assisted extraction. J Agric Food Chem 61:5888–5892

Li J, Wang X, Lin X, Yan G, Liu L, Zheng H, Zhao B, Tang J, Guo YD (2018) Alginate-derived oligosaccharides promote water stress tolerance in cucumber (*Cucumis sativus* L.). Plant Physiol Biochem 130:80–88

Li S, Liu S, Zhang Q, Cui M, Zhao M, Li N, Wang S, Wu R, Zhang L, Cao Y, Wang L (2022) The interaction of ABA and ROS in plant growth and stress resistances. Front Plant Sci 13:1050132

Li Z, Duan S, Lu B, Yang C, Ding H, Shen H (2023) Spraying alginate oligosaccharide improves photosynthetic performance and sugar accumulation in citrus by regulating antioxidant system and related gene expression. Front Plant Sci 13:1108848

Liew CY, Li WL, Ni CK (2024) Structural determination of fructooligosaccharides and raffinose family oligosaccharides using logically derived sequence tandem mass spectrometry. Analyst 149:5714–5727

Liu H, Zhang YH, Yin H, Wang WX, Zhao XM, Du YG (2013) Alginate oligosaccharides enhanced *Triticum aestivum* L. tolerance to drought stress. Plant Physiol Biochem 62:33–40

Liu R, Jiao T, Zhang Z, Yao Z, Li Z, Wang S, Xin H, Li Y, Wang A, Zhu J (2022) Ectopic expression of the *Allium cepa* 1-SST gene in cotton improves drought tolerance and yield under drought stress in the field. Front Plant Sci 12:783134

Lu Y, Fricke W (2023) Salt stress – regulation of root water uptake in a whole-plant and diurnal context. Int J Mol Sci 24:8070

Mabrok H, Soliman M, Mohammad M, Husein L (2018) HPLC profiles of onion fructooligosaccharides and inulin and their prebiotic effects on modulating key markers of colon function, calcium metabolism and bone mass in rat model. J Biochem Physiol 1:1

Maeda T, Watanabe A, Wambrauw DZ, Osanai S, Honda K, Oku S, Shimura H, Suzuki T, Yamasaki A, Okabe Y, Ueno K, Onodera S (2017) Analysis of varietal differences in the fructo-oligosaccharide accumulation profile among onion (*Allium cepa* L.) cultivars grown by spring-sown cultivation. J Hortic 86:501–510

Mansoor S, Ali A, Kour N, Bornhorst J, AlHarbi K, Rinklebe J, Abd El Moneim D, Ahmad P, Chung YS (2023) Heavy metal induced oxidative stress mitigation and ROS scavenging in plants. Plants 12:3003

Mishra S, Spaccarotella K, Gido J, Samanta I, Chowdhary G (2023) Effects of heat stress on plant-nutrient relations: an update on nutrient uptake, transport, and assimilation. Int J Mol Sci 24:15670
Mittler R, Zandalinas SI, Fichman Y, Van Breusegem F (2022) Reactive oxygen species signaling in plant stress responses. Nat Rev Mol Cell Biol 23:663–679
Munns R (2011) Plant adaptations to salt and water stress: differences and commonalities. Adv Bot Res 57:1–32
Mussatto SI, Mancilha IM (2007) Non-digestible oligosaccharides: a review. Carbohydr Polym 68:387–616
Nievola CC, Carvalho CP, Carvalho V, Rodrigues E (2017) Rapid responses of plants to temperature changes. Temperature 4:371–405
Nishizawa-Yokoi A, Yabuta Y, Shigeoka S (2008) The contribution of carbohydrates including raffinose family oligosaccharides and sugar alcohols to protection of plant cells from oxidative damage. Plant Signal Behav 3:1016–1018
Noronha H, Silva A, Silva T, Frusciante S, Diretto G, Gerós H (2022) VviRafS5 is a raffinose synthase involved in cold acclimation in grapevine woody tissues. Front Plant Sci 12:754537
Páez-Watson T, Álvarez-Suárez JM, Rivas-Romero F, Estrada L, López D, Pérez Pelea L, Cabrera JC, González-Pérez L (2020) Increased salinity stress tolerance of *Nicotiana tabacum* L. *in vitro* plants with the addition of xyloglucan oligosaccharides to the culture medium. In Vitro Cell Dev Biol Plant 56:325–334
Parthasarathi T, Firdous S, David EM, Lesharadevi K, Djanaguiraman M (2022) Chapter 2 – Effects of high temperature on crops. In: Ngui Kimatu J (ed) Advances in plant defense mechanisms, 1st edn. IntechOpen, pp 1–19
Peng C, Xu W, Wang X, Meng F, Zhao Y, Wang Q, Wang X, Lodi RS, Dong X, Zhu C, Peng L (2025) Alginate oligosaccharides trigger multiple defense responses in tobacco and induce resistance to *Phytophthora infestans*. Front Plant Sci 16:1506873
Pepe O, Ventorino V, Cavella S, Fagnano M, Brugno R (2013) Prebiotic content of bread prepared with flour from immature wheat grain and selected dextran-producing lactic acid bacteria. Appl Environ Microbiol 79:3779–3785
Phukon LC, Singh SP, Pandey A, Rai AK (2022) Chapter 1 – Microbial bioprocesses for production of nutraceuticals and functional foods. In: Rai AK, Singh SP, Pandey A, Larroche C, Soccol CR (eds) Current developments in biotechnology and bioengineering, 1st edn. Elsevier, pp 1–29
Pintó-Marijuan M, Munné-Bosch S (2013) Ecophysiology of invasive plants: osmotic adjustment and antioxidants. Trends Plant Sci 18:660–666
Rahim MA, Saeed F, Khalid W, Hussain M, Anjum FM (2021) Functional and nutraceutical properties of fructo-oligosaccharides derivatives: a review. Int J Food Prop 24:1588–1602
Reichelt N, Korte A, Krischke M, Mueller MJ, Maag D (2023) Natural variation of warm temperature-induced raffinose accumulation identifies TREHALOSE-6-PHOSPHATE SYNTHASE 1 as a modulator of thermotolerance. Plant Cell Environ 46:3392–3404
Resentini F, Orozco-Arroyo G, Cucinotta M, Mendes MA (2023) The impact of heat stress on plant reproduction. Front Plant Sci 14:1271644
Rohin MAK, Abu Bakar CA, Ali AM (2014) Isolation and characterization of oligosaccharides composition in organically grown red pitaya, white pitaya and papaya. Int J Pharm Pharm Sci 6:131–136
Sánchez-Martínez MJ, Soto-Jover S, Antolinos V, Martínez-Hernández GB, López-Gómez A (2020) Manufacturing of short-chain fructooligosaccharides: from laboratory to industrial scale. Food Eng Rev 12:149–172
Sangeetha PT, Ramesh MN, Prapulla SG (2012) Production of fructosyl transferase by *Aspergillus oryzae* CFR 202 in solid-state fermentation using agricultural by-products. Appl Microbiol Biotechnol 65:530–537
Sanyal R, Kumar S, Pattanayak A, Kar A, Bishi SK (2023) Optimizing raffinose family oligosaccharides content in plants: a tightrope walk. Front Plant Sci 14:1134754

Serrano N, Ling Y, Bahieldin A, Mahfouz MM (2019) Thermopriming reprograms metabolic homeostasis to confer heat tolerance. Sci Rep 9:181
Sophonputtanaphoca S, Pridam C, Chinnak J, Nathong M, Juntipwong P (2018) Production of non-digestible oligosaccharides as value-added by-products from rice straw. Agric Nat Resour 52:169–175
Soualiou S, Duan F, Li X, Zhou W (2022) Crop production under cold stress: an understanding of plant responses, acclimation processes, and management strategies. Plant Physiol Biochem 190:47–61
Sridevi V, Sumathi V, Guru Prasad M, Kumar SM (2014) Fructooligosaccharides-type prebiotic: a review. J Pharm Res 8:321–330
Sunu P, Sunarti D, Mahfudz LD, Yunianto VD (2019) Prebiotic activity of garlic (*Allium sativum*) extract on *Lactobacillus acidophilus*. Vet World 12:2046–2051
Takahashi F, Kuromori T, Urano K, Yamaguchi-Shinozaki K, Shinozaki K (2020) Drought stress responses and resistance in plants: from cellular responses to long-distance intercellular communication. Front Plant Sci 11:556972
Thakuria A (2025) Chapter 15 – Xylooligosaccharide: a potential prebiotic. In: Sheth MK, Dwivedi S (eds) Prebiotic foods for a healthy gut: scientific studies to manage metabolic disorders, 1st edn. Academic Press, pp 1–10
Van den Ende W, Valluru R (2009) Sucrose, sucrosyl oligosaccharides, and oxidative stress: scavenging and salvaging? J Exp Bot 60:9–18
Vivodová Z, Hačkuličová D, Bačovčinová M, Šípošová K, Labancová E, Kollárová K (2023) Galactoglucomannan oligosaccharides alleviate cadmium toxicity by improving physiological processes in maize. Ecotoxicol Environ Saf 255:114777
Wang F, Feng G, Chen K (2009) Burdock fructooligosaccharide induces resistance to tobacco mosaic virus in tobacco seedlings. Physiol Mol Plant Pathol 74:34–40
Wang K, Xu Y, Xuan Z, Xiao X, Gu G, Lu L (2023) Enzymatic synthesis of prebiotic galactooligosaccharides from galactose derived from gum arabic. Food Chem 429:136987
Weinhold B (2006) Epigenetics: the science of change. Environ Health Perspect 114:A160–A167
Wu Y, Yang Z, How J, Xu H, Chen L, Li K (2017) Overexpression of a peroxidase gene (AtPrx64) of *Arabidopsis thaliana* in tobacco improves plant's tolerance to aluminum stress. Plant Mol Biol 95:157–168
Yamauchi T, Yoshioka M, Fukazawa A, Mori H, Nishizawa NK, Tsutsumi N, Yoshioka H, Nakazono M (2017) An NADPH oxidase RBOH functions in rice roots during lysigenous aerenchyma formation under oxygen-deficient conditions. Plant Cell 29:775–790
Yan S, Liu Q, Li W, Yan J, Fernie AR (2022) Raffinose family oligosaccharides: crucial regulators of plant development and stress responses. Crit Rev Plant Sci 41:286–303
Yang C, Li W, Cao J, Meng F, Yu Y, Huang J, Jiang L, Liu M, Zhang Z, Chen X, Miyamoto K, Yamane H, Zhang J, Chen S, Liu J (2017) Activation of ethylene signaling pathways enhances disease resistance by regulating ROS and phytoalexin production in rice. Plant J 89:338–353
Ye N, Zhu G, Liu Y, Li Y, Zhang J (2011) ABA controls H2O2 accumulation through the induction of OsCATB in rice leaves under water stress. Plant Cell Physiol 52:689–698
Yigang Y, Ya'nan X, Yeran B, Yuanpei Z, Wei H, Saito M, Guohua L, Jiqing S, Wenbo B (2024) Mixed-oligosaccharides promote seedling growth of direct-seeded rice under salt and alkaline stress. Rice Sci 31:712–724
You J, Chan Z (2015) ROS regulation during abiotic stress responses in crop plants. Front Plant Sci 6:1092–1092
Yue P, Hu Y, Tian R, Bian J, Peng F (2022) Hydrothermal pretreatment for the production of oligosaccharides: a review. Bioresour Technol 343:126075
Zandi P, Schnug E (2022) Reactive oxygen species, antioxidant responses and implications from a microbial modulation perspective. Biology 11:155
Zeng M, Mohapatra S, Oh JH, Astmann T, van Pijkeren JP, Pan X (2023) Novel galacto-oligosaccharides from lactose: chemical synthesis, structural characterization, and *in vitro* assessment of prebiotic activity. ACS Sustain Chem Eng 11:14031–14045

Zhao Q, Zhou L, Liu J, Cao Z, Du X, Huang F, Pan G, Cheng F (2018) Involvement of CAT in the detoxification of HT-induced ROS burst in rice anther and its relation to pollen fertility. Plant Cell Rep 37:741–757

Zhao J, Lu Z, Wang L, Jin B (2020) Plant responses to heat stress: physiology, transcription, non-coding RNAs, and epigenetics. Int J Mol Sci 22:117

Zhao S, Zhang Q, Liu M, Zhou H, Ma C, Wang P (2021) Regulation of plant responses to salt stress. Int J Mol Sci 22:4609

Zhou H, Shi H, Yang Y, Feng X, Chen X, Xiao F, Lin H, Guo YI (2024) Insights into plant salt stress signaling and tolerance. J Genet Genomics 51:16–34

Zong H, Li K, Liu S, Song L, Xing R, Chen X, Li P (2017) Improvement in cadmium tolerance of edible rape (*Brassica rapa* L.) with exogenous application of chitooligosaccharide. Chemosphere 181:92–100

Chapter 10
Ultraviolet Defense Mechanisms in Higher Plants Against Environmental Stress

Sayon Mukherjee, Sourav Das, Gorantla Prathap Reddy, Dibyajyoti Panda, Partha Mondal, Sabyasachi Koley, Madhurya Ray, Soumen Chakraborty, Debdut Manna, and Abhik Patra

Abstract Sunlight is indispensable for the photosynthesis of plants. Ultraviolet (UV) light is an integral part of sunlight. Due to the diminution of stratospheric ozone, UVB (280–315 nm) light is reaching the Earth's surface and damaging the crucial components of cells such as membrane lipids, nucleic acid, and proteins, producing reactive oxygen species (ROS) and hampering the photosynthesis process. There is also a substantial alteration in the growth cum development of plants due to changes at the molecular, cytological, and morphological strata by the UVB light. Plants have managed to develop various defense mechanisms to counter UVB stress, which include the accumulation of secondary metabolites (flavonoids and phenolic compounds), generation of antioxidant factors (enzymes like superoxide

S. Mukherjee (✉) · D. Panda · S. Koley
Department of Soil Science and Agricultural Chemistry, Institute of Agricultural Sciences, Banaras Hindu University, Varanasi, Uttar Pradesh, India
e-mail: sayon.m@bhu.ac.in

S. Das
Division of Soil Science and Agricultural Chemistry, Indian Agricultural Research Institute, New Delhi, India

G. Prathap Reddy · S. Chakraborty
Department of Agronomy, Institute of Agricultural Sciences, Banaras Hindu University, Varanasi, Uttar Pradesh, India

P. Mondal
Department of Agricultural Biochemistry, Bidhan Chandra Krishi Viswavidyalaya, Mohanpur, West Bengal, India

M. Ray
Department of Plant Physiology, Institute of Agricultural Sciences, Banaras Hindu University, Varanasi, Uttar Pradesh, India

D. Manna
Division of Biochemistry, Indian Agricultural Research Institute, New Delhi, India

A. Patra
Krishi Vigyan Kendra, Narkatiaganj, West Champaran, Dr. Rajendra Prasad Central Agricultural University, Pusa, Bihar, India

H. I. Mohamed (ed.), *Elicitors for Sustainable Crop Production*, Sustainability Sciences in Asia and Africa, https://doi.org/10.1007/978-981-95-3533-0_10

dismutase (SOD) and catalase (CAT)), quick repairing of DNA (photorepair and dark repair), and signal transduction pathways (UV response locus 8 (UVR8)).

Keywords Photosynthesis · UVB · DNA · Reactive oxygen species · Secondary metabolites

10.1 Introduction

One of the constituents of sunshine is ultraviolet (UV) light. The UV part of the spectrum is separated into three sections: "UVA (315–400 nm), UVB (280–315 nm), and UVC (100–280 nm)." The wavelength at which UV radiation passes through the Earth's atmosphere determines its transmission. The gases in the atmosphere completely absorb UVC radiation and most of the UVB radiation (absorbed by the ozone layer). The radiation that reaches the Earth's surface consists of UVA and a small portion of UVB, due to which these two radiations have biological importance. Due to the high energy content of UVB, it shows potential negative effects on plants (Ueda and Nakamura 2011; Pathak et al. 2019).

Because of the artificial pollutants like methyl bromide, halons, and chlorofluorocarbons that are released into the atmosphere, the ozone layer's ability to filter UVB radiation has been reduced, and the quantity of UVB radiation reaching Earth has increased. The responses of plants to UVB radiation depend on the intensity, duration, and wavelength of UVB radiation (Chen et al. 2022). Other considerations like solar angle, cloud cover, altitude, and shading effect also determine the UV influx into the Earth. Equatorial regions are highly exposed to UVB radiation due to the narrow solar zenith angles (Pathak et al. 2019).

Plants exposed to UVB radiation suffer changes in physiological (formation of reactive oxygen, lipid peroxidation, reduction in the ribulose-1,5-bisphosphate carboxylase/oxygenase (RuBisCO) activity, and hormonal imbalance), morphological (reduced root growth and leaf area, inhibition of hypocotyl growth, and yield reduction), and molecular (downregulation of the gene associated with metabolic pathways) transformations (Kataria et al. 2014). UVB radiation affects the plant by the production of reactive oxygen species (ROS), which further oxidize proteins and lipids, causing oxidative stress in plants, and damage the DNA. In plants, UVB light is received through the photoreceptor "UV resistance locus 8" (UVR8), a principal gene, to elicit the light signal for acclimatization and photomorphogenic responses (Yadav et al. 2020). The other key proteins acting along with UVR8 include "elongated hypocotyl 5 (HY5) and constitutive photomorphogenic 1 (COP1)." UVR8 controls the genes involved in DNA repair, the synthesis of sunscreen, the function of chloroplasts, and the protection against oxidative stress (Rizzini et al. 2011).

The survival of plants under UVB stress is achieved by an amalgamation of repair mechanisms, exposure limit, and protection. To maintain the photosynthetic activity, chloroplasts alter their position to have minimum impact on photosystem II (PSII), and if it is damaged, then it is repaired by the neo-synthesis of D1 subunit (Takahashi and Badger 2011). There is an accretion of secondary metabolites like polyphenols, flavonoids, sinapate esters, and hydroxycinnamic acid in the epidermal cell due to the activation of the enzyme chalcone synthase (CHS) and phenylalanine ammonia-lyase (PAL) upon the exposure of plants to UVB irradiation (Ryan et al. 2002; Podolec and Ulm 2018).

The DNA damage is repaired in two ways, viz., photorepair (involving photolyases) and dark repair (removal of the photoproduct through nucleotide excision repair (NER)). The enzyme cyclobutane pyrimidine dimer (CPD) photolyase is induced by UVR8 signaling pathways, UVA, and blue light (Li et al. 2015). The nonenzymatic antioxidants (carotenoids, glutathione, ascorbate, and tocopherol) and enzymatic antioxidants (ascorbate peroxidase (APX), catalase (CAT), superoxide dismutase (SOD), and peroxidase (POX)) participate in critical functions in detoxifying reactive oxygen species (Mittler 2017). The development of plants resistant to UV stress is a key concept and one of the promising technologies, along with the application of nanomaterials (TiO_2, Si, and Ag nanoparticles) for the development of tolerance toward UV stress (Ueda and Nakamura 2011; Chen et al. 2022; Soni et al. 2022). More insights into UVR8–dependent and independent gene transduction pathways and other pathways that regulate gene expression for photomorphogenesis is responsible for UVB stress resistance mechanisms in plants. This chapter delves into the fascinating world of plant UV defense, exploring the various strategies employed by these green sentinels to shield themselves from sun's harmful rays.

10.2 Shielding the Green: Ultraviolet Defense Mechanisms in Plants by Antioxidant Agents

Sunlight nourishes plants by fueling photosynthesis. However, sunlight also packs a punch in the form of ultraviolet (UV) radiation (UVR). While essential for vitamin D synthesis in animals, UV radiation can wreak havoc on plant cells, damaging DNA, proteins, and membranes. To thrive in this seemingly contradictory environment, plants have evolved a sophisticated arsenal of UV defense mechanisms.

10.2.1 Sun's Double-Edged Sword: Understanding UV Radiation

Sunlight is a spectrum of electromagnetic radiation, with visible light representing a small portion. The invisible UV radiation, further classified into UVA, UVB, and UVC. While the atmosphere filters out most UVC, some UVB and UVA reach the Earth's surface. Although beneficial in small doses for plant growth, excessive UVB exposure can be detrimental.

10.2.2 The UV Defense Toolbox: A Multipronged Approach

Plants combat UV radiation through a multilayered defense system. Here, we explore some key mechanisms.

10.2.2.1 Antioxidant Arsenal

UV exposure can generate harmful free radicals and reactive oxygen molecules that damage cells. Plants counter these threats with a potent arsenal of antioxidants. Enzymes like superoxide dismutase (SOD) and catalase work tirelessly to neutralize free radicals, protecting cellular structures. Additionally, plants synthesize antioxidant compounds like carotenoids and glutathione, further bolstering their defense (Hasanuzzaman et al. 2020).

10.2.2.2 Enzymatic Antioxidants

Enzymatic antioxidants are essential for scavenging ROS and maintaining cellular homeostasis (Hasanuzzaman et al. 2020):

- Superoxide Dismutase (SOD): SOD breaks down the conversion of superoxide radicals ($O_2^{\cdot-}$) to hydrogen peroxide (H_2O_2). It is a critical enzyme in the antioxidant defense system.
- Ascorbate Peroxidase (APX): APX detoxifies H_2O_2 by converting it into water (H_2O) using ascorbate (vitamin C) as a co-substrate.
- Glutathione Peroxidase (GPX): GPX reduces H_2O_2 and lipid hydroperoxides, protecting cell membranes from oxidative damage.
- Glutathione Reductase (GR): GR maintains the reduced form of glutathione (GSH), which is crucial for ROS scavenging.
- Dehydroascorbate Reductase (DHAR): DHAR regenerates ascorbate from its oxidized form (dehydroascorbate), ensuring a continuous supply of this potent antioxidant.
- Catalase (CAT): CAT efficiently decomposes H_2O_2 into water and oxygen.

10.2.2.3 Free Radical Scavengers

UVB rays trigger harmful molecules (ROS) in cells, damaging fats and proteins (oxidative stress) (Frohnmeyer and Staiger 2003; Wei et al. 2024). The antioxidant defense system comprises a multifaceted network of enzymatic and nonenzymatic components (Blokhina et al. 2003). Cellular defense against reactive oxygen species employs a multienzyme system. SOD promotes the conversion of $O_2^{\cdot-}$ to H_2O_2 and molecular oxygen. Catalase then facilitates the decomposition of H_2O_2 into water and O_2. Peroxidase offers an alternative pathway for H_2O_2 detoxification, utilizing co-substrates like ascorbate and glutathione of phenolic compounds and antioxidants, which undergo oxidation in the process (Blokhina et al. 2003; Aslan et al. 2024). These antioxidant enzymes are also triggered when exposed to environmental stressors other than UVB irradiation (Nair et al. 2008). One common outcome of all these unfavorable environmental circumstances is the development of oxidative damage in plants. Nonenzymatic antioxidants, characterized by their small size, exhibit distinct compartmentalization within the cell. Ascorbate and glutathione are primarily situated in the solution phase, the intracellular fluid, while tocopherol and carotenoids exert their antioxidant activity within cell membranes. The *Arabidopsis* vtc1 (vitamin C-1) mutants, which lack ascorbate, are exceptionally vulnerable to UVB rays.

10.2.3 The Ever-Changing Environment: The Future of Plant UV Defense

As the ozone layer thins, allowing more UV to reach Earth, understanding plant UV defense mechanisms becomes increasingly crucial (Lisboa et al. 2024). Plant breeders can leverage this understanding to produce crops with increased UV resistance, ensuring food security in a changing environment. Furthermore, studying plant UV defense mechanisms can inspire the generation of novel sunscreen technologies and techniques for safeguarding against the detrimental impacts of UV radiation on other organisms.

10.3 Ultraviolet Defense Mechanisms in Higher Plants Against Environmental Stress

10.3.1 UV Defense Mechanisms in Plants: UV Sunscreens

To alleviate the detrimental impacts of UV irradiation, plants have evolved various defense systems, including the manufacturing of UV-absorbing compounds that act as natural sunscreens, like phenolic compounds, flavonoids, anthocyanins, lignins, tannins, specialized metabolites, alkaloids, glucosinolates, and hormones, which help them protect themselves from photosynthetic damage (Elbouzidi et al. 2025).

10.3.2 Phenolic Compounds

Phenolic compounds include a wide range of secondary metabolites that serve important functions in plant defense against UV radiation. These compounds, which include flavonoids and phenolic acids, contribute to UV tolerance. Phenolic compounds act as powerful antioxidants, scavenging the ROS generated by UV exposure (Rai and Agrawal 2017). Their ability to donate electrons and stabilize free radicals makes them effective in reducing oxidative stress and preventing cellular damage.

10.3.3 Flavonoids

Flavonoids are a subclass of phenolic compounds, which are particularly important in UV defense. They are found in high concentrations in the epidermal layers of leaves, absorb UV radiation, and protect underlying tissues from damage. Additionally, flavonoids strengthen the defense of plant against free radicals by enhancing the efficiency of enzymes like APX and SOD. These compounds not only protect plants from UV-induced DNA damage but also play a role in signaling pathways that regulate stress responses (Rai and Agrawal 2017).

10.3.4 Anthocyanins

Anthocyanins, composed of glycosylated flavonoids, are responsible for the purple, red, and blue pigmentation in plants. They accumulate under UV stress and provide photoprotection by absorbing UV radiation and quenching ROS (Soni et al. 2022).

10.3.5 Lignin

Lignin comprises a multifaceted polymer found inside the secondary cell walls of vascular plants. Although lignin is not primarily involved in UV screening, it plays a vital role in strengthening the cell walls and reducing UV transmittance. The biosynthesis of lignin involves the polymerization of monolignols, which are derived from phenylpropanoids. Increased lignin content in response to UV stress enhances the mechanical strength of plant tissues and imparts a physical blockage against UV penetration (Rai and Agrawal 2017).

10.3.6 Tannins

Tannins are another group of phenolic compounds with significant UV-absorbing properties. The abundance of tannins in the epidermal layers confirms their significance in UV protection by reducing oxidative damage (Rai and Agrawal 2017).

10.3.7 Specialized Metabolites

Specialized metabolites, including alkaloids and glucosinolates, also contribute to plant defense against UV radiation. Alkaloids comprise nitrogen-containing molecules that have multiple configurations and activities, including UV protection. Glucosinolates, found in *Brassicaceae* species, are sulfur-containing compounds that can be converted into bioactive products with antioxidative properties (Mmbando 2023).

10.3.8 Alkaloids and Glucosinolates

Alkaloids and glucosinolates are nitrogen-containing compounds that perform specific roles in plant defense. Although their primary function is often associated with deterring herbivores and pathogens, they also contribute to UV protection. These compounds can absorb UV radiation and have antioxidative characteristics that reduce the effects of UV-induced oxidative stress (Mmbando 2023).

10.3.9 Hormones

Plant hormones, such as abscisic acid (ABA), salicylic acid (SA), and jasmonic acid (JA), serve regulatory functions in plant responses to UV stress by modulating the expression of genes interrelated with antioxidant defenses, phenolic biosynthesis, and stress signaling pathways (Rai and Agrawal 2017; Mmbando 2023).

10.4 Other UV Defense Strategies

10.4.1 Structural Adaptations

In addition to biochemical defenses, plants also exhibit structural adaptations to minimize UV damage. These include the development of thicker cuticles, increased trichome density, and altered leaf morphology. These physical barriers help reduce the penetration of UV rays into deeper tissues (Zhang et al. 2024b).

10.4.2 DNA Repair Mechanisms

Plants possess a potent DNA repair process to reduce UV-prompted DNA damage. Photoreactivation, nucleotide excision repair, and homologous recombination (HR) are some of the pathways employed to maintain genome integrity under UV stress.

10.4.3 Interaction with Microbes

The symbiotic relationship between plants and certain soil microbes can also enhance UV resistance.

10.5 DNA Repair Mechanisms

The chromatin structure and sequence composition affect the relative proportion and nonrandom dissemination of the two primary forms of DNA lesions instigated by UV radiation in the plant genome, namely, "cyclobutane pyrimidine dimers" (CPDs) and "6-4 photolyase" (6-4 PP). Even though DNA damage is widespread and frequently occurs, it needs to be fixed right away to guarantee the precise genetic information transfer that occurs during cell division (Law et al. 2013; Vijg 2021). The kinds and locations of lesions in the genome primarily determine which repair pathway a cell uses. There is a dearth of well-supported research on the dynamics of accumulated DNA damage and the radiation injury-mediated molecular mechanisms concerning dosage rate. However, evidence so far in the present context advocates that plants have very complex techniques for repairing DNA damage that involve multiple loci (Esnault et al. 2010). It is important to note that there are several ways to repair UV radiation-mediated DNA damage, including direct repair, photorepair (photoreactivation), "base excision repair" (BER), "nucleotide excision repair" (NER), or "mismatch repair" (MMR), in plants, where the aforementioned lesion-specific repair pathways can revert UV-induced DNA impairment to maintain genomic integrity (Kunz et al. 2005).

The main mechanisms that underlie the coping mechanisms for UV-induced DNA damage are briefly reviewed below.

10.5.1 Light-Dependent Repair

The primary mechanism by which plants repair low-frequency DNA impairment spawned by UVB radiation, particularly UVB-induced CPDs, is called photorepair (light-dependent repair). The photoreactivating enzyme "photolyase," which is highly preserved and present in all three realms of life, is used to carry out the photoreactivation process (Zgłobicki et al. 2024). The enzyme utilizes the energy from visible/blue light to selectively bind to CPD photolyase or 6-4 PPs and immediately break down the cyclobutene ring pyrimidine dimer. This serves as a safeguard for the genome against the detrimental impacts of ultraviolet radiation (UVR) (Essen and Klar 2006). Because photoreactivation is error-free, plants use it as the preferred and fastest mechanism to lessen the harmful effects of DNA photodimers that are produced during regular solar radiation exposure. The quantity of the photolyase protein that is present in the injured cell determines whether plants can photoreactivate UV-induced pyrimidine dimers. In *Arabidopsis*, the photomorphogenic repressors AtDET1 and AtCOP1 sustain the dark suppression of photorepair gene expression until de-etiolation, when their negative effect is mitigated. However, the transcription factors Elongated Hypocotyl 5 (HY5) and Elongated Hypocotyl 5-Homolog (HYH) further enhance the transcriptional activation of photolyase genes, including "AtPHR1" and "AtUVR3" (Garcia-Caparros et al. 2021). The recently discovered UVB-specific photoreceptor "AtUVR8" is a major constituent of the plant's response to UVB. It controls a "downstream signaling cascade" that includes "COP1" and "HY5," which regulate the numerous genetic expressions linked to plant UV defense and photomorphogenic response (Tilbrook et al. 2013).

10.5.2 Light-Independent Repair

Light-independent repair is also known as dark repair. NER, BER, and MMR are the major light-independent repair mechanisms. Genomic consistency and survival of organisms under the exposure of UV light are central to both base and excision repair of DNA (Bray and West 2005). Light-independent repair (dark repair) mechanisms are simpler than light-dependent repair, as in the case of light-dependent repair, the recognition, targeting, and subsequent removal of damaged product consist of 24–32 base and gap filling via the synthesis of DNA and ligation of the nicks happens (Maillard et al. 2008; Huang and Zhou 2021). Nuclear excision repair concedes structural alterations in the DNA duplex rather than a particular type of DNA injury, and it can restore multiple types of DNA injury. This is why it is considered to be the most versatile technique of DNA repair (Boubriak et al. 1997; Krasikova et al. 2021). NER encompasses the two sub-pathways, which are "global genomic repair" (GGR), which is interrelated with the entire DNA damage repair, and "transcription-coupled repair" (TCR), which is restricted to repairing the DNA of the expressed gene (Li et al. 2022).

The NER mechanism successfully repairs major "helix-distorting lesions," such as CPDs and 6-4 photoproducts, by means of the sequential involvement of DNA impairment, followed by incision on the impaired DNA strand and excision of damaged oligonucleotides, and finally, DNA synthesis, followed by ligation (Li et al. 2022).

Like NER, BER also consists of two sub-pathways, "DNA polymerase (beta-dependent) short-patch BER and DNA polymerase (delta/epsilon-dependent) long-patch BER." BER reforms the oxidized or hydrated bases and Single-stranded binding protein (SSB). The process begins with the liberation of the injured base and DNA glycosylases by cleaving the sugar phosphate chain. This is followed by the removal of the residue carrying "oligonucleotides" and subsequent DNA synthesis and ligation. The initiation of short-patch BER occurs when a DNA glycosylase enzyme removes the impaired base (Kumar et al. 2022). On the other hand, long-patch BER involves a repair process where a nick translation reaction takes place, escorted by strand expulsion in the 5′–3′ direction (Harrington and Lieber 1994).

Mismatched bases can arise due to the insertion and/or deletion (frame shift mutation) of nucleotide or during recombination. The MMR systems are used to rectify a significant proportion of these faults, resulting in a further reduction of the error rate to a range of 10^{-9}–10^{-10} (Putnam 2021). The highly conserved MMR systems encourage genomic sturdiness by rectifying mistakes in DNA replication, inhibiting homologous recombination, and reacting to various types of DNA injury (Pećina-Šlaus et al. 2020). The MMR process is remarkably consistent across different species and is responsible for eliminating the bulk (99.9%) of mistakes that persist even after polymerase proofreading. In addition, MMR may also detect discrepancies at recombination sites between DNA sequences, thus decreasing the frequency of recombination events (Leonard et al. 2003). Some key highlights of DNA repair mechanisms in plants against UV defense are outlined in Table 10.1.

Table 10.1 DNA repair mechanism in plants as a UV defense mechanism

Pathways	Organism targeted	Mechanism of repair	Reference
Photoreactivation/ photorepair	Rice (*Oryza sativa*)	Increases the activity of photolyase enzymes	Teranishi et al. (2012), Otake et al. (2024)
	Arabidopsis thaliana (model plant)	Cyclobutane pyrimidine dimer (CPD) photolyase repair mechanism	Kaiser et al. (2009)
	Rice (*Oryza sativa*) transgenic	CPD repair mechanism	Otake et al. (2024)
Dark repair	Carrot (*Daucus carota*)	Excision of pyrimidine dimers	Hays (2002), Lestari et al. (2024)
	Wheat (*Triticum aestivum* L. cv. Chinese spring)	CPD repair mechanism	Taylor et al. (1996), Nakamura et al. (2021)
	Rice (*Oryza sativa* L. cv. Sasanishiki)	CPD repair mechanism	Hidema et al. (1997)
	Soybean (*Glycine max*)	Removal of CPD by both excision and photoreactivation	Sutherland et al. (1996)
Homologous recombination (HR)	Rice (*Oryza sativa*)	The interaction between plant NBS1 and MRE11 orthologs supports the idea that the Mre11-Rad50-Nbs1 complex (MRN) complex functions in the repair of double-strand breaks	Syed and Tainer (2018)
Nonhomologous recombination (NHR)	Human cells	During the process of nonhomologous end joining (NHEJ), the DNA ends are brought into proximity by a complex consisting of many proteins and then directly joined together	Lieber (2010)

10.6 Signal Transduction and UV Defense Mechanism in Plants

The process by which a cell translates an external stimulus into an operational response is known as signal transduction. Through a sequence of molecular events, including phosphorylation and protein activation, signals from a cell surface receptor are relayed. This process eventually results in a particular cellular response, such as changes in gene expression or metabolism. When plants encounter UV radiation, signal transduction takes place involving several signaling pathways to combat the damage that occurs. Table 10.2 enumerates some of its mechanisms administered by plants as a response to UV stress.

UVB radiation specifically impacts plant growth and development. The damage due to UVB and UVC light ranges from cellular membrane damage, lipid peroxidation, and generation of ROS to, ultimately, DNA damage and death of the plant. In order to sense and react to UV light, plants have developed complex processes that enable them to adjust to shifting environmental conditions, which are elaborated by the following points with reference to UVB and in Fig. 10.1.

Table 10.2 Signal transduction in plants in the presence of UV radiation

Mechanism	Description	Key components	Reference
UV perception	Plants recognize UVB radiation using specific photoreceptors	UVR8	Jenkins (2017)
Signal transduction pathways	Initiation of signaling cascades upon detection of ultraviolet radiation	UVR8, COP1, HY5, ROS	Lee (2016)
Flavonoid biosynthesis	Production of UV-absorbing substances and antioxidants that provide protection against UV-induced damage	Chalcone synthase (CHS), flavanol synthase (FLS)	Falcone Ferreyra et al. (2012)
DNA repair mechanisms	Strategies like nucleotide excision repair and photoreactivation to repair the DNA damage caused by ultraviolet (UV) radiation	Photolyase, nucleotide excision repair (NER)	Britt (1996)
Cuticle formation	Building a shielding barrier to minimize the infiltration of UV radiation	Cuticular waxes, cutin	Yeats and Rose (2013)
Hormonal regulation	Plant hormones regulate the UV reactions	SA, JA, ABA	Kazan and Manners (2011)
Cross talk with other pathways	Integration of ultraviolet (UV) response pathways with other pathways involved in environmental and stress responses	Light, pathogens, abiotic stresses	Ballaré (2014)

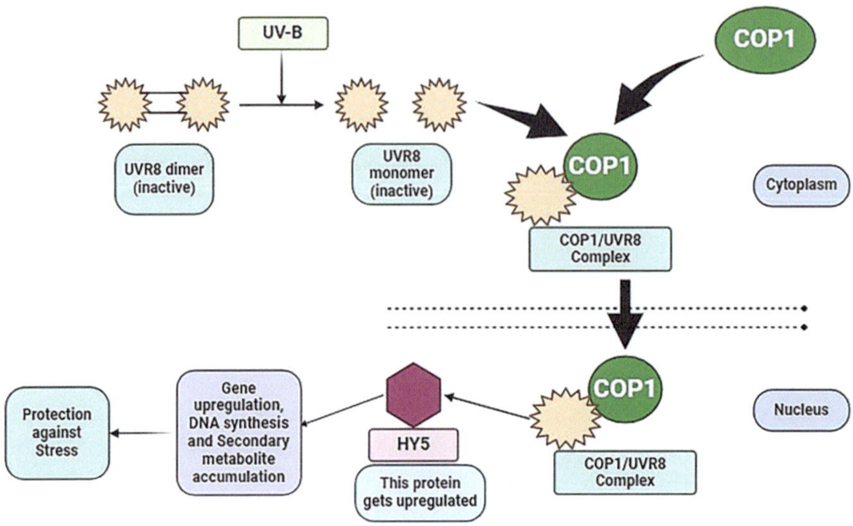

Fig. 10.1 Signal transduction pathway in response to UVB radiation (stress) in plants

10.6.1 UVB Perception

10.6.1.1 UV Resistance Locus 8 (UVR8)

In plants, UVR8 is a specific type of UVB photoreceptor. When UVB is not present, it exists as a homodimer. UVR8 monomers split apart in response to UVB radiation, which triggers downstream signaling cascades (Depaepe et al. 2023).

10.6.2 Downstream Signaling Pathways

10.6.2.1 Elongated Hypocotyl 5 (HY5)

In order to stabilize and accumulate HY5, a transcription factor that governs the gene expression of implicated UVB responses, "UVR8" interacts with "COP1," a pivotal regulator of photomorphogenesis. The primary mechanism by which UVR8 binds to COP1 is through its WD40 domain, which is in charge of identifying its substrates. Certain amino acids within UVR8 that are accessible during its monomerization aid in the interchange between UVR8 and COP1. The activity of COP1 is inhibited when UVR8 binds to it. This suppression is important because COP1 breaks down positive regulators of photomorphogenesis (like transcription factors HY5 and HYH) when UVB is absent. UVR8 binds to COP1 to stop these transcription factors from degrading, allowing them to accumulate and trigger the expression of UVB-responsive genes, thus protecting the plant (Zhang et al. 2024a).

10.7 Role of UV Light in Alleviating Environmental Stress

10.7.1 Heat and Cold Stress

Climate change poses previously unheard-of risks to global food yield because of abiotic factors like temperature that harm plant development and metabolism. Because plants are sessile, they have evolved sophisticated communication systems to sense temperature variations in their surroundings. This then triggers a series of molecular alterations that support plant reproduction and survival in challenging environments. Upon exposure to UV radiation, plants suffer from oxidative stress, DNA injury, photosynthesis impairment, and modification of the metabolic pathways. Similarly, plants under heat and cold stress suffer from oxidative stress, impairment in respiration, photosynthesis, cellular homeostasis, and membrane stability (Fig. 10.2). The defense mechanism plants develop to combat UV radiation also helps them combat high- and low-temperature stress (Qian et al. 2024).

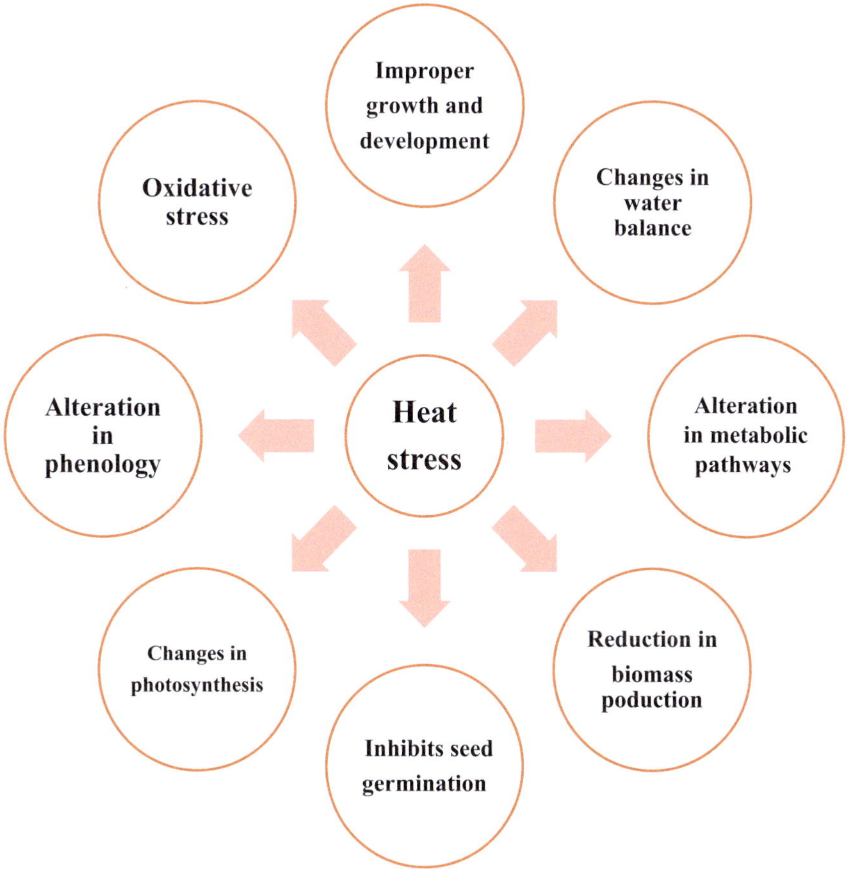

Fig. 10.2 Impacts of heat stress (HS) on plants

10.7.1.1 Responses of Plants to Heat Stress

Heat stress (HS) changes the effectiveness of enzymatic reactions within the cell, which is impeded by the main physiological process and results in metabolic disparity. By disrupting thylakoid membranes, heat stress lowers the photosynthetic rate by preventing the activity of enzymes and electron carriers linked to the membrane. In particular, there is a significant decrease in photosystem II (PSII) activity. Furthermore, HS affects the structure of chloroplasts and the thermostability of photosynthetic system building blocks by reducing the number of photosynthetic pigments, the carbon fixing ability, and the activity of “ribulose-1,5-bisphosphate carboxylase/oxygenase” (RuBisCO), impacting the yield (Zhao et al. 2020).

The primary physiological impact of plant susceptibility to HS is membrane disruption. The disintegration of membrane lipids and a rise in membrane fluidity that result from the loosened chemical bonds caused by the enhanced kinetic energy and mobilization of biomolecules thwart membranes under severe HS. Heat stress can decouple enzymes and metabolic pathways, resulting in the buildup of undesirable and damaging ROS. These include “hydrogen peroxide” (H_2O_2), “superoxide radical” (O^{-2}), “hydroxyl radical” (OH^-), and “singlet oxygen” ($_1O^2$). PSI and PSII are the primary sources of ROS. Surplus energy in PSII leads chlorophylls to transition into the triplet state, transferring excitation energy to O_2 to produce singlet oxygen. When PSI is reduced too much, the superoxide anion is made, which encourages the creation of H_2O_2. The antioxidant shielding mechanism of plants, which includes nonenzymatic antioxidants, including “ascorbate,” “glutathione,” “tocopherol,” and “carotenoids,” and numerous enzymes like “superoxide dismutase” (SOD), “catalase” (CAT), “peroxidase” (POD), and “ascorbate peroxidase” (APX), helps them become more thermotolerant (Hossain et al. 2020).

10.7.1.1.1 Transcriptional Regulation of Genes

Under heat stress, a set of heat shock transcription factors (HSFs) quickly triggers the expression of heat shock protein (HSP) genes, which modulate ROS production and thermotolerance in plants.

10.7.1.1.2 Regulation of the Thermotolerance Gene by Noncoding RNAs

Noncoding RNAs (ncRNAs) have a significant role in reacting to HS. They trigger the targeted gene and transcription factor for thermotolerance in plants. These RNAs downregulate the free radical scavenger, which, in turn, accumulates the ROS and increases the concentration of HSF and HSP (Guan et al. 2013).

10.7.1.1.3 Epigenetic Regulation of Genes

The term "epigenetics" describes the heritable modifications in gene expression that are essential to the plant HS response but do not result from alterations in DNA sequence. Heat-induced DNA methylation, histone modification, chromatin remodeling, "small interfering RNAs" (sRNAs), and "long noncoding RNAs" (lncRNAs) are all components of the epigenetic regulatory system that changes the pattern of gene expression and/or epigenetic memory in plants (Liu et al. 2015).

Long-term evolutionary adaptations in phenology and morphology, along with heat stress responses in plants, are illustrated in Fig. 10.3.

10.7.1.2 Adaptation of Plants to Cold Stress

Cold stress tolerance in plants is improved by the accumulation of soluble sugar, protein, and proline and also due to higher activities of antioxidant enzymes. These essential changes at the physiological level help maintain membrane stability, cell homeostasis, and protection from ROS. The activation of these antioxidant enzymes is facilitated by the induction of transcription factors like basic leucine zipper (bZIP) and myeloblastosis (MYB). The increasing level of ROS also triggers the activation of "C-repeat binding factor" (CBF) signaling pathways. Increased levels of cold stress activate signaling pathways of various plant hormones like ABA, salicylic acid, ethylene, and jasmonic acid. The activation of the ABA signaling pathway is

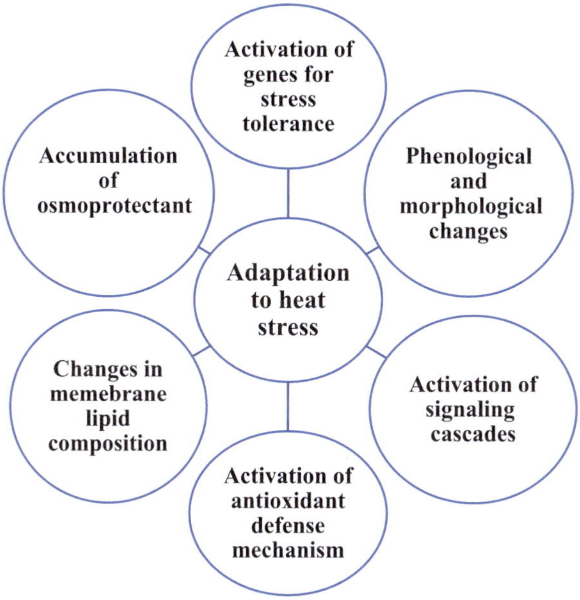

Fig. 10.3 Morphological and biochemical adaptation of plants to heat stress

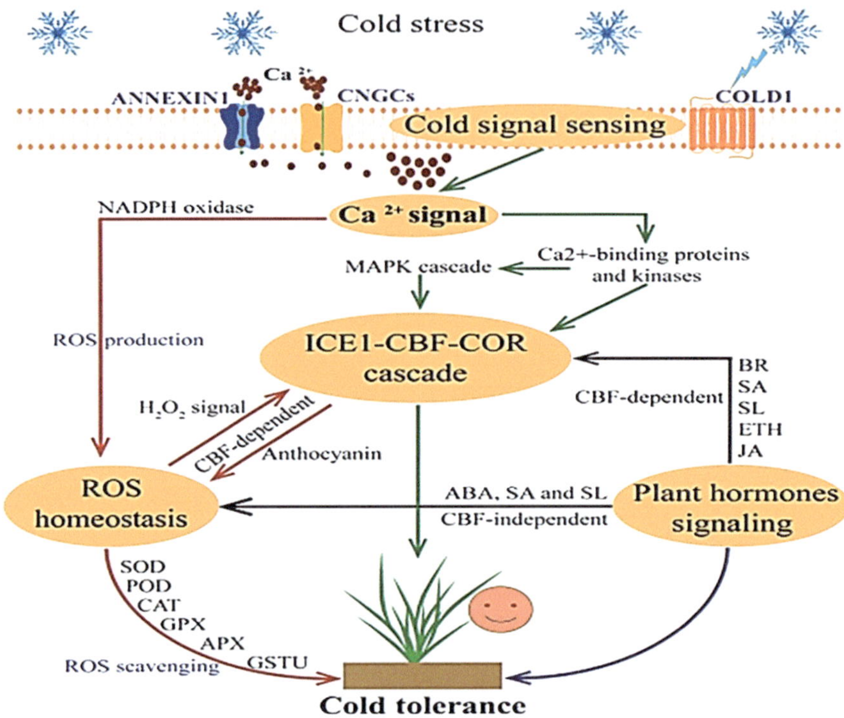

Fig. 10.4 Cold stress mechanism in plants (Qian et al. 2024)

CBF-independent, while the other hormones are independent. In a general way, we can say that the cold stress response of plants is an interaction between the hormonal signaling pathway, ROS homeostasis, and CBF transcription (Fig. 10.4) (Qian et al. 2024).

10.7.2 Role of UV in Alleviating Drought Stress

Throughout their life cycle, crop plants face certain environmental and anthropogenic stresses, like drought, heat, low temperature, salinity, and heavy metal contamination. Among these stresses, drought, which is characterized by a reduction in the available water and moisture and is coupled with heat, is considered to be the most devastating form of abiotic stress hampering crop productivity (Seleiman et al. 2021).

The intense abiotic stress of soil salinization impairs plant growth and development, causing physiological anomalies and severely endangering the food supply. The problem results from massive salt buildup in the soil, mostly from human activities, including overfertilization, inappropriate land use, and irrigation. Elevated levels of Na^+, Cl^-, and other ions in the soil can interfere with plants' cellular processes and change vital metabolic processes like photosynthesis and seed germination, resulting in serious tissue damage and, in the worst cases, plant death (Balasubramaniam et al. 2023).

UV radiation is omnipotent with sunlight, and prolonged exposure negatively impacts the plant's health, just like any other abiotic stress, altering plant morphology, biomass, photosynthesis, stomatal conductance, transpiration, and ROS generation. It also paves the way for the accumulation of malonaldehyde (MDA) and membrane damage. Plants perceive both UVA and UVB light through the "UVB" photoreceptor UVR8. Through UVR8 and crystal protein gene (CRYs), respectively, plants are able to detect short and long wavelengths within the UVA wave band (Rai et al. 2021). Lower degree of UVB radiation imposes restorative impacts on plant growth and salinity tolerance; this phenomenon is known as hormesis. However, different studies have shown that a controlled subjection to low-level UVB radiation plays a complex role and has potential drought- and salinity stress-mitigating effects (Roeber et al. 2021).

Controlled, low-level exposure to UV in drought- and salinity-affected plants has a multidimensional effect and stress acclimating responses that are evident at the morphological, physiological, biochemical, and molecular levels (Jansen et al. 2019). UV has a potential role in particularly hinting at morphological changes owing to stress, initiation of antioxidant defense techniques, osmotic regulation, and augmentation of stress markers. The role of UV in mitigating drought and salinity stress in plants is discussed below and is depicted in Fig. 10.5.

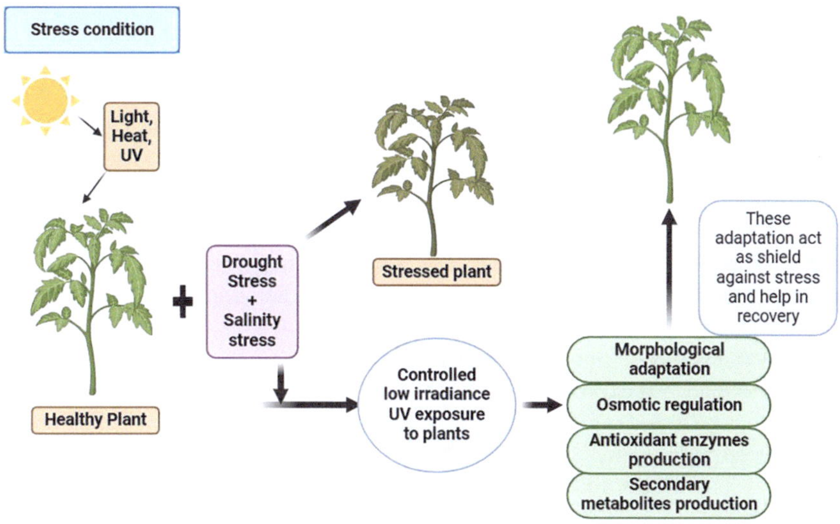

Fig. 10.5 Schematic diagram of the influence of UV on drought and salinity stress recovery in plants

10.7.2.1 Morphological Ways Influenced by UV to Alleviate Drought and Salinity Stress (Jansen et al. 2022)

(a) Reduction in Leaf Area Index to reduce photosynthesis and conserve water
(b) Proliferation of root growth and reduction in shoot-to-root ratio for enhancing water uptake
(c) Reduction in specific leaf area and leaf thickness

10.7.2.2 Physiological Ways Influenced by UV to Alleviate Drought and Salinity Stress

(a) Decrease in stomatal conductance by stomatal closure due to the accumulation of ABA (Li et al. 2021).
(b) Controlled UV exposure renders the formation of compatible osmolytes like proline and glycine betaine, which have positive impacts in maintaining the tissue water status of the drought- and salinity-exposed plants (Dikilitas et al. 2020).
(c) Drought and salinity stress together correspond to low turgor pressure and the collapse of the cell wall, thereby jeopardizing cellular integrity. UVB is known to stimulate lignin biosynthesis and associated proteins, thereby strengthening the cell wall (Sadhna et al. 2023).
(d) The upregulation of different cation and anion transporter genes for the effective exclusion and avoidance of salts is an important physiological mechanism to withstand salinity stress, as evident from the studies on *Conyza blinii*, a traditional Chinese medicinal plant (Zhan et al. 2021).

10.7.2.3 Biochemical Ways Influenced by UV to Alleviate Drought Stress

(a) UVB exposure enhances certain acclimation-associated metabolites, such as proline, flavonoids, anthocyanins, unsaturated fatty acids, and antioxidants, and secondary metabolites like monoterpenes and diterpenes, which are significantly upregulated in plants in response to UVB radiation, along with drought stress. These metabolites are essential for providing defense against drought and salinity stress (Shoaib et al. 2024; Pant et al. 2021).
(b) Controlled UVB exposure also enhances antioxidant defense practice in plants, which is necessary to withstand drought. Superoxide dismutase, catalase, ascorbate peroxidase, etc., are essentially upregulated. They scavenge the ROS produced due to stress and act as a shield against drought and salinity stress (Nawaz et al. 2020).

10.7.2.4 Molecular Pathways Influenced by UV to Alleviate Drought Stress

In its ground form, UVR8 is a homodimer that, when exposed to UVB photons, monomerizes through certain tryptophan residues. Monomeric UVR8 inhibits COP1's E3 ubiquitin ligase activity, opposing target proteins, encasing transcriptional regulators like HY5, by competitively interacting with its substrate-binding site. Members of the Phytochrome-Interacting Factor basic helix-loop-helix factor (PIF bHLH) factor family are similarly destabilized as a result of the UVR8–COP1 interaction. Furthermore, UVR8 directly interacts with another group of transcription factors and prevents them from binding to the DNA. Each of these UVR8 signaling pathways initiates changes in nuclear gene expression that result in photomorphogenesis and acclimation to drought brought on by UVB (Podolec et al. 2021). UVB initiates UVR8, which regulates osmotic potential, tissue water status, anthocyanin accumulation, and downward leaf curling to curtail down transpiration and promoting thermotolerance, which become essential to combat drought and salinity stress (Burko et al. 2020; Bursch et al. 2020). UVB radiation and exposure upregulate the gene involving "ABA response," "jasmonic acid response," and "salicylic acid response," critical for mitigating drought and salinity stress (Shan et al. 2022).

10.7.3 Role of UV in Alleviating Heavy Metal Stress

One of the primary environmental issues faced by the modern world is heavy metal contamination. Heavy metals like cadmium (Cd), lead (Pb), mercury (Hg), chromium (Cr), and arsenic (As) are released into the environment by industrial processes, urbanization, agricultural practices, and fossil fuel combustion. Ecological systems and human health are at risk because of the toxicity of these metals to humans, animals, and plants. The presence of these metals in soil, water, and air has grown to be a major problem since it impacts plant development, productivity, and metabolic functions. It is noteworthy that ultraviolet (UV) radiation has manifested as an effective tool for reducing heavy metal stress in plants.

10.8 Effects of UVB on Phenology and Growth

10.8.1 Growth Parameters

Research indicates that growth variables like the height of the plant and the leaf area are surprisingly lowered in the UVB-susceptible plants, with the extent of this reduction varying notably by species and cultivar (Lydon et al. 1986). For instance, in controlled experiments, elevated UVB levels have been shown to delay seedling emergence and the onset of flowering in various crop plants (Liu et al. 2023).

10.8.2 Phytohormonal Impact

Mark and Tevini (1996) provided evidence that UVB might inhibit stem elongation by changing the levels of crucial phytohormones, particularly indole-3-acetic acid (IAA), as UVB exposure accelerates the breakdown of IAA, resulting in reduced cell length and decreased plant height (Liaqat et al. 2024). Furthermore, UVB exposure can cause coiling in both attached and detached tendrils of plants.

10.8.3 Leaf Area and Growth

Leaf area, a highly receptive growth parameter, typically decreases under increased UVB radiation, leading to smaller leaves. This reduction in leaf size is often a preventative strategy attributed to a decline in the cell count, cell division, and cell expansion (González et al. 1998). Interestingly, some studies, like those by Nedunchezhian and Kulandaivelu (1997), report that marginally raised UVB levels can increase the leaf area in cowpea. Additionally, broad beans and wheat plants subjected to increased UVB levels have demonstrated increased leaf area.

10.8.4 Biomass Production

When examining biomass production, research findings are mixed. Approximately 54% of studies report a reduction in biomass (Mathur et al. 2024; Sherin et al. 2022) with upgraded UVB levels, while 35% show no impact on the dry weight, along with a few studies indicating an enhancement in the crop dry weight (Lee et al. 2019). These variations are likely due to genotypic differences, UVB doses, and the ratio of photosynthetically active radiation (PAR) to UVA and UVB. The dosage of UVB plays a critical role in determining the crop biomass production. Higher total irradiance levels generally result in leaves more tolerant to UVB injury, whereas lower PAR makes plants more receptive to UVB radiation.

10.9 Effects of UVB on Crop Yield and Biomass Partitioning

10.9.1 Photosynthesis and Carbon Assimilation

Exposure to UVB radiation substantially reduces photosynthesis levels, leading to alterations in carbon assimilation processes within plants (Mathur et al. 2024). When foliage is exposed to UVB, there is an accretion of carbohydrates in the source leaves, while a translocation to distant sinks is notably reduced. This is because phloem loading, an energy-dependent process, is impaired under increased UVB levels (Kumar et al. 2023). UVB radiation oxidizes sensitive proteins entangled with phloem loading, thereby modifying the distribution of stored carbohydrates in crops such as wheat and barley. This disruption in carbohydrate distribution directly affects biomass partitioning across different parts of the plant.

10.9.2 Translocation and Biomass Partitioning

UVB inhibits the migration of photosynthates from the source leaves to the sink tissues, which, in turn, impacts the growth of various parts of the plant. Photosynthate allocation patterns are crucial for overall plant growth and reproductive success. The most UV-sensitive period for plants is during the vegetative-to-reproductive transition phase (Teramura and Sullivan 1994). During the flowering and fruiting stages, reproductive organs require higher concentrations of photosynthates, diverting essential resources away from leaves and roots.

10.9.3 Impact on Cereals and Flag Leaves

In cereal crops, flag leaves play a vital role during anthesis, significantly contributing to grain filling. Ozone fumigation at anthesis, which simulates the UVB impact, increases carbohydrate concentrations in wheat flag leaves due to disrupted transport processes due to membrane injury in mesophyll or companion cells (Brewster et al. 2024). UVB exposure instigates senescence in flag leaves, leading to the loss of chlorophyll and soluble proteins, which are critical for sustaining photosynthesis and assimilate production amid the grain-filling period.

10.9.4 Yield Reduction

The timescale of assimilate production and the grain-filling period are critical factors in determining the crop yield. UVB radiation alters these processes, often resulting in a reduced yield. UVB exposure can decrease the grain-filling rate, ultimately affecting the final yield. Additionally, the stress induced by UVB can lead to premature senescence, further reducing the effective grain-filling period and the overall yield of the crop.

10.10 Potential Strategies for Breeding UVB-Resistant Plants

10.10.1 Inherent Defense Mechanisms

Plants possess inherent defense strategies, including ROS scavenging and the DNA repair process, that enable them to resist subordinate levels of UVB radiation. However, these repair techniques might not be sufficient to counteract the high induction rates of UVB damage, particularly in regions with elevated UVB levels, such as the tropics. For instance, in the tropical regions of Africa and the Bengal area, UVB radiation inhibited growth and reduced the number of tillers, leading to decreased rice productivity (Mmbando et al. 2020).

10.10.2 Strategies for Adaptation

Several strategies were proposed to assist plants in acclimatizing to the increased UVB vulnerability. These include the removal, repair, reversal, and tolerance of UVB-induced damage (Soni et al. 2022). Enhancing these approaches through molecular techniques can significantly improve the resilience of the crops to elevated UVB stress.

10.10.3 Molecular Techniques and Breeding Efforts

Although many UVB-sensitive crops, such as African rice, happen to be resistant to tissue culture, other strategies to promote revitalization and alteration are needed to boost breeding efforts (Mmbando 2022). The following approaches have been suggested for improving UVB protection in vegetation:

- *Enhancing DNA Repair Mechanisms*: One promising strategy is to genetically engineer plants to increase their DNA repair mechanisms. By increasing the expression of photolyase, an enzyme that specifically repairs UVB-induced DNA damage, researchers successfully developed UVB resistance in rice plants (Hidema et al. 2007).
- *Improving Antioxidant Capacity*: Another approach involves boosting the antioxidative capability of plants to scavenge the ROS generated by UVB exposure. This can be achieved by overexpressing genes connected with the synthesis of antioxidants, such as SOD and CAT.
- *Selective Breeding and Genetic Screening*: Traditional breeding methods combined with modern genetic screening techniques can be used to identify and select naturally occurring UVB-resistant traits. This approach involves screening large populations of plants for UVB tolerance and selecting the most resistant individuals for further breeding.

- *Gene-Editing Technologies*: Advanced gene-editing technologies, such as Clustered Regularly Interspaced Short Palindromic Repeats (CRISPR) and CRISPR-associated protein 9 (Cas9), can be employed to introduce specific mutations that confer UVB resistance. This method makes specific modifications to the plant genome to enhance UVB tolerance.
- *Developing UVB-Protective Compounds*: Research is also focused on developing and applying UVB-protective compounds, such as sunscreens for plants. These compounds can be applied to crops to provide an additional layer of protection against harmful UVB radiation.

By integrating these approaches, researchers aim to produce crop varieties that are more resilient to the adverse impacts of UVB exposure, thereby ensuring sustainable agricultural productivity in regions with high UVB exposure. The following methods have been suggested to develop UVB resistance in crops, as shown in Table 10.3.

These methods represent promising avenues for enhancing UVB resistance in crops, potentially leading to improved growth and productivity in regions with high UVB radiation.

Table 10.3 Crop-breeding methods to counter the stress of UVB radiation exposure

Method	Description	Role	Reference
Enhancement of DNA-repairing enzymatic activity	Overexpression of the CPD photolyase enzyme gene in UVB-resistant genotypes	This method aims to boost the plant's natural ability to repair UVB-induced DNA impairment by increasing the levels of photolyase	Hidema et al. (2007)
Controlling microRNA396 and also growth-regulating factors	Regulating the endogenous expression levels of miRNA396 or the transcript levels and activity of the principal growth-regulating agents	MicroRNA396 and growth regulators take part in modulating the plant's response to UVB stress, impacting growth and development	Fina et al. (2017)
Improvement of the antioxidative mechanism	Enhancing ROS-scavenging mechanisms in plants	This approach focuses on increasing the plant's ability to neutralize the ROS generated by UVB exposure	Zhang et al. (2022)
Alterations of the UVR8 signaling pathway	Overexpression of UVR8 associated with flavonoid and anthocyanin accumulation	Enhancing UVR8 signaling can boost the production of flavonoids and anthocyanins, which provide UV protection	Jenkins (2017)
DNA methylation modification	Altering the DNA methylation status of specific genes or regulatory regions	DNA methylation can influence gene expression related to UVB stress responses, affecting the overall plant tolerance	Shaikh et al. (2022)
Controlling Mitogen-Activated Protein Kinase (MAPK) phosphatase	Using gene-editing methods to maintain low levels of MKP3 and MKP6 expression	MAPK phosphatases such as MKP3 and MKP6 are involved in stress signaling; controlling their levels can enhance UVB resistance	Köhler et al. (2017)

10.11 Conclusion

UV radiation, especially UVB radiation, exerts detrimental impacts on plant growth and development and subsequently reduces the economic yield. Plants, through their remarkable biochemical and physiological UV defense strategies, like the production of phenolic compounds, flavonoids, anthocyanins, lignins, tannins, specialized metabolites, alkaloids, glucosinolates, and hormones and overall DNA repair and signal transduction mechanisms, demonstrate a deep-rooted resilience that protects plants from photosynthetic damage. By understanding these sophisticated strategies, we gain a newfound appreciation for the green world's ability to adapt and thrive in a dynamic environment. This knowledge not only holds promise for safeguarding plant life but also offers valuable insights for the development of modern strategies to protect ourselves from food insecurities. Thus, understanding these UV defense mechanisms is crucial for creating ways to improve crop resilience to UV stress, specifically in the realm of climate change and increasing UV radiation levels in the near and far future.

References

Aslan D, Alan B, Yay NÖ et al (2024) Neuroprotective effect of Myrtus communis against ionizing radiation-induced brain injury: insights from histopathological and biochemical analysis in rats. J Radiat Res Appl Sci 17(4):101082. https://doi.org/10.1016/j.jrras.2024.101082

Balasubramaniam T, Shen G, Esmaeili N, Zhang H (2023) Plants' response mechanisms to salinity stress. Plants 12(12):2253. https://doi.org/10.3390/plants12122253

Ballaré CL (2014) Light regulation of plant defense. Annu Rev Plant Biol 65(1):335–363. https://doi.org/10.1146/annurev-arplant-050213-040145

Blokhina O, Virolainen E, Fagerstedt KV (2003) Antioxidants, oxidative damage and oxygen deprivation stress: a review. Ann Bot 91(2):179–194. https://doi.org/10.1093/aob/mcf118

Boubriak I, Kargiolaki H, Lyne L et al (1997) The requirement for DNA repair in desiccation tolerance of germinating embryos. Seed Sci Res 7(2):97–106. https://doi.org/10.1017/S0960258500003433

Bray CM, West CE (2005) DNA repair mechanisms in plants: crucial sensors and effectors for the maintenance of genome integrity. New Phytol 168(3):511–528. https://doi.org/10.1111/j.1469-8137.2005.01548.x

Brewster C, Fenner N, Hayes F (2024) Chronic ozone exposure affects nitrogen remobilization in wheat at key growth stages. Sci Total Environ 908:168288. https://doi.org/10.1016/j.scitotenv.2023.168288

Britt AB (1996) DNA damage and repair in plants. Annu Rev Plant Biol 47(1):75–100. https://doi.org/10.1146/annurev.arplant.47.1.75

Burko Y, Seluzicki A, Zander M, Pedmale UV, Ecker JR, Chory J (2020) Chimeric activators and repressors define HY5 activity and reveal a light-regulated feedback mechanism. Plant Cell 32(4):967–983. https://doi.org/10.1105/tpc.19.00772

Bursch K, Toledo-Ortiz G, Pireyre M, Lohr M, Braatz C, Johansson H (2020) Identification of BBX proteins as rate-limiting cofactors of HY5. Nat Plants 6(8):921–928. https://doi.org/10.1038/s41477-020-0725-0

Chen Z, Dong Y, Huang X (2022) Plant responses to UVB radiation: signaling, acclimation and stress tolerance. Stress Biol 2(1):51. https://doi.org/10.1007/s44154-022-00076-9

Depaepe T, Vanhaelewyn L, Van Der Straeten D (2023) UVB responses in the spotlight: dynamic photoreceptor interplay and cell-type specificity. Plant Cell Environ 46(11):3194–3205. https://doi.org/10.1111/pce.14680

Dikilitas M, Simsek E, Roychoudhury A (2020) Role of proline and glycine betaine in overcoming abiotic stresses. In: Protective chemical agents in the amelioration of plant abiotic stress: biochemical and molecular perspectives. Wiley, Newark, pp 1–23. https://doi.org/10.1002/9781119552154.ch1

Elbouzidi A, Haddou M, Baraich A et al (2025) Biochemical insights into specialized plant metabolites: advancing cosmeceutical applications for skin benefits. J Agric Food Res 19:101651. https://doi.org/10.1016/j.jafr.2025.101651

Esnault MA, Legue F, Chenal C (2010) Molecular mechanisms of ultraviolet radiation-induced DNA damage and repair. Environ Exp Bot 68(3):231–232. https://doi.org/10.4061/2010/592980

Essen LO, Klar T (2006) Light-driven DNA repair by photolyases. Cell Mol Life Sci 63:1266–1277. https://doi.org/10.1007/s00018-005-5447-y

Falcone Ferreyra ML, Rius SP, Casati P (2012) Flavonoids: biosynthesis, biological functions, and biotechnological applications. Front Plant Sci 3:222. https://doi.org/10.3389/fpls.2012.00222

Fina J, Casadevall R, AbdElgawad H et al (2017) UVB inhibits leaf growth through changes in growth regulating factors and gibberellin levels. Plant Physiol 174(2):1110–1126. https://doi.org/10.1104/pp.17.00365

Frohnmeyer H, Staiger D (2003) Ultraviolet-B radiation-mediated responses in plants. Balancing damage and protection. Plant Physiol 133(4):1420–1428. https://doi.org/10.1104/pp.103.030049

Garcia-Caparros P, De Filippis L, Gul A, Hasanuzzaman M et al (2021) Oxidative stress and antioxidant metabolism under adverse environmental conditions: a review. Bot Rev 87:421–466. https://doi.org/10.1007/s12229-020-09231-1

González R, Mepsted R, Wellburn AR, Paul ND (1998) Non-photosynthetic mechanisms of growth reduction in pea (*Pisum sativum* L.) exposed to UVB radiation. Plant Cell Environ 21(1):23–32. https://doi.org/10.1046/j.1365-3040.1998.00243.x

Guan Q, Lu X, Zeng H, Zhang Y, Zhu J (2013) Heat stress induction of mi R 398 triggers a regulatory loop that is critical for thermotolerance in Arabidopsis. Plant J 74(5):840–851. https://doi.org/10.1111/tpj.12169

Harrington JJ, Lieber MR (1994) Functional domains within FEN-1 and RAD2 define a family of structure-specific endonucleases: implications for nucleotide excision repair. Genes Dev 8(11):1344–1355. https://doi.org/10.1101/gad.8.11.1344

Hasanuzzaman M, Bhuyan MB, Zulfiqar F et al (2020) Reactive oxygen species and antioxidant defense in plants under abiotic stress: revisiting the crucial role of a universal defense regulator. Antioxidants 9(8):681. https://doi.org/10.3390/antiox9080681

Hays JB (2002) Arabidopsis thaliana, a versatile model system for study of eukaryotic genome-maintenance functions. DNA Repair 1(8):579–600. https://doi.org/10.1016/S1568-7864(02)00093-9

Hidema J, Kumagai T, Sutherland JC, Sutherland BM (1997) Ultraviolet B-sensitive rice cultivar deficient in cyclobutyl pyrimidine dimer repair. Plant Physiol 113(1):39–44. https://doi.org/10.1104/pp.113.1.39

Hidema J, Taguchi T, Ono T et al (2007) Increase in CPD photolyase activity functions effectively to prevent growth inhibition caused by UVB radiation. Plant J 50(1):70–79. https://doi.org/10.1111/j.1365-313X.2007.03041.x

Hossain MA, Liu F, Burritt D, Fujita M, Huang B (eds) (2020) Priming-mediated stress and cross-stress tolerance in crop plants. Academic Press

Huang R, Zhou PK (2021) DNA damage repair: historical perspectives, mechanistic pathways and clinical translation for targeted cancer therapy. Signal Transduct Target Ther 6(1):254. https://doi.org/10.1038/s41392-021-00648-7

Jansen MA, Bilger W, Hideg É et al (2019) Interactive effects of UVB radiation in a complex environment. Plant Physiol Biochem 134:1–8. https://doi.org/10.1016/j.plaphy.2018.10.021

Jansen MA, Ač A, Klem K, Urban O (2022) A meta-analysis of the interactive effects of UV and drought on plants. Plant Cell Environ 45(1):41–54. https://doi.org/10.1111/pce.14221
Jenkins GI (2017) Photomorphogenic responses to ultraviolet-B light. Plant Cell Environ 40(11):2544–2557. https://doi.org/10.1111/pce.12934
Kaiser G, Kleiner O, Beisswenger C, Batschauer A (2009) Increased DNA repair in Arabidopsis plants overexpressing CPD photolyase. Planta 230:505–515. https://doi.org/10.1007/s00425-009-0962-y
Kataria S, Jajoo A, Guruprasad KN (2014) Impact of increasing Ultraviolet-B (UVB) radiation on photosynthetic processes. J Photochem Photobiol B 137:55–66. https://doi.org/10.1016/j.jphotobiol.2014.02.004
Kazan K, Manners JM (2011) The interplay between light and jasmonate signalling during defence and development. J Exp Bot 62(12):4087–4100. https://doi.org/10.1093/jxb/err142
Köhler H, Contreras RA, Pizarro M et al (2017) Antioxidant responses induced by UVB radiation in Deschampsia antarctica Desv. Front Plant Sci 8:921. https://doi.org/10.3389/fpls.2017.00921
Krasikova Y, Rechkunova N, Lavrik O (2021) Nucleotide excision repair: from molecular defects to neurological abnormalities. Int J Mol Sci 22(12):6220. https://doi.org/10.3390/ijms22126220
Kumar A, Reed AJ, Zahurancik WJ et al (2022) Interlocking activities of DNA polymerase β in the base excision repair pathway. Proc Natl Acad Sci USA 119(10):e2118940119. https://doi.org/10.1073/pnas.2118940119
Kumar S, Bhushan B, Wakchaure GC et al (2023) Unveiling the impact of heat stress on seed biochemical composition of major cereal crops: implications for crop resilience and nutritional value. Plant Stress 9:100183. https://doi.org/10.1016/j.stress.2023.100183
Kunz BA, Anderson HJ, Osmond MJ et al (2005) Components of nucleotide excision repair and DNA damage tolerance in Arabidopsis thaliana. Environ Mol Mutagen 45(2–3):115–127. https://doi.org/10.1002/em.20094
Law YK, Forties RA, Liu X et al (2013) Sequence-dependent thymine dimer formation and photoreversal rates in double-stranded DNA. Photochem Photobiol Sci 12(8):1431–1439. https://doi.org/10.1039/c3pp50078k
Lee JH (2016) UVB signal transduction pathway in Arabidopsis. J Plant Biol 59:223–230. https://doi.org/10.1007/s12374-016-0155-8
Lee JH, Oh MM, Son KH (2019) Short-term ultraviolet (UV)-A light-emitting diode (LED) radiation improves biomass and bioactive compounds of kale. Front Plant Sci 10:1042. https://doi.org/10.3389/fpls.2019.01042
Leonard JM, Bollmann SR, Hays JB (2003) Reduction of stability of Arabidopsis genomic and transgenic DNA-repeat sequences (microsatellites) by inactivation of AtMSH2 mismatch-repair function. Plant Physiol 133(1):328–338. https://doi.org/10.1104/pp.103.023952
Lestari K, Babikian H, Kulsum ID et al (2024) The combination of *Gardenia jasminoides*, *Boswellia serrata*, *Commiphora myrrha*, *Foeniculum vulgarae*, and *Daucus carota* essential oil blend improved the inflammatory and clinical status in respiratory tract infection of COVID-19 patients: a multicentre, randomized, open-label, controlled trial. Indones Biomed J 16(3):237–247. https://doi.org/10.18585/inabj.v16i3.3023
Li N, Teranishi M, Yamaguchi H et al (2015) UVB-induced CPD photolyase gene expression is regulated by UVR8-dependent and-independent pathways in Arabidopsis. Plant Cell Physiol 56(10):2014–2023. https://doi.org/10.1093/pcp/pcv121
Li S, Liu J, Liu H, Qiu R, Gao Y, Duan A (2021) Role of hydraulic signal and ABA in decrease of leaf stomatal and mesophyll conductance in soil drought-stressed tomato. Front Plant Sci 12:653186. https://doi.org/10.3389/fpls.2021.653186
Li W, Jones K, Burke TJ, Hossain MA, Lariscy L (2022) Epigenetic regulation of nucleotide excision repair. Front Cell Dev Biol 10:847051. https://doi.org/10.3389/fcell.2022.847051
Liaqat W, Altaf MT, Barutçular C et al (2024) Ultraviolet-B radiation in relation to agriculture in the context of climate change: a review. Cereal Res Commun 52(1):1–24. https://doi.org/10.1007/s42976-023-00375-5

Lieber MR (2010) The mechanism of double-strand DNA break repair by the nonhomologous DNA end-joining pathway. Annu Rev Biochem 79(1):181–211. https://doi.org/10.1146/annurev.biochem.052308.093131

Lisboa HM, Pasquali MB, dos Anjos AI et al (2024) Innovative and sustainable food preservation techniques: enhancing food quality, safety, and environmental sustainability. Sustainability 16(18):8223. https://doi.org/10.3390/su16188223

Liu J, Feng L, Li J, He Z (2015) Genetic and epigenetic control of plant heat responses. Front Plant Sci 6:267. https://doi.org/10.3389/fpls.2015.00267

Liu X, Shi R, Gao M, He R, Li Y, Liu H (2023) Growth of tomato and cucumber seedlings under different light environments and their development after transplanting. Front Plant Sci 14:1164768. https://doi.org/10.3389/fpls.2023.1164768

Lydon J, Teramura AH, Summers EG (1986) Effects of ultraviolet-B radiation on the growth and productivity of field grown soybean. In: Worrest RC, Caldwell MM (eds) Stratospheric ozone reduction, solar ultraviolet radiation and plant life, NATO ASI series, vol 8. Springer, Berlin, Heidelberg. https://doi.org/10.1007/978-3-642-70090-3_24

Maillard O, Camenisch U, Blagoev KB et al (2008) Versatile protection from mutagenic DNA lesions conferred by bipartite recognition in nucleotide excision repair. Mutat Res Rev Mutat Res 658(3):271–286. https://doi.org/10.1016/j.mrrev.2008.01.007

Mark U, Tevini M (1996) Combination effects of UVB radiation and temperature on sunflower (*Helianthus annuus* L., cv. Polstar) and maize (*Zea mays* L, cv. Zenit 2000) seedlings. J Plant Physiol 148(1–2):49–56. https://doi.org/10.1016/S0176-1617(96)80293-8

Mathur S, Bheemanahalli R, Jumaa SH, Kakar N et al (2024) Impact of ultraviolet-B radiation on early-season morpho-physiological traits of indica and japonica rice genotypes. Front Plant Sci 15:1369397. https://doi.org/10.3389/fpls.2024.1369397

Mittler R (2017) ROS are good. Trends Plant Sci 22:11–19. https://doi.org/10.1016/j.tplants.2016.08.00

Mmbando GS (2022) Challenges and prospects in using biotechnological interventions in *O. glaberrima*, an African cultivated rice. GM Crops Food 13(1):372–387. https://doi.org/10.1080/21645698.2022.2149212

Mmbando GS (2023) The recent relationship between ultraviolet-B radiation and biotic resistance in plants: a novel non-chemical strategy for managing biotic stresses. Plant Signal Behav 18(1):2191463. https://doi.org/10.1080/15592324.2023.2191463

Mmbando GS, Teranishi M, Hidema J (2020) Very high sensitivity of African rice to artificial ultraviolet-B radiation caused by genotype and quantity of cyclobutane pyrimidine dimer photolyase. Sci Rep 10:3158. https://doi.org/10.1038/s41598-020-59720-x

Nair AS, Abraham TK, Jaya DS (2008) Studies on the changes in lipid peroxidation and antioxidants in drought stress induced cowpea (*Vigna unguiculata* L.) varieties. J Environ Biol 29(5):689–691

Nakamura M, Nunoshiba T, Hiratsu K (2021) Detection and analysis of UV-induced mutations in the chromosomal DNA of Arabidopsis. Biochem Biophys Res Commun 554:89–93. https://doi.org/10.1016/j.bbrc.2021.03.087

Nawaz M, Anjum SA, Ashraf U, Azeem F, Wang Z (2020) Antioxidant defense system and reactive oxygen species (ROS) interplay in plants under drought condition. In: Handbook of climate change management: research, leadership, transformation. Springer, Cham, pp 1–25. https://doi.org/10.1007/978-3-030-22759-3_121-1

Nedunchezhian N, Kulandaivelu G (1997) Changes induced by ultraviolet-B (280–320 nm) radiation to vegetative growth and photosynthetic characteristics in field grown Vigna unguiculata L. Plant Sci 123(1–2):85–92. https://doi.org/10.1016/S0168-9452(97)04583-4

Otake M, Teranishi M, Komatsu C, Hara M et al (2024) Poaceae plants transfer cyclobutane pyrimidine dimer photolyase to chloroplasts for ultraviolet-B resistance. Plant Physiol 195(1):326–342. https://doi.org/10.1093/plphys/kiae060

Pant P, Pandey S, Dall'Acqua S (2021) The influence of environmental conditions on secondary metabolites in medicinal plants: a literature review. Chem Biodivers 18(11):e2100345. https://doi.org/10.1002/cbdv.202100345

Pathak J, Rajneesh Ahmed H, Singh DK et al (2019) Oxidative stress and antioxidant defense in plants exposed to ultraviolet radiation. In: Reactive oxygen, nitrogen and sulfur species in plants: production, metabolism, signaling and defense mechanisms. Wiley, Newark, pp 371–420. https://doi.org/10.1002/9781119468677

Pećina-Šlaus N, Kafka A, Salamon I, Bukovac A (2020) Mismatch repair pathway, genome stability and cancer. Front Mol Biosci 7:122. https://doi.org/10.3389/fmolb.2020.00122

Podolec R, Ulm R (2018) Photoreceptor-mediated regulation of the COP1/SPA E3 ubiquitin ligase. Curr Opin Plant Biol 45:18–25. https://doi.org/10.1016/j.pbi.2018.04.018

Podolec R, Demarsy E, Ulm R (2021) Perception and signaling of ultraviolet-B radiation in plants. Annu Rev Plant Biol 72(1):793–822. https://doi.org/10.1146/annurev-arplant-050718-095946

Putnam CD (2021) Strand discrimination in DNA mismatch repair. DNA Repair 105:103161. https://doi.org/10.1016/j.dnarep.2021.103161

Qian ZF, He LL, Li F (2024) Understanding cold stress response mechanisms in plants: an overview. Front Plant Sci 15:1443317. https://doi.org/10.3389/fpls.2024.1443317

Rai K, Agrawal SB (2017) Effects of UVB radiation on morphological, physiological and biochemical aspects of plants: an overview. J Sci Res 61:87–113

Rai N, Morales LO, Aphalo PJ (2021) Perception of solar UV radiation by plants: photoreceptors and mechanisms. Plant Physiol 186(3):1382–1396. https://doi.org/10.1093/plphys/kiab162

Rizzini L, Favory JJ, Cloix C et al (2011) Perception of UVB by the Arabidopsis UVR8 protein. Science 332(6025):103–106. https://doi.org/10.1126/science.1200660

Roeber VM, Bajaj I, Rohde M, Schmülling T, Cortleven A (2021) Light acts as a stressor and influences abiotic and biotic stress responses in plants. Plant Cell Environ 44(3):645–664. https://doi.org/10.1111/pce.13948

Ryan KG, Swinny EE, Markham KR, Winefield C (2002) Flavonoid gene expression and UV photoprotection in transgenic and mutant Petunia leaves. Phytochemistry 59(1):23–32. https://doi.org/10.1016/s0031-9422(01)00404-6

Sadhna PBD, Tripathi S, Devi RS, Singh R, Bhadouria R (2023) Chapter 7 – Ecophysiological responses of tropical plants to rising air pollution. In: Ecophysiology of tropical plants: recent trends and future perspectives. CRC Press, Boca Raton. https://doi.org/10.1201/9781003335054-9

Seleiman MF, Al-Suhaibani N, Ali N et al (2021) Drought stress impacts on plants and different approaches to alleviate its adverse effects. Plants 10(2):259. https://doi.org/10.3390/plants10020259

Shaikh AA, Chachar S, Chachar M et al (2022) Recent advances in DNA methylation and their potential breeding applications in plants. Horticulturae 8(7):562. https://doi.org/10.3390/horticulturae8070562

Shan T, Gul B, Rafique M, Jabeen A, Gurmani AR, Ortas I (2022) Cereals and phytohormones under UV stress. In: Sustainable remedies for abiotic stress in cereals. Springer Nature Singapore, Singapore, pp 425–441. https://doi.org/10.1007/978-981-19-5121-3_17

Sherin G, Aswathi KR, Puthur JT (2022) Photosynthetic functions in plants subjected to stresses are positively influenced by priming. Plant Stress 4:100079. https://doi.org/10.1016/j.stress.2022.100079

Shoaib N, Pan K, Mughal N et al (2024) Potential of UVB radiation in drought stress resilience: a multidimensional approach to plant adaptation and future implications. Plant Cell Environ 47(2):387–407. https://doi.org/10.1111/pce.14774

Soni S, Jha AB, Dubey RS, Sharma P (2022) Application of nanoparticles for enhanced UVB stress tolerance in plants. Plant Nano Biol 2:100014. https://doi.org/10.1016/j.plana.2022.100014

Sutherland BM, Takayanagi S, Sullivan JH, Sutherland JC (1996) Plant responses to changing environmental stress: cyclobutyl pyrimidine dimer repair in soybean leaves. Photochem Photobiol 64(3):464–468. https://doi.org/10.1111/j.1751-1097.1996.tb03092.x

Syed A, Tainer JA (2018) The MRE11–RAD50–NBS1 complex conducts the orchestration of damage signaling and outcomes to stress in DNA replication and repair. Annu Rev Biochem 87(1):263–294. https://doi.org/10.1146/annurev-biochem-062917-012415

Takahashi S, Badger MR (2011) Photoprotection in plants: a new light on photosystem II damage. Trends Plant Sci 16(1):53–60. https://doi.org/10.1016/j.tplants.2010.10.001

Taylor RM, Nikaido O, Jordan BR et al (1996) Ultraviolet-B-induced DNA lesions and their removal in wheat (Triticum aestivum L.) leaves. Plant Cell Environ 19(2):171–181. https://doi.org/10.1111/j.1365-3040.1996.tb00238.x

Teramura AH, Sullivan JH (1994) Effects of UVB radiation on photosynthesis and growth of terrestrial plants. Photosynth Res 39:463–473. https://doi.org/10.1007/BF00014599

Teranishi M, Taguchi T, Ono T, Hidema J (2012) Augmentation of CPD photolyase activity in japonica and indica rice increases their UVB resistance but still leaves the difference in their sensitivities. Photochem Photobiol Sci 11:812–820. https://doi.org/10.1039/c2pp05392f

Tilbrook K, Arongaus AB, Binkert M et al (2013) The UVR8 UVB photoreceptor: perception, signaling and response. Arabidopsis Book 11:e0164. https://doi.org/10.1199/tab.0164

Ueda T, Nakamura C (2011) Ultraviolet-defense mechanisms in higher plants. Biotechnol Biotechnol Equip 25(1):2177–2182. https://doi.org/10.5504/BBEQ.2011.0001

Vijg J (2021) From DNA damage to mutations: all roads lead to aging. Ageing Res Rev 68:101316. https://doi.org/10.1016/j.arr.2021.101316

Wei M, He X, Liu N, Deng H (2024) Role of reactive oxygen species in ultraviolet-induced photodamage of the skin. Cell Div 19(1):1. https://doi.org/10.1186/s13008-024-00107-z

Yadav A, Singh D, Lingwan M, Yadukrishnan P, Masakapalli SK, Datta S (2020) Light signaling and UVB-mediated plant growth regulation. J Integr Plant Biol 62(9):1270–1292. https://doi.org/10.1111/jipb.12932

Yeats TH, Rose JK (2013) The formation and function of plant cuticles. Plant Physiol 163(1):5–20. https://doi.org/10.1104/pp.113.222737

Zgłobicki P, Hermanowicz P, Kłodawska K, Łabuz J et al (2024) The photoreactivation of 6–4 photoproducts in chloroplast and nuclear DNA depends on the amount of the Arabidopsis UV repair defective 3 protein. BMC Plant Biol 24(1):723. https://doi.org/10.1186/s12870-024-05439-0

Zhan J, Yang Q, Lin Z et al (2021) Enhanced antioxidant capacity and upregulated transporter genes contribute to the UVB-induced increase in blinin in *Conyza blinii*. Environ Sci Pollut Res Int 28:13275–13287. https://doi.org/10.1007/s11356-020-11502-8

Zhang F, Huang J, Guo H et al (2022) *OsRLCK160* contributes to flavonoid accumulation and UVB tolerance by regulating *OsbZIP48* in rice. Sci China Life Sci 65:1380–1394. https://doi.org/10.1007/s11427-021-2036-5

Zhang L, Wang L, Fang Y et al (2024a) Phosphorylated transcription factor PuHB40 mediates ROS-dependent anthocyanin biosynthesis in pear exposed to high light. Plant Cell 36:3562–3583. https://doi.org/10.1093/plcell/koae167

Zhang M, Ming Y, Wang HB, Jin HL (2024b) Strategies for adaptation to high light in plants. aBIOTECH 5(3):381–393. https://doi.org/10.1007/s42994-024-00164-6

Zhao J, Lu Z, Wang L, Jin B (2020) Plant responses to heat stress: physiology, transcription, non-coding RNAs, and epigenetics. Int J Mol Sci 22(1):117. https://doi.org/10.3390/ijms22010117

Chapter 11
Peptide Elicitors for Defense Against Abiotic Stress

Rania Francis, Nahed M. Rashed, Saleh M. AlTurki, Randa A. Zarban, and Tony Kevork Sajyan

Abstract Peptide elicitors have emerged as critical components in enhancing plant defense mechanisms against abiotic stress. These small signaling molecules, which can be either endogenous (plant-derived) or exogenous (synthetically and microbially derived), play a pivotal role in stress perception, signal transduction, and adaptive responses.

Peptide elicitors contribute to abiotic stress tolerance through various mechanisms, including receptor activation, hormonal cross talk (abscisic acid (ABA), jasmonic acid (JA), and ethylene), and stress-specific responses. They regulate osmoprotectant production under drought stress, maintain ionic homeostasis in saline conditions, and induce heat shock proteins (HSPs) for thermal stress tolerance. Additionally, they activate cold-responsive genes (COR) for freezing stress adaptation, chelate heavy metals to reduce toxicity, and enhance antioxidant enzyme activity to combat oxidative stress.

The practical applications of peptide elicitors in agriculture are also discussed, focusing on foliar sprays, root applications, and seed priming as innovative strategies for improving crop resilience. The integration of peptide-based treatments with biostimulants and fertilizers offers promising avenues for sustainable agricultural

---

R. Francis (✉)
Department of Agriculture and Food Engineering, School of Engineering, Holy Spirit University of Kaslik, Jounieh, Lebanon
e-mail: rania.francis@usek.edu.lb

N. M. Rashed · S. M. AlTurki
Department of Arid Land Agriculture, College of Agricultural and Food Science, King Faisal University, Al-Ahsa, Saudi Arabia

R. A. Zarban
Department of Plant Biotechnology, College of Agricultural and Food Science, King Faisal University, Al-Ahsa, Saudi Arabia

T. K. Sajyan
Department of Agriculture and Food Engineering, School of Engineering, Holy Spirit University of Kaslik, Jounieh, Lebanon

Faculty of Agricultural and Food Sciences, American University of Beirut, Beirut, Lebanon

H. I. Mohamed (ed.), *Elicitors for Sustainable Crop Production*, Sustainability Sciences in Asia and Africa, https://doi.org/10.1007/978-981-95-3533-0_11

practices. Despite their potential, challenges such as cost-effectiveness, stability, and large-scale application remain key concerns.

This chapter explores the chemical nature, classification, and structural diversity of peptide elicitors and provides insights into the role of peptide elicitors in plant defense against abiotic stress, emphasizing their potential in improving crop productivity and stress resilience. Future research directions, including advancements in peptide engineering and formulation strategies, will be essential for optimizing their agricultural use.

Keywords Peptide elicitors · Abiotic stress tolerance · Plant immunity · Oxidative stress · Sustainable agriculture

11.1 Introduction

In the quest for sustainable agricultural practices, the exploration of biological alternatives has gained remarkable momentum. Among these, peptide elicitors, particularly plant elicitor peptides (Peps), have emerged as pivotal agents for enhancing plant defense mechanisms against a variety of stresses. Discovered approximately two decades ago in *Arabidopsis thaliana*, Peps are small protein fragments that act as signaling molecules, initiating a cascade of defense responses throughout plant tissues (Huffaker et al. 2006; Bartels and Boller 2015). They play a crucial role not only in bolstering immunity against a wide range of biotic stressors, such as bacteria, fungi, oomycetes, nematodes, and herbivorous insects, but also in enhancing resilience to abiotic stresses like drought, salinity, and extreme temperatures (Zhang et al. 2023).

Abiotic stresses, such as extreme temperatures, salinity, drought, and flooding, impose significant challenges to plant growth, resulting in a considerable reduction in crop yields worldwide. Addressing the adverse effects of these environmental stresses is critical to meeting the growing global food demand. Plants have developed a variety of defense mechanisms to mitigate these stresses. These mechanisms include structural adaptations such as biochemical responses involving reactive oxygen species (ROS) scavengers, molecular chaperones, and compatible solutes. The regulation of these stress responses is governed by a complex signaling network that involves upstream molecules such as stress hormones, ROS, polyamines, gasotransmitters, and calcium, which activate downstream regulatory factors like transcription factors (He et al. 2018).

Recent research underscores the broad potential of peptides in agricultural applications, including their use as antimicrobial agents, immune boosters, plant growth regulators, and biostimulants (Pankaew et al. 2023; Saini et al. 2024). Due to their high bioactivity, eco-friendliness, and low toxicity, peptides offer significant promise as alternatives to conventional agrochemicals (Moreno-Hernández et al. 2020; Mendoza-Mendoza et al. 2023). Advancements in formulation technologies, biosynthesis, and delivery systems have further propelled their integration into modern farming practices, aligning with global efforts toward environmentally sustainable and climate-resilient agriculture (Ruiz et al. 2018; Sun et al. 2024).

Moreover, the application methods for peptide elicitors have expanded to include foliar sprays, root treatments, and seed priming, each offering unique advantages in enhancing plant defense and growth (Huffaker et al. 2013; Lee et al. 2018). Additionally, integrating peptides with biostimulants and fertilizers supports a holistic approach to sustainable farming by improving nutrient uptake and plant productivity (Dash and Ghosh 2018; Mendoza-Mendoza et al. 2023).

Given these promising attributes, peptide elicitors are poised to become central players in the future of sustainable agriculture. In this chapter, the practical applications of peptide elicitors in agriculture will be explored, specifically their use through foliar sprays and root applications, seed-priming techniques, and their integration with biostimulants and fertilizers.

11.2 Chemical Nature and Classification of Peptide Elicitors

Peptide elicitors are small signaling molecules that trigger plant defense mechanisms and initiate immune responses. Upon stress detection, elicitors activate the synthesis of specific secondary metabolites that induce the expression of a set of genes involved in plant stress response (Meena et al. 2022; Mubeen et al. 2025). These elicitors can be classified according to their nature as biotic, from plants or microbial sources, and abiotic molecules, which could be physical or chemical compounds. They can also be classified according to their origin as endogenous, which are released from the plant itself, or exogenous, which are perceived from external agents such as microbes or synthetic sources. Elicitors can also be natural or artificial molecules (Moreno-Escamilla et al. 2018). Figure 11.1 summarizes the different classifications of elicitors. This section will focus on the types of peptide elicitors as endogenous and exogenous molecules, in addition to the diversity in their structure and function in abiotic stress mitigation.

11.2.1 Types of Peptide Elicitors

Plants, though immobile, have evolved highly sophisticated immune systems to detect and respond to environmental stressors and threats posed by different microbial attacks, diseases, and harsh conditions (Jing et al. 2023; Martínez-Lorente et al. 2024). One essential element of this immune system involves elicitors' molecular cues perceived by pattern recognition receptors (PRRs) that trigger defense responses (Steinbrenner 2020). Among them, peptide elicitors are particularly noteworthy due to their defined structure, specificity, and rapid mode of action in modulating plants' innate immunity (Albert 2013; Bartels and Boller 2015). These peptides can either originate endogenously as an internal alarm, acting as

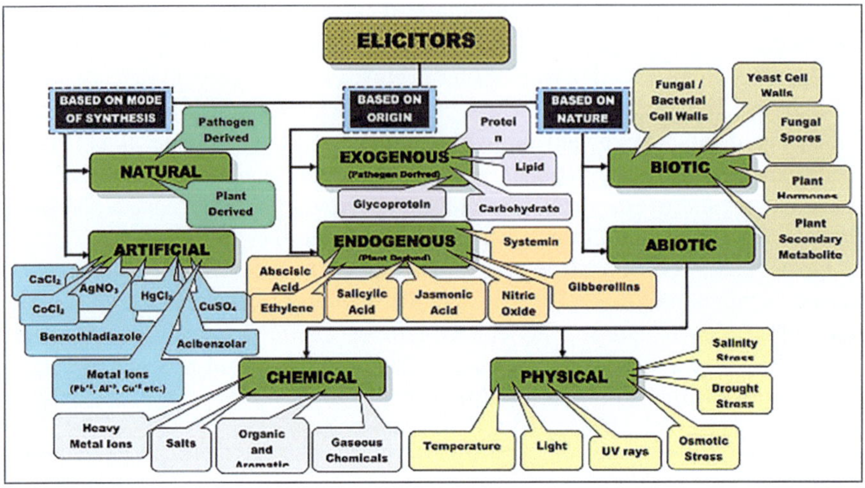

Fig. 11.1 Extensive classification of elicitors based on the mode of synthesis, origin, and nature. (Reprinted from Lone et al. (2023). Open access journal)

damage-associated molecular patterns (DAMPs), or be introduced from external sources such as pathogenic microbes, functioning as pathogen-associated molecular patterns (PAMPs) (Bartels and Boller 2015; Zelman et al. 2024). Their detection typically initiates PAMP-triggered immunity (PTI) or DAMP-triggered immunity (DTI), both of which are essential first layers of plant defense. PTI activation leads to cellular responses, including oxidative burst, mitogen-activated protein kinase (MAPK) cascade activation, transcriptional reprogramming, and cell wall reinforcement. As research on peptide elicitors expands, their utility in agrobiotechnological applications becomes increasingly evident. Leveraging this natural defense activation mechanism allows for reduced chemical pesticide and fertilizer use and more resilient crop varieties (Albert 2013).

11.2.1.1 Endogenous Peptide Elicitors (Plant-Derived)

Endogenous peptide elicitors are small, bioactive peptide molecules that are naturally produced within plant cells that play significant roles in regulating plant growth, development, and responses to environmental stimuli, including biotic and abiotic stresses (Moreno-Escamilla et al. 2018). Endogenous peptide elicitors typically function as part of plant defense mechanisms against pathogens, herbivores, and abiotic stresses, such as drought, salinity, extreme temperatures, heavy metal contamination, and UV radiation. These peptides are involved in processes like immune responses, cell division, root growth, organ abscission, and cell-to-cell communication. They are typically cleaved from larger precursor proteins (PROPEPs) and serve as DAMPs to amplify defense signaling locally and systemically. The defense responses of plants are initiated by the recognition of the conserved molecules released by the invading organism, known as microbe-associated molecular patterns (MAMPs), which, in return, are bound specifically to PRRs and induce the production of peptide elicitors. The specific mechanism by which proteolytic enzymes process and release active peptide elicitors remains under investigation (Hou et al. 2019; Steinbrenner 2020).

Endogenous peptide elicitors have been identified in several plant species via sequence homology search in most angiosperm genomes like maize, wheat, rice, and soybean (Lee et al. 2018; Lone et al. 2023).

Several classes and families of endogenous elicitors have been identified (Yadav 2024). The *A. thaliana* plant elicitor peptide (*At*Peps) family is a well-characterized endogenous elicitor composed of eight paralogs. *At*Pep1, the first isolated peptide elicitor from *A. thaliana*, is a 23-amino-acid peptide in length and released from the C-terminal of 92-amino-acid precursor proteins (AtproPep1). Their perception triggers rapid transcriptional activation of defense-related genes, increases cytosolic calcium levels, and induces reactive oxygen species (ROS) (Huffaker et al. 2006).

C-terminally encoded peptides (CEPs), initially identified in *A. thaliana*, regulate root system architecture by modulating root hair formation, lateral root initiation, and primary root elongation (Delay et al. 2013). They also play a pivotal role in nitrogen signaling by acting as long-distance root-to-shoot signals under nitrogen-limited conditions, leading to the upregulation of nitrate transporter genes such as NRT2.1 via CEP receptor (CEPR/XIP1) and downstream effectors CEPD1/2 (Tabata et al. 2014). CEP signaling contributes not only to nutrient acquisition and root morphology but also to plant adaptability under abiotic stress by modulating nitrate uptake and integrating nutrient status into developmental processes (De Coninck and De Smet 2016).

Casparian strip integrity factor (CIF) signaling regulates the deposition of lignin and suberin, important for barrier formation and selective solute transport, as well as the transcription of genes related to oxidative stress, water transport, and nutrient flux (Shiono et al. 2014; Roppolo et al. 2014). Thus, CIF peptides are critical for maintaining apoplastic barrier function, which can modulate abiotic stress responses by affecting water retention and ion homeostasis under stress conditions.

The CLAVATA3/Embryo Surrounding Region-Related (CLE) peptide family, particularly CLE9 and CLE25, is involved in regulating drought responses. CLE25 acts as a mobile signal, moving from roots to leaves to induce stomatal closure by enhancing ABA synthesis through the activation of NCED3 (Takahashi et al. 2018). Conversely, CLE9 enhances drought tolerance by regulating stomatal closure through signaling cascades involving MAPK3/6 and ROS (Zhang et al. 2019).

The inflorescence deficient in abscission (IDA) peptide, involved in floral organ abscission, has been shown to play a role in drought-induced leaf shedding. IDA signaling is necessary for the abscission of cauline leaves in response to dehydration stress (Kim et al. 2021). Additionally, the phytosulfokine (PSK) family, particularly PSK-α, has been implicated in promoting cell elongation during drought stress. The overexpression of PSK-related genes has been shown to improve drought tolerance by regulating processes like flower drop in tomatoes (Reichardt et al. 2020).

The PSK1 peptide has been identified as a key player in cold stress responses in loquat fruits. Under chilling conditions, PSK1 treatment not only reduced injury symptoms but also maintained energy balance and cell integrity, thus enhancing cold tolerance. This suggests that PSK1 could be a useful tool in improving the cold tolerance of crops (Song et al. 2017).

Rapid alkalinization factor (RALF) peptides, a group of cysteine-rich molecules, play a role in regulating various physiological processes, including stress responses. RALF peptides interact with the FERONIA (FER) receptor to modulate salt tolerance by influencing the cell wall's interaction with the plasma membrane. For example, the RALF22/23-FER signaling module controls salt tolerance by regulating cell wall extensins (Zhao et al. 2018). Mutants with disrupted RALF or FER pathways exhibit reduced growth and hypersensitivity to salt, suggesting that RALF peptides are integral to salt stress responses (Kim et al. 2021). Furthermore, the FER receptor interacts with RALF peptides to modulate root cell protection against oxidative damage under salt stress (Feng et al. 2018).

One prominent peptide elicitor, miPEP408, has been shown to play a significant role in plant adaptation to heavy metal stress, particularly arsenic toxicity. Research by Datta et al. demonstrated that miPEP408 modulates plant responses to arsenic and sulfur deficiencies by regulating the expression of miR408, a microRNA that affects stress tolerance (Datta et al. 2024).

The peptide elicitor PIP-1 has been studied for its role in mediating plant defense responses to UV irradiation. In tobacco cells, PIP-1 induces a variety of immunological responses, including the synthesis of the phytoalexin capsidiol (Kim et al. 2015).

11.2.1.2 Exogenous Peptide Elicitors (Microbially and Synthetically Derived)

Exogenous peptide elicitors in plants are small compounds that originate from external sources and are typically derived from microbial pathogens or artificially synthesized peptide analogs that mimic naturally occurring elicitors. Exogenous peptide elicitors are recognized by plants as nonself signals, and they induce plant basal immunity by acting like PAMPs (Albert 2013; Hou et al. 2019; Moreno-Escamilla et al. 2018).

The application of exogenous peptide elicitors in plants is not limited to strengthening basal immunity and enhancing resistance to pathogens. These exogenous elicitors also serve as tools to stimulate the production of secondary metabolite compounds that often possess significant pharmacological, nutraceutical, or therapeutic value. By activating specific signaling pathways, exogenous peptide elicitors can upregulate metabolic biosynthetic genes, resulting in an increased synthesis of alkaloids, phenolics, terpenoids, and other bioactive molecules (Aitouguinane et al. 2023; Mubeen et al. 2025). Such secondary metabolites are widely recognized for their diverse roles as herbal remedies and natural health products. Their enhanced accumulation in plant tissues following elicitor treatment provides an additional avenue for producing high-value compounds with medicinal properties. This effect supports the use of elicitor treatments beyond plant protection, extending to the enrichment of extracts used in pharmaceuticals, functional foods, and cosmetics (Baenas et al. 2014).

Consequently, the strategy of utilizing exogenous peptide elicitors has gained interest not only in plant health and disease resistance but also in industrial and biotechnological applications. Their potential to increase the yield and diversity of natural compounds may reduce dependence on synthetic chemicals, offering more sustainable and potentially safer alternatives for both agriculture and industry. This dual benefit, improved plant defense and elevated production of valuable secondary metabolites, highlights the expanding utility of peptide elicitors in integrated plant management and natural product biosynthesis (Baenas et al. 2014; Aitouguinane et al. 2023; Mubeen et al. 2025).

Synthetic peptide elicitors are small, artificially engineered peptides designed to replicate or modify the activity of naturally occurring signaling molecules recognized by plant immune systems. Unlike their natural counterparts, which are typically derived from microbial pathogens or external biological sources, synthetic peptides can be rationally tailored in terms of sequence and functional properties (Guo and Cheng 2022; Montesinos 2023). Producing synthetic peptides aims to (1) enhance their immunogenic potential by extracting the active sequence, (2) eliminate negative effects like toxicity and low activity, (3) enhance their resistance to proteolysis, and (4) enable large-scale production for practical agricultural applications (Lima et al. 2021).

Synthetic peptide elicitors can be engineered to exhibit enhanced biological activity compared to their natural analogs and may be effective at significantly lower concentrations.

Researchers employ strategies like computational modeling, alanine scanning, incorporation of non-natural amino acids, and alterations in peptide length to improve important characteristics such as stability, binding affinity to immune receptors, resistance to enzymatic degradation, and specificity (Li et al. 2021; Montesinos 2023). For instance, chemical modifications like cyclization or backbone methylation are frequently introduced to extend peptide durability within the plant while maintaining or enhancing immune activation (Badosa et al. 2022). These peptides can be synthesized either de novo or by modifying existing natural peptide frameworks. They may be used individually or integrated into chimeric peptides that combine different functional motifs (Montesinos 2023). The arrangement and structural context of these domains are crucial, as they significantly affect the efficacy of the immune response and the overall safety profile of the construction (Badosa et al. 2022; Pearce et al. 2010).

Experimental evidence has shown that synthetic peptide elicitors can trigger typical plant immune responses, including bursts of ROS, changes in pH, the activation of MAPK, and the upregulation of defense-related genes (Lee et al. 2018; Pearce et al. 2010; Zelman and Berkowitz 2023). Importantly, these synthetic molecules also offer the ability to fine-tune the intensity and timing of immune responses, which can help mitigate potential negative impacts on plant growth and development.

In the context of sustainable agriculture, synthetic peptide elicitors hold considerable promise as environmentally friendly biopesticides and biofertilizers. Their customizable nature, scalability, improved stability, and low toxicity make them strong candidates for future crop protection strategies in eco-agriculture. Continued advances in synthetic design may enable the development of crop- or pathogen-specific elicitors, supporting enhanced plant resistance while reducing dependence on conventional chemical inputs (Badosa et al. 2022; Li et al. 2021; Zhang et al. 2023).

11.2.2 Structural and Functional Diversity of Peptide Elicitors

Peptide elicitors play multifaceted roles in plant immunity through structurally and functionally diverse mechanisms. The distinction between endogenous and exogenous peptides highlights the plant's ability to detect both internal damage and external threats. Owing to their specificity, adaptability, and biotechnological potential, these peptides represent promising tools for sustainable agriculture.

Peptide elicitors exhibit considerable structural heterogeneity in terms of sequence length, charge distribution, secondary structure, and post-translational modifications. They can be classified based on their charge into cationic peptides (rich in arginine or lysine) and anionic peptides (rich in aspartic or glutamic acid) (Dashper et al. 2005; Li et al. 2021). Peptide elicitors are structurally small in size, between 2 and 50 amino acids. Endogenous peptides can range from short chains of 10–20 amino acids to larger peptides. Structural traits and confirmation influence binding specificity to receptors and stability. Structural motifs determine the specificity, enabling distinct immune responses and receptor–ligand interactions. Some peptides exhibit cyclic or beta sheet structures, which confer protease resistance (Guo and Cheng 2022; Li et al. 2021; Lima et al. 2021; Zhang et al. 2023).

Post-translational modifications such as hydroxylation, glycosylation, and sulfation influence peptide solubility, binding affinity, and half-life. These features are crucial for their biological function. Moreover, synthetic variants often include D-amino acids, hydrophobic and cationic residues, cyclization, or terminal protection to increase resistance to enzymatic degradation, extending their utility in agricultural applications (Almaaytah et al. 2014; Lima et al. 2021).

Functionally, peptide elicitors mediate diverse biological roles beyond the mere activation of immunity. They can enhance basal defense, regulate hormonal cross talk, and prime the plant for faster and stronger responses upon subsequent stress by engaging specific receptors and initiating cascades that culminate in enhanced resistance, thus serving as a key component in the plant's defense arsenal against diverse environmental stressors.

Elicitors trigger plant defense responses upon recognition by specific plant receptors. In the context, endogenous peptides like Systemin (SYS) and AtPeps are released following tissue damage, acting as DAMPs, while exogenous peptides like flg22 and elf18, derived from microbial components, function as PAMPs that initiate PTI. DAMP and PAMP molecules are recognized by PRRs, such as leucine-rich repeat receptor-like kinases (LRR-RLKs), on the plant cell surface and activate a cascade of intracellular signaling events. These include the phosphorylation of receptor proteins, the activation of kinase cascades, the rapid production of ROS and nitric oxide (NO), Ca^{2+}-dependent signaling, and the activation of MAPKs. These signaling pathways subsequently lead to the activation of transcription factors like W-box transcription factors (WRKY), which induce the overexpression of defense-related genes (Huffaker et al. 2006; Albert 2013; Moreno-Escamilla et al. 2018; Hou et al. 2019; Montesinos 2023).

11.3 Applications of Peptide Elicitors in Agriculture

An alternative that can support other sustainable farming methods is the use of naturally occurring chemicals. Since their discovery in the model plant *A. thaliana* 20 years ago, several intriguing studies have revealed that Peps can protect various agricultural plants from biotic and abiotic stresses. Peptides are small protein fragments that play a crucial role in enhancing plant immunity against various abiotic stresses, including salinity, drought, extreme temperatures, heavy metal contamination, and UV radiation (Fig. 11.2). They are a type of signaling molecule produced by plants to spread danger signals around the organs of an individual plant and initiate defensive maneuvers. They are considered promising tools in agriculture for promoting plant health and reducing the reliance on chemicals. Table 11.1 summarizes the function of the main small peptide elicitors in abiotic stress responses.

Recent research underscores the expanding role of peptides in plant protection strategies. These include their use as stress-mitigating factors, antimicrobial and immune-inducing agents, plant growth regulators, biofertilizers, biopesticides, and bioherbicides owing to their broad resource base, potent bioactivity, and favorable ecological profile. Over the past century, there has been substantial progress in peptide-based agrochemical development, driven by the global need for high-efficiency, low-toxicity, and environmentally safe alternatives. The field of peptide agrochemicals has matured considerably, with advancements in structure optimization, formulation technologies, delivery systems, and biosynthesis contributing to its growth (Huffaker et al. 2006; Zhang et al. 2023; Zelman and Berkowitz 2024).

Over the past 120 years, there has been a notable emergence of peptides alongside the swift advancement of peptide-based pharmaceuticals and agrochemicals, reflecting a global trend toward products characterized by high efficiency, low toxicity, and environmental safety. The domain of peptide agrochemicals is advancing in various dimensions, encompassing sustained and comprehensive research on the application, production, and identification of multiple agricultural peptides as viable candidates for farming practices. Various advanced technologies in structure optimization, formulation, delivery systems, and biosynthesis will persist in

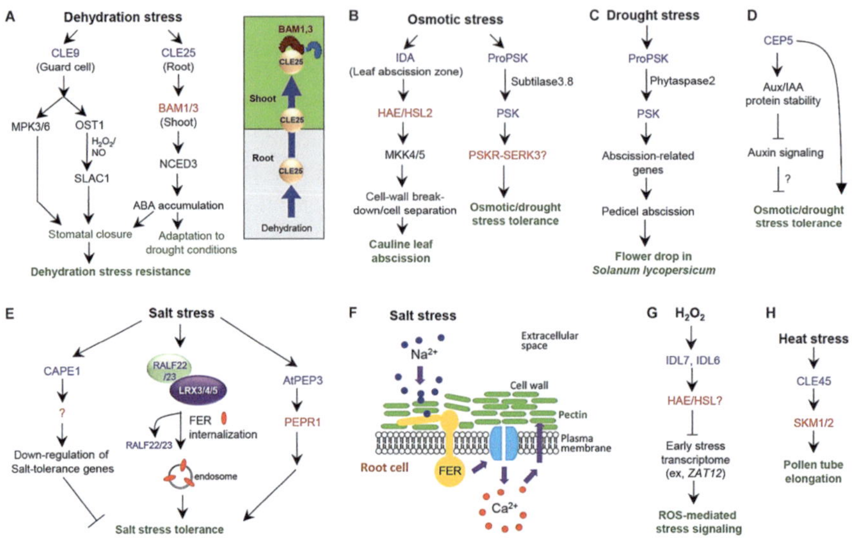

Fig. 11.2 Signaling peptides control abiotic stress responses in plants. Blue and red letters indicate peptides and the cognate receptors, respectively. The plant not specified in the subtitle indicates *Arabidopsis*. (**a**) Roles and signaling pathways of CLE9/25 in dehydration stress response. (**b**) Roles and signaling pathways of IDA and PSK in osmotic stress response. (**c**) Roles of PSK in flower abscission in tomatoes under drought stress. (**d**) Role of CEP5 in osmotic and drought stress tolerance. (**e**) Roles and signaling pathways in high salinity stress response. (**f**) Role of FER receptor kinase during high salinity stress response to maintain cell wall integrity in root cells. (**g**) Role of IDL6/7 in ROS-mediated stress signaling. (**h**) Role of CLE45 in pollen tube elongation under heat stress. (Reprinted from Kim et al. (2021). Open access journal)

fostering the development of peptide agrochemicals (Bartels and Boller 2015; Moreno-Hernández et al. 2020; Zhang et al. 2023).

Peptide elicitors offer sustainable, eco-friendly solutions for enhancing plant immunity and crop protection by activating natural defense mechanisms, reducing the reliance on chemicals. Peptides with bioactive properties may serve as potential candidates for application in the pharmaceutical, agricultural, and food industries (Pankaew et al. 2023; Saini et al. 2024; Setty et al. 2024). These molecules, applied via foliar, root, or seed-priming treatments, improve disease resistance and stress resilience while promoting growth and nutrient uptake (Ruiz et al. 2018). Integrated with biostimulants and fertilizers, they support holistic agricultural practices, aligning with the goals of high-yield, climate-resilient farming. Research highlights their role in sustainable and eco-friendly crop protection, emphasizing their potential as alternatives to conventional agrochemicals (Mendoza-Mendoza et al. 2023; Sun et al. 2024).

Protein hydrolysates (PHs), particularly those enriched with antioxidant peptides, are derived from natural sources. Many researchers regard them as biostimulants due to their minimal side effects, ease of absorption, cost-effectiveness, high

Table 11.1 Typical small peptides involved in plant abiotic stress responses

Peptide	Plant	Stress	Precursor	Key receptor	Function	References
CEP	*Arabidopsis* and *M. truncatula*	Drought N deficiency	proCEP	CEPR1 CEPR2	Enhances drought tolerance by downregulating auxin-related genes; upregulates nitrate transporter NRT2.1 to enhance nitrate uptake under low N	Smith et al. (2020); Luo et al. (2022)
PSK	*Arabidopsis*, *Solanum lycopersicum*	Drought Immunity Heat stress	proPSK	PSKR1 PSKR2	Regulates drought-induced flower drop; triggers calcium influx, activating auxin-mediated pathways to enhance immunity; helps plants tolerate heat stress	Reichardt et al. (2020); Stührwohldt et al. (2021)
CAPE1	*Arabidopsis*, *S. lycopersicum*, *Triticum aestivum*	Salinity Immunity	proCAPE1	Unknown	Negatively regulates salt tolerance by downregulating abscisic acid (ABA)–dependent and salinity-responsive genes	Sung et al. (2021)
RALF	*Arabidopsis*	Salinity Immunity	proRALF	FER	Enhances salt tolerance by maintaining cell wall integrity and suppressing ABA signaling; regulates immunity by modulating receptor-mediated signaling	Feng et al. (2018)
PEP	*Arabidopsis*	Salinity Immunity	proPEP	PEPR1	Activates immune responses under extracellular alkaline conditions; deprotonation enhances receptor binding and immune activation; links salt tolerance and immunity	Nakaminami et al. (2018); He et al. (2023)
PIP1 PIP3	*Arabidopsis*, *S. pimpinellifolium*	Salinity Immunity	proPIP1 proPIP3	RLK7	Regulates salt tolerance via activating the MPK3/6 pathway; regulates immunity via the salicylic acid (SA) and jasmonic acid (JA) signaling pathways	Yang et al. (2023)
CLE25	*Arabidopsis*	Drought	proCLE25	BAM1 BAM3	Induces stomatal closure to prevent water loss under drought conditions, enabling dehydration resistance	Takahashi et al. (2018)
CLE45	*Arabidopsis*	Temperature stress	proCLE45	SKM1 SKM2	Enhances pollen tube growth under heat stress, supporting reproductive success and seed production	Endo et al. (2013)
CLE9 CLE10	*Arabidopsis*	Drought	proCLE9 proCLE10	HSL1 BAM1	Enhances drought tolerance by inducing stomatal closure to reduce water loss	Zhang et al. (2019)

(continued)

Table 11.1 (continued)

Peptide	Plant	Stress	Precursor	Key receptor	Function	References
CLE1 CLE3 CLE4 CLE7	*Arabidopsis*	N deficiency	proCLE1 proCLE3 proCLE4 proCLE7	CLV1	Modulates root system architecture by inhibiting lateral root growth and promoting primary root elongation under N-deficient conditions	Araya et al. (2014)
EPFL6	*Arabidopsis*	Temperature stress	proEPFL6	ER	Coordinates stamen elongation at low temperatures, enabling proper self-pollination for reproductive success	Negoro et al. (2023)

Reprinted from Xiao et al. (2025). Open access journal

activity, and lower molecular weight. PHs and peptides serve as significant components in the formulation of functional foods (Balan et al. 2022).

11.3.1 Foliar Sprays and Root Applications

Peptide application via foliar sprays and root treatments has gained significant attention due to its efficiency and sustainability. Foliar application of peptide elicitors involves spraying solutions directly onto plant leaves, where they are absorbed and activate defense pathways, which can be effectively applied to trigger defense mechanisms and enhance plant resilience against abiotic stresses.

PHs, particularly those with antioxidant peptides, can be utilized as foliar sprays or via drip irrigation systems, with amino acids being absorbed through both the leaves and roots (Dash and Ghosh 2018; Santi et al. 2017).

PHs, composed of polypeptides, oligopeptides, and amino acids derived from plant and animal protein sources, offer promising applications in mitigating abiotic stress in plants. They are typically produced via enzymatic, chemical, or thermal hydrolysis, with enzymatic hydrolysis being preferred for plant-based PHs due to its lower salinity and consistent composition. PHs can be applied as foliar sprays or root treatments to alleviate various stressors, including water and temperature stress. The application of PHs, particularly those derived from legume and soybean biomass, has shown positive effects on plant stress tolerance. For example, in grapevine production, PH treatments from lupin, soybean, and casein significantly improved the yield, grapevine performance, and secondary metabolite synthesis, especially in response to environmental stress. PHs reduced stomatal conductance and transpiration under water stress conditions, likely through the action of abscisic acid (ABA), which enhances stress tolerance by regulating calcium signaling. In addition to their effects on water stress, PHs have been shown to modify plant metabolomes and proteomes, improving plant resilience. For instance, legume-derived PHs delayed physiological maturity in vines and altered metabolite concentrations, with notable increases in stress-related compounds and antioxidants like stigmasterol and plastoquinol-9. These metabolic changes are thought to help plants better manage environmental stress, including summer heat, by dissipating excess energy (Cataldo et al. 2022).

Foliar application of okara enzymatic extract (OEE), a biostimulant rich in isoflavones and protein hydrolysates, significantly mitigated the deleterious effects of acute ozone exposure on pepper plants. Key photosynthetic parameters—including net photosynthetic rate (AN), electron transport rate (ETR), effective quantum yield of photosystem II (PSII) (ΦPSII), and delayed fluorescence (DF)—were maintained at higher levels in OEE-treated plants compared to untreated controls. Furthermore, a transcriptomic analysis revealed that OEE modulated the expression of genes associated with photosystem II, indicating protection at the molecular level (Orts et al. 2024).

While specific studies on Peps directly enhancing photosynthesis are limited, their role in activating defense mechanisms can lead to overall improved plant health, potentially supporting better growth and photosynthetic efficiency. Research by Zhang et al. revealed that the application of Pep1 in *A. thaliana* roots triggers systemic immune signaling to shoots, enhancing the plant's ability to withstand various stresses. This systemic response suggests potential applications in improving plant resilience through targeted treatments (Zhang et al. 2022). Jing et al. studied the effect of Peps on root hair development in *A. thaliana*. Their findings indicate that Peps enhance root hair growth, which is crucial for nutrient and water uptake, thereby promoting overall root development. While direct evidence of Peps enhancing nutrient uptake is still under investigation, their role in promoting root hair development suggests a potential for improved nutrient absorption due to increased root surface area (Jing et al. 2024). Additional specialized applications can be anticipated. For instance, antitranspirants applied to bedding plants and floral crops effectively delay wilting by inducing stomatal closure, allowing bedding plants to withstand water deficit (Park et al. 2016). Given the significance of Pep signaling in salinity stress, osmopriming with a Pep may prove beneficial for cultivating crops in saline soil conditions. In *A. thaliana*, 17 small peptide-encoding genes were upregulated under salinity stress. Four of these genes, particularly *At*PROPEP3, were found to improve root growth and survival in transgenic plants under salt stress. The *At*PROPEP3 peptide, when applied exogenously, protected plants from salt-induced chlorophyll bleaching (Nakaminami et al. 2018; Kim et al. 2021). Nakaminami et al. showed that silencing *At*PROPEP3 expression results in increased sensitivity to salt stress, whereas overexpressing *At*PROPEP3 or applying synthetic *At*PEP3 peptide enhances plant resistance to salinity (Nakaminami et al. 2018). Additionally, Peps like *At*PEP1–6 bind to receptor kinases (PEPR1 and PEPR2), triggering immune responses and conferring resistance to salinity. Notably, plants lacking PEPR receptors exhibited severe hypersensitivity to salt, which could be mitigated by peptide treatments (Kim et al. 2021).

11.3.2 Seed Priming with Peptides

Seed priming with peptide elicitors is an innovative and sustainable approach to enhancing seed vigor and stress tolerance. Peptide elicitors, small protein fragments that mimic natural signaling molecules, can activate plant defense mechanisms even before germination. This pretreatment strengthens seedlings against abiotic stresses, leading to improved crop performance. Peptide elicitors function by activating plant immune responses: Primed seeds exhibit early expression of defense-related genes. Priming seeds with peptide elicitors can also enhance tolerance to abiotic stresses. Studies have shown that seed priming with elicitors leads to increased total phenolic content and antioxidant activity in rice plants under drought conditions. This biochemical enhancement correlates with improved drought tolerance and better overall plant health (Samota et al. 2017).

Preplanting seed treatment is standard practice but necessitates drying before insertion into planting machinery; various seeds may undergo treatment for differing durations before germination rates are compromised. The requisite treatment durations for mitigating abiotic stress will be essential for all Pep seed treatments. Seeds cultivated for commercial purposes are frequently subjected to aqueous solutions of growth regulators (hydropriming) or osmolyte solutions (osmopriming) to enhance germination and growth rates (Weissmann et al. 2023). Their application can notably influence nitrogen metabolism in plants and enhance productivity, especially when used as a seed pretreatment (Dash and Ghosh 2018).

An innovative application of Peps is in the regeneration of tissue culture. Tomato Pep, identified as a regeneration factor, has been demonstrated to facilitate shoot regeneration from callus (Yang et al. 2023). Shoot regeneration is a crucial process in the application of tissue culture for plant breeding and micropropagation. Cannabis exemplifies a crop requiring enhanced shoot regeneration protocols, given the low regeneration rates of transgenic cannabis plants (Hesami et al. 2021).

Matsubayashi emphasizes the potential of using synthetic or recombinant peptides to enhance tissue culture regeneration meristem activity. This study demonstrates how CLV3 signaling controls cell proliferation and differentiation in meristematic tissues. The findings suggest that manipulating CLV3 or related peptides could improve shoot regeneration in tissue culture (Matsubayashi 2014). The CEP family of peptides regulates root growth and development. A study conducted by Nikolau et al. identifies a specific CEP peptide that promotes root regeneration in *A. thaliana* (Nikolau et al. 2008). The CLE45 peptide plays a pivotal role in heat stress tolerance, especially in reproductive organs like flowers. In *A. thaliana*, CLE45 interacts with the SKM1 and SKM2 receptors in the pollen tube, promoting pollen tube growth under high-temperature conditions. The CLE45-SKM signaling pathway is essential for seed production under heat stress, making it a potential target for improving crop resilience to elevated temperatures (Endo et al. 2013; Xie et al. 2022). The tracheary element differentiation inhibitory factor (TDIF) peptide is a member of the CLE family and plays a critical role in vascular tissue development. A study shows that TDIF can suppress premature differentiation of vascular cells, promoting the proliferation of undifferentiated cells in tissue culture. Such peptides could be applied to enhance callus formation and regeneration efficiency (Ito et al. 2006). Ikeuchi et al. discuss the molecular mechanisms underlying callus formation and shoot regeneration. The potential of peptide hormones, such as those involved in cytokinin signaling, to overcome recalcitrance in tissue culture suggests that the exogenous application of peptides could improve regeneration rates in difficult-to-propagate species (Ikeuchi et al. 2013). Phytosulfokine (PSK) is a sulfated peptide hormone that promotes cell proliferation and differentiation. A study demonstrates that PSK enhances callus formation and shoot regeneration in tissue culture. The results propose that PSK could be used as a supplement in tissue culture media to improve regeneration efficiency (Wang et al. 2015).

11.3.3 Integration with Biostimulants and Fertilizers

Biostimulants have emerged as vital agents in enhancing crop resilience to abiotic stresses, which are intensifying due to climate change and increasingly unpredictable environmental conditions (Goud et al. 2022). By modulating plant physiological and metabolic responses, biostimulants help maintain productivity under stress conditions, such as drought, salinity, temperature extremes, ozone, and nutrient deficiencies. The impact of various biostimulants on crops encompasses enhancements in visual quality of produce, stimulation of plant immune systems, induction of biosynthesis of defensive biomolecules, remediation of heavy metals in contaminated soils, enhancement of crop performance, reduction of leaching, improvement of root development and seed germination, induction of tolerance to abiotic stressors, promotion of crop establishment, and increased nutrient use efficiency.

Protein hydrolysates consist of polypeptides and free amino acids derived from the enzymatic and chemical hydrolysis of agro-industrial protein by-products from both animal and plant sources. They have been shown to mitigate environmental stress, enhance growth, and increase crop productivity (Sun et al. 2024). Protein hydrolysates and glycine betaine-based products have demonstrated priming effects that stabilize photosynthetic activity, enhance carotenoid accumulation, and strengthen plant responses to water limitation (Antonucci et al. 2021). Nonhormonal biostimulants, including formulations based on peptides and antioxidants, have been shown to alleviate osmotic stress in crops such as cowpea and maize by supporting metabolic homeostasis and improving water use efficiency (Atta-Boateng and Berlyn 2021; Ambrosini et al. 2021).

Orts et al. evaluated the protective role of a novel biostimulant derived from okara (okara enzymatic extract) against ozone-induced oxidative stress in *Capsicum annuum* (pepper) plants. OEE, a water-soluble extract produced through the enzymatic hydrolysis of okara (a soybean by-product), is enriched with bioactive compounds, notably isoflavones and protein hydrolysates. These molecules exhibit strong antioxidant activities and may function synergistically to mitigate oxidative stress. The predominance of hydrophobic and sulfur-containing amino acids further enhances the antioxidative potential of the peptides. These constituents are known for their antioxidant properties, which are crucial in counteracting oxidative stress induced by environmental factors such as ozone. Results showed that OEE reduced ozone-induced lipid peroxidation, as indicated by lower malondialdehyde accumulation, a reliable marker of oxidative membrane damage. In parallel, OEE application elevated the activities of key antioxidant enzymes—catalase, superoxide dismutase, and guaiacol peroxidase—in both ozone-exposed and nonexposed plants. These results support the extract's role in bolstering the antioxidant defense system (Orts et al. 2024).

The integration of peptide elicitors with biostimulants and fertilizers has emerged as a promising strategy for sustainable agriculture. Peptide elicitors, which are small signaling molecules, enhance plant defense mechanisms and stress tolerance. When combined with biostimulants (e.g., microbial inoculants and seaweed extracts) and fertilizers, they improve nutrient use efficiency, crop productivity, and resilience to abiotic stresses. Recent studies have demonstrated that this synergy optimizes plant growth, reduces fertilizer input, and promotes eco-friendly agricultural practices. Microbial biostimulants, especially plant growth-promoting rhizobacteria (PGPR), arbuscular mycorrhizal fungi (AMF), and endophytes, significantly enhance plant tolerance to multiple abiotic stresses. These include drought, nutrient limitation, and salinity, primarily by improving nutrient availability, enhancing antioxidant responses, and facilitating root development (Kalozoumis et al. 2021; Toubali et al. 2022; Mellidou et al. 2021). Microbial inoculants, including strains of *Pseudomonas*, *Bacillus*, and halotolerant fungi, have also been effective in mitigating salt stress through improved root growth, enhanced secondary metabolism, and upregulation of defense-related enzymes (Mellidou et al. 2021; Xiao et al. 2022; Tarroum et al. 2022). Studies also highlight the synergistic effects of combining microbial and nonmicrobial biostimulants to amplify stress resilience and yield outcomes (Lalarukh et al. 2022; Kang et al. 2021). Integrating peptide elicitors with biostimulants and fertilizers offers a promising strategy to enhance plant growth, nutrient uptake, and stress resilience. This combined approach leverages the synergistic effects of these substances, leading to more sustainable and efficient agricultural practices. Peptide elicitors, such as protein hydrolysates, contain bioactive peptides that exhibit phytohormone-like activities, stimulating root growth and overall plant development. When combined with biostimulants like seaweed extracts or humic substances, these peptides can further enhance plant biomass and vigor. For instance, enzyme-based biostimulants have been shown to influence physiological and biochemical processes in plants, promoting growth and improving stress tolerance (Tamburino et al. 2023). The integration of peptide elicitors with fertilizers can lead to more efficient nutrient uptake and utilization. Peptides can modulate root architecture, increasing the surface area for nutrient absorption. Additionally, biostimulants have been reported to enhance the uptake of essential minerals such as phosphorus, potassium, and micronutrients, thereby improving overall nutrient use efficiency (Martínez-Lorente et al. 2024). Combining peptide elicitors with biostimulants can bolster a plant's defense mechanisms against abiotic stresses like drought and salinity. This combination can activate antioxidant pathways and stress-responsive genes, leading to improved resilience. For example, the application of protein hydrolysates has been associated with increased tolerance to water deficit in grapevines, suggesting potential benefits in stress mitigation (Jindo et al. 2022) (Fig. 11.3).

Protein hydrolysates serve as biostimulants, presenting a novel approach to enhance crop nutrition. Their application may decrease the reliance on chemical fertilizers across various crops, particularly in horticulture. Commercial protein-based biostimulants are primarily derived from vegetable protein sources; however, animal proteins are also utilized in their production, providing additional functionalities such as enhancing plant abiotic tolerance. Food by-products present a viable option for the production of protein hydrolysate biostimulants, given the substantial volumes of high-protein effluents discarded by food-processing industries. The conversion of these compounds into bioactive substances for agricultural purposes will

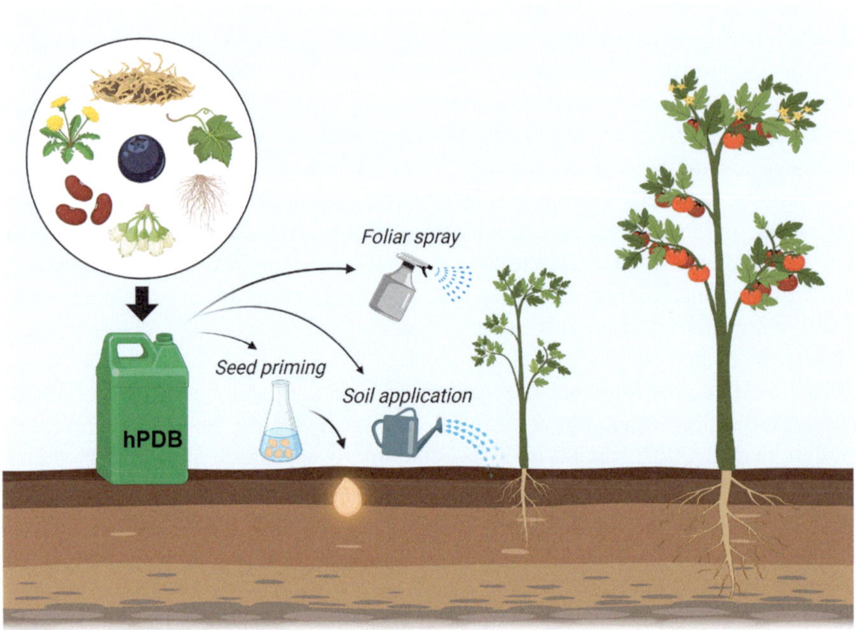

Fig. 11.3 Higher plant-derived biostimulants (hPDB) main sources and application method. (Reprinted from Martínez-Lorente et al. (2024). Open access journal)

not only mitigate the environmental impacts associated with their discharge but also enhance agricultural practices through the production of plant biostimulants. In the formulation of biostimulants, protein by-products may be chosen based on the peptides encoded in their primary sequence and their potential bioactivity (Moreno-Hernández et al. 2020). The use of gelatin derived from animals can enhance the root nitrogen assimilation process and increase the shoot dry weight in crops such as corn, cucumbers, tomatoes, arugula, peppers, and field peas (Wilson et al. 2018). Rouphael et al. demonstrated that the use of microgranules derived from vegetal-protein hydrolysates can enhance the concentration of carotenoids and total chlorophyll (Rouphael et al. 2021). Increased mineral-nutrient and water use efficiency, better plant output, and tolerance to adverse conditions are all possible outcomes of protein hydrolysate's beneficial effects on the root length and the total root area. Additionally, it has the ability to increase the productivity and leaf area of fruit trees and horticulture plants (Pankaew et al. 2023). Table 11.2 summarizes a few examples related to the impact of different protein hydrolysates on various plants regarding abiotic stress mitigation.

Table 11.2 The impacts of different protein hydrolysates on various plants

Plant	Plant family	Protein hydrolysate	Key point	Reference
Common bean (*Phaseolus vulgaris* L.)	Fabaceae	Pumpkin seed protein hydrolysate	Application of 2000 μL L^{-1} to obtain an appropriate yield and growth of plants under salt stress	Sitohy et al. (2020)
Grape tomatoes (*Solanum lycopersicum* var. cerasiforme)	Solanaceae	Fish-derived protein hydrolysates	Application of fish-derived protein hydrolysates could reduce the negative impacts of drought and improve total plant biomass yield, leaf dry weight, and fruit number	Garcia-Santiago et al. (2022)
Lettuce (*Lactuca sativa* L.)	Asteraceae	Fish-derived protein hydrolysate	It contains amino acids and peptides, which can improve root biomass and leaf number and enhance the photosynthetic rate and chlorophyll content	Xu and Mou (2017)
		A *Graminaceae*-derived protein hydrolysate	It can improve the growth and yield of plants and improve the resistance of plants under mild salinity conditions	El-Nakhel et al. (2023)
Passion fruit (*Passiflora edulis*)	Passifloraceae	Commercial preparation of peptides and amino acids	It has peptides and amino acids, which can increase the photosynthetic process in plants and increase transplanting success	Ertani et al. (2009)
Persimmon (*Diospyros kaki*)	Ebenaceae	Protein hydrolysate	Increases the biosynthesis of salt stress response proteins	Aktsoglou et al. (2021)
Spinach (*Spinacia oleracea* L.)	Amaranthaceae	*Xcell Boost*, a mixture of fish protein hydrolysates and kelp extract	It is highly beneficial for promoting the tolerance of spinach to water shortage stress	Liatile et al. (2022)
Sweet pepper (*Capsicum annuum* L.)	Solanaceae	Organic fertilizer based on hydrolyzed proteins	It could improve the performance in nitrogen uptake, increase resistance to tolerance, and mitigate the negative impacts of toxic elements	Testani et al. (2017)
Tomato (*Solanum lycopersicum* L.)	Solanaceae	Protein hydrolysates	It can be considered an important biostimulant to improve plant resilience to abiotic stresses	Francesca et al. (2022)

(continued)

Table 11.2 (continued)

Plant	Plant family	Protein hydrolysate	Key point	Reference
		Pig blood-derived protein hydrolysate	It can increase salt tolerance in tomatoes and improve photosynthetic efficiency, chlorophyll levels, and plant growth	Zhou et al. (2022)
		Pig blood-derived protein hydrolysate	It can increase yield and mitigate the negative impacts of drought stress by regulating chloroplast ultrastructure, antioxidant systems, stomatal aperture, and osmotic changes	Wang et al. (2022)
Wheat (*Triticum aestivum* L.)	Poaceae	Protein hydrolysate	It can improve wheat grain seed germination and the final production	Trakselyte-Rupsiene et al. (2021)

Reprinted from Sun et al. (2024). Open access journal

11.4 Challenges and Future Opportunities in the Application of Peptide Elicitors

The use of peptide elicitors and biostimulants can reduce the reliance on chemical fertilizers and pesticides, promoting environmentally friendly farming practices. Studies have highlighted the potential of biostimulants to improve crop performance while minimizing environmental impact (Ruzzi et al. 2024). By enhancing natural plant processes and, more importantly, enhancing crop tolerance against environmental stresses and improving nutrient use efficiency, this integrated approach supports sustainable agriculture.

Peptide elicitors have shown immense potential in agriculture through applications in foliar sprays, root treatments, seed priming, and integration with biostimulants and fertilizers. However, several challenges hinder their widespread adoption. Addressing these issues could unlock new opportunities for improving crop resilience, productivity, and sustainability (Bartels and Boller 2015). The main challenge is the elicitors' stability in field conditions. Peptide elicitors are susceptible to degradation from environmental factors such as UV radiation and microbial activity. Ensuring their stability during application is crucial for optimal effectiveness. Strategies like nano-encapsulation or the use of carrier molecules are being explored to protect peptides from rapid degradation (Zelman and Berkowitz 2024). The efficacy of peptide elicitors can vary, depending on the plant species and environmental conditions. Research indicates that Peps exhibit conserved activity within related plant families but may have limited functionality across distant families. This specificity necessitates tailored approaches for different crops (Huffaker et al. 2013). Additionally, the lack of standardized guidelines for the optimal use of peptide elicitors, including dosages and application methods, affects the efficacy of these elicitors. Establishing clear protocols through extensive field trials is necessary to ensure consistent and reliable outcomes. The long-term impacts of peptide elicitor applications on soil health, nontarget organisms, and overall ecosystem dynamics remain under-researched. Comprehensive studies are needed to assess any potential unintended consequences of their use (Zelman and Berkowitz 2023). Moreover, the synthesis and extraction of peptide elicitors can be cost-prohibitive, limiting their widespread use in agriculture. Developing cost-effective production methods, possibly through biotechnological advancements, is essential to make peptide-based solutions economically viable for farmers (Zelman and Berkowitz 2024). Peptide elicitors often occupy a regulatory gray area, complicating their approval and commercialization. Clear regulatory frameworks are required to facilitate their integration into mainstream agricultural practices.

11.5 Conclusion

Peptide elicitors hold significant promise as tools for promoting plant health and productivity in sustainable agriculture. They can be used in various forms, such as foliar sprays, seed priming, and integration with other agricultural inputs. By stimulating natural defense pathways and supporting growth, they offer an environmentally friendly alternative to conventional agrochemicals. Despite their potential, several technical and practical barriers, such as stability, delivery efficiency, and production costs, currently limit their broader adoption.

Ongoing innovations are addressing these constraints. Progress in peptide design and biotechnological approaches, including Clustered Regularly Interspaced Short Palindromic Repeats (CRISPR)-based methods, could enhance the efficiency and affordability of elicitor production. Novel delivery systems, particularly nano- and microencapsulation technologies, improve peptide stability, bioavailability, and uptake by plants. Additionally, combining peptide elicitors with beneficial microbes and biostimulants may lead to synergistic improvements in nutrient acquisition, stress tolerance, and soil health.

The integration of digital agriculture tools and precision agriculture, such as artificial intelligence (AI) and automated application technologies, presents new opportunities to optimize the application of peptide elicitors. AI and machine learning models can optimize peptide dosages, application timing, and crop-specific responses based on real-time data. Drones and automated sprayers can improve precision in foliar applications, reducing waste and enhancing efficiency.

Furthermore, as agriculture faces increasing climatic pressures, peptide elicitors could support the development of resilient crops. Peptide elicitors can be integrated into climate-smart agricultural strategies to help crops withstand temperature extremes, drought, and salinity. Their role in reducing chemical fertilizer dependence aligns with global sustainability goals and regenerative agriculture. Research into their role as natural biopesticides and biofungicides also highlights their potential to reduce chemical inputs and enhance food safety.

In summary, while challenges remain, peptide elicitors are poised to play a pivotal role in the transition toward more resilient, efficient, and ecologically sound agricultural systems. Continued research and cross-disciplinary innovation will be essential to fully harness their benefits.

References

Aitouguinane M, El Alaoui-Talibi Z, Rchid H, Fendri I, Abdelkafi S, El-Hadj MDO, Boual Z, Le Cerf D, Rihouey C, Gardarin C, Dubessay P, Michaud P, Pierre G, Delattre C, El Modafar C (2023) Elicitor activity of low-molecular-weight alginates obtained by oxidative degradation of alginates extracted from *Sargassum muticum* and *Cystoseira myriophylloides*. Mar Drugs 21:301

Aktsoglou DC, Kasampalis DS, Sarrou E, Tsouvaltzis P, Chatzopoulou P, Martens S, Siomos AS (2021) Protein hydrolysates supplement in the nutrient solution of soilless grown fresh peppermint and spearmint as a tool for improving product quality. Agronomy 11:317

Albert M (2013) Peptides as triggers of plant defence. J Exp Bot 64:5269–5279

Almaaytah A, Tarazi S, Abu-Alhaijaa A, Altall Y, Alshar'i N, Bodoor K, Al-Balas Q (2014) Enhanced antimicrobial activity of AamAP1-Lysine, a novel synthetic peptide analog derived from the scorpion venom peptide Aam AP1. Pharmaceuticals 7:502–516

Ambrosini S, Sega D, Santi C, Zamboni A, Varanini Z, Pandolfini T (2021) Evaluation of the potential use of a collagen-based protein hydrolysate as a plant multi-stress protectant. Front Plant Sci 12:600623

Antonucci G, Croci M, Miras-Moreno B, Fracasso A, Amaducci S (2021) Integration of gas exchange with metabolomics: high-throughput phenotyping methods for screening biostimulant-elicited beneficial responses to short-term water deficit. Front Plant Sci 12:678925

Araya T, Miyamoto M, Wibowo J, Suzuki A, Kojima S, Tsuchiya YN, Sawa S, Fukuda H, Von Wirén N, Takahashi H (2014) CLE-CLAVATA1 peptide-receptor signaling module regulates the expansion of plant root systems in a nitrogen-dependent manner. Proc Natl Acad Sci USA 111:2029–2034

Atta-Boateng A, Berlyn GP (2021) Limiting-stress-elimination hypothesis: using non-hormonal biostimulant to reduce stress and increase savanna cowpea [*Vigna unguiculata* (L.) Walp.] productivity. Front Plant Sci 12:732279

Badosa E, Planas M, Feliu L, Montesinos L, Bonaterra A, Montesinos E (2022) Synthetic peptides against plant pathogenic bacteria. Microorganisms 10:1784

Baenas N, García-Viguera C, Moreno DA (2014) Elicitation: a tool for enriching the bioactive composition of foods. Molecules 19:13541–13563

Balan D, Luţă G, Stanca M, Jerca O, Niculescu M, Gaidau C, Jurcoane S, Mihalcea A (2022) Effect of protein gel treatments on biometric and biochemical attributes of tomato seedlings in greenhouse condition. Agriculture 13:54

Bartels S, Boller T (2015) Quo vadis, Pep? Plant elicitor peptides at the crossroads of immunity, stress, and development. J Exp Bot 66:5183–5193

Cataldo E, Fucile M, Mattii GB (2022) Biostimulants in viticulture: a sustainable approach against biotic and abiotic stresses. Plants 11:162

Dash P, Ghosh G (2018) Amino acid profiling and antimicrobial activity of *Cucurbita moschata* and *Lagenaria siceraria* seed protein hydrolysates. Nat Prod Res 32:2050–2053

Dashper SG, O'Brien-Simpson NM, Cross KJ, Paolini RA, Hoffmann B, Catmull DV, Malkoski M, Reynolds EC (2005) Divalent metal cations increase the activity of the antimicrobial Peptide kappacin. Antimicrob Agents Chemother 49:2322–2328
Datta T, Kumar RS, Sinha H, Trivedi PK (2024) Small but mighty: peptides regulating abiotic stress responses in plants. Plant Cell Environ 47:1207–1223
De Coninck B, De Smet I (2016) Plant peptides–taking them to the next level. J Exp Bot 67:4791–4795
Delay C, Imin N, Djordjevic MA (2013) CEP genes regulate root and shoot development in response to environmental cues and are specific to seed plants. J Exp Bot 64:5383–5394
El-Nakhel C, Cristofano F, Colla G, Pii Y, Lucini L, Rouphael YA (2023) *Graminaceae*-derived protein hydrolysate and its fractions provide differential growth and modulate qualitative traits of lettuce grown under non-saline and mild salinity conditions. Sci Hortic 319:112130
Endo S, Shinohara H, Matsubayashi Y, Fukuda H (2013) A novel pollen pistil interaction conferring high-temperature tolerance during reproduction via CLE45 signaling. Curr Biol 23:1670–1676
Ertani A, Cavani L, Pizzeghello D, Brandellero E, Altissimo A, Ciavatta C, Nardi S (2009) Biostimulant activity of two protein hydrolyzates in the growth and nitrogen metabolism of maize seedlings. J Plant Nutr Soil Sci 172:237–244
Feng W, Kita D, Peaucelle A, Cartwright HN, Doan V, Duan Q, Liu MC, Maman J, Steinhorst L, Schmitz-Thom I, Yvon R, Kudla J, Wu HM, Cheung AY, Dinneny JR (2018) The FERONIA receptor kinase maintains cell-wall integrity during salt stress through Ca2+ signaling. Curr Biol 28:666–675
Francesca S, Najai S, Zhou R, Decros G, Cassan C, Delmas F, Ottosen CO, Barone A, Rigano MM (2022) Phenotyping to disset the biostimulant action of a protein hydrolysate in tomato plants under combined abiotic stress. Plant Physiol Biochem 179:32–43
Garcia-Santiago JC, Cavazos CJ, Gonzalez-Fuentes J, Zermeno-Gonzalez A, Alvarado ER, Duarte AR, Preciado-Rangel P, Troyo-dieguez E, Ramos FMP, Valdez-Aguilar LA, Alvarado-Camarillo D, Hernández Liatile PC, Potgieter G, Moloi MJ (2022) A natural biostimulant consisting of a mixture of fish protein hydrolysates and kelp extract enhances the physiological, biochemical, and growth responses of Spinach under different water levels. Plants 11:3374
Goud EL, Singh J, Kumar P (2022) Chapter 19 – Climate change and their impact on global food production. In: Kumar A, Singh J, Ferreira LFR (eds) Microbiome under changing climate: implications and solutions, 1st edn. Elsevier, pp 415–436
Guo J, Cheng Y (2022) Advances in fungal elicitor-triggered plant immunity. Int J Mol Sci 23:12003
He M, He CQ, Ding NZ (2018) Abiotic stresses: general defenses of land plants and chances for engineering multistress tolerance. Front Plant Sci 9:1771
He YH, Chen SY, Chen XY, Xu YP, Liang Y, Cai XZ (2023) RALF22 promotes plant immunity and amplifies the Pep3 immune signal. J Integr Plant Biol 65:2519–2534
Hesami M, Baiton A, Alizadeh M, Pepe M, Torkamaneh D, Jones AMP (2021) Advances and perspectives in tissue culture and genetic engineering of cannabis. Int J Mol Sci 22:5671
Hou S, Liu Z, Shen H, Wu D (2019) Damage-associated molecular pattern-triggered immunity in plants. Front Plant Sci 10:646
Huffaker A, Pearce G, Ryan CA (2006) An endogenous peptide signal in *Arabidopsis* activates components of the innate immune response. Proc Natl Acad Sci USA 103:10098–10103
Huffaker A, Pearce G, Veyrat N, Erb M, Turlings TCJ, Sartor R, Shen Z, Briggs SP, Vaughan MM, Alborn HT (2013) Plant elicitor peptides are conserved signals regulating direct and indirect antiherbivore defense. Proc Natl Acad Sci USA 110:5707–5712
Ikeuchi M, Sugimoto K, Iwase A (2013) Plant callus: mechanisms of induction and repression. Plant Cell 25:3159–3173
Ito Y, Nakanomyo I, Motose H, Iwamoto K, Sawa S, Dohmae N, Fukuda H (2006) Dodeca-CLE peptides as suppressors of plant stem cell differentiation. Science 313:842–845
Jindo K, Goron TL, Pizarro-Tobías P, Sánchez-Monedero MÁ, Audette Y, Deolu-Ajayi AO, van der Werf A, Goitom Teklu M, Shenker M, Pombo Sudré C (2022) Application of biostimu-

lant products and biological control agents in sustainable viticulture: a review. Front Plant Sci 13:932311

Jing Y, Zheng X, Sharifi R, Chen J (2023) Plant elicitor peptide induces endocytosis of plasma membrane proteins in *Arabidopsis*. Front Plant Sci 14:1328250

Jing Y, Zhao F, Lai K, Sun F, Sun C, Zou X, Xu M, Fu A, Sharifi R, Chen J (2024) Plant elicitor peptides regulate root hair development in *Arabidopsis*. Front Plant Sci 15:1336129

Kalozoumis P, Savvas D, Aliferis K, Ntatsi G, Marakis G, Simou E, Tampakaki A, Karapanos I (2021) Impact of plant growth-promoting rhizobacteria inoculation and grafting on tolerance of tomato to combined water and nutrient stress assessed via metabolomics analysis. Front Plant Sci 12:670236

Kang Y, Kim M, Shim C, Bae S, Jang S (2021) Potential of algae–bacteria synergistic effects on vegetable production. Front Plant Sci 12:656662

Kim Y, Miyashita M, Miyagawa H (2015) Photocontrol of elicitor activity of PIP-1 to investigate temporal factors involved in phytoalexin biosynthesis. J Agric Food Chem 63:5894–5901

Kim JS, Jeon BW, Kim J (2021) Signaling peptides regulating abiotic stress responses in plants. Front Plant Sci 12:704490

Lalarukh I, Al-Dhumri SA, Al-Ani LKT, Hussain R, Al Mutairi KA, Mansoora N, Amjad SF, Abbas MHH, Abdelhafez AA, Poczai P, Meena KR, Galal TM (2022) A combined use of rhizobacteria and moringa leaf extract mitigates the adverse effects of drought stress in wheat (*Triticum aestivum* L.). Front Microbiol 13:813415

Lee MW, Huffaker A, Crippen D, Robbins RT, Goggin FL (2018) Plant elicitor peptides promote plant defences against nematodes in soybean. Mol Plant Pathol 19:858–869

Li J, Hu S, Jian W, Xie C, Yang X (2021) Plant antimicrobial peptides: structures, functions, and applications. Bot Stud 62:5

Liatile PC, Potgieter G, Moloi MJ (2022) A natural biostimulant consisting of a mixture of fish protein hydrolysates and kelp extract enhances the physiological, biochemical, and growth responses of Spinach under different water levels. Plants 11:3374

Lima PG, Oliveira JTA, Amaral JL, Freitas CDT, Souza PFN (2021) Synthetic antimicrobial peptides: characteristics, design, and potential as alternative molecules to overcome microbial resistance. Life Sci 278:119647

Lone R, Hassan N, Bashir B, Rohela GK, Malla NA (2023) Role of growth elicitors and microbes in stress management and sustainable production of Sorghum. Plant Stress 9:100179

Luo Z, Wang J, Li F, Lu Y, Fang Z, Fu M, Mysore KS, Wen J, Gong J, Murray JD, Xie F (2022) The small peptide CEP1 and the NIN-like protein NLP1 regulate NRT2.1 to mediate root nodule formation across nitrate concentrations. Plant Cell 35:776–794

Martínez-Lorente SE, Martí-Guillén JM, Pedreño MÁ, Almagro L, Sabater-Jara AB (2024) Higher plant-derived biostimulants: mechanisms of action and their role in mitigating plant abiotic stress. Antioxidants 13:318

Matsubayashi Y (2014) Posttranslationally modified small-peptide signals in plants. Annu Rev Plant Biol 65:385–413

Meena M, Yadav G, Sonigra P, Nagda A, Mehta T, Swapnil P, Harish, Marwal A (2022) Role of elicitors to initiate the induction of systemic resistance in plants to biotic stress. Plant Stress 5:100103

Mellidou I, Ainalidou A, Papadopoulou A, Leontidou K, Genitsaris S, Karagiannis E, Van de Poel B, Karamanoli K (2021) Comparative transcriptomics and metabolomics reveal an intricate priming mechanism involved in PGPR-mediated salt tolerance in tomato. Front Plant Sci 12:713984

Mendoza-Mendoza A, Serrano M, Schwan-Estrada KRF, Ferreira JFS, Marchese JA (2023) Eliciting plant defense responses: from basic to applied science for sustainable agriculture. Front Plant Sci 14:1129305

Montesinos E (2023) Functional peptides for plant disease control. Annu Rev Phytopathol 61:301–324

Moreno-Escamilla JO, Alvarez-Parrilla E, de la Rosa LA, Núñez-Gastélum JA, González-Aguilar GA, Rodrigo-García J (2018) Chapter 3 – Effect of elicitors in the nutritional and sensorial quality of fruits and vegetables. In: Siddiqui MW (ed) Preharvest modulation of postharvest fruit and vegetable quality, 1st edn. Elsevier, pp 71–91

Moreno-Hernández JM, Benítez-García I, Mazorra-Manzano MA, Ramírez-Suárez JC, Sánchez E (2020) Strategies for production, characterization and application of protein-based biostimulants in agriculture: a review. Chil J Agric Res 80:274–289

Mubeen B, Murtaza S, Yaqoob S, Aregbe AY, Boasiako TA, Xiong YQ, Ma YK (2025) Harnessing plant defense: elicitors, hormones and immunity-driven production of medicinally valuable secondary metabolites. S Afr J Bot 179:280–292

Nakaminami K, Okamoto M, Higuchi-Takeuchi M, Yoshizumi T, Yamaguchi Y, Fukao Y, Shimizu M, Ohashi C, Tanaka M, Matsui M (2018) AtPep3 is a hormone-like peptide that plays a role in the salinity stress tolerance of plants. Proc Natl Acad Sci USA 115:5810–5815

Negoro S, Hirabayashi T, Iwasaki R, Torii KU, Uchida N (2023) EPFL peptide signalling ensures robust self-pollination success under cool temperature stress by aligning the length of the stamen and pistil. Plant Cell Environ 46:451–463

Nikolau BJ, Perera MADN, Brachova L, Shanks B (2008) Platform biochemicals for a biorenewable chemical industry. Plant J 54:536–545

Orts A, Navarro-Torre S, Macías-Benítez S, Orts JM, Naranjo E, Castaño A, Parrado J (2024) A new biostimulant derived from soybean by-products enhances plant tolerance to abiotic stress triggered by ozone. BMC Plant Biol 24:580

Pankaew C, Supdensong K, Tothong C, Roytrakul S, Phaonakrop N, Kongbangkerd A, Limmongkon A (2023) Combining elicitor treatment of chitosan, methyl jasmonate, and cyclodextrin to induce the generation of immune response bioactive peptides in peanut hairy root culture. Plant Sci 331:111670

Park S, Mills SA, Moon Y, Waterland NL (2016) Evaluation of antitranspirants for enhancing temporary water stress tolerance in bedding plants. HortTechnology 26:444–452

Pearce G, Yamaguchi Y, Barona G, Ryan CA (2010) A subtilisin-like protein from soybean contains an embedded, cryptic signal that activates defense-related genes. Proc Natl Acad Sci USA 107:14921–14925

Reichardt S, Piepho HP, Stintzi A, Schaller A (2020) Peptide signaling for drought-induced tomato flower drop. Science 367:1482–1485

Roppolo D, Boeckmann B, Pfister A, Boutet E, Rubio MC, Dénervaud-Tendon V, Vermeer JE, Gheyselinck J, Xenarios I, Geldner N (2014) Functional and evolutionary analysis of the CASPARIAN STRIP MEMBRANE DOMAIN PROTEIN family. Plant Physiol 165:1709–1722

Rouphael Y, Colla G, Hoagland L, Giordano M, El-Nakhel C, Cardarelli M (2021) Vegetal-protein hydrolysates based microgranule enhances growth, mineral content, and quality traits of vegetable transplants. Sci Hortic 290:110554

Ruiz C, Nadal A, Foix L, Montesinos L, Montesinos E, Pla M (2018) Diversity of plant defense elicitor peptides within the *Rosaceae*. BMC Genet 19:1–12

Ruzzi M, Colla G, Rouphael Y (2024) Biostimulants in agriculture II: towards a sustainable future. Front Plant Sci 15:1427283

Saini DR, Prakash P, Jangde S, Kumar K, Maiti I (2024) Chapter 8 – Plant elicitor peptides: mechanism of action and its applications in agriculture. In: Singh S, Mehrotra R (eds) Plant elicitor peptides: new tool for sustainable agriculture, 1st edn. Springer, pp 171–197

Samota MK, Sasi M, Awana M, Yadav OP, Amitha Mithra SV, Tyagi A, Kumar S, Singh A (2017) Elicitor-induced biochemical and molecular manifestations to improve drought tolerance in rice (*Oryza sativa* L.) through seed-priming. Front Plant Sci 8:934

Santi C, Zamboni A, Varanini Z, Pandolfini T (2017) Growth stimulatory effects and genome-wide transcriptional changes produced by protein hydrolysates in maize seedlings. Front Plant Sci 8:433

Setty J, Singh P, Tantuway G, Vijai P (2024) Chapter 12 – Recent advancement on peptide research and their application in eco-agriculture. In: Singh S, Mehrotra R (eds) Plant elicitor peptides: new tool for sustainable agriculture, 1st edn. Springer, pp 269–281

Shiono K, Ando M, Nishiuchi S, Takahashi H, Watanabe K, Nakamura M, Matsuo Y, Yasuno N, Yamanouchi U, Fujimoto M, Takanashi H, Ranathunge K, Franke RB, Shitan N, Nishizawa NK, Takamure I, Yano M, Tsutsumi N, Schreiber L, Yazaki K, Nakazono M, Kato K (2014) RCN1/OsABCG5, an ATP-binding cassette (ABC) transporter, is required for hypodermal suberization of roots in rice (*Oryza sativa*). Plant J 80:40–51

Sitohy MZ, Desoky EM, Osman A, Rady MM (2020) Pumpkin seed protein hydrolysate treatment alleviates salt stress effects on *Phaseolus vulgaris* by elevating antioxidant capacity and recovering ion homeostasis. Sci Hortic 271:109495

Smith S, Zhu S, Joos L, Roberts I, Nikonorova N, Vu LD, Stes E, Cho H, Larrieu A, Xuan W, Goodall B, van de Cotte B, Waite JM, Rigal A, Ramans Harborough S, Persiau G, Vanneste S, Kirschner GK, Vandermarliere E, Martens L, Stahl Y, Audenaert D, Friml J, Felix G, Simon R, Bennett MJ, Bishopp A, De Jaeger G, Ljung K, Kepinski S, Robert S, Nemhauser J, Hwang I, Gevaert K, Beeckman T, De Smet I (2020) The CEP5 peptide promotes abiotic stress tolerance, as revealed by quantitative proteomics, and attenuates the AUX/IAA equilibrium in *Arabidopsis*. Mol Cell Proteomics 19:1248–1262

Song H, Wang X, Hu W, Yang X, Diao E, Shen T, Qiang Q (2017) A cold-induced phytosulfokine peptide is related to the improvement of loquat fruit chilling tolerance. Food Chem 232:434–442

Steinbrenner AD (2020) The evolving landscape of cell surface pattern recognition across plant immune networks. Curr Opin Plant Biol 56:135–146

Stührwohldt N, Bühler E, Sauter M, Schaller A (2021) Phytosulfokine (PSK) precursor processing by subtilase SBT3.8 and PSK signaling improve drought stress tolerance in *Arabidopsis*. J Exp Bot 72:3427–3440

Sun W, Shahrajabian MH, Kuang Y, Wang N (2024) Amino acids biostimulants and protein hydrolysates in agricultural sciences. Plants 13:210

Sung YC, Outram MA, Breen S, Wang C, Dagvadorj B, Winterberg B, Kobe B, Williams SJ, Solomon PS (2021) PR1-mediated defence via C-terminal peptide release is targeted by a fungal pathogen effector. New Phytol 229:3467–3480

Tabata R, Sumida K, Yoshii T, Ohyama K, Shinohara H, Matsubayashi Y (2014) Perception of root-derived peptides by shoot LRR-RKs mediates systemic N-demand signaling. Science 346:343–346

Takahashi F, Suzuki T, Osakabe Y, Betsuyaku S, Kondo Y, Dohmae N, Fukuda H, Yamaguchi-Shinozaki K, Shinozaki K (2018) A small peptide modulates stomatal control via abscisic acid in long-distance signaling. Nature 556:235–238

Tamburino R, Docimo T, Sannino L, Gualtieri L, Palomba F, Valletta A, Ruocco M, Scotti N (2023) Enzyme-based biostimulants influence physiological and biochemical responses of *Lactuca sativa* L. Biomolecules 13:1765

Tarroum M, Romdhane WB, Al-Qurainy F, Ali AAM, Al-Doss A, Fki L, Hassairi A (2022) A novel PGPF *Penicillium olsonii* isolated from the rhizosphere of *Aeluropus littoralis* promotes plant growth, enhances salt stress tolerance, and reduces chemical fertilizers inputs in hydroponic system. Front Microbiol 13:996054

Testani E, Campanelli G, Leteo F, Ciaccia C, Canali S, Tittarelli F (2017) Sweet pepper (*Capsicum annuum* L.) organic seedling production: the role of compost, cultivar, and protein hydrolyzate. Compost Sci Util 25:112–119

Toubali S, Ait-El-Mokhtar M, Boutasknit A, Anli M, Ait-Rahou Y, Benaffari W, Ben-Ahmed H, Mitsui T, Baslam M, Meddich A (2022) Root reinforcement improved performance, productivity, and grain bioactive quality of field-droughted quinoa (*Chenopodium quinoa*). Front Plant Sci 13:860484

Trakselyte-Rupsiene K, Juodeikiene G, Cernauskas D, Bartkiene E, Klupsaite D, Zadeike D, Bendoraitiene J, Damasius J, Ignatavicius J, Sikorskaite-Gudziuniene S (2021) Integration of

ultrasound into the development of plant-based protein hydrolysate and its bio-stimulatory effect for growth of wheat grain seedlings *in vivo*. Plants 10:1319
Wang C, Yu H, Zhang Z, Yu L, Xu X, Hong Z, Luo L (2015) Phytosulfokine is involved in positive regulation of *Lotus japonicus* nodulation. Mol Plant-Microbe Interact 28:847–855
Wang W, Zheng W, Lv H, Liang B, Jin S, Li J, Zhou W (2022) Animal-derived plant biostimulant alleviates drought stress by regulating photosynthesis, osmotic adjustment, and antioxidant systems in tomato plants. Sci Hortic 305:111365
Weissmann EA, Raja K, Gupta A, Patel M, Buehler A (2023) Chapter 16 – Seed quality enhancement. In: Dadlani M, Yadava DK (eds) Seed science and technology: biology, production, quality, 1st edn. Springer, pp 391–414
Wilson HT, Amirkhani M, Taylor AG (2018) Evaluation of gelatin as a biostimulant seed treatment to improve plant performance. Front Plant Sci 9:1006
Xiao X, Wang Q, Ma X, Lang D, Guo Z, Zhang X (2022) Physiological biochemistry-combined transcriptomic analysis reveals mechanism of *Bacillus cereus* G2 improved salt-stress tolerance of *Glycyrrhiza uralensis* fisch. Seedlings by balancing carbohydrate metabolism. Front Plant Sci 12:712363
Xiao F, Zhou H, Lin H (2025) Decoding small peptides: regulators of plant growth and stress resilience. J Integr Plant Biol 67:596–631
Xie H, Zhao W, Li W, Zhang Y, Hajný J, Han H (2022) Small signaling peptides mediate plant adaptions to abiotic environmental stress. Planta 255:72
Xu C, Mou B (2017) Drench application of fish-derived protein hydrolysates affects lettuce growth, chlorophyll content, and gas exchange. HortTechnology 27:539–543
Yadav V (2024) Chapter 5 – Endogenous peptides involved in plant growth and development. In: Singh S, Mehrotra R (eds) Plant elicitor peptides: new tool for sustainable agriculture, 1st edn. Springer, pp 113–130
Yang R, Liu J, Wang Z, Zhao L, Xue T, Meng J, Luan Y (2023) The SlWRKY6-SlPROPEP-SlPep module confers tomato resistance to *Phytophthora infestans*. Sci Hortic 318:112117
Zelman AK, Berkowitz GA (2023) Plant elicitor peptide (Pep) signaling and pathogen defense in tomato. Plants 12:2856
Zelman AK, Berkowitz GA (2024) Chapter 15 – Peps, pathogens, and pests: challenges and opportunities for usage of pep signaling in sustainable farming. In: Singh S, Mehrotra R (eds) Plant elicitor peptides: new tool for sustainable agriculture, 1st edn. Springer, pp 335–351
Zelman AK, Ma Y, Berkowitz GA (2024) Pathogen elicitor peptide (pep), systemin, and their receptors in tomato: sequence analysis sheds light on standing disagreements about biotic stress signaling components. BMC Plant Biol 24:728
Zhang L, Shi X, Zhang Y, Wang J, Yang J, Ishida T, Jiang W, Han X, Kang J, Wang X, Pan L, Lv S, Cao B, Zhang Y, Wu J, Han H, Hu Z, Cui L, Sawa S, He J, Wang G (2019) CLE9 peptide induced stomatal closure is mediated by abscisic acid, hydrogen peroxide, and nitric oxide in *Arabidopsis thaliana*. Plant Cell Environ 42:1033–1044
Zhang J, Li Y, Bao Q, Wang H, Hou S (2022) Plant elicitor peptide 1 fortifies root cell walls and triggers a systemic root-to-shoot immune signaling in *Arabidopsis*. Plant Signal Behav 17:2034270
Zhang YM, Ye DX, Liu Y, Zhang XY, Zhou YL, Zhang L, Yang XL (2023) Peptides, new tools for plant protection in eco-agriculture. Adv Agrochem 2:58–78
Zhao C, Zayed O, Yu Z, Jiang W, Zhu P, Hsu CC et al (2018) Leucine-rich repeat extensin proteins regulate plant salt tolerance in *Arabidopsis*. Proc Natl Acad Sci USA 115:13123–13128
Zhou W, Zheng W, Lv H, Wang Q, Liang B, Li J (2022) Foliar application of pig blood-derived protein hydrolysates improves antioxidant activities in lettuce by regulating phenolic biosynthesis without compromising yield production. Sci Hortic 291:110602

Chapter 12
Nano-elicitors: A Promising Strategy for Sustainable Crop Production Under Abiotic Stress Conditions

Alia Telli, Aicha Mouane, Manal Bencheikh, and Heba I. Mohamed

Abstract This chapter presents an overview of nano-elicitors, an innovative method for influencing plant responses to abiotic stress, along with their agricultural applications. Abiotic stress is a substantial risk to worldwide agricultural output, influencing plant growth, development, and yield. The rising prevalence and intensity of abiotic stresses require the formulation of novel ways to improve crop resilience. Nano-elicitors have emerged as a promising instrument in agriculture, providing an innovative method to regulate plant responses to abiotic stress. This chapter examines the mechanisms of action of nano-elicitors, their agricultural applications, and the problems and constraints related to their use. The application of nano-elicitors in agriculture may enhance crop yields, diminish environmental impact, and foster sustainable agricultural methods.

Keywords Abiotic stress · Crop production · Nano-elicitors · Crop resilience · Sustainable agriculture

A. Telli (✉)
Laboratory of Preservation of Ecosystems in Arid and Semi-Arid Areas, Faculty of Nature and Life Sciences, University of Kasdi Merbah-Ouargla, Ouargla, Algeria
e-mail: telli.alia@univ-ouargla.dz

A. Mouane
Department of Biology, Faculty of Natural and Life Sciences, University of El Oued, El Oued, Algeria

M. Bencheikh
Laboratory of Phoeniciculture Research "Phoenix," Faculty of Nature and Life Sciences, University of Kasdi Merbah, Ouargla, Algeria

H. I. Mohamed
Biological and Geological Sciences, Faculty of Education, Ain Shams University, Cairo, Egypt

H. I. Mohamed (ed.), *Elicitors for Sustainable Crop Production*, Sustainability Sciences in Asia and Africa, https://doi.org/10.1007/978-981-95-3533-0_12

12.1 Introduction

Food security is a fundamental requirement for survival. Given the growing population and the limited availability of resources, such as land and water for agriculture, it is imperative to improve agricultural yield in the face of abiotic pressures. Environmental stressors, including salinity, drought, thermal extremes, heavy metal contamination, and pesticide residues in the soil, significantly impede the growth and development of many crop species (Rivero-Montejo et al. 2021).

Abiotic stress is well-documented as a significant factor in the reduction of grain yield in many crops, and the current climate change is anticipated to exacerbate this issue in agricultural production. Consequently, it is essential to devise innovative strategies to mitigate these adverse effects on our crops and thereby enhance their output. The application of agricultural techniques in cropping systems utilizing plants already acclimated to their environment, along with specific genotypes or soil enhancement treatments that promote greater resilience to existing abiotic stressors, is widely acknowledged to benefit the agricultural sector. Furthermore, it is essential to focus on the enhancement of specific biotic and abiotic stress-related traits in currently significant crop species (Kopecká et al. 2023; Hossain et al. 2021; Yadav et al. 2020; Chaudhry and Sidhu 2022).

The incorporation of nanotechnology in agriculture offers numerous advantages compared to traditional agrochemicals, including regulated release kinetics, improved permeability and solubility, and mitigation of premature degradation in adverse environmental circumstances (Parra-Torrejón et al. 2021). Nevertheless, nano-agronomy, a concept that recognizes the optimized application of nanoparticles (NPs), offers innovative prospects for improving crop protection. Nanotechnology has shown significant potential in agricultural research and the agri-food sector. Nano-materials, especially solid nanoparticles and unconventional structures, offer a novel approach to sustainable crop protection, characterized by the reduced usage of control substances, diminished drift issues, and rapid efficacy against pathogenic and parasitic microorganisms. Nanotechnology offers potential for enhancing plant responses to abiotic stress and for identifying stress variables as environmental risk indicators. The utilization of innovative nano-elicitors, including metal-based nanoparticles, silicon nanoparticles, carbon-based nanoparticles, functional seed priming, and nano-carriers in agricultural science, has led to the emergence of novel insights and ongoing advancements due to their exceptional properties and efficacy (Mariyam et al. 2025; Gómez-Llorente et al. 2022).

Contemporary farming methods primarily rely on fertilizers and chemical pesticides, which, although beneficial in the short term, can lead to major environmental impacts such as soil degradation, water pollution, and biodiversity loss (Garnett et al. 2013). With the transition to sustainable agricultural techniques, the incorporation of nano-elicitors offers an alternative method to enhance resilience and reduce reliance on chemical inputs. Through the application of nanotechnology, these elicitors can boost stress tolerance features in plants such as improved root development, increased photosynthetic efficiency, and improved nutrient acquisition (Rhaman et al. 2022).

In addition, the processes behind the effect of nano-materials comprise the activation of several stress-responsive signaling pathways, like positive antioxidant enzyme regulation, osmolyte synthesis, and gene expression variations (Wahab et al. 2024). These responses are essential for developing a resilient defense mechanism in plants, enabling them to withstand physical and physiological stressors. For example, it has been shown that the use of silica nanoparticles promotes drought tolerance in several crops, facilitating osmotic adjustment and enhancing water usage efficiency (Sharf-Eldin et al. 2023).

So, besides boosting plant resistance, nano-elicitors additionally provide an opportunity to improve agricultural productivity. These nanoscale compounds can enhance biomass accumulation and performance in stressed plants by modulating growth-regulatory hormones and boosting metabolic pathways. The dual functionality of nano-elicitors—stress tolerance and growth promotion—highlights their promise as a sustainable agricultural solution (Khan et al. 2021).

Ultimately, tackling the intricate issues posed by abiotic stresses necessitates the formulation and execution of creative solutions that enhance resilience and sustain productivity. Nano-elicitors represent a revolutionary method for attaining these objectives, embodying a convergence of plant science and nanotechnology that may significantly impact the future of sustainable agriculture. This developing topic demands further examination into specific action mechanisms, appropriate application settings, and long-term environmental implications of nano-elicitor use in various agricultural contexts. Nano-elicitors are a novel category of molecules, defined by their nanometric dimensions, that are pivotal in regulating plant responses to abiotic stressors, including salt, drought, and severe temperatures (Inam et al. 2023).

The rising incidence of abiotic stressors presents a considerable risk to worldwide agricultural output, highlighting the necessity for new and sustainable strategies to improve crop resilience. Recently, nano-elicitors have emerged as a viable approach to alleviate the detrimental impacts of abiotic stress on crops. These minuscule molecules have demonstrated the ability to engage with plant physiology, activating defense mechanisms that allow crops to endure environmental challenges. These compounds encompass nanoparticles of metals and metal oxides, as well as natural biomolecules such as chitosan, humic acids, and different plant extracts. Their distinctive characteristics, including enhanced surface area, heightened responsiveness, and the capacity to infiltrate plant tissues, confer substantial advantages over conventional elicitors. Nano-elicitors can trigger signaling cascades at the cellular and molecular levels, resulting in physiological and biochemical alterations in plants, thus enhancing their resilience and productivity under adverse conditions. This work seeks to deliver an exhaustive analysis of nano-elicitors and their function in enhancing sustainable crop production under abiotic stress situations. This research also aims to enhance crop resilience and tackle urgent agricultural concerns by investigating the processes of action, obstacles, and constraints related to nano-elicitors.

12.2 Negative Impacts of Abiotic Stress on Crop Production

Plants are perpetually subjected to various environmental pressures, encompassing both biotic and abiotic factors. Abiotic stress notably impedes agricultural productivity, positioning it as a primary focus of research in plant biology. Abiotic stress denotes environmental variables that adversely impact plant growth and output. Light, water, temperature, humidity, gases, soil constituents, wind, and other factors are responsible for abiotic stress (Fig. 12.1). The likelihood of plant exposure to these elements has increasingly escalated due to global climate change. The primary kinds of abiotic stress are drought, salt, and thermal stress. Among the three, soil drought is the most harmful stress condition and constitutes a global problem (Dhaliwal et al. 2021; Dhaliwal and Williams 2022; Devi et al. 2024; Khalid et al. 2025).

Drought is a prevalent abiotic stressor that impedes plants from acquiring sufficient water necessary for their optimal growth and metabolic processes. The remaining three abiotic stresses—elevated temperature, inundation, and reduced temperature—significantly affect plant growth and agricultural yield. Elevated temperatures restrict plant growth, disrupt normal physiological functions, and induce cellular necrosis. Reduced temperature modifies standard physiological functions, membrane fluidity, and oxidative stress and compromises the structural integrity of cells (Kopecká et al. 2023; Kumari et al. 2022). Drought alters water allocation

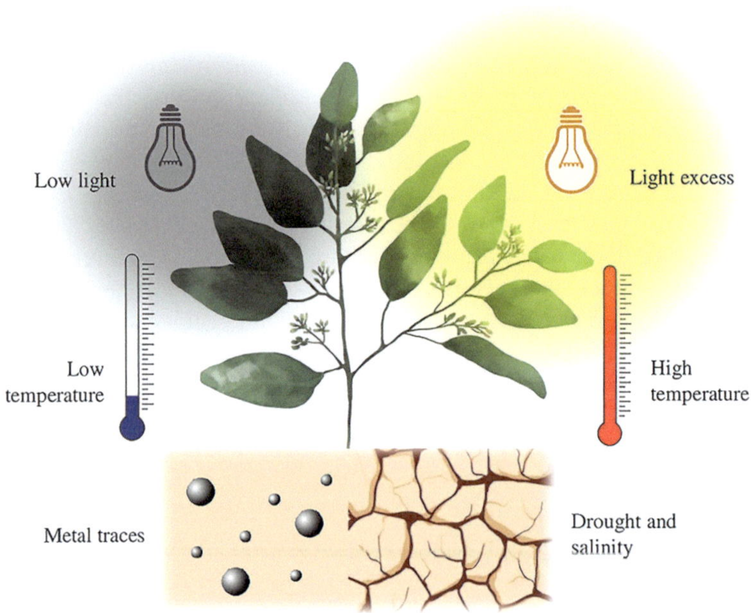

Fig. 12.1 Different abiotic stress impacts on plants (Aguirre-Becerra et al. 2022)

within plants, with roots and stems showing reduced water retention under stress. It inhibits biomass accumulation in seeds while promoting root growth as a survival response (Morris et al. 2024). In addition, drought stress severely impairs photosynthesis in cereals by causing stomatal closure, which limits CO_2 availability and triggers photorespiration, reducing carbon fixation efficiency. It also damages photosystems and blocks electron transport chains. This results in the accumulation of reactive oxygen species (ROS), which cause oxidative damage to chloroplasts and other cellular structures, further decreasing photosynthetic productivity. Plants respond by activating protective mechanisms like non-photochemical quenching to dissipate excess light energy and alternative electron pathways to maintain energy balance. Hormones such as abscisic acid modulate stomatal conductance and osmotic adjustment to enhance drought tolerance. Structural changes like altered leaf orientation and root architecture also occur to mitigate stress effects (Qiao et al. 2024; Fang et al. 2023).

Salinity stress mostly impacts the lenticel region, resulting in fruit rot. This abiotic stress situation additionally fosters the proliferation of infections. In contrast to other abiotic pressures, flooding can inhibit seed germination and reduce the likelihood of successful germination. Excessive water content and insufficient aeration disrupt the usual regulation mechanisms of plant hormones, ultimately leading to the plant's demise. The direct impact of these abiotic stresses on sustainability and the inhibition of normal plant growth would certainly result in a significant decline in agricultural production (Chaudhry and Sidhu 2022; Khan et al. 2021). High sodium ion (Na^+) concentrations in saline soils disrupt potassium (K^+) uptake, essential for enzyme activation and osmoregulation, leading to stunted root growth and inhibited seed germination. In maize, salt stress downregulates microRNA *ZmmiR169q*, which results in increased auxin production that halts primary root growth. Salt stress also activates ethylene signaling and upregulates tryptophan biosynthesis genes (*ASA1* and *ASA2*), further increasing auxin levels that inhibit root elongation (Rane et al. 2021; Caverzan et al. 2016).

Heat stress is a major abiotic factor that significantly reduces crop productivity worldwide. Its effects span from physiological disruptions at the cellular level to large-scale yield losses in the field. The severity and timing of heat events, especially during sensitive growth stages, are critical determinants of their impact. High temperatures disrupt plant growth by causing morphological, biochemical, and physiological changes. These include reduced germination rates, poor stand establishment, accelerated leaf senescence, and inhibited biomass accumulation. In crops like maize and sugarcane, heat stress leads to shorter internodes and lower overall biomass (Peverelli and Rogers 2013; Fahad et al. 2017; Morris et al. 2024). Heat stress impairs photosynthesis by damaging chloroplast membranes, reducing chlorophyll content, and causing stomatal closure. This leads to lower CO_2 assimilation and energy production. For example, in sorghum and soybean, heat stress significantly decreases photosynthetic rates, chlorophyll content, and stomatal conductance (Hasanuzzaman et al. 2013; Peverelli and Rogers 2013). It also increases oxidative stress via the production of reactive oxygen species, further damaging cellular structures (Peverelli and Rogers 2013). The reproductive phase is particularly sensitive to heat. Elevated temperatures during flowering (anthesis) and grain filling can cause flower and pod abortion, reduced pollen viability, and poor seed set, leading to substantial yield losses. For instance, in sweet corn, each additional degree day above 30 °C during anthesis reduces the yield by 2% in rainfed fields. In cereals, heat stress during flowering can result in floret sterility, poor pollen germination, and reduced grain number and weight (Fahad et al. 2017; Dhaliwal and Williams 2022). Heat stress negatively affects the quality of harvested crops. In cereals and oilseeds, it reduces starch, protein, and oil content. In barley, heat stress alters amino acid profiles and lowers carbohydrate levels, impacting malting and nutritional quality (Peverelli and Rogers 2013; Fahad et al. 2017). Severe heat waves can cause yield losses of up to 50% in affected croplands. For example, in 2021, about 4% of European cropland experienced such losses due to heat during flowering. Projections suggest that with global warming, up to 80% of European maize area could be affected, potentially reducing yields by 20% or more (Martin et al. 2024). Rainfed crops are generally more vulnerable than irrigated ones, but even irrigation cannot fully mitigate the negative effects of extreme heat (Dhaliwal and Williams 2022).

Consequently, it is imperative to understand the molecular and physiological responses of plants to abiotic stress to produce crop cultivars with enhanced tolerance to unfavorable climatic conditions. The part below discusses the molecular and genetic foundations of plant responses to drought, salinity, and temperature stress. Considering the widespread parallels in the mechanisms governing plant responses to various stresses, we will first examine generic stress response pathways, followed by a detailed analysis of the distinct pathways related to drought, salinity, and temperature stress.

12.3 Responses of Plants to Abiotic Stressors

Abiotic stressful circumstances, characterized as nonliving environmental conditions that adversely impact biological organisms, are essential determinants of plant development and survival. Drought, salinity, and temperature extremes are particularly serious stresses owing to their prevalence and the profound physiological challenges they present to plants, especially in the context of global climate change. Plants' responses to stressful conditions are intricate and multifarious, encompassing several physiological, biochemical, and molecular systems that facilitate short-term survival and long-term adaptation (Dos Santos et al. 2022).

Drought stress arises from inadequate water availability, resulting in diminished turgor pressure and impaired physiological activities, including photosynthesis, nutrient absorption, and growth. In drought situations, plants frequently undergo stomatal closure to reduce transpiration, which conserves water but dramatically restricts carbon dioxide supply, ultimately diminishing photosynthetic efficiency (Qiao et al. 2024). This response pertains to the regulation of abscisic acid (ABA), a hormone that prompts the plant to initiate drought response pathways, which may encompass alterations in root growth, osmotic adjustments via the accumulation of compatible solutes, and the activation of stress-responsive genes. The relationship between water availability and physiological response mechanisms underscores the significant impact of drought as an abiotic stress influencing plant adaptation methods across diverse settings (Priya et al. 2025).

SOS (salt overly sensitive) genes, pivotal genes in the SOS pathway, are intricately linked to the regulation of salt tolerance in plants. Huang et al. (2024) delineated the tomato SOS gene family and discovered that salt stress elevated the expression levels of most SOS genes in the leaves of 'LA0716' (*Solanum pennellii* L.), 'LA2093' (*Solanum pimpinellifolium* L.), and "Heinz 1706" (*Solanum lycopersicon* L.). The expression levels of SOS1-1, SOS1-2, SOS2-2, SOS3-3, SOS4-1, and SOS5-2 in the leaves of salt-tolerant tomato varieties ('LA0516' and 'LA1598') were elevated compared to those in salt-sensitive varieties ('LA1698' and 'LA0012') in response to salinity. Tang et al. (2024) illustrated that the trends in superoxide dismutase (SOD), peroxidase (POD), proline, and total soluble content varied in pomegranate (*Punica granatum*) under low- and high-salinity stress after 3 days. Transcription factors (TFs) are crucial in plant defense against stress by controlling the expression of pertinent genes (Yu et al. 2023). Aside from physiological characteristics, Tang et al. (2024) discovered 6571 differentially expressed genes (DEGs), including 374 transcription factors (e.g., AP2/ERF, HD-ZIP, MYB, and WRKY), in pomegranate leaves subjected to low (0.5%) and high (0.8%) salt concentrations for one and 2 days, using transcriptome sequencing.

Crop plants employ a complex array of responses to cope with heat stress, spanning morphological, physiological, biochemical, and molecular levels. These mechanisms work together to minimize damage, maintain productivity, and ensure survival under high-temperature conditions. Plants may alter leaf angle and orientation to reduce direct exposure to sunlight, thereby minimizing heat absorption and water loss. High temperatures can hasten leaf senescence and reduce the duration of grain filling, impacting the yield (Hasanuzzaman et al. 2013). Heat often inhibits root and shoot elongation, but in some cases, plants may redirect resources to root growth to access deeper soil moisture (Khanzada et al. 2025). Heat stress triggers the biosynthesis of stress-related hormones such as abscisic acid (ABA), ethylene, salicylic acid (SA), jasmonic acid, brassinosteroids, and cytokinins, which regulate stress responses and adaptation. The balance and interplay of these hormones determine the plant's ability to withstand heat. Compatible solutes like proline, glycine betaine, and trehalose accumulate to stabilize proteins and membranes and maintain cell turgor. Heat stress stimulates the production of protective enzymes and may also inhibit or denature others, impacting metabolic pathways. Plants under heat stress upregulate antioxidant enzymes (superoxide dismutase, catalase (CAT), ascorbate peroxidase (APX), and glutathione reductase) and nonenzymatic antioxidants (ascorbate, glutathione, flavonols, and α-tocopherol) to scavenge ROS and protect cellular structures (Hasanuzzaman et al. 2013). The synthesis of heat shock proteins (HSPs), regulated by heat shock transcription factors (HSFs), helps refold denatured proteins and protect cellular machinery (Khanzada et al. 2025; Zhou et al. 2022). Heat stress induces the expression of genes involved in protection, repair, and signaling, including those for osmoprotectants, detoxifying enzymes, and regulatory proteins. Signal transduction pathways involving calcium-dependent protein kinases (CDPKs), mitogen-activated protein kinases (MAPKs), nitric oxide (NO), and sugars activate stress-responsive genes (Hasanuzzaman et al. 2013). Epigenetic modifications and microRNAs (miRNAs) modulate gene expression, contributing to acquired thermotolerance and adaptation (Khanzada et al. 2025). Evolutionary adaptations, such as development of C4 or CAM photosynthetic pathways, improved antioxidant systems, and stabilized HSP expression, enhance plant resilience to environmental stress (Hasanuzzaman et al. 2013).

12.4 Nanotechnology and Agriculture

The agricultural sector faces substantial hurdles in satisfying the escalating global food demands due to climate change, water scarcity, and soil degradation. In recent years, nanotechnology has emerged as a promising instrument to improve agricultural methods and tackle these urgent challenges. Utilizing the distinctive characteristics of nano-materials, researchers and agriculturists can create novel strategies to enhance crop productivity, minimize ecological consequences, and foster sustainable agriculture (Khan et al. 2022).

Nanotechnology is a multidisciplinary domain of study and invention that entails the manipulation of materials at the nanoscale, generally characterized as ranging from 1 to 100 nanometers. At this scale, materials display distinct physical, chemical, and biological properties that vary dramatically from their bulk counterparts. Nanotechnology has been progressively utilized across multiple areas, like medicine, electronics, energy, and agriculture, to produce creative solutions and products. The distinctive characteristics of nanoparticles render them exceptionally beneficial for numerous applications. Nanoparticles can be designed to possess certain optical, electrical, or magnetic characteristics, facilitating their application in biomedical imaging, diagnostics, and therapy. Likewise, nanostructured materials can be engineered to exhibit improved mechanical strength, thermal conductivity, or catalytic activity, rendering them appropriate for energy storage, conversion, and transmission. In agriculture, nanotechnology presents several prospects for enhancing crop yields, disease resistance, and water management. Nano-materials can enhance the efficacy of fertilizers, insecticides, and irrigation systems, thereby mitigating environmental contamination and fostering sustainable farming practices. Nanoscale sensors and monitoring systems can identify crop stress, infections, and nutrient deficiencies, allowing farmers to act swiftly and reduce losses (Gupta et al. 2025).

Nano-materials can be applied as follows:

(a) *Nano-fertilizers*: Nano-fertilizers are engineered to gradually release nutrients, thus decreasing the frequency of applications and minimizing waste. This method can enhance agricultural productivity while diminishing environmental contamination.

(b) *Nano-pesticides*: Nano-pesticides are designed to specifically target pests, thereby diminishing the adverse effects on beneficial insects and lessening the emergence of pesticide-resistant pests.

(c) *Nano-biosensors*: Nano-biosensors can identify crop stress, illnesses, and nutritional deficiencies, allowing farmers to respond swiftly and reduce losses.

(d) *Nano-enhanced irrigation systems*: Nano-enhanced irrigation systems optimize water distribution, decreasing waste and enhancing crop water use efficiency.

The implementation of nanotechnology in agriculture has resulted in various novel applications and success narratives. This section presents several instances of nanotechnology's application in enhancing crop yields, disease resistance, and water management globally (Rautela et al. 2021; Neme et al. 2021).

(a) *Nano-fertilizers in India*: Indian researchers have created nano-fertilizers that enhance agricultural yields by as much as 30% while decreasing fertilizer usage by 50% (Patel et al. 2023).

(b) *Nano-pesticides in the United States*: Researchers in the United States have created nano-pesticides that specifically target pests, thereby diminishing the adverse effects on beneficial insects and minimizing environmental contamination (Xu et al. 2022).

(c) *Nano-enhanced irrigation systems in Australia*: Australian farmers are implementing nano-enhanced irrigation systems that optimize water distribution, minimizing waste and enhancing crop water productivity by as much as 25% (Saritha et al. 2022).

(d) *Nano-biosensors in China*: Chinese researchers have created nano-biosensors capable of detecting agricultural stress, illnesses, and nutritional deficits, allowing farmers to respond swiftly and reduce losses (Luo et al. 2025).

Nanotechnology possesses the capacity to transform agriculture by offering novel solutions to enhance crop yields, disease resistance, and water management. Nonetheless, challenges (scalability, regulatory frameworks, toxicity and environmental impacts, public acceptance, etc.) must be confronted to guarantee the secure and extensive implementation of nanotechnology in agriculture.

12.5 Elicitors and Nano-elicitors

Elicitor is a natural or artificially generated factor capable of enhancing the secondary metabolite accumulation in plants by the stimulation of strong activation of metabolite production pathways (Baenas et al. 2014). Nano-elicitors are substances that trigger plant responses, encompassing defense mechanisms, at the nanoscale. They have garnered interest in recent years for their prospective applications in agriculture. The word "nano-elicitor" denotes a nanoparticle or nano-material that functions as an elicitor, inducing a particular reaction in plants. Nano-elicitors can originate from diverse sources, including natural substances like plant extracts and manufactured compounds such as nanoparticles. They can be engineered to target specific plant receptors, eliciting a desired reaction. Nano-elicitors provide numerous benefits compared to conventional elicitors, such as increased efficiency, diminished toxicity, and enhanced stability. The advancement of nano-elicitors is a burgeoning domain, with current research concentrating on elucidating their mechanisms of action, refining their design, and investigating their prospective applications in agriculture. Potential applications of nano-elicitors comprise augmenting crop yields, bolstering plant resilience to diseases and pests, and mitigating the environmental impact of agricultural methods (Darbahani et al. 2024; Subhash et al. 2024; Abou El-Ela et al. 2022; Rivero-Montejo et al. 2021).

12.5.1 Definition and Characteristics

Nano-elicitors are nanoscale substances that stimulate plants to evoke advantageous responses for the improvement of growth and resistance (Lala 2021). Nano-elicitors are defined as nanoparticles that elicit a defense response when applied to organisms they are in a close proximity with, considering distance across the interface triggers the phenomenon. In the case of agriculture, the responding organisms are plants (Fares et al. 2024; Murthy et al. 2024).

In spite of the considerable amount of knowledge on the plant defense mechanisms and the characterization of a wide array of nano-elicitors already discovered, their triggering processes are still imperfectly known. Both features of the nano-elicitor and plant cells influence their interaction, which eventually triggers a defense response. Therefore, an understanding of the physicochemical characteristics that determine the defense activity of nano-elicitors is crucial to establishing the relation between the features of nano-elicitors and their agronomic effect (Murthy et al. 2024).

A nano-elicitor, which is within a given size range, normally 1–100 nm, is a particle with specific chemical, physical, and biological properties that interacts with plant recognition mechanisms and modulates defense responses (Lala 2021). The first direct imaging of a nano-elicitor in interaction with a plant surface membrane receptor was conducted in 1999. The atomic force microscopy (AFM) imaged Lateral Test Band–Gold Nanoparticle (LTB-GNP) aggregation on the plasma membrane of soybean, which enables G protein activation. Plasma membrane localization is indicated by the AFM images of nano-elicitors, which disperse extensively but induce unrestrained responses. The result demonstrates that nano-elicitors associate with generic recognition sites in the plasma membrane of a plant. Size–activity fingerprinting of nano-elicitors involves generic recognition at the plant plasma membrane. The study uncovers the nanostructure–function relationship in the molecular dialogue between plants and nano-elicitors (Anjum et al. 2019; Parra-Torrejón et al. 2021).

Nano-elicitors play a crucial role in designing applications, with size being a key parameter. Studies have shown that phytotoxicity is associated with NPs of <70 nm, while NPs are toxic to bacteria at <5 nm. Smaller NPs are more efficient in translocating across plant plasma membranes and exhibit greater reactivity. The Lennard-Jones potential between the NP and the lipid bilayer determines a maximum active radius at 5.1 nm for NPs. The phytotoxicity increases with decreasing NP size at <70 nm. The plant–NP interaction is size-dependent, with uptake and translocation being NP-size-dependent. The plant–NP interface and size dependence of the NP diffusion coefficient in plants remains unknown. The development of nano-elicitors with tailored surface chemistry and novel shapes is expanding. This information is essential for developing smart nanoparticular delivery systems for translation research and agri-biotechnology (Fares et al. 2024; Nazeer et al. 2023; Zhang and Su 2024; Zhou et al. 2023).

12.5.2 Types of Nanoparticles

Nano-elicitors are nanoscale substances (often <100 nm) employed to provoke plant defense mechanisms and enhance stress resilience. They exhibit distinctive physicochemical characteristics, including a high surface area–to-volume ratio, increased reactivity, and the capacity to infiltrate plant cells and tissues. Nano-elicitors have been categorized broadly according to their composition and function in Table 12.1.

12.5.3 Synthesis Methods of Nanoparticles

Nano-elicitors are synthesized via biological (green) and chemical approaches, each offering distinct advantages in terms of scalability, cost, and nanoparticle properties. Below is a detailed breakdown of these methods, supported by experimental evidence from recent studies.

12.5.3.1 Green Synthesis

- *Plant-Mediated Synthesis*
  - *Mechanism*: Phytochemicals (e.g., polyphenols and flavonoids) in plant extracts reduce metal ions and stabilize nanoparticles.
  - *Example*
    - *Selenium Nanoparticles (SeNPs)*: SeNPs are synthesized using garlic (*Allium sativum*) extract. Sodium selenite solution is reduced by garlic phytochemicals, forming brick-red SeNPs (20–100 nm) after 4–6 days of incubation (Ali et al. 2023).

Table 12.1 Categorization of nano-elicitors according to their composition and function

Nano-elicitor type	Example	Properties	Applications	Plant response	References
Metal-based nano-elicitors	Silver (Ag), gold (Au), copper (Cu), zinc (Zn), iron (Fe), and titanium (Ti) nanoparticles	High catalytic activity and antimicrobial properties Ability to generate reactive oxygen species (ROS) at controlled levels to trigger plant defense Facilitate nutrient delivery and uptake	AgNPs and CuNPs act as nano-pesticides against fungal and bacterial pathogens ZnO and FeO NPs improve nutrient efficiency and photosynthesis	Enhanced secondary metabolite production, improved antioxidant enzyme activities, and pathogen resistance	Anjum et al. (2019); Hatami et al. (2019); Lala (2021)
Metal oxide nano-elicitors	Zinc oxide (ZnO), titanium dioxide (TiO_2), iron oxide (Fe_2O_3/Fe_3O_4), silicon dioxide (SiO_2), and cerium oxide (CeO_2)	Photocatalytic and antioxidant capabilities Stability under environmental conditions Ability to modulate ROS signaling and protect chlorophyll	TiO_2 NPs enhance photosynthetic efficiency and drought tolerance SiO_2 NPs strengthen cell walls and improve salinity tolerance CeO_2 NPs mimic antioxidant enzymes, scavenging ROS	Improved stress tolerance via enhanced antioxidant defense and structural reinforcement	Anjum et al. (2019); Hatami et al. (2019); Lala (2021)
Carbon-based nano-elicitors	Carbon nanotubes (CNTs), graphene oxide, and fullerenes	High mechanical strength and electrical conductivity Facilitate water and nutrient transport Can absorb and immobilize heavy metals	CNTs improve seed germination and root elongation under drought Graphene-based materials aid in pollutant remediation and stress mitigation	Enhanced growth, water uptake, and detoxification capacity	Anjum et al. (2019); Hatami et al. (2019); Salem et al. (2023)

(continued)

Table 12.1 (continued)

Nano-elicitor type	Example	Properties	Applications	Plant response	References
Polymeric nano-elicitors	Chitosan nanoparticles, alginate nanoparticles, cellulose nanofibers, and lignin nano-capsules	Biodegradable, biocompatible, and nontoxic Capable of controlled release of agrochemicals (fertilizers, pesticides, and elicitors) Can stimulate plant immune responses	Chitosan NPs induce systemic acquired resistance (SAR) and osmolyte accumulation Cellulose-based hydrogels improve soil water retention	Enhanced defense gene expression, improved drought and cold tolerance	Parra-Torrejón et al. (2021); Salem et al. (2023)
Hybrid and functionalized nano-elicitors	Metal or metal oxide NPs functionalized with bioactive molecules (e.g., methyl jasmonate (MEJ) and salicylic acid) or coated with polymers	Targeted delivery and sustained release of elicitors Enhanced stability and bioavailability	Calcium phosphate NPs loaded with methyl jasmonate for improved phytoalexin production Levan-coated CeO_2 NPs for enhanced antioxidant activity	Amplified stress signaling and metabolite production with lower doses	Spanos et al. (2021); Easwaran et al. (2024)

  - *Silver Nanoparticles (AgNPs)*: *Foeniculum vulgare* (fennel) extracts reduce Ag^+ ions to AgNPs, enhancing secondary metabolite production (Hatami et al. 2019).

- *Microbial Synthesis*
  - *Bacteria*
    - *Rhodopseudomonas capsulata* and sulfate-reducing bacteria (*Desulfovibrio desulfuricans*) produce intracellular/extracellular nanoparticles via enzymatic reduction.
    - *Lactobacillus* spp. synthesize selenium and silver nanoparticles with antimicrobial properties (Karunakaran et al. 2023).
  - *Fungi*
    - *Fusarium oxysporum* extracellularly synthesizes Au/Ag nanoparticles (15–30 nm) using reductase enzymes (Karunakaran et al. 2023).
    - *Penicillium fellutanum* produces spherical AgNPs (5–25 nm) in mangrove sediments (Karunakaran et al. 2023).

- *Yeast*
    - *Saccharomyces cerevisiae* synthesizes selenium (30–100 nm) and palladium (10–100 nm) nanoparticles via glutathione-mediated detoxification pathways (Karunakaran et al. 2023).

12.5.3.2 Chemical Synthesis

(a) *Coprecipitation*

- *Calcium Phosphate Nano-elicitors*: Methyl jasmonate (MeJ) is functionalized onto calcium phosphate nanoparticles (CaP NPs) via coprecipitation. This method ensures sustained release and prolonged leaf retention, enhancing phytoalexin production in grapes (Parra-Torrejón et al. 2021).

(b) *Functionalization*

- *Polymer-Based NPs*: Chitosan or alginate nanoparticles are functionalized with elicitors (e.g., salicylic acid) to improve stability and targeted delivery.

12.5.3.3 Hybrid and Emerging Methods

- *Nanocomposites*
    - *Nano-MeJ*: Calcium phosphate nanoparticles loaded with MeJ show ten times higher efficiency than free MeJ, enabling low-dose applications in vineyards (Parra-Torrejón et al. 2021).
- *Metal-Doped NPs*
    - Yttrium-doped Fe_2O_3 NPs enhance root architecture for drought resilience (Karunakaran et al. 2023).

Despite the considerable advantages of green nanoparticle manufacturing processes compared to conventional methods, they nevertheless exhibit certain limitations. The benefits and constraints of green synthesis are depicted in Table 12.2.

12.6 Nano-elicitors' Mechanisms of Action in Alleviating Abiotic Stress in Plants

Nano-elicitors, including metal, metalloid, carbon-based, and polymer-based nanoparticles (NPs), mitigate abiotic stress in plants through multifaceted mechanisms. These mechanisms span molecular, biochemical, and physiological levels, enhancing stress tolerance while maintaining crop productivity.

Table 12.2 Advantages and limitations of nano-elicitor synthesis methods

Method	Advantages	Limitations	References
Plant-mediated (green)	Eco-friendly, cost-effective, scalable, biocompatible, minimally toxic by-products	Limited control over size/shape, variability in plant extract composition	Hano and Abbasi (2021); Dutta et al. (2021); Ali et al. (2023)
Microbial (green)	High enzymatic specificity, low toxicity, sustainability	Slow synthesis, complex downstream processing, potential pathogenicity	Hano and Abbasi (2021); Ali et al. (2023)
Chemical	Precise control over size/shape, high reproducibility, rapid synthesis	Toxic reagents (e.g., sodium borohydride), environmental hazards, high cost	Hano and Abbasi (2021); Omran and Baek (2021); Dutta et al. (2021)
Physical (top-down)	No chemical solvents, suitable for industrial-scale production	High energy consumption, broad size distribution, equipment-intensive	Dutta et al. (2021)
Hybrid (e.g., coated NPs)	Enhanced stability, targeted delivery, improved biocompatibility	Higher production costs, potential reduction in therapeutic efficacy	Omran and Baek (2021)

12.6.1 Reactive Oxygen Species (ROS) Scavenging and Antioxidant Defense

Abiotic stresses induce oxidative damage via ROS accumulation. Nano-elicitors counteract this by the following actions:

- *Direct ROS Neutralization*: CeO_2 NPs mimic antioxidant enzymes, while Fe_2O_3 and ZnO NPs reduce ROS levels by enhancing superoxide dismutase (SOD), catalase (CAT), and ascorbate peroxidase (APX) activity (Aguirre-Becerra et al. 2022; Sarraf et al. 2022; Tortella et al. 2023).
- *Lipid Peroxidation Reduction*: Silica NPs lower malondialdehyde (MDA) levels, protecting membrane integrity (Tortella et al. 2023).
- *ROS Signaling Priming*: Low-dose AgNPs and chitosan NPs activate early ROS signaling, triggering systemic acquired resistance (SAR) without causing oxidative damage (Singh et al. 2024, 2025).
- *Modulation of the Content and/or Composition of Secondary Metabolites*: The increase in secondary metabolite content is typically associated with the activity of enzymes involved in oxidative defense and secondary metabolism, as well as with the transcription of their respective genes (Lala 2021) (Fig. 12.2).

12.6.2 Gene Regulation and Stress Signaling

NPs modulate stress-responsive genetic pathways:

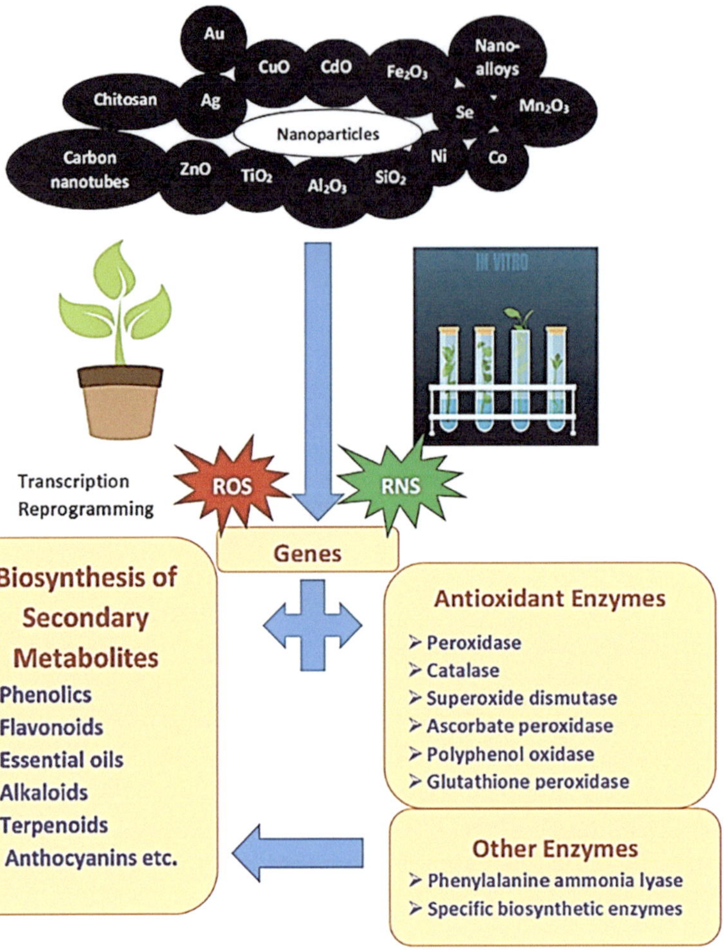

Fig. 12.2 Schema illustrating the elicitation of secondary metabolites by currently known nanoparticles in plants

- *Upregulation of Defense Genes*: SiO_2 NPs enhance SOS1 expression for Na^+/K^+ homeostasis, while TiO_2 NPs increase RuBisCO and starch synthase activity under drought (Sarraf et al. 2022; Singh et al. 2025).
- *miRNA and Phytohormone Regulation*:
  - ZnNPs downregulate *ZmmiR169q* in maize to stabilize root growth under salinity (Aguirre-Becerra et al. 2022).
  - AgNPs and Al_2O_3 NPs regulate ABA and ethylene pathways, improving stomatal closure and osmotic adjustment (Sarraf et al. 2022; Singh et al. 2024).

12.6.3 Nutrient Uptake and Ion Homeostasis

NPs improve nutrient efficiency and ion balance:

- *Na^+ Exclusion*: SiO_2 and FeO NPs enhance K^+ uptake while reducing Na^+ accumulation in roots and shoots under salinity (Sarraf et al. 2022; Singh et al. 2025).
- *Heavy Metal Detoxification*: ZnO NPs and Fe NPs reduce arsenic (As) and lead (Pb) uptake by immobilizing metals in the soil (Sarraf et al. 2022; Singh et al. 2025).

12.6.4 Enhanced Photosynthesis and Stomatal Regulation

- *Chlorophyll Protection*: TiO_2 NPs prevent chlorophyll degradation under heat and drought, while CuO NPs stabilize photosynthetic electron transport (Sarraf et al. 2022; Singh et al. 2025).
- *Stomatal Optimization*: Fe_3O_4 NPs improve stomatal conductance and CO_2 assimilation by modulating ABA signaling (Sarraf et al. 2022).

12.6.5 Osmoprotectant Synthesis

NPs stimulate osmolyte production to maintain cellular turgor:

- *Proline and Glycine Betaine*: Chitosan and carbon-based NPs upregulate proline synthesis genes (*P5CS* and *P5CR*) under drought and salinity (Singh et al. 2025).
- *Soluble Sugars*: ZnO NPs enhance UDP Uridine DiPhospahte-glucose pyrophosphorylase activity, boosting sucrose accumulation for osmotic balance (Sarraf et al. 2022).

12.6.6 Structural Modifications

- *Cell Wall Fortification*: SiO_2 NPs accumulate in the cell wall matrix, altering wall structure and plant stress responses. SiO_2 NPs deposition in cell walls could improve mechanical strength or change permeability. SiO_2 NPs deposit in cell walls, reducing water loss and pathogen entry under drought (Singh et al. 2025; Tortella et al. 2023).
- *Root Architecture Remodeling*: Yttrium-doped Fe_2O_3 NPs promote lateral root growth for deeper water access (Sarraf et al. 2022).

12.6.7 Phytohormone Modulation

NPs interact with hormonal pathways to coordinate stress responses:

- *Salicylic Acid (SA) Pathway*: Silica NPs activate SA-dependent *PR* genes, enhancing systemic resistance (Tortella et al. 2023).
- *Brassinosteroid Signaling*: CeO_2 NPs upregulate *BZR1* transcription factors, improving heat tolerance (Singh et al. 2024).

Table 12.3 summarizes the mechanisms of some nano-elicitors in alleviating abiotic stress in the production of some strategic crops.

Table 12.3 Some nano-elicitors and their roles in alleviating abiotic stress

Nanoparticles	Stress targeted	Mechanism	Example crops	References
SiO_2	Drought, salinity	Enhances cell wall integrity, ROS scavenging, Na^+/K^+ homeostasis	Rice, wheat	Aguirre-Becerra et al. (2022); Rehman et al. (2024); Singh et al. (2025)
ZnO	Drought, heat	Activates antioxidant enzymes (SOD and CAT), stimulates osmolyte synthesis (proline)	Maize, soybean	Aguirre-Becerra et al. (2022); Rehman et al. (2024); Singh et al. (2025)
FeO/Fe_3O_4	Salinity, heavy metal contamination	Improves Fe nutrition, reduces ROS, detoxifies heavy metals via chelation	Tomato, rapeseed	Rehman et al. (2024); Singh et al. (2025)
TiO_2	Drought, UV	Protects chlorophyll, enhances photosynthetic efficiency, regulates stress-responsive genes	Wheat, sorghum	Rehman et al. (2024); Cao et al. (2025)
CeO_2	Heat, oxidative stress	Mimics antioxidant enzymes (CAT), neutralizes ROS	General crops	Rehman et al. (2024)
Chitosan	Drought, cold	Regulates stress-responsive genes (e.g., *PR* genes), enhances osmoprotectant synthesis	Tea, pepper	Lala (2021); Rehman et al. (2024); Singh et al. (2025)
Carbon nanotubes (CNTs)	Drought, heavy metal contamination	Enhances water transport, reduces metal toxicity via adsorption	Various crops	Rehman et al. (2024); Singh et al. (2025)
Hydrogel NPs	Drought	Improves water retention, ensures gradual water release	General crops	Rehman et al. (2024)

12.7 Applications of Nano-elicitors in Abiotic Stress Resilience in Some Strategic Crop Cultures

Crop plants often encounter biotic and abiotic stressors, resulting in substantial reductions in their production potential. The utilization of phyto-immune activators, such as nano-elicitors, has garnered significant interest in addressing these restrictions. These nano-elicitors are secure, sustainable, regenerative, and an environmentally beneficial method for plant growth and defense against various pressures. They are essential for developing resistance to various challenges, thereby modulating phytohormones, mineral elements, and defense-related genes. The application of nano-elicitors activates the host's immune system for several signaling molecules, leading to changes in the cell wall and inducing lignin phenolic buildup, thus boosting systemic acquired resistance and promoting early active defense responses. The expression induced by phytohormones activates downstream defense-related genes, thus establishing a fundamental defense system that may serve as a method for resisting diverse biotic and abiotic challenges. Moreover, the utilization of nano-elicitors enhanced the expression patterns in both primary and secondary metabolite pathways (Jadoon et al. 2024; Humbal and Pathak 2023).

In defense signaling, the secondary metabolite medicarpin, generated from chalcone, and stilbenes were synthesized in response to pathogen attacks. The administration of exogenous nano-elicitors markedly enhanced photosynthetic activities, correlated with the response of reactive oxygen species, and considerably elevated antioxidant-related enzymes to mitigate oxidative stress induced by diverse abiotic stressors. The intracellular antioxidant defense system was effectively activated, resulting in improved cell membrane stability, photosynthetic efficiency, and carbon balance, which may facilitate natural product biosynthesis and enhance plant productivity under diverse environmental constraints. In terms of plant nutrition, the nanoparticle-treated plants exhibited significant mineral buildup, including manganese, copper, and iron, alongside enhancements in photosynthesis, glucose metabolism, and nitrogen metabolism. By integrating phytohormone signaling, enhancing the accumulation of vital mineral elements, and promoting the production of defense-related secondary metabolites, nanoparticles played a proactive role in the defense response within strategies that enhance systemic acquired resistance, thereby activating several stress-resistance genes with physiological and molecular significance (Xu et al. 2022).

Seed priming is a technique involving regulated seed hydration succeeded by desiccation. It is a pre-sowing treatment that initiates germination processes while inhibiting radicle emergence. This treatment enhances the consistency and speed of germination. It enhances seedling establishment, stress resilience, and metabolic activity and augments nutrient utilization efficiency. The application of nano-elicitors in seed priming augments the activity of defense enzymes throughout the seedling phase, resulting in heightened resistance to abiotic stress. Priming seeds may prolong the duration until harvest, although this delay is compensated by enhanced seedling establishment, earlier blooming, accelerated phenological development, uniform flowering and fruit maturation, and vigorous plants. Seed coating can modify the seed microbiome by promoting the colonization of advantageous bacteria from the rhizosphere on the seed and, subsequently, the young seedlings. The application of an external chemical to the seed serves as a safeguard against infections. Seed coat encapsulation technology is a novel and eco-friendly method that envelops individual seeds with a precise layer of diverse beneficial chemicals using natural binders. The seed entity, comprising the embryonic axis and cotyledons, is enveloped in a nanocomposite that offers protection and nourishment to the developing seedling in unfavorable environmental conditions. The application of nano-elicitors in seed coating has demonstrated the ability to safeguard ionic concentration inside cells, maintain seed vitality, support post-germination seedling development, and augment root hair density and microvillus characteristics in roots, thus enhancing resistance to abiotic stress (Murthy et al. 2024).

Foliar applications exhibit considerable potential. This is an advantageous method to enhance water availability while simultaneously supplying tissues with the necessary strength in adequate quantities and at the appropriate time. These applications are cost-effective, eco-friendly, user-friendly, and compatible with supplementary items utilized in the field. The efficacy of nano-elicitors is enhanced by the superior absorption of these particles by the leaf. Similar to their larger counterparts, an environment devoid of wind and characterized by elevated relative humidity after the application of the spray facilitates the efficacy of these particles. The root-involving alternative is beneficial, featuring a structure that is well-adapted to the bioactive chemical. The direct delivery of nano-elicitors to specific roots may facilitate precise reactions among them. While the drawbacks of transport must be addressed, the association of the nano-elicitor/compound is not particularly advantageous due to the plants' internal isolation at the site. Dosage is not governed by the same protective mechanisms. The formulation of mixes and suspensions by combining the drug with solid particles occupying a well-defined internal space will enhance this condition. Regular irrigation to maintain root separation from the compound must be implemented. It is also necessary to optimize systems by integrating them into broader operations (Darbahani et al. 2024; Jadoon et al. 2024; Inam et al. 2023).

Seven distinct crop species, specifically *Brassica campestris*, *Brassica juncea*, *Robinia pseudoacacia*, *Oryza sativa*, *Sorghum bicolour*, *Vigna radiata*, and *Vigna unguiculata*, have been examined for their growth and biochemical, molecular, and physiological responses to the application of BiNPs at varying dosages under controlled experimental conditions. The findings indicated that Bi-NPs generally elicited a favorable effect on growth indices by improving the photosynthetic efficiency of the plants. The *Vigna* species exhibited superior growth and metabolic responses to these nanoparticles. The gene expression profiling of some biomass-related genes indicated that expression was downregulated at elevated concentrations of nanoparticles on the seventh day after sowing. Nonetheless, the expression was elevated when the seedlings were exposed to these two examined nanoparticles (Dousari et al. 2025; Kaur et al. 2024; Sarwar et al. 2024; Zacchini 2024; Gupta et al. 2018; Syu et al. 2014).

Given that the application of nano-elicitors on the majority of agronomic crops has not been documented, four supplementary field experiments were conducted to expand and corroborate the findings and enhance the understanding of the mechanisms behind these induced resistances. The primary conclusion was that the impact of TMZn on the 'Monte Verde' series tomato commercial varieties was consistent across varying circumstances and locations. The results were particularly promising for two leguminous commercial crops: faba bean and pea, demonstrating their stimulating and biostimulant properties. All of these practical experiences demonstrate significant possibilities for the application of TMZn in agriculture. A pilot experiment on the heat resistance of tomatoes was conducted; however, no definitive results were achieved due to a significant weather event that transpired prior to the initiation of treatment with TMZn (Sajyan et al. 2025; Khan et al. 2021; Moghadas et al. 2021).

The recent reduction in the technical implementation of comprehensive research methodologies utilized in this experiment on sesame has led to an initial investigation of the impact of nanoparticles on this drought-sensitive and commercially valuable crop. The application of nanoparticles to sesame seeds at approved dosages before sowing significantly improved the crop's tolerance to water stress, particularly during the flowering and reproductive phases of plant development, which are known to be highly sensitive to drought. The noted improvement in the physicochemical condition of sesame plants, linked to the elicited adaptive responses to drought through induced resistance, strongly endorses the subsequent field trials for the comprehensive implementation of these promising practices in commercial sesame cultivation (Hasani et al. 2023; Mahdavi Khorami et al. 2020).

Nano-elicitors applied at various crop growth stages improve stress tolerance by enhancing physiological functions, antioxidant defenses, nutrient homeostasis, and grain or fruit quality, offering a promising approach for sustainable crop production under abiotic stress conditions. Table 12.4 summarizes the effects of the application of nano-elicitors on abiotic stress resilience at different stages of growth during the production of some strategic crops.

12.8 Challenges and Future Directions

With the advent of modern biotechnology, the application of nano-elicitors in food production raises concerns over their potential impact on the environment and human health. The extensive manufacturing of metallic nanoparticles, including silver and copper, can significantly impact the environment. Consequently, it is prudent to synthesize nanoparticles with eco-friendly plant extracts. Phytosynthesized nanoparticles exhibit catalytic behavior comparable to that of chemogenically synthesized nanoparticles, making them suitable for elicitor activity, perhaps offering a safer alternative for the environment. Concentration is a significant issue since

Table 12.4 The application of nano-elicitors for abiotic stress resilience at different growth stages of several strategic crops

Crop type	Growth stage	Type of nano-elicitor	Abiotic stress targeted	Effects	References
Wheat	Seedling stage	Zinc oxide (ZnO) NPs, carbon nanotubes (CNTs)	Drought, salinity	Enhanced seed germination, root elongation, antioxidant enzyme activation, improved water uptake	Khan and Upadhyaya (2019); Muhammad et al. (2022); Abbas et al. (2023); Zaman et al. (2025)
	Vegetative stage	Titanium dioxide (TiO_2) NPs, ZnO NPs, iron oxide (Fe_3O_4) NPs, silicon (Si) NPs	Drought, heat, salinity	Maintained chlorophyll content, improved photosynthesis, enhanced antioxidant defense, nutrient uptake	Aguirre-Becerra et al. (2022); Banerjee and Roychoudhury (2022); Grewal et al. (2024)
	Reproductive stage	ZnO NPs, Fe_3O_4 NPs, chitosan NPs	Heat, drought, salinity	Stabilized pollen viability, improved grain filling, biofortification (zinc enrichment), stress tolerance	Aguirre-Becerra et al. (2022); Banerjee and Roychoudhury (2022); Muhammad et al. (2022)
	All growth stages	Cerium oxide (CeO_2) NPs, polymer-based NPs (chitosan, hydrogels)	Drought, salinity, heat	ROS scavenging, osmoprotectant synthesis, enhanced drought and salinity tolerance	Banerjee and Roychoudhury (2022); Muhammad et al. (2022); Zaman et al. (2025)
Barley	Seedling stage	Zinc oxide (ZnO) NPs	Salinity, oxidative stress	Enhanced seed germination, shoot/root elongation, modulation of miRNA expression, increased ROS signaling	Plaksenkova et al. (2020)

elevated levels of nanoparticles can interact with flora and fauna, potentially resulting in toxicity at the cellular and molecular levels. Conducting more comprehensive research on the stability of nanoparticles and their toxicological effects is crucial, as minimal investigation has been undertaken to assess the fate and behavior of nanoparticles during their uptake by plants. Investigating the molecular characteristics of nano-elicitors is crucial for comprehending their mechanisms of action within plant cells. To optimize their effects, a comprehensive understanding of the interaction between nanoparticles and the plant is essential. Thorough research is crucial to elucidate the role of nanoparticles in signaling pathways, identify the constituents inside the nanoparticles that elicit plant responses, and assess their stability within a cell. Following the design of the nanoparticles, it is crucial to assess their impact on the subsequent postharvest processes that a plant may experience.

Meticulous characterization of nanoparticle properties is essential for assessing their dispersal effects, ensuring that they do not adhere to the consumable portions of plants. Furthermore, preliminary studies may be conducted to evaluate the prospective applications of nanotechnology in agriculture. Despite the numerous challenges and obligations that must be addressed to fully leverage nano-elicitors in agricultural crop management, it is evident that these substances could enable a shift toward more sustainable crop management by enhancing plant tolerance to stress (Kumari et al. 2022; Pandita 2022).

The potential of nanotechnology in agriculture is rapidly increasing globally, leading to numerous formulations with desirable qualities for mitigating abiotic stresses in plants. The utilized materials comprise metallic nanoparticles, including silver, gold, zinc, and nano-clays, which are predominantly applied by the foliar spray method. The efficacy and potential of nanoparticles in agricultural applications mostly hinge on their compliance with regulatory standards, safety considerations, ethical implications, environmental impact, and societal difficulties. As of now, the utilization of agricultural nanoproducts is limited; numerous regulations have been established by various nations to oversee and manage the application of nanotechnology in this sector, although few are actively implemented. The toxicological impacts of nanoceuticals that may accumulate in soil–aquatic systems on human health and the environment due to primary and secondary releases are significant problems that require more evaluation. Essential factors for sustainability in the agricultural market include regulations that address the primary international concerns regarding safety, evaluation, and risk characterization of nano-materials utilized in nano–agricultural formulations; thus, regulatory agencies must actively consider these aspects (Azameti and Imoro 2023; Ernst et al. 2023; Cao and Wang 2022).

Various problems may hinder the extensive implementation of nano-elicitors in agricultural output. Addressing these factors is essential for surmounting the problems linked to the extensive implementation of nano-elicitors in agricultural production. The challenges encompass the following:

- Insufficient comprehension of the long-term implications of nano-elicitors on environmental and human health.
- Deficient regulatory frameworks overseeing the application of nano-elicitors in agriculture.
- Elevated production costs, rendering nano-elicitors less competitive compared to traditional agricultural products.
- Restricted market availability of nano-elicitors, obstructing access to these products.
- Absence of standardization in the production and application of nano-elicitors, which may result in inconsistent outcomes.
- Limited awareness and understanding of nano-elicitors among farmers and agricultural professionals, impeding adoption.
- Potential risks linked to the use of nano-elicitors, including unintended consequences on nontarget organisms or the emergence of resistance in target pests or pathogens.

- Further research and development are necessary to comprehensively elucidate the mechanisms of action of nano-elicitors and enhance their efficacy and safety (Rani et al. 2025; Kumar et al. 2022, 2024; Rajpal et al. 2024; Khan et al. 2021).

The use of nano-elicitors in crop production may pose potential risks to the environment, human health, and the crops themselves. Some of the possible risks associated with the use of nano-elicitors include the following:

- Unintended effects on nontarget organisms, such as beneficial insects or microorganisms
- Accumulation of nanoparticles in soil, water, or air, potentially leading to environmental pollution
- Potential toxicity to humans or animals through ingestion or inhalation of nanoparticles
- Development of resistance in target pests or pathogens, reducing the efficacy of nano-elicitors over time
- Unknown long-term effects on soil health and fertility
- Potential disruption to ecosystems and biodiversity
- Inadequate regulation and oversight of nano-elicitor use, potentially leading to misuse or overuse (Kalia and Sreelakshmi 2025; Sajyan et al. 2025; Jadoon et al. 2024; Upadhyay et al. 2022; Anjum et al. 2019; Jain et al. 2018)

Multiple study domains require additional investigation to comprehensively clarify the potential of nano-elicitors in sustainable crop production. These comprise the following:

1. *Examining the Mechanisms of Action*: Clarifying the molecular mechanisms that govern the interaction between nano-elicitors and plant physiology is essential for comprehending their mode of action and pinpointing possible areas for enhancement.
2. *Optimizing Nano-elicitor Formulations*: Creating nano-elicitor formulations customized for certain crop kinds and stress conditions is crucial for enhancing their effectiveness and minimizing potential hazards.
3. *Evaluating Environmental Fate and Toxicity*: Examining the environmental fate and potential toxicity of nano-elicitors is essential for guaranteeing their safe and responsible application in agriculture.
4. *Assessing Nano-elicitor Interactions with Additional Agricultural Inputs*: Investigating the interactions between nano-elicitors and other agricultural inputs, including fertilizers and pesticides, is essential for comprehending their potential effects on agricultural ecosystems.
5. *Formulating Scalable and Economical Production Techniques*: Formulating scalable and economical techniques for the manufacturing of nano-elicitors will be crucial for their feasible integration into agriculture on a large scale.

6. *Exploring the Possibilities of Nano-elicitor-Based Precision Agriculture*: Investigating the application of nano-elicitors to precisely target specific stress responses in plants will be a significant focus of future research.
7. *Investigating the Potential of Nano-elicitors in Integrated Pest Management*: Analyzing the role of nano-elicitors in integrated pest control methods is essential for comprehending their contribution to sustainable agricultural practices.
8. *Evaluating the Socioeconomic Impacts of Nano-elicitor Adoption*: Analyzing the socioeconomic effects of nano-elicitor adoption, particularly their potential influence on farmer livelihoods and rural communities, is crucial for comprehending their wider implications for sustainable agriculture (Darbahani et al. 2024; Pandita 2022).

12.9 Conclusion

The incorporation of nano-elicitors into sustainable agriculture offers an effective approach to alleviate the detrimental impacts of abiotic conditions, including severe temperatures, drought, salinity, and nutrient deficits. By modulating plant physiological processes and activating defense systems, nano-elicitors present a promising avenue for improving agricultural resilience and output. Notwithstanding their potential, issues, including comprehending their long-term environmental impact and overcoming regulatory concerns, persist as substantial barriers to their widespread implementation. This highlights the need for focused research to clarify the mechanics of nano-elicitors and investigate their wider uses in agriculture. Ongoing innovation and advancement are essential to maximize the potential of nano-elicitors, guaranteeing their effective contribution to global food security and sustainable agricultural practices under abiotic stress conditions. Recent studies emphasise that fostering interdisciplinary collaboration is essential for overcoming complex challenges and maximising the impact of disruptive technologies. Effective collaboration brings together diverse expertise, addresses technical, social, and regulatory barriers, and enables innovation across sectors.

References

Abbas SF, Bukhari MA, Raza MAS, Abbasi GH, Ahmad Z, Alqahtani MD, Almutairi KF, Abd Allah EF, Iqbal MA (2023) Enhancing drought tolerance in wheat cultivars through nano-ZnO priming by improving leaf pigments and antioxidant activity. Sustainability 15(7):5835

Abou El-Ela AS, Ntiri ES, Munawar A, Shi XX, Zhang C, Pilianto J et al (2022) Silver and copper-oxide nanoparticles prepared with GA3 induced defense in rice plants and caused mortalities to the brown planthopper, Nilaparvata lugens (Stål). NanoImpact 28:100428

Aguirre-Becerra H, Feregrino-Pérez AA, Esquivel K, Perez-Garcia CE, Vazquez-Hernandez MC, Mariana-Alvarado A (2022) Nanomaterials as an alternative to increase plant resistance to abiotic stresses. Front Plant Sci 13:1023636

Ali A, Mashwani Z-u-R, Raja NI, Mohammad S, Luna-Arias JP, Ahmad A, Kaushik P (2023) Phytomediated selenium nanoparticles and light regimes elicited *in vitro* callus cultures for biomass accumulation and secondary metabolite production in *Caralluma tuberculata*. Front Plant Sci 14:1253193

Anjum S, Anjum I, Hano C, Kousar S (2019) Advances in nanomaterials as novel elicitors of pharmacologically active plant specialized metabolites: current status and future outlooks. RSC Adv 9(69):40404–40423

Azameti MK, Imoro AW (2023) Nanotechnology: a promising field in enhancing abiotic stress tolerance in plants. Crop Design 2(2):100037

Baenas N, García-Viguera C, Moreno DA (2014) Elicitation: a tool for enriching the bioactive composition of foods. Molecules 19(9):13541–13563

Banerjee A, Roychoudhury A (2022) Explicating the cross-talks between nanoparticles, signaling pathways and nutrient homeostasis during environmental stresses and xenobiotic toxicity for sustainable cultivation of cereals. Chemosphere 286(Pt 3):131827

Cao X, Wang Z (2022) Application of nano-agricultural technology for biotic stress management: mechanisms, optimization, and future perspectives. Environ Sci Nano 9(12):4336–4353

Cao Y, Turk K, Bibi N, Ghafoor A, Ahmed N, Azmat M, Ahmed R, Ghani MI, Ahanger MA (2025) Nanoparticles as catalysts of agricultural revolution: enhancing crop tolerance to abiotic stress: a review. Front Plant Sci 15:1510482

Caverzan A, Casassola A, Brammer SP (2016) Antioxidant responses of wheat plants under stress. Genet Mol Biol 39(1):1–6

Chaudhry S, Sidhu GP (2022) Climate change regulated abiotic stress mechanisms in plants: a comprehensive review. Plant Cell Rep 41(1):1–31

Darbahani M, Ghiyasi MR, Rahaie M (2024) Nanoparticles as new elicitors for the production of bioactive and phytochemicals in vitro and in vivo plant culture. Phytochem Rev 24:3179–3203. https://doi.org/10.1007/s11101-024-10022-4

Devi L, Kushwaha P, Ansari TM, Kumar A, Rao A (2024) Recent trends in biologically synthesized metal nanoparticles and their biomedical applications: a review. Biol Trace Elem Res 202(7):3383–3399

Dhaliwal DS, Williams MM (2022) Evidence of sweet corn yield losses from rising temperatures. Sci Rep 12:18218

Dhaliwal SS, Sharma V, Shukla AK, Verma V, Behera SK, Singh P, Alotaibi SS, Gaber A, Hossain A (2021) Comparative efficiency of mineral, chelated and nano forms of zinc and iron for improvement of zinc and iron in chickpea (Cicer arietinum L.) through biofortification. Agronomy 11(12):2436

Dos Santos TB, Ribas AF, de Souza SG, Budzinski IG, Domingues DS (2022) Physiological responses to drought, salinity, and heat stress in plants: a review. Stress 2(1):113–135

Dousari A, Shakibaie M, Adeli M (2025) Biosynthesis and properties of bismuth nanoparticles: a review. Biol Trace Elem Res 203:3949–3966

Dutta S, Bhadara R, Kalam M, Tripathi S (2021) Green synthesis is preferred over chemical synthesis of nanomaterials. J Agric Aquacult 3(1):1–18

Easwaran C, Christopher SR, Moorthy G, Mohan P, Marimuthu R, Koothan V, Nallusamy S (2024) Nano hybrid fertilizers: a review on the state of the art in sustainable agriculture. Sci Total Environ 929:172533

Ernst D, Kolenčík M, Šebesta M, Ďurišová Ľ, Ďúranová H, Kšiňan S, Illa R, Safarik I, Černý I, Kratošová G, Žitniak Čurná V (2023) Agronomic investigation of spray dispersion of metal-based nanoparticles on sunflowers in real-world environments. Plants 12(9):1789

Fahad S, Bajwa AA, Nazir U, Anjum SA, Farooq A, Zohaib A, Sadia S, Nasim W, Adkins S, Saud S, Ihsan MZ, Alharby H, Wu C, Wang D, Huang J (2017) Crop production under drought and heat stress: plant responses and management options. Front Plant Sci 8:1147

Fang L, Martre P, Jin K, Du X, van der Putten PEL, Yin X, Struik PC (2023) Neglecting acclimation of photosynthesis under drought can cause significant errors in predicting leaf photosynthesis in wheat. Glob Chang Biol 29(2):505–521

Fares A, Mahdy A, Ahmed G (2024) Unraveling the mysteries of silver nanoparticles: synthesis, characterization, antimicrobial effects and uptake translocation in plant—a review. Planta 206:7
Garnett T, Appleby MC, Balmford A, Bateman IJ, Benton TG, Bloomer P, Burlingame B, Dawkins M, Dolan L, Fraser D, Herrero M (2013) Sustainable intensification in agriculture: premises and policies. Science 341(6141):33–34
Gómez-Llorente H, Hervás P, Pérez-Esteve É, Barat JM, Fernández-Segovia I (2022) Nanotechnology in the agri-food sector: consumer perceptions. NanoImpact 26:100399
Grewal S, Boora R, Kumari S, Thakur R, Goel S (2024) Fascinating aspects of nanosilicon enabled plant stress tolerance – a comprehensive review. Plant Nano Biol 23:100077
Gupta SD, Agarwal A, Pradhan S (2018) Phytostimulatory effect of silver nanoparticles (AgNPs) on rice seedling growth: an insight from antioxidative enzyme activities and gene expression patterns. Ecotoxicol Environ Saf 161:624–633
Gupta C, Gupta MK, Tripathi S (2025) Nanotechnology: a promising technology in sustainable agriculture. Discov Biotechnol 2(1):12
Hano C, Abbasi BH (2021) Plant-based green synthesis of nanoparticles: production, characterization and applications. Biomolecules 12(1):31
Hasani M, Faryabi S, Ehsanzadeh P, Vaghar M (2023) Foliar-applied zinc enhances sesame physiological and yield attributes under different irrigation regimes. S Afr J Bot 162:451–460
Hasanuzzaman M, Nahar K, Alam MM, Roychowdhury R, Fujita M (2013) Physiological, biochemical, and molecular mechanisms of heat stress tolerance in plants. Int J Mol Sci 14(5):9643–9684
Hatami M, Naghdi Badi H, Ghorbanpour M (2019) Nano-elicitation of secondary pharmaceutical metabolites in plant cells: a review. J Med Plants 18(71):6–36
Hossain A, Skalicky M, Brestic M, Maitra S, Ashraful Alam M, Syed MA, Hossain J, Sarkar S, Saha S, Bhadra P, Shankar T (2021) Consequences and mitigation strategies of abiotic stresses in wheat (*Triticum aestivum* L.) under the changing climate. Agronomy 11(2):241
Huang J, Liu J, Jiang F, Liu M, Chen Z, Zhou R, Wu Z (2024) Identification and expression pattern analysis of the SOS gene family in tomatoes. Agronomy 14:773
Humbal A, Pathak B (2023) Influence of exogenous elicitors on the production of secondary metabolite in plants: a review ("VSI: secondary metabolites"). Plant Stress 8:100166
Inam M, Attique I, Zahra M, Khan AK, Hahim M, Hano C, Anjum S (2023) Metal oxide nanoparticles and plant secondary metabolism: unraveling the game-changer nano-elicitors. Plant Cell Tissue Organ Cult 155(2):327–344
Jadoon L, Gul A, Fatima H, Babar MM (2024) Nano-elicitation and hydroponics: a synergism to enhance plant productivity and secondary metabolism. Planta 259(4):80
Jain A, Ranjan S, Dasgupta N, Ramalingam C (2018) Nanomaterials in food and agriculture: an overview on their safety concerns and regulatory issues. Crit Rev Food Sci Nutr 58(2):297–317
Kalia A, Sreelakshmi MV (2025) Nanoelicitors: a promising strategy for sustainable crop production under biotic stress conditions. In: Elicitors for sustainable crop production: overcoming biotic stress challenges in plants. Springer Nature Singapore, Singapore, pp 125–142
Karunakaran G, Sudha KG, Ali S, Cho EB (2023) Biosynthesis of nanoparticles from various biological sources and its biomedical applications. Molecules 28(11):4527
Kaur S, Garg T, Joshi A, Awasthi A, Kumar V, Kumar A (2024) Potential effects of metal oxide nanoparticles on leguminous plants: practical implications and future perspectives. Sci Hortic 331:113146
Khan Z, Upadhyaya H (2019) Impact of nanoparticles on abiotic stress responses in plants: an overview. In: Kumar Tripathi D et al (eds) Nanomaterials in plants, algae and microorganisms. Academic Press, pp 305–322. https://doi.org/10.1016/B978-0-12-811488-9.00015-9
Khan AK, Kousar S, Tungmunnithum D, Hano C, Abbasi BH, Anjum S (2021) Nano-elicitation as an effective and emerging strategy for in vitro production of industrially important flavonoids. Appl Sci 11(4):1694

Khalid F, Rasheed Y, Ashraf H et al. (2025) Nanoparticle-Mediated Phytohormone Interplay: Advancing Plant Resilience to Abiotic Stresses. J Crop Health 77:56. https://doi.org/10.1007/s10343-025-01121-z

Khan F, Pandey P, Upadhyay TK (2022) Applications of nanotechnology-based agrochemicals in food security and sustainable agriculture: an overview. Agriculture 12(10):1672

Khanzada A, Yan K, Hu W, Malko M, Khan KA, Bao Y, Elboughdiri N, Li Y (2025) Heat stress response mechanisms and resilience strategies in wheat. J Agron Crop Sci 211:e70023

Kopecká R, Kameniarová M, Černý M, Brzobohatý B, Novák J (2023) Abiotic stress in crop production. Int J Mol Sci 24(7):6603

Kumar G, Puthusseri B, Peethambaran D, Savanur MA (2022) Application of nanotechnology in plant secondary metabolites production. In: Phytonanotechnology. Springer Nature Singapore, Singapore, pp 99–120

Kumar DA, Parasurama DS, Sankar D (2024) Plant bioactive compounds: crucial pharmacological properties and role of elicitors in enhancing production. Indian J Pharm Educ Res 58(4):1034–1044

Kumari S, Khanna RR, Nazir F, Albaqami M, Chhillar H, Wahid I, Khan MI (2022) Bio-synthesized nanoparticles in developing plant abiotic stress resilience: a new boon for sustainable approach. Int J Mol Sci 23(8):4452

Lala S (2021) Nanoparticles as elicitors and harvesters of economically important secondary metabolites in higher plants: a review. IET Nanobiotechnol 15(1):28–57

Luo X, Li J, Guo S, Yu H, Zeng X, Zhou Z, Shangguan Y, He M, Ouyang Y, Chen K, Chen Z (2025) Analysis of research status and trends on nano-agricultural application: a bibliometric study. Front Plant Sci 16:1530629

Mahdavi Khorami A, Masoud Sinaki J, Amini Dehaghi M, Rezvan S, Damavandi A (2020) Sesame (*Sesame indicum* L.) biochemical and physiological responses as affected by applying chemical, biological, and nano-fertilizers in field water stress conditions. J Plant Nutr 43(3):456–475

Mariyam MS, Subburayan RR, Dhanraj G (2025) Harnessing the potential of TiO2 nanoparticles from marine macroalgae *Ulva lactuca* for antibacterial, anti-inflammatory, and cytotoxic properties. J Appl Pharm Sci 15(4):135–141

Martin L, Smerald A, Haas E, Klimiuk T, Sánchez-Benítez A, Scheer C (2024) Assessing the vulnerability of agricultural areas under climate change in Europe through a heat stress index approach. EGU General Assembly 2024, Vienna, Austria, 14–19 Apr 2024, EGU24-15603. https://doi.org/10.5194/egusphere-egu24-15603

Moghadas MRS, Naghavi M, Sabokdast M, Motemadi E, Nasiri J (2021) The effect of nano elicitors on the expression of the genes involved in alkaloids biosynthetic pathway in Papaver orientale L. suspension culture. Iran J Field Crop Sci 52(1):133–141

Morris PC, Reuveni M, Ortiz R (2024) Editorial: impact and mitigation of abiotic stress in cereals. Front Plant Sci 15:1374631

Muhammad F, Raza MAS, Iqbal R, Zulfiqar F, Aslam MU, Yong JWH, Altaf MA, Zulfiqar B, Amin J, Ibrahim MA (2022) Ameliorating drought effects in wheat using an exclusive or 2 co-applied rhizobacteria and ZnO nanoparticles. Biology 11(11):1564

Murthy HN, Joseph KS, Paek KY, Park SY (2024) Nanomaterials as novel elicitors of pharmacologically active plant specialized metabolites in cell and organ cultures: current status and future outlooks. Plant Growth Regul 104(1):5–30

Nazeer A, Ahmad F, Verma N, Ahmad S (2023) Nanomaterials and plant biomolecules: basics of interactions. In: Targeted delivery of nanopesticides and nanofertilizers in sustainable agricultural farming. Springer Nature Switzerland, Cham, pp 9–49

Neme K, Nafady A, Uddin S, Tola YB (2021) Application of nanotechnology in agriculture, postharvest loss reduction and food processing: food security implication and challenges. Heliyon 7(12):e08539

Omran B, Baek KH (2021) Nanoantioxidants: pioneer types, advantages, limitations, and future insights. Molecules 26(22):7031

Pandita D (2022) Role of nano-biotechnology in medicinal plant production. In: Environmental challenges and medicinal plants: sustainable production solutions under adverse conditions. Springer International Publishing, Cham, pp 355–384

Parra-Torrejón B, Ramírez-Rodríguez GB, Giménez-Bañón MJ, Moreno-Olivares JD, Paladines-Quezada DF, Gil-Muñoz R, Delgado-López JM (2021) Nanoelicitors with prolonged retention and sustained release to produce beneficial compounds in wines. Environ Sci Nano 8(12):3524–3535

Patel C, Singh J, Karunakaran A, Ramakrishna W (2023) Evolution of nano-biofertilizer as a green technology for agriculture. Agriculture 13(10):1865

Peverelli MC, Rogers WJ (2013) Heat stress effects on crop performance and tools for tolerance breeding. Rev Fac Cienc Agrar 45(2):349–368

Plaksenkova I, Kokina I, Petrova A, Jermaļonoka M, Gerbreders V, Krasovska M (2020) The impact of zinc oxide nanoparticles on cytotoxicity, genotoxicity, and miRNA expression in barley (Hordeum vulgare L.) seedlings. ScientificWorldJournal 2020(1):6649746

Priya M, Farooq M, Siddique KH (2025) Enhancing tolerance to combined heat and drought stress in cool-season grain legumes: mechanisms, genetic insights, and future directions. Plant Cell Environ. https://doi.org/10.1111/pce.15382

Qiao M, Hong C, Jiao Y, Hou S, Gao H (2024) Impacts of drought on photosynthesis in major food crops and the related mechanisms of plant responses to drought. Plants 13(13):1808

Rajpal VR, Dhingra Y, Khungar L, Mehta S, Minkina T, Rajput VD, Husen A (2024) Current research in biotechnology. Curr Res Biotechnol 7:100219

Rane J, Singh AK, Kumar M, Boraiah KM, Meena KK, Pradhan A, Prasad PV (2021) The adaptation and tolerance of major cereals and legumes to important abiotic stresses. Int J Mol Sci 22(23):12970

Rani N, Boora R, Singh Y, Choudhary Y, Devi S, Mohan N, Kumari S, Grewal S (2025) Nano-elicitation: an emerging and potential technique for enhanced production of pharmaceutically important secondary metabolites in plants. Nanotechnol Environ Eng 10:42

Rautela I, Dheer PA, Thapliyal PR, Shah DH, Joshi M, Upadhyay S, Gururani P, Sinha VB, Gaurav NA, Sharma MD (2021) Current scenario and future perspectives of nanotechnology in sustainable agriculture and food production. Plant Cell Biotechnol Mol Biol 22:99–121

Rehman A, Khan S, Sun F, Peng Z, Feng K, Wang N, Jia Y, Pan Z, He S, Wang L, Qayyum A, Du X, Li H (2024) Exploring the nano-wonders: unveiling the role of nanoparticles in enhancing salinity and drought tolerance in plants. Front Plant Sci 14:1324176

Rhaman MS, Tania SS, Imran S, Rauf F, Kibria MG, Ye W, Hasanuzzaman M, Murata Y (2022) Seed priming with nanoparticles: an emerging technique for improving plant growth, development, and abiotic stress tolerance. J Soil Sci Plant Nutr 22(4):4047–4062

Rivero-Montejo SDJ, Vargas-Hernandez M, Torres-Pacheco I (2021) Nanoparticles as novel elicitors to improve bioactive compounds in plants. Agriculture 11(2):134

Sajyan TK, Ahmad SA, El Ghandour I, Al Ahad LA, Francis R (2025) Innovative elicitors in plant science. In: Elicitors for sustainable crop production: overcoming biotic stress challenges in plants. Springer Nature Singapore, Singapore, pp 297–321

Salem SS, Hammad EN, Mohamed AA, El-Dougdoug W (2023) A comprehensive review of nanomaterials: types, synthesis, characterization, and applications. Biointerface Res Appl Chem 13(1):41

Saritha GNG, Anju T, Kumar A (2022) Nanotechnology—big impact: how nanotechnology is changing the future of agriculture? J Agric Food Res 10:100457

Sarraf M, Vishwakarma K, Kumar V, Arif N, Das S, Johnson R, Janeeshma E, Puthur JT, Aliniaeifard S, Chauhan DK, Fujita M, Hasanuzzaman M (2022) Metal/metalloid-based nanomaterials for plant abiotic stress tolerance: an overview of the mechanisms. Plants 11(3):316

Sarwar S, Ali S, Khan SR, Jamil S, Farooq S, Jafar T, Latif MJ, Shehroz H (2024) Marriage of Green Technology and Chemistry – bismuth-based nanoparticles: a remarkable combination. Environ Technol Rev 13(1):544–571

Sharf-Eldin AA, Alwutayd KM, El-Yazied AA, El-Beltagi HS, Alharbi BM, Eisa MA, Alqurashi M, Sharaf M, Al-Harbi NA, Al-Qahtani SM, Ibrahim MF (2023) Response of maize seedlings to silicon dioxide nanoparticles (SiO_2NPs) under drought stress. Plants 12(14):2592

Singh R, Choudhary P, Kumar S, Daima HK (2024) Mechanistic approaches for crosstalk between nanomaterials and plants: plant immunomodulation, defense mechanisms, stress resilience, toxicity, and perspectives. Environ Sci Nano 11:2324–2351

Singh J, Singh VK, Mishra S (2025) Nanotechnology-driven strategies for abiotic stress alleviation in plants. Int J Adv Biochem Res 9(4):75–81

Spanos A, Athanasiou K, Ioannou A, Fotopoulos V, Krasia-Christoforou T (2021) Functionalized magnetic nanomaterials in agricultural applications. Nanomaterials 11(11):3106

Subhash RS, Meshram JH, Gogle DP, Chaudhary RG, Mahakhode RH, Patil VN, Kamble RB, Sawant RC, Gadpayale JV, Chaudhary RR (2024) Nanoelicitors mediated abiotic stresses in plant defense response mechanisms: current review and future perspectives. In: Plant secondary metabolites and abiotic stress. Wiley/Scrivener Publishing LLC, Hoboken/Beverly, pp 591–621. https://doi.org/10.1002/9781394186457.ch19

Syu YY, Hung JH, Chen JC, Chuang HW (2014) Impacts of size and shape of silver nanoparticles on Arabidopsis plant growth and gene expression. Plant Physiol Biochem 83:57–64

Tang H, Wang C, Mei J, Feng L, Wu Q, Yin Y (2024) Transcriptome analysis revealed the response mechanism of pomegranate to salt stress. Agronomy 14(10):2261

Tortella G, Rubilar O, Pieretti JC, Fincheira P, de Melo Santana B, Fernández-Baldo MA, Benavides-Mendoza A, Seabra AB (2023) Nanoparticles as a promising strategy to mitigate biotic stress in agriculture. Antibiotics 12(2):338

Upadhyay SK, Kumar S, Rout C, Vashistha G, Aggarwal D (2022) Risks and concerns of use of nanoparticles in agriculture. In: The role of nanoparticles in plant nutrition under soil pollution: nanoscience in nutrient use efficiency. Springer International Publishing, Cham, pp 371–394

Wahab A, Muhammad M, Ullah S, Abdi G, Shah GM, Zaman W, Ayaz A (2024) Agriculture and environmental management through nanotechnology: eco-friendly nanomaterial synthesis for soil-plant systems, food safety, and sustainability. Sci Total Environ 23:171862

Xu Z, Tang T, Lin Q, Yu J, Zhang C, Zhao X, Kah M, Li L (2022) Environmental risks and the potential benefits of nanopesticides: a review. Environ Chem Lett 20(3):2097–2108

Yadav S, Modi P, Dave A, Vijapura A, Patel D, Patel M (2020) Effect of abiotic stress on crops. In: Sustainable crop production. IntechOpen, pp 5–16. https://doi.org/10.5772/intechopen.88434

Yu Y, Wu Y, He L (2023) A wheat WRKY transcription factor TaWRKY17 enhances tolerance to salt stress in transgenic Arabidopsis and wheat plant. Plant Mol Biol 113(4):171–191

Zacchini M (2024) Bismuth interaction with plants: uptake and transport, toxic effects, tolerance mechanisms – a review. Chemosphere 22:142414

Zaman W, Ayaz A, Park S (2025) Nanomaterials in agriculture: a pathway to enhanced plant growth and abiotic stress resistance. Plants 14(5):716

Zhang HY, Su WH (2024) Classification, uptake, translocation, and detection methods of nanoparticles in crop plants: a review. Environ Sci Nano 11:1847–1870

Zhou Y, Xu F, Shao Y, He J (2022) Regulatory mechanisms of heat stress response and thermomorphogenesis in plants. Plants 11(24):3410

Zhou Y, Liu X, Yang X, Du Laing G, Yang Y, Tack FM, Bank MS, Bundschuh J (2023) Effects of platinum nanoparticles on rice seedlings (*Oryza sativa* L.): size-dependent accumulation, transformation, and ionomic influence. Environ Sci Technol 57(9):3733–3745

Chapter 13
The Effect of Saccharides on Plant Response to Stress and Their Regulatory Function During Development

Karima Guehaz, Fatma Zohra Chakou, Matilde Ciani, Zakaria Boual, Alessandra Adessi, Alia Telli, and Heba I. Mohamed

Abstract Saccharides, encompassing simple sugars and complex polysaccharides, are integral to plant stress responses and developmental regulation. They function as protective agents and signaling molecules, augmenting stress tolerance and affecting growth and gene expression. The present work examines the several functions of saccharides in plant stress responses and developmental control. It offers an extensive examination of the roles of saccharides in plant systems, highlighting their essential participation in stress response mechanisms and developmental processes. The study investigates the fundamental molecular mechanisms that enable these tasks and explores the techniques utilized to analyze saccharides' roles in plants. The study underlines the significance of saccharides in plant resilience and growth through an analysis of both biochemical and physiological factors. The constraints of existing approaches and the prospects for future research lines are critically reviewed to improve the comprehension of saccharides' roles in plant biology.

Keywords Saccharides · Developmental regulation · Stress responses · Gene expression · Tolerance

---

K. Guehaz · F. Z. Chakou · Z. Boual · A. Telli (✉)
Laboratory of Preservation of Ecosystems in Arid and Semi-Arid Areas, Faculty of Nature and Life Sciences, University of Kasdi Merbah-Ouargla, Ouargla, Algeria
e-mail: telli.alia@univ-ouargla.dz

M. Ciani · A. Adessi
Department of Agriculture, Food Environment and Forestry (DAGRI), University of Florence, Florence, Italy

H. I. Mohamed
Biological and Geological Sciences, Faculty of Education, Ain Shams University, Cairo, Egypt

H. I. Mohamed (ed.), *Elicitors for Sustainable Crop Production*, Sustainability Sciences in Asia and Africa, https://doi.org/10.1007/978-981-95-3533-0_13

13.1 Introduction

Saccharides, or carbohydrates, are essential molecules in plants, serving both as energy sources and as key structural and regulatory components. Their diverse roles extend from basic metabolism and growth to stress responses and cellular communication, making them fundamental to plant life (Lubbe et al. 2021). Acknowledged for their functions as energy sources and structural elements, saccharides have significant roles in plant biology that surpass their conventional purposes. Recent studies emphasize their crucial role in plant stress responses and developmental control, highlighting their importance in enhancing plant resilience and adaptability (Jeandet et al. 2022; Poonam et al. 2016).

Sugars are essential in modulating physiological processes in plants via a complex network of signaling pathways. The identification of sugar transporters and signaling systems has been crucial to research, revealing the significant roles of sugars in controlling plant development and responses to diverse abiotic stressors (Martínez-Noël and Tognetti 2018).

Carbohydrates, including monosaccharides and their derivatives, disaccharides, and oligosaccharides, collectively referred to as saccharides, serve as structural components of living cells and supply energy, construction materials, and metabolic substrates for synthesis and catabolism. These sugars participate in several biological processes and signaling mechanisms related to plant growth and development, stress response and adaptation, as well as interactions between plants and their environment. During abiotic stress, fluctuations in the quantities and compositions of several saccharides are noted. Various saccharide types have been discovered to enhance plant tolerance to drought, salinity, extreme temperatures, and heavy metals (Qi et al. 2025; Alagoz et al. 2023; Akensous et al. 2022; Morcillo and Manzanera 2021).

Saccharides are carbohydrates, or sugars, that form the basis of the carbohydrate group. They are classically categorized into three groups based on their degree of polymerization: monosaccharides, oligosaccharides, and polysaccharides; however, their diversity is, in fact, much greater. The simplest monosaccharides contain a single hydrophilic carbonyl (C=O) group and several hydroxyl (−OH) groups. The presence of an asymmetric carbon atom creates a chiral center, determining whether the molecule is in the D or L form. Recently, researchers have been studying their involvement in signaling pathways that regulate development and responses to biotic and abiotic stresses (Jeandet et al. 2022; Kaur et al. 2021).

Plants must respond to both abiotic and biotic stresses through a range of reactions, including the readjustment of growth and metabolism, changes in gene expression to activate defense genes, and reprogramming of developmental patterns. A key survival strategy involves the production of specialized metabolites, such as compatible solutes and osmoregulatory compounds like proline, glycine betaine, and certain saccharides. These saccharides help delay water loss by maintaining cellular turgor and provide primary defense mechanisms against stress-induced damage (Kaur et al. 2021; Ma et al. 2023).

Plants use sugars not only as energetic compounds but also as signaling molecules regulating key physiological processes. During photosynthesis, plants produce D-glucose, which can be converted into D-fructose, sucrose, or starch via different metabolic pathways. Sucrose, due to its nonreducing properties, enables stable osmotic/chemical signaling in phloem transport. Oligosaccharides like α-galactosides (e.g., raffinose) protect proteins during desiccation. Raffinose family oligosaccharides (RFOs) modulate defense responses and development. Both fructans and raffinose family oligosaccharides (RFOs) serve as reserve carbohydrates, membrane stabilizers, and mediators of stress tolerance. Additionally, fructan metabolism may play a significant role in osmoregulation (such as in flower opening) and in the dynamics of source–sink relationships. Fructans and RFOs may help maintain cellular reactive oxygen species (ROS) homeostasis through specific ROS-scavenging mechanisms near organellar membranes (e.g., vacuoles and chloroplasts). It is proposed that small fructans and RFOs function as phloem-mobile signaling molecules during stress conditions. α-galactosides play a similar protective role. These sugars, as well as glycosylation events, are involved in modulating plant defense responses, growth, and development (Van den Ende 2013; Afzal et al. 2021; Choudhary et al. 2022).

Additionally, several carbohydrates, including glucose, fructose, sucrose, trehalose, and sugar alcohols like mannitol, have been demonstrated to play regulatory roles in plant development by modulating cell division and growth. These saccharides influence cell number and size by integrating phytohormonal and environmental signals. While phytochrome (PHY) signaling coordinates developmental responses to light, the interplay between sugar signaling and PHY pathways contributes to the fine-tuning of plant growth and adaptation (Salvi et al. 2022).

This chapter examines the intricate functions of saccharides, investigating their impact on a plant's ability to endure environmental challenges and facilitate its growth and development. The study seeks to clarify the complex interactions of saccharides inside plant systems by investigating the molecular mechanisms that govern these processes. Moreover, the approaches employed to examine these interactions are evaluated, providing insights into existing limits and facilitating future research avenues. This investigation broadens our comprehension of saccharides in plant biology and paves the way for advancements in agricultural methods and the enhancement of crop resistance amid escalating environmental problems.

13.2 Types of Saccharides and Their Roles

Carbon, hydrogen, and oxygen atoms combine to form organic molecules known as carbohydrates. They are divided into simple and complicated carbohydrates according to their structure and complexity. Energy storage, structural support, cell–cell recognition, and cell communication are just a few of the biological functions in which carbohydrates are essential. In the following subsections, there is a discussion of all the different types of carbohydrates, including monosaccharides, disaccharides, oligosaccharides, and polysaccharides (Fig. 13.1).

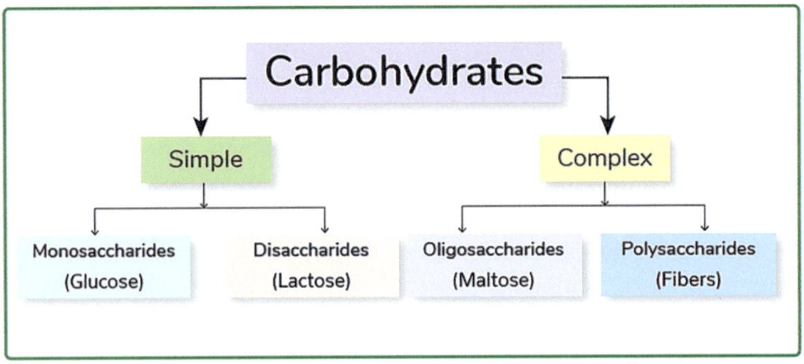

Fig. 13.1 Different types of carbohydrates

13.2.1 Monosaccharides

Monosaccharides, also known as simple sugars, are the most basic form of carbohydrates, consisting of single sugar units with the general formula $(CH_2O)_n$, where n typically ranges from three to seven carbon atoms. Common examples of monosaccharides include glucose, fructose, and galactose (which are hexoses), as well as ribose (a pentose) and glyceraldehyde (a triose). These simple sugars play several crucial roles in biological systems: They act as the primary energy source for cellular metabolism, serve as building blocks for larger carbohydrates, and function as signaling molecules that regulate growth and stress responses in organisms (Ruan 2014).

Glucose is a crucial metabolite produced during photosynthesis and plays a vital role in cellular respiration, where it is utilized to generate ATP, providing the energy necessary for various metabolic processes essential for plant growth and maintenance. It also serves as a precursor for the biosynthesis of structural carbohydrates, such as cellulose and starch, which support biomass accumulation. Plants dynamically adjust glucose metabolism to meet their energy demands, particularly in response to changing environmental conditions. For instance, the overexpression of sucrose synthase genes can increase Uridine Diphosphate (UDP)-glucose pools, thereby enhancing cellulose biosynthesis and photosynthetic rates. This illustrates glucose's significant role in carbon allocation and overall plant growth (Li et al. 2024; Zhou et al. 2024a).

Glucose functions as a significant signaling molecule that regulates gene expression and influences various physiological processes, including seed germination, root and shoot development, flowering, senescence, and stress responses. It operates through both hexokinase (HXK)-dependent and independent pathways to modulate transcription factors and hormonal signaling networks, such as those involving abscisic acid (ABA), ethylene (ET), and brassinosteroids (BRs). Additionally, glucose signaling plays a critical role in nitrogen metabolism and the expression of stress-responsive genes, interacting with other nutrient signals to orchestrate plant adaptation. Recent advancements in fluorescent biosensors have enabled real-time monitoring of glucose dynamics, providing insights into its spatial and temporal signaling roles within living plants (Zhou et al. 2024a; Pottier et al. 2023).

Glucose plays a significant role in plant growth and development, notably inducing a delay in the transition from the juvenile to the vegetative phase. It stimulates the synthesis of chlorophyll, ribulose-1,5-bisphosphate carboxylase/oxygenase (RuBisCO), and various photoprotective pigments, all of which are essential for photosynthesis and plant vitality. Furthermore, glucose helps alleviate the harmful effects of abiotic stress by increasing the levels of antioxidants and sugars within the plant. It also interacts with key hormones, such as auxins, cytokinins (CKs), ethylene, abscisic acid (ABA), and brassinosteroids (BRs) to modulate growth and metabolic processes. Notably, lower concentrations of glucose can relieve germination inhibition caused by ABA catabolism, highlighting its multifaceted role in promoting plant resilience and development (Siddiqui et al. 2020).

Fructose plays multifaceted roles in plant development and signaling, serving both as a metabolic substrate and a regulatory molecule under normal and abiotic stress conditions. Recent research underscores its involvement in growth regulation, stress adaptation, and hormonal cross talk. Under normal conditions, fructose signaling is crucial for seedling establishment, where it modulates early development through the action of FRUCTOSE INSENSITIVE1 (FINS1), a cytosolic fructose-1,6-bisphosphatase. FINS1 interacts with WUSCHEL (WUS) transcription factors to regulate stem cell function in meristems, thereby controlling cell division and growth. Additionally, fructose influences primary root development by regulating auxin biosynthesis and accumulation through the action of tyrosylprotein sulfotransferase (TPST). Moreover, fructose signaling engages in hormonal cross talk, interacting with abscisic acid (ABA) and ethylene pathways. It acts downstream of ABA synthesis and negatively regulates ethylene signaling, thus fine-tuning developmental transitions in response to varying environmental conditions (Fañanás-Pueyo et al. 2025; Li et al. 2023; Cho and Yoo 2011).

Under abiotic stress conditions, fructose plays a vital role in enhancing plant resilience through various mechanisms. One of its key functions is membrane stabilization; fructose polymers, known as fructans, protect cellular membranes during drought, cold, or salinity stress by forming hydrogen bonds with lipids. High–degree polymerization (DP) fructans contribute to structural integrity, effectively reducing water loss. In addition to membrane protection, fructose acts as a stress signaling molecule by triggering the expression of stress-responsive genes, such as NAC089, and modulating reactive oxygen species (ROS) scavenging. This process promotes the accumulation of osmoprotectants, including raffinose family oligosaccharides (RFOs), which enhance drought and cold tolerance. Furthermore, under stress conditions, fructose amplifies abscisic acid (ABA) signaling, prioritizing resource allocation for stress adaptation rather than growth. This hormonal integration involves the repression of ethylene pathways, effectively delaying senescence and allowing plants to better cope with adverse environmental conditions (Fañanás-Pueyo et al. 2025; Gentile et al. 2024; Valluru 2015; Valluru and Van den Ende 2008).

Galactose plays multifaceted roles in plant development and signaling under both normal and abiotic stress conditions, functioning as a structural component, metabolic intermediate, and signaling molecule. In terms of its role in normal development, galactose is a vital component of cell wall polymers, such as pectins and hemicellulose. During seedling development, cell wall remodeling releases galactose, which is then recycled into nucleotide sugars via galactokinase (GALK) for reuse in biosynthesis. Additionally, galactose contributes to the synthesis of raffinose family oligosaccharides (RFOs), such as raffinose and stachyose, through the action of galactinol synthase (GolS). These RFOs play a crucial role in protecting embryos during desiccation and serve as transport carbohydrates in certain plant families (Höftberger et al. 2022; Wang et al. 2022; Sengupta et al. 2015).

In the context of signaling, excess galactose can inhibit primary root growth and pollen germination in *Arabidopsis* by disrupting cell elongation and meristem function. This inhibition is mediated by GALK, which converts galactose into toxic intermediates. Furthermore, galactose signaling engages in sugar–hormone cross talk, interacting with abscisic acid (ABA) and ethylene pathways. For example, the inhibition of root growth induced by galactose requires ABA synthesis, while ethylene signaling modulates the plant's response to galactose levels, highlighting its complex role in regulating growth and development (Höftberger et al. 2022; Wang et al. 2022; Eveland and Jackson 2012).

Under abiotic stress conditions, galactose plays a crucial role in enhancing plant resilience through various mechanisms. One of its primary functions is osmoprotection; during periods of drought, salinity, or cold stress, plants accumulate raffinose family oligosaccharides (RFOs) derived from galactose. These RFOs act as compatible solutes, stabilizing cellular membranes and proteins during dehydration. Moreover, the regulation of stress-responsive genes is tightly linked to galactose metabolism; for instance, GolS genes, such as OsGolS1 and OsGolS2 in rice, are upregulated under stress conditions. Their expression is posttranscriptionally regulated through intron retention, establishing a connection between galactose metabolism and osmotic stress adaptation. In addition, galactose influx via transporters like Sugars Will Eventually be Exported Transporter 5 (SWEET5) modulates metabolic flux dynamics, influencing overall stress tolerance. Notably, mutants lacking SWEET5 exhibit reduced sensitivity to galactose during pollen germination under stress conditions. Furthermore, RFOs help mitigate oxidative damage by enhancing the plant's antioxidant capacity, although their precise role remains a topic of debate due to functional redundancy observed in loss-of-function mutants. This multifaceted involvement of galactose and its derivatives underscores their importance in plant responses to abiotic stress (Mukherjee et al. 2019; Sengupta et al. 2015).

Ribose, a pentose sugar primarily recognized as a structural component of nucleotides and nucleic acids (RNA and DNA), plays an essential but mostly indirect role in plant development and signaling. Although ribose itself is not commonly described as a direct signaling molecule like glucose or fructose, its metabolic derivatives and its involvement in nucleotide biosynthesis are critical for various developmental processes and stress responses. In terms of its structural and metabolic role, ribose is a fundamental constituent of nucleotides, including ATP, NADH,

NADPH, and FAD, which are vital for all cellular processes such as DNA replication, transcription, and energy metabolism. The availability of ribose and its phosphate derivatives, such as ribose-5-phosphate in the pentose phosphate pathway (PPP), supports nucleotide synthesis, which is crucial for cell division and growth during development. Moreover, ribose has an indirect role in signaling through nucleotide-derived molecules. Nucleotides like ATP and NAD(P)H function as energy carriers and cofactors in signaling pathways that regulate development and stress responses. Additionally, ribose-containing molecules, such as cyclic Adenosine Diphosphate-ribose (cADPR), act as secondary messengers in calcium signaling, influencing processes like stomatal movement and stress adaptation. Furthermore, ribosomal RNA and ribosomal proteins, which depend on ribose-containing nucleotides, regulate the translation of key developmental regulators, such as auxin response factors (ARFs), thereby integrating hormonal and environmental signals in plant development (Rosado et al. 2012).

Ribose plays a significant role in plant stress responses, particularly through its involvement in the pentose phosphate pathway (PPP), which generates ribose-5-phosphate and provides reducing power in the form of NADPH. This reducing power is critical for the detoxification of reactive oxygen species (ROS) generated under abiotic stress conditions. Enhanced activity of the PPP supports nucleotide repair and synthesis during instances of stress-induced DNA damage, thereby facilitating recovery and adaptation in plants. Recent insights into ribosome-mediated regulation further underscore the importance of ribose in plant development. Mutations in ribosomal proteins can adversely affect the translation of auxin-signaling components, leading to developmental defects. This highlights the crucial role of ribose-derived RNA in modulating plant growth and responses to environmental cues. Additionally, translational regulation by ribosomes serves as a key control point for decoding complex auxin signals, translating them into specific developmental programs that enable plants to adapt to their changing environments effectively (Noctor and Foyer 2020; Xiong and Sheen 2014).

Mannitol is a polyol derived from the monosaccharide mannose. It plays significant roles in plant development and signaling under both normal and abiotic stress conditions, primarily acting as a compatible solute, osmoprotectant, and modulator of stress responses. As an osmoprotectant, mannitol helps plants tolerate salt, drought, and osmotic stresses by stabilizing proteins and membranes, thereby maintaining cellular osmotic balance. Research has shown that transgenic plants engineered to produce mannitol exhibit improved tolerance to salt and temperature stresses, which is linked to the activation of genes involved in hormone signaling, reactive oxygen species (ROS) scavenging, and ion homeostasis (Rathor et al. 2020).

Mannitol is also widely used in experimental settings to impose osmotic stress and study the mechanisms of growth repression. For instance, in *Arabidopsis*, mild mannitol-induced stress triggers rapid changes in the leaf phosphoproteome, affecting proteins that regulate shoot growth and leaf size. Additionally, mannitol-induced osmotic stress influences calcium signaling pathways, which are crucial for transmitting stress signals and regulating growth responses (Nikonorova et al. 2018).

Moreover, mannitol can effectively quench ROS, protecting plants from oxidative damage during stress conditions. Some phytopathogenic fungi produce mannitol to suppress plant defenses mediated by ROS, while plants induce mannitol dehydrogenase to catabolize fungal mannitol, counteracting this suppression. Mannitol also plays a role in hormonal cross talk, particularly influencing abscisic acid (ABA) homeostasis and signaling, which are vital for stress adaptation and shoot growth regulation under osmotic stress. This interaction helps balance growth and stress responses. Furthermore, mannitol treatments have been shown to enhance photosynthetic pigments, biomass, and seed yield in crops such as faba bean, indicating a positive impact on growth and productivity, even under challenging environmental conditions (Meena et al. 2015; Jennings et al. 1998).

13.2.2 Disaccharides

Disaccharides are carbohydrates formed by the linkage of two monosaccharides through a glycosidic bond, a process that releases a molecule of water. This class of sugars plays important roles in various biological functions. Notable examples of disaccharides include sucrose, which is composed of glucose and fructose and serves as the main transport sugar in plants, facilitating the distribution of energy. Another example is lactose, formed from glucose and galactose, commonly known as milk sugar, which is essential for providing energy to young mammals. Maltose, consisting of two glucose molecules, is another important disaccharide that plays a role in energy metabolism, particularly during the digestion of starch. Trehalose is a nonreducing disaccharide present in diverse organisms, ranging from bacteria and fungi to invertebrates, in which it serves as an energy source as well as an osmolyte and/or protein/membrane protectant. Together, these disaccharides contribute significantly to energy storage and transport in living organisms (Szilagyi 2019; Lemoine et al. 2013; López-Gómez and Lluch 2012; Zeeman et al. 2010).

Disaccharides, particularly sucrose and trehalose, play dual roles as metabolic substrates and signaling molecules in plants, regulating development and stress responses through complex molecular networks. Their functions under normal and abiotic stress conditions are essential for plant health and resilience. In terms of plant development, sucrose acts as a key developmental regulator by controlling gene expression related to phloem loading, such as BvSUT1, carbon partitioning, and critical developmental transitions, including flowering and senescence. Notably, sucrose exerts its effects independently of hexose-derived signals, with nonmetabolizable analogs capable of mimicking its actions. Meanwhile, trehalose-6-phosphate (T6P), an intermediate in trehalose biosynthesis, functions as a growth sensor by integrating sucrose availability with developmental programs. T6P promotes growth under favorable conditions by inhibiting Sucrose Non-Fermenting-1-Related Protein Kinase (SnRK1) kinases, which normally repress energy-intensive processes like protein synthesis (Hernández-Bernal et al. 2022; Jeandet et al. 2022; Sakr et al. 2018; Delorge et al. 2014).

Regarding their signaling mechanisms, sucrose-specific pathways involve plasma membrane transporters, such as AtSUT2, which facilitate sucrose perception and trigger downstream signaling involving calcium-dependent protein kinases (CDPKs) and protein phosphatases. This process regulates source–sink dynamics and the expression of stress-responsive genes. On the other hand, trehalose-mediated signaling is characterized by T6P's modulation of SnRK1 activity, effectively linking carbon status to stress adaptation. Additionally, T6P interacts with Basic Leucine Zipper (bZIP) transcription factors, such as bZIP11, to balance growth and stress responses, further illustrating the intricate role of disaccharides in plant physiology (Han et al. 2024; Ramel et al. 2009).

13.2.3 Oligosaccharides

Oligosaccharides, consisting of three to ten monosaccharide units, are short chains of sugars that often exhibit specific biological functions. A well-known example of an oligosaccharide is raffinose, which is composed of galactose, glucose, and fructose. Oligosaccharides play several important roles in plant physiology and human health. They act as compatible solutes, helping protect cells from various stress conditions, such as drought and cold, by stabilizing cellular structures and maintaining osmotic balance. Additionally, in the human diet, oligosaccharides serve as prebiotics, promoting the growth of beneficial gut bacteria and enhancing digestive health. Furthermore, these molecules play a crucial role in plant defense mechanisms and signaling pathways, thereby enhancing the plant's ability to respond to environmental challenges and interact with other organisms. Overall, oligosaccharides are vital compounds that support both plant resilience and human health (Wayne 2024; Peters and Keller 2023; Yan et al. 2022).

Oligosaccharides play significant roles in plant development and signaling under both normal and abiotic stress conditions by acting as signaling molecules, modulators of growth, and triggers of defense responses. Various oligosaccharides, including those derived from the plant cell wall, such as oligogalacturonides (OGs), function as endogenous signals that regulate plant growth, development, and defense mechanisms. These oligosaccharides are released during cell wall remodeling or degradation and can activate signaling pathways that modulate gene expression and cellular responses (Ferrari et al. 2013; Etzler 1998).

Oligogalacturonides are oligomers of galacturonic acid released from pectin in the cell wall during stress or pathogen attack. They elicit defense responses, including the accumulation of reactive oxygen species (ROS) and the expression of pathogenesis-related proteins. OGs are perceived by wall-associated kinases (WAKs) and share signaling components with microbe-associated molecular patterns (MAMPs), thereby activating innate immunity and wound responses. In addition to their role in defense, OGs also influence developmental processes by antagonizing auxin responses, which affects root formation and overall growth (Ferrari et al. 2013).

Moreover, oligosaccharides can modulate hormone signaling and growth by antagonizing auxin-induced elongation and root formation, highlighting their role in balancing growth and defense mechanisms. For instance, OGs inhibit auxin-induced root formation, suggesting a regulatory mechanism wherein oligosaccharides adjust hormone signaling to facilitate development under stress or damage (Guarnizo et al. 2020; Fry et al. 1993).

In the context of symbiotic signaling, certain lipochitin oligosaccharides, known as Nod factors, produced by rhizobia, act as key signaling molecules that trigger root nodule formation in legumes. These oligosaccharides are recognized by specific plant receptors with high affinity, initiating developmental programs necessary for symbiosis and nitrogen fixation (Broghammer et al. 2012).

During abiotic stress conditions such as drought, cold, or mechanical wounding, oligosaccharides like OGs serve as local signals that activate defense and repair pathways. Their limited mobility confines their action to damaged tissues, where they induce localized responses to maintain cellular integrity and promote adaptation, further emphasizing the critical role of oligosaccharides in plant resilience and development (Ferrari et al. 2013).

13.2.4 Polysaccharides

Polysaccharides are large polymers of monosaccharides linked by glycosidic bonds, which can be either linear or branched in structure. They are categorized into two main types: homopolysaccharides, which consist of a single type of monosaccharide (such as starch, cellulose, and glycogen), and heteropolysaccharides, which contain different monosaccharides (like pectin and hemicellulose) (Zhang et al. 2021; Zeeman et al. 2010).

Starch is the primary energy storage polysaccharide in plants, composed of two components: amylose, which is linear, and amylopectin, which is branched. This structure allows plants to efficiently store and mobilize energy as needed. Cellulose, another important polysaccharide, serves as a crucial structural component of plant cell walls, providing rigidity and strength to the plant structure. In contrast, glycogen is the energy storage polysaccharide found in animals, allowing for rapid energy release when needed (Cho and Kang 2020; Zeeman et al. 2010).

Additionally, inulin acts as a storage polysaccharide in some plants and is known for its prebiotic effects, promoting beneficial gut bacteria in humans. Pectin and hemicellulose are also critical components of plant cell walls; they contribute to the structural integrity and gel-forming properties of the wall, playing essential roles in plant growth and development. Together, these polysaccharides are vital for energy storage, structural support, and overall plant health and functionality (Mensink et al. 2015).

Polysaccharides serve as crucial structural components, metabolic regulators, and signaling molecules in plants, significantly influencing growth, development, and stress adaptation. Their diverse roles under normal and abiotic stress conditions highlight their importance in plant physiology, as supported by recent research. Under normal conditions, polysaccharides provide essential structural support; cellulose and hemicellulose form the primary matrix of the cell wall, imparting rigidity and enabling cell expansion during growth. Pectin plays a vital role in cell adhesion and wall flexibility, facilitating the development of various plant organs. Additionally, polysaccharides derived from plant growth-promoting rhizobacteria (PGPR), such as *Burkholderia* sp. BK01, enhance root and shoot elongation and biomass by stimulating nutrient uptake and hormone signaling, thus promoting overall growth (Chen et al. 2024a; Hoffmann et al. 2021).

In terms of energy storage, starch, composed of amylose and amylopectin, serves as a crucial reserve carbohydrate that supports metabolic processes during the germination and growth phases (Chen et al. 2024a).

Polysaccharides also function as signaling molecules under normal conditions. Oligosaccharins, which are fragments of cell wall polysaccharides like oligogalacturonides, act as signaling compounds that modulate gene expression related to morphogenesis, flowering, and root development. Furthermore, exopolysaccharides (EPSs) produced by rhizobacteria trigger symbiotic signaling pathways that promote root nodulation and enhance nutrient exchange between plants and their microbial partners. Collectively, these roles underscore the multifaceted contributions of polysaccharides to plant health and development (Mensink et al. 2015).

Under abiotic stress conditions, polysaccharides play vital roles in helping plants adapt, particularly in response to salt stress. Exopolysaccharides (EPS) derived from plant growth-promoting rhizobacteria (PGPR), such as those from *Burkholderia* sp. BK01, have been shown to increase the potassium (K^+)-to-sodium (Na^+) ratio, thereby enhancing the activity of antioxidant enzymes like superoxide dismutase (SOD), catalase (CAT), and peroxidase (POD). This response reduces oxidative damage in plants like *Arabidopsis*. Additionally, natural polysaccharides extracted from halophytes, such as *Lobularia maritima*, improve membrane stability and ion homeostasis in durum wheat when exposed to saline environments (Chen et al. 2024a, b; Morcillo and Manzanera 2021; Bouteraa et al. 2022).

Polysaccharides also contribute to the mitigation of oxidative stress by scavenging reactive oxygen species (ROS). They achieve this by boosting proline synthesis and enhancing antioxidant activity, leading to lower levels of lipid peroxidation and hydrogen peroxide (H_2O_2). Furthermore, PSLm has been found to upregulate stress-responsive genes, including NHX1, SOS1, and SOD, which collectively enhance salt tolerance in plants. These mechanisms illustrate the essential roles that polysaccharides play in promoting plant resilience and adaptation to challenging environmental conditions (Zha et al. 2023; Zhao et al. 2022).

13.3 Saccharides' Biochemical Properties

Plant saccharides, including monosaccharides, oligosaccharides, and polysaccharides, exhibit diverse biochemical properties that are crucial for plant structure, metabolism, signaling, and interaction with the environment. Recent studies have provided a detailed understanding of their composition, structural features, and functional roles. In terms of key biochemical properties, plant polysaccharides are macromolecules composed of identical or different monosaccharides linked via α- or β-glycosidic bonds, which determine their structure and function. The nature of these glycosidic linkages, such as α-1,4 in starch and β-1,4 in cellulose, influences solubility, digestibility, and mechanical properties. Additionally, strong intermolecular hydrogen bonding and other interactions confer high stability to plant polysaccharides, making them resistant to pH changes and enzymatic degradation. This stability is essential for maintaining cell wall integrity and resistance to environmental stresses (Salehi and Rashidinejad 2025; Chen et al. 2020).

The molecular weight, degree of branching, and substitution patterns of polysaccharides (such as sulfation and acetylation) significantly affect their physicochemical properties and bioactivities. For instance, the branching in amylopectin influences starch granule architecture and energy mobilization. Furthermore, functional groups such as hydroxyl groups and reducing ends enable chemical modifications—like sulfation and phosphorylation—that can modulate solubility, bioactivity, and interactions with other biomolecules. These modifications can enhance various biological activities, including antioxidant, immunomodulatory, and antitumor effects (Li et al. 2025a; Niu et al. 2023).

Last, plant polysaccharides play a vital role in the microbiome and metabolism by promoting the formation of short-chain fatty acids (SCFAs) by gut microbiota, which are essential for host health. They also regulate carbohydrate-active enzymes (CAZymes), enhancing polysaccharide metabolism and associated biological effects. Collectively, these properties and activities underscore the importance of plant saccharides in plant physiology and their beneficial impacts on human health (Chen et al. 2024b; Xie et al. 2016).

13.4 Plant Stress Responses

Recent advances in understanding plant abiotic stress responses have highlighted the complex molecular, physiological, and developmental adaptations that enable plants to survive and acclimate to challenging environments. An overview of plant abiotic stress responses reveals that plants often encounter multiple and combined stresses, such as heat coupled with drought or flooding followed by drought, which elicit unique phenotypic and transcriptomic responses distinct from those triggered by single stresses. Key regulators, including EARLY FLOWERING 6 (ELF6) and ARABIDOPSIS TÓXICOS EN LEVADURA 80 (ATL80), play crucial roles in modulating growth and survival under these combined stress conditions (Jiang et al. 2025).

Hormonal regulation is central to mediating tolerance to osmotic stresses, such as drought and salinity, with abscisic acid (ABA) playing a pivotal role by regulating stomatal closure and gene expression. Other hormones, including jasmonic acid (JA), salicylic acid (SA), and brassinosteroids (BRs), interact in complex ways, depending on the type of stress and plant species involved (Kashtoh et al. 2025).

Signal transduction pathways are also vital, with mitogen-activated protein kinases (MAPKs), receptor-like kinases (RLKs), and calcium-dependent protein kinases (CDPKs) transducing stress signals to activate downstream protective mechanisms. Reactive oxygen species (ROS) function as signaling molecules that integrate multiple stress signals and trigger antioxidant defenses (Pinski et al. 2025).

Additionally, epigenetic and transcriptional regulation through histone modifications, such as H3K4me3, and specific transcription factors shape gene expression patterns during stress acclimation. Stress-responsive genes include those encoding ion transporters like NHX1 and SOS1, enzymes involved in osmoprotectant biosynthesis, and antioxidant proteins such as superoxide dismutase (SOD) and catalase (CAT) (Moustakas 2025).

Physiologically and morphologically, plants exhibit adaptations to cope with abiotic stresses: During drought, they promote root elongation to access deeper water sources, while salt stress typically restricts root growth. Stomatal regulation is the key to balancing transpiration and photosynthesis under heat and drought conditions, and flooding can inhibit photosynthesis and induce leaf senescence. Last, secondary metabolites significantly contribute to osmotic adjustment, ROS scavenging, and membrane stabilization, thereby enhancing stress tolerance in forage crops and other plant species (Mishra et al. 2025).

13.5 Roles of Saccharides in Stress Responses

Saccharides play multifaceted and crucial roles in plant abiotic stress responses, acting as osmoprotectants, signaling molecules, and metabolic regulators to enhance plant tolerance and adaptation, as shown in Table 13.1. Recent research highlights the dynamic involvement of sugars in stress perception, signal transduction, and protective mechanisms. Under stresses such as drought, salinity, and cold, plants accumulate soluble sugars (e.g., glucose, fructose, sucrose, and trehalose) that function as osmolytes to maintain cell turgor and protect cellular structures from dehydration and ionic imbalance. Dynamic starch–sugar interconversion between source and sink tissues supports energy supply and stress adaptation (Jiang et al. 2025).

Sugars act as signaling entities regulating gene expression and hormonal pathways. For example, sucrose and trehalose-6-phosphate (T6P) modulate abscisic acid (ABA) signaling, which is central to osmotic stress tolerance. Sugar signaling pathways interact with calcium-dependent protein kinases, MAP kinases, and SnRKs to orchestrate stress responses (Kashtoh et al. 2025; Pinski et al. 2025).

Table 13.1 Roles of saccharides in abiotic stress responses in plants

Saccharide type	Example	Stress type	Role in stress response	Mechanism	Reference
Monosaccharide	Glucose	Drought	Osmoprotection, signaling	Maintains turgor; activates stress-responsive genes	Kaur et al. (2021)
Disaccharide	Trehalose	Heat	Cryoprotection, energy supply	Stabilizes membranes, enhances ROS scavenging	Zheng et al. (2024)
Oligosaccharide	Raffinose	Cold,	Antioxidant, osmoprotection	Inhibits lipid peroxidation, scavenges ROS	Gu et al. (2018)
Polysaccharide	Fructans	Drought	Stabilization of membranes and proteins	Binds water, prevents cellular damage	Zhou et al. (2023)
Polysaccharide	Exopolysaccharide	Salinity	Structural barrier and ion binding	Chelates toxic ions, maintains cell wall integrity	Li et al. (2025a, b)

Sugars influence antioxidant defense by modulating enzymes such as superoxide dismutase (SOD), catalase (CAT), and peroxidase (POD), helping balance ROS production and scavenging during stress. This antioxidative role reduces lipid peroxidation and protects membrane integrity (Mishra et al. 2023; Zhang et al. 2023).

Sugar accumulation affects expression of stress-responsive genes (e.g., NHX1 and SOS1) involved in ion homeostasis and osmoprotection. Cross talk between sugar signaling and hormones like ABA, jasmonic acid (JA), and salicylic acid (SA) fine-tunes plant acclimation to combined stresses (Jiang et al. 2025; Zhang et al. 2025b).

Microbial exopolysaccharides (EPS) and plant-derived polysaccharides enhance stress tolerance by modulating ion transport, antioxidant systems, and osmotic balance, highlighting saccharides' role in plant–microbe interactions under stress (Mishra et al. 2025; Pinski et al. 2025).

13.5.1 Saccharides as Signaling Molecules

Saccharides function as important signaling molecules in plant responses to drought stress, modulating physiological and molecular pathways to enhance tolerance. Recent research has highlighted their roles in osmotic adjustment, hormone cross talk, and the activation of stress-responsive genes. In terms of osmotic adjustment and energy supply, soluble sugars such as sucrose, glucose, and trehalose accumulate in response to drought conditions, acting as osmoprotectants that help maintain cell turgor and protect cellular membranes. These sugars also serve as vital energy sources to support the increased metabolic demands associated with stress (Ali et al. 2025).

Regarding sugar signaling and hormone interaction, sugars play a critical role in regulating drought responses by interacting with abscisic acid (ABA) signaling, which is the central hormone involved in drought adaptation. For example, the accumulation of sugars can modulate ABA biosynthesis and influence signaling components such as Pyrabactin Resistance/Pyrabactin Resistance-Like (PYR/PYL) receptors, protein phosphatase 2Cs (PP2Cs), and SnRK2 kinases, all of which control stomatal closure and gene expression. Moreover, sugars also influence cross talk with other hormones, including salicylic acid (SA), ethylene (ET), and auxin, thereby fine-tuning stomatal behavior and overall stress responses (Ali et al. 2025; Wang et al. 2020; Supriya et al. 2025).

Additionally, sugar signaling triggers the expression of drought-responsive genes that are crucial for osmoprotection, antioxidant defense, and ion transport. For instance, these signaling pathways can induce the production of antioxidant enzymes that mitigate the effects of reactive oxygen species (ROS) generated during drought stress (Du et al. 2025).

Finally, sugars interact with calcium-dependent protein kinases (CDPKs) and are integrated into ROS signaling networks, which amplify drought signals and regulate downstream responses such as stomatal closure and the activation of stress-related genes. This multifaceted role of saccharides underscores their importance in plant resilience under drought conditions (Li et al. 2023).

Saccharides act as critical signaling molecules in plant responses to salt stress by modulating osmotic balance, hormone signaling, ion homeostasis, and antioxidant defenses. Recent studies provide insights into how sugar metabolism and signaling integrate with calcium, reactive oxygen species (ROS), and hormonal pathways to enhance salt tolerance. Salt stress alters sugar levels, such as those of sucrose and fructose, affecting source–sink sugar transport and glycolytic fluxes. This metabolic remodeling supports osmotic adjustment and the energy supply needed for stress adaptation (Abuslima et al. 2025; Peng et al. 2025a, b).

Salt stress triggers rapid increases in cytosolic Ca^{2+}, which activates calcium sensors (e.g., CBL–CIPK complexes) that regulate ion transporters like SOS1 to maintain Na^+/K^+ homeostasis. ROS production via NADPH oxidases (e.g., AtrbohF) is modulated by sugar signaling and calcium, forming a feedback loop that fine-tunes stress responses (Xiao and Zhou 2023; Zhou et al. 2024b).

Sugars interact with abscisic acid (ABA) signaling pathways, influencing stomatal closure and gene expression. ABA and Ca^{2+} signaling converge to regulate ROS homeostasis and activate salt tolerance mechanisms, including phosphorylation cascades involving SnRK2 kinases (Xiao and Zhou 2023; Zhang et al. 2025b). Sugar signaling modulates the expression of ion transporters (e.g., SOS1 and NHX1), antioxidant enzymes (SOD and CAT), and osmoprotectant biosynthesis genes, enhancing cellular protection against salt-induced oxidative and ionic stress (Zhang et al. 2025b).

Exopolysaccharides from beneficial microbes and natural polysaccharides act as biostimulants, promoting osmolyte accumulation and ROS scavenging, thereby improving salt tolerance. Saccharides serve as important signaling molecules in plant heat stress responses, modulating physiological and molecular pathways that enhance thermotolerance. Recent studies reveal their involvement in heat perception, signal transduction, and the activation of protective mechanisms, often interacting with calcium, ROS, and hormone signaling. Under heat stress, plants accumulate soluble sugars such as sucrose and trehalose, which stabilize proteins and membranes, acting as osmoprotectants and molecular chaperones to maintain cellular homeostasis. Heat stress induces cytosolic Ca^{2+} influx, activating calcium sensors and kinases (e.g., CALMODULIN3 and CDPKs) that interact with sugar signaling pathways. Reactive oxygen species (ROS), particularly H_2O_2, function as signaling molecules mediating heat shock responses and thermomemory, with sugars modulating antioxidant defenses to balance ROS levels (Naaz et al. 2025; Peng et al. 2025a, b).

Sugars influence the expression and activity of Heat Shock Factor (HSFs), which regulate heat shock proteins (HSPs) crucial for protein folding and protection under heat. Trehalose and sucrose signaling can modulate HSFs and downstream gene expression to confer thermotolerance (Fortunato et al. 2023). Heat stress alters polysaccharide metabolism in the cell wall (e.g., pectin demethylesterification), affecting wall elasticity and cell expansion. Sugar-derived signaling molecules participate in these modifications, facilitating adaptation to heat (Lu et al. 2025). Sugars contribute to heat stress memory by regulating gene expression patterns and chromatin modifications, enabling plants to respond more effectively to repeated heat stress events (Naaz et al. 2025; Jiang et al. 2025).

13.5.2 Metabolic Adjustment During Stress

Plants dynamically adjust saccharide metabolism under abiotic stresses such as drought, salinity, and wounding, to maintain energy balance, osmotic homeostasis, and signaling functions. This involves coordinated synthesis, degradation, transport, and signaling of sugars like sucrose, glucose, fructose, and trehalose. Stress conditions trigger both the synthesis and degradation of soluble sugars, balancing energy supply and osmoprotection. For example, drought induces starch breakdown into soluble sugars to maintain the osmotic balance and provide energy (Deng et al. 2025).

Abiotic stress affects sugar partitioning between source and sink tissues. Salt stress alters the expression of sucrose transporters such as SWEET5 (Sugars Will Eventually be Exported Transporter 5) and SUT (Sucrose Transporter), modulating sugar unloading in roots or grains to optimize resource allocation under stress (Abuslima et al. 2025). Sugars influence reactive oxygen species (ROS) levels by fueling NADPH production via the pentose phosphate pathway and directly scavenging ROS. This dual role helps mitigate oxidative damage during stress (Keunen et al. 2013). Plants sense sugars via hexokinase-dependent and independent pathways, triggering downstream signaling cascades involving calcium, cyclic nucleotides, and G-proteins that regulate growth and stress responses (Li 2022).

13.6 Regulatory Functions of Saccharides

Sugars, such as sucrose, serve as primary transport forms of carbon and energy, fueling various metabolic processes essential for plant growth and development. Beyond their metabolic roles, these sugars also act as signaling molecules that regulate gene expression, enzyme activities, and developmental transitions. For example, glucose, sucrose, and trehalose-6-phosphate (T6P) are pivotal in regulating growth phases, flowering time, and stress responses by modulating transcription factors and signaling kinases (Jeandet et al. 2022; Ciereszko 2018).

In terms of growth and development regulation, sugars influence key developmental processes, including vegetative phase change, apical dominance, flowering, and fruiting. T6P, in particular, signals the plant's sucrose status, allowing it to coordinate growth with carbon availability. Additionally, sugar signaling interacts with phytohormones and protein kinases, such as SnRK1 and target of rapamycin (TOR) kinase, which mediate growth and stress responses. High sugar levels activate TOR to promote growth, while sugar deprivation activates SnRK1, prompting the plant to conserve resources (Zhai et al. 2021; Lastdrager et al. 2014).

Sugars also play a crucial role in regulating protein synthesis and metabolic enzyme activities, including those involved in starch and fatty acid biosynthesis. For instance, T6P inhibits SnRK1 to promote anabolic processes, while sucrose activates enzymes that facilitate starch and anthocyanin synthesis. Furthermore, oligosaccharides derived from cell wall polysaccharides can function as regulatory molecules, known as oligosaccharins, which trigger defense responses against pathogens and regulate morphogenesis, influencing cell growth rate, flowering, and rooting. Collectively, these diverse functions of sugars underscore their integral role in plant physiology and development (Cosgrove 2024; Jeandet et al. 2022).

13.7 Influence of Saccharides on Development

13.7.1 *Germination Stage*

Saccharides play multifaceted regulatory roles during seed germination, primarily by interacting with hormonal pathways and signaling networks to control the timing and progression of germination. One key regulatory function of saccharides during germination is the modulation of abscisic acid (ABA) signaling. Sugars, particularly glucose, influence seed germination by regulating cellular ABA concentrations and response pathways. High sugar levels can enhance ABA signaling, which inhibits germination, allowing sugars to act as negative regulators under certain conditions by maintaining dormancy or delaying germination (Wang et al. 2020; Price et al. 2003).

In addition, sugar signaling involves interaction with ABA-responsive transcription factors such as ABI4 and ABI5, which are crucial for regulating ABA responses during germination. For instance, ABI5 helps maintain the embryo in a quiescent state post-germination, and sugar signaling can influence these transcription factors to control the timing of germination. Moreover, saccharides facilitate cross talk with gibberellic acid (GA) signaling pathways, which promote germination. The protein gibberellic acid-stimulated Arabidopsis6 (GASA6) functions as an integrator of glucose, ABA, and GA signaling, coordinating the balance between dormancy and germination (Wang et al. 2020).

Trehalose-6-phosphate (T6P) serves as a key signal during this process, acting as a proxy for sucrose availability. T6P promotes germination by reducing seed sensitivity to glucose and ABA, favoring germination by antagonizing ABA signaling and supporting the transition to seedling growth. Furthermore, sugar sensing via hexokinase 1 (HXK1) is critical, as glucose sensing through HXK1-dependent and independent pathways affects gene expression linked to germination. Glucose can repress positive regulators of germination, such as AtGASA6, through HXK1, modulating the hormonal balance necessary for radicle protrusion (Wang et al. 2020).

After germination, the regulation of seedling establishment is also influenced by saccharides. Sugars released from the hydrolysis of storage compounds provide essential carbon and osmotic support for radicle expansion. They also signal to coordinate early seedling growth, including hypocotyl elongation and root meristem development, often in conjunction with cytokinins (CKs) and auxin signaling (Wang et al. 2020).

Last, the sucrose transporter SUT4/Cyb5-2 mediates the sugar-dependent inhibition of germination through pathways independent of ABA signaling, indicating that multiple sugar-regulated mechanisms are involved in controlling the germination process (Wang et al. 2022).

13.7.2 Seedling Stage

Saccharides (sugars) play a crucial role in regulating seedling development by functioning as signaling molecules and energy sources that coordinate growth, metabolic transitions, and stress responses. Their diverse functions involve interactions with hormonal pathways, transporters, and metabolic enzymes, all aimed at optimizing early plant establishment. One of the key regulatory functions of saccharides during seedling development is growth stimulation through sugar signaling. Optimal sugar availability accelerates seedling growth by activating the target of rapamycin (TOR) kinase pathway, which promotes protein synthesis and cell division. Elevated levels of sucrose and glucose stimulate hypocotyl elongation and root development, contributing to robust seedling architecture (Wingler and Henriques 2022; Yuan and Wysocka-Diller 2006).

In addition to direct growth stimulation, sugars also facilitate hormonal cross talk. They interact with gibberellin (GA) and cytokinin pathways to promote growth, with glucose enhancing GA signaling while simultaneously suppressing responses mediated by abscisic acid (ABA). This dynamic interaction facilitates resource allocation toward root and shoot expansion, ensuring that seedlings can effectively utilize the available nutrients (Ciereszko 2018).

Trehalose-6-phosphate (Tre6P) serves as a critical metabolic sensor during seedling development. As a sucrose-derived signal, Tre6P reflects the carbon status of the plant. High levels of Tre6P activate anabolic processes, such as starch synthesis, while inhibiting the SnRK1 kinase, thereby ensuring that growth aligns with energy availability (Wingler and Henriques 2022).

The role of sugar transporters in resource partitioning is also essential. Transporters such as SWEETs and SUTs mediate the movement of sucrose to sink tissues. In seedlings, CoSWEET1b (symplasmic) and CoSUT2 (apoplasmic) ensure efficient sugar delivery to roots and shoots, supported by invertases that maintain hexose gradients necessary for cell division (Guo et al. 2023).

Finally, sugars contribute to stress adaptation in seedlings. For instance, compounds like raffinose accumulate under abiotic stress conditions to stabilize cellular membranes and scavenge reactive oxygen species (ROS), thereby enhancing seedling resilience. This accumulation integrates with osmotic adjustment and antioxidant defense mechanisms, further supporting the plant's ability to withstand environmental challenges. Overall, the multifaceted roles of saccharides are vital for successful seedling development and establishment.

13.8 Future Directions

Recent advances and emerging trends indicate that saccharides will play increasingly pivotal roles in enhancing plant growth, development, and resilience to abiotic stresses. One key future direction involves deciphering the intricate sugar signaling networks and their cross talk with hormonal, epigenetic, and environmental signals. A deeper understanding of how saccharides such as sucrose, trehalose, and various polysaccharides modulate gene expression, influence developmental phase transitions, and activate stress-responsive networks is essential for advancing agricultural practices.

By elucidating these complex interactions, researchers can develop targeted strategies to manipulate plant development and improve stress tolerance. This knowledge will not only enhance our understanding of plant biology but also provide practical applications in breeding and biotechnology aimed at creating crops that can better withstand environmental challenges. As the demand for sustainable agriculture grows, harnessing the regulatory potential of saccharides will be critical in promoting resilient and high-yielding plant varieties.

Exploiting saccharides as biostimulants and priming agents represents a promising approach to enhance plant tolerance to various abiotic stresses, including drought, salinity, heat, and oxidative stress. The development of saccharide-based biostimulants can effectively prime plants, preparing them to better withstand environmental challenges. Natural and microbial polysaccharides can be harnessed to activate key physiological mechanisms, such as antioxidant defenses, osmoprotection, and ion homeostasis. These mechanisms help mitigate the adverse effects of abiotic stress, ultimately reducing the impact on plant growth and productivity.

By leveraging the beneficial properties of saccharides, researchers and agricultural practitioners can create innovative solutions that promote resilience in crops, contributing to sustainable agriculture and food security in the face of climate change and environmental degradation. This area of research not only highlights the versatility of saccharides in plant biology but also opens up new avenues for developing effective agricultural inputs that enhance plant performance under stress conditions.

The integration of genomics and metabolomics for trait improvement offers a powerful approach to accelerate the breeding of stress-resilient crops by leveraging genomic tools and pan-genome datasets to identify sugar metabolism and signaling genes linked to desirable traits. Multi-omics approaches that combine transcriptomics, metabolomics, and phenotyping will provide a comprehensive understanding of saccharide functions across various developmental stages and stress conditions, enabling the identification of key regulatory pathways.

Moreover, advancements in the sustainable production and utilization of saccharides are essential for supporting agriculture and industry while reducing reliance on fossil resources. Innovations in bio-based production from renewable feedstocks, coupled with low-energy conversion technologies and the valorization of agricultural residues, will expand the availability of saccharides for crop improvement and the development of bioproducts.

In addition, engineering sugar metabolism through genetic engineering and genome editing presents a promising route to enhance carbon allocation, growth regulation, and stress adaptation in plants. By targeting sugar biosynthesis, transporters, and signaling components, researchers can create crops with improved resilience to climate challenges. Furthermore, modulating sugar-related epigenetic marks may provide a means to establish durable stress memory and enhance developmental plasticity.

Finally, understanding the roles of saccharides in microbiome interactions will reveal novel strategies to improve plant health and stress tolerance through beneficial microbes. Exploring how plant saccharides influence root and phyllosphere microbiomes, as well as optimizing the interplay between microbial exopolysaccharides and plant polysaccharides, can lead to synergistic effects that bolster abiotic stress resilience. Collectively, these integrative strategies highlight the potential of saccharides in advancing sustainable agriculture and enhancing crop performance in changing environments.

13.9 Conclusion

Plants must acclimate to altered environmental circumstances to guarantee life and reproduction. Stressors, including salinity, drought, heat, flooding, and cold, impede plant growth. This stress can result in premature mortality in severe instances. The perception of sugar signals initiates several reactions, encompassing alterations in gene expression, protein accumulation, and metabolite accumulation. This then affects physiological factors, including osmotic adjustments, turgor pressure, blooming period, and cell division and elongation. A variety of signal transduction pathways participate in sugar signaling. Gene expression reliant on sugar transpires within seconds after sugar introduction and is governed by transcription factors, which are influenced by protein phosphorylation, chromatin remodeling, and the recruitment of RNA.

Higher plants synthesize both simple and complex carbohydrates to guarantee energy availability, metabolic equilibrium, and fundamental physiological functions. These sugars function as vital sources of energy, carbon, and structural components for plants. They function as organic osmolytes, regulating cellular osmolarity, safeguarding proteins from denaturation, and neutralizing harmful free radicals. Monosaccharides serve as crucial signaling molecules that modulate gene expression, metabolic processes, developmental pathways, and plant responses to environmental stimuli. Numerous sugar-sensing and signal transduction mechanisms have been elucidated. Recent years have seen an enhanced comprehension of the functions of complex sugars in plant development and stress responses. Plant sugars are also involved in the application of biostimulants for crops under abiotic stresses.

References

Abuslima E, Kanbar A, Ismail A, Raorane ML, Eiche E, El-Sharkawy I, Junker BH, Riemann M, Nick P (2025) Salt stress-induced remodeling of sugar transport: a role for promoter alleles of SWEET13. Sci Rep 15:7580. https://doi.org/10.1038/s41598-025-90432-2

Afzal S, Chaudhary N, Singh NK (2021) Role of soluble sugars in metabolism and sensing under abiotic stress. In: Aftab T, Hakeem KR (eds) Plant growth regulators. Springer, Cham. https://doi.org/10.1007/978-3-030-61153-8_14

Akensous F-Z, Anli M, Meddich A (2022) Biostimulants as innovative tools to boost date palm (*Phoenix dactylifera* L.) performance under drought, salinity, and heavy metal(oid)s' stresses: a concise review. Sustainability 14(23):15984. https://doi.org/10.3390/su142315984

Alagoz SM, Lajayer BA, Ghorbanpour M (2023) Proline and soluble carbohydrates biosynthesis and their roles in plants under abiotic stresses. In: Ghorbanpour M, Shahid MA (eds) Plant stress mitigators. Academic Press, Elsevier, pp 169–185. https://doi.org/10.1016/B978-0-323-89871-3.00027-6

Ali S, Mir RA, Haque MA, Danishuddin, Almalki MA, Alfredan M, Khalifa A, Mahmoudi H, Shahid M, Tyagi A, Mir ZA (2025) Exploring physiological and molecular dynamics of drought stress responses in plants: challenges and future directions. Front Plant Sci 16:1565635. https://doi.org/10.3389/fpls.2025.1565635

Bouteraa MT, Mishra A, Romdhane WB, Hsouna AB, Siddique KHM, Saad RB (2022) Biostimulating effect of natural polysaccharides from Lobularia maritima on durum wheat seedlings: improved plant growth, salt stress tolerance by modulating biochemical responses and ion homeostasis. Plants 11(15):1991. https://doi.org/10.3390/plants11151991
Broghammer A, Krusell L, Blaise M, Sauer J, Sullivan JT, Maolanon N, Vinther M, Lorentzen A, Madsen EB, Jensen KJ, Roepstorff P (2012) Legume receptors perceive the rhizobial lipochitin oligosaccharide signal molecules by direct binding. Proc Natl Acad Sci USA 109(34):13859–13864. https://doi.org/10.1073/pnas.1205171109
Chen Y, Jia Y, Guo Q (2020) Polysaccharides and polysaccharide complexes as potential sources of antidiabetic compounds: a review. In: Atta-ur-Rahman (ed) Studies in natural products chemistry, vol 67. Elsevier, pp 199–220. https://doi.org/10.1016/B978-0-12-819483-6.00006-0
Chen E, Yang C, Tao W, Li S (2024a) Polysaccharides produced by plant growth-promoting rhizobacteria strain *Burkholderia* sp. BK01 enhance salt stress tolerance to *Arabidopsis thaliana*. Polymers 16(1):145. https://doi.org/10.3390/polym16010145
Chen Y, Li H, Lai F, Min T, Wu H, Zhan Q (2024b) The influence and mechanisms of natural plant polysaccharides on intestinal microbiota-mediated metabolic disorders. Foods 13(23):3882. https://doi.org/10.3390/foods13233882
Cho YG, Kang KK (2020) Functional analysis of starch metabolism in plants. Plants 9(9):1152. https://doi.org/10.3390/plants9091152
Cho YH, Yoo SD (2011) Signaling role of fructose mediated by FINS1/FBP in *Arabidopsis thaliana*. PLoS Genet 7(1):e1001263. https://doi.org/10.1371/journal.pgen.1001263
Choudhary A, Kumar A, Kaur N, Kaur H (2022) Molecular cues of sugar signaling in plants. Physiol Plant 174(1):e13630. https://doi.org/10.1111/ppl.13630
Ciereszko I (2018) Regulatory roles of sugars in plant growth and development. Acta Soc Bot Pol 87(2):3583. https://doi.org/10.5586/asbp.3583
Cosgrove DJ (2024) Structure and growth of plant cell walls. Nat Rev Mol Cell Biol 25(5):340–358. https://doi.org/10.1038/s41580-023-00691-y
Delorge I, Janiak M, Carpentier S, Van Dijck P (2014) Fine tuning of trehalose biosynthesis and hydrolysis as novel tools for the generation of abiotic stress tolerant plants. Front Plant Sci 5:147. https://doi.org/10.3389/fpls.2014.00147
Deng H, Xie P, Han Y, Song X, Ge Y (2025) Starch-sugar metabolism in response to wound stress in apple fruit. Plant Physiol Biochem 227:110142. https://doi.org/10.1016/j.plaphy.2025.110142
Du W, Huang H, Wang Z, Yang G, Pang Y (2025) The calcium-regulated kinase MsCIPK4 confers drought and salt stress tolerance in alfalfa by enhancing ROS scavenging. Plant Sci 359:112646. https://doi.org/10.1016/j.plantsci.2025.112646
Etzler ME (1998) Oligosaccharide signaling of plant cells. J Cell Biochem 30–31:123–128
Eveland AL, Jackson DP (2012) Sugars, signalling, and plant development. J Exp Bot 63(9):3367–3377. https://doi.org/10.1093/jxb/err379
Fañanás-Pueyo I, Carrera-Castaño G, Pernas M, Oñate-Sánchez L (2025) Signalling and regulation of plant development by carbon/nitrogen balance. Physiol Plant 177(2):e70228. https://doi.org/10.1111/ppl.70228
Ferrari S, Savatin DV, Sicilia F, Gramegna G, Cervone F, De Lorenzo G (2013) Oligogalacturonides: plant damage-associated molecular patterns and regulators of growth and development. Front Plant Sci 4:49. https://doi.org/10.3389/fpls.2013.00049
Fortunato S, Lasorella C, Dipierro N, Vita F, de Pinto MC (2023) Redox signaling in plant heat stress response. Antioxidants 12(3):605. https://doi.org/10.3390/antiox12030605
Fry SC, Suzanne Aldington P, Hetherington R, Aitken J (1993) Oligosaccharides as signals and substrates in the plant cell wall. Plant Physiol 103(1):1–5. http://www.jstor.org/stable/4275348
Gentile D, Serino G, Frugis G (2024) CRF transcription factors in the trade-off between abiotic stress response and plant developmental processes. Front Genet 15:1377204. https://doi.org/10.3389/fgene.2024.1377204

Gu H, Lu M, Zhang Z, Xu J, Cao W, Miao M (2018) Metabolic process of raffinose family oligosaccharides during cold stress and recovery in cucumber leaves. J Plant Physiol 224:112–120

Guarnizo N, Oliveros D, Murillo-Arango W, Bermúdez-Cardona MB (2020) Oligosaccharides: defense inducers, their recognition in plants, commercial uses and perspectives. Molecules 25(24):5972. https://doi.org/10.3390/molecules25245972

Guo W-J, Pommerrenig B, Neuhaus HE, Keller I (2023) Interaction between sugar transport and plant development. J Plant Physiol 288:154073. https://doi.org/10.1016/j.jplph.2023.154073

Han Y, Liang A, Xu D, Zhang Y, Shi J, Li M, Liu T, Qi H (2024) Versatile roles of trehalose in plant growth and development and responses to abiotic stress. Veg Res 4:e007. https://doi.org/10.48130/vegres-0024-0007

Hernández-Bernal AF, Gregorio-Jorge J, León P (2022) The role of sugars as signaling molecules in plants. TIP Rev Esp Cienc Quim Biol 25(1):1–20

Hoffmann N, King S, Samuels AL, McFarlane HE (2021) Subcellular coordination of plant cell wall synthesis. Dev Cell 56(7):933–948. https://doi.org/10.1016/j.devcel.2021.03.004

Höftberger M, Althammer M, Foissner I, Tenhaken R (2022) Galactose induces formation of cell wall stubs and cell death in Arabidopsis roots. Planta 256(2):26. https://doi.org/10.1007/s00425-022-03919-x

Jeandet P, Formela-Luboińska M, Labudda M, Morkunas I (2022) The role of sugars in plant responses to stress and their regulatory function during development. Int J Mol Sci 23(9):5161. https://doi.org/10.3390/ijms23095161

Jennings DB, Ehrenshaft M, Pharr DM, Williamson JD (1998) Roles for mannitol and mannitol dehydrogenase in active oxygen-mediated plant defense. Proc Natl Acad Sci USA 95(25):15129–15133. https://doi.org/10.1073/pnas.95.25.15129

Jiang Z, van Zanten M, Sasidharan R (2025) Mechanisms of plant acclimation to multiple abiotic stresses. Commun Biol 8:655. https://doi.org/10.1038/s42003-025-08077-w

Kashtoh H, Rabbee MF, Baek KH (2025) New insights into plant signaling mechanisms in biotic and abiotic stress. Plants 14(13):1953. https://doi.org/10.3390/plants14131953

Kaur H, Manna M, Thakur T, Gautam V, Salvi P (2021) Imperative role of sugar signaling and transport during drought stress responses in plants. Physiol Plant 171(4):833–848. https://onlinelibrary.wiley.com/doi/epdf/10.1111/ppl.13364

Keunen E, Peshev D, Vangronsveld J, Van den Ende W, Cuypers A (2013) Plant sugars are crucial players in the oxidative challenge during abiotic stress: extending the traditional concept. Plant Cell Environ 3636:1242–1255. https://doi.org/10.1111/pce.12061

Lastdrager J, Hanson J, Smeekens S (2014) Sugar signals and the control of plant growth and development. J Exp Bot 65(3):799–807. https://doi.org/10.1093/jxb/ert474

Lemoine R, La Camera S, Atanassova R, Dédaldéchamp F, Allario T, Pourtau N, Bonnemain J-L, Laloi M, Coutos-Thévenot P, Maurousset L, Faucher M, Girousse C, Lemonnier P, Parrilla J, Durand M (2013) Source-to-sink transport of sugar and regulation by environmental factors. Front Plant Sci 4:272. https://doi.org/10.3389/fpls.2013.00272

Li ZG (2022) Sugar signaling in plants under physiological and stress conditions. In: Biostimulants for crop production and sustainable agriculture. CABI, pp 372–385

Li Y, Sun W, Yao Y, Zhang L, Xu S, Zhang Q, Huang T (2023) FRUCTOSE INSENSITIVE1 regulates stem cell function in Arabidopsis in response to fructose signalling. J Exp Bot 74(10):3060–3073. https://doi.org/10.1093/jxb/erad067

Li J, Hu Y, Hu J, Xie Q, Chen X, Qi X (2024) Sucrose synthase: an enzyme with multiple roles in plant physiology. J Plant Physiol 303:154352. https://doi.org/10.1016/j.jplph.2024.154352

Li Z, Zhang M, Liu Y, Li M, Qi J (2025a) Advances in the preparation, structure-activity relationship, and biological mechanisms of Agaricus polysaccharides. Food Sci Human Wellness 15:1–26. https://doi.org/10.26599/FSHW.2025.9250590

Li Y, Zhang J, Wang X, Feng Z, Yang E, Wu M, Jiang Y, Huang J, Gao Z, Du Y (2025b) The synergistic effect of extracellular polysaccharide-producing salt-tolerant bacteria and biochar promotes grape growth under saline-alkaline stress. Environ Technol Innov 38:104070

López-Gómez M, Lluch C (2012) Trehalose and abiotic stress tolerance. In: Ahmad P, Prasad M (eds) Abiotic stress responses in plants. Springer, New York. https://doi.org/10.1007/978-1-4614-0634-1_14
Lu C, Li W, Feng X, Chen J, Hu S, Tan Y, Wu L (2025) The dynamic remodeling of plant cell wall in response to heat stress. Genes 16(6):628. https://doi.org/10.3390/genes16060628
Lubbe FC, Klimeš A, Doležal J, Jandová V, Mudrák O, Janeček Š, Bartušková A, Klimešová J (2021) Carbohydrate storage in herbs: the forgotten functional dimension of the plant economic spectrum. Ann Bot 127(6):813–825. https://doi.org/10.1093/aob/mcab014
Ma L, Han R, Yang Y, Liu X, Li H, Zhao X, Li J, Fu H, Huo Y, Sun L, Yan Y, Zhang H, Li Z, Tian F, Li J, Guo Y (2023) Phytochromes enhance SOS2-mediated PIF1 and PIF3 phosphorylation and degradation to promote Arabidopsis salt tolerance. Plant Cell 35(8):2997–3020. https://doi.org/10.1093/plcell/koad117
Martínez-Noël GMA, Tognetti JA (2018) Sugar signaling under abiotic stress in plants. In: Parvaiz A et al (eds) Plant metabolites and regulation under environmental stress. Academic Press, Elsevier, pp 397–406. https://doi.org/10.1016/B978-0-12-812689-9.00022-4
Meena M, Prasad V, Zehra A, Gupta VK, Upadhyay RS (2015) Mannitol metabolism during pathogenic fungal–host interactions under stressed conditions. Front Microbiol 6:1019. https://doi.org/10.3389/fmicb.2015.01019
Mensink MA, Frijlink HW, van der Voort Maarschalk K, Hinrichs WL (2015) Inulin, a flexible oligosaccharide I: review of its physicochemical characteristics. Carbohydr Polym 130:405–419. https://doi.org/10.1016/j.carbpol.2015.05.026
Mishra N, Jiang C, Chen L, Paul A, Chatterjee A, Shen G (2023) Achieving abiotic stress tolerance in plants through antioxidative defense mechanisms. Front Plant Sci 14:1110622. https://doi.org/10.3389/fpls.2023.1110622
Mishra UN, Chauhan J, Singhal RK, Anuragi H, Dey P, Lal D, Pandey S, Gupta NK, Nayak JK, Tripathi A, Singh M, Yadav M, Sajeevan RS (2025) Abiotic stress responses in forage crops and grasses: the role of secondary metabolites and biotechnological interventions. Front Plant Sci 16:1542519. https://doi.org/10.3389/fpls.2025.1542519
Morcillo RJL, Manzanera M (2021) The effects of plant-associated bacterial exopolysaccharides on plant abiotic stress tolerance. Metabolites 11(6):337. https://doi.org/10.3390/metabo11060337
Moustakas M (2025) Molecular mechanisms of plant abiotic stress tolerance. Int J Mol Sci 26(6):2731. https://doi.org/10.3390/ijms26062731
Mukherjee S, Sengupta S, Mukherjee A, Basak P, Majumder AL (2019) Abiotic stress regulates expression of galactinol synthase genes post-transcriptionally through intron retention in rice. Planta 249(3):891–912. https://doi.org/10.1007/s00425-018-3046-z
Naaz S, Pande A, Laxmi A (2025) Nitric oxide-mediated thermomemory: a new perspective on plant heat stress resilience. Front Plant Sci 16:1525336. https://doi.org/10.3389/fpls.2025.1525336
Nikonorova N, Van den Broeck L, Zhu S, van de Cotte B, Dubois M, Gevaert K, Inzé D, De Smet I (2018) Early mannitol-triggered changes in the Arabidopsis leaf (phospho)proteome reveal growth regulators. J Exp Bot 69(19):4591–4607. https://doi.org/10.1093/jxb/ery261
Niu Y, Liu W, Fan X, Wen D, Wu D, Wang H, Liu Z, Li B (2023) Beyond cellulose: pharmaceutical potential for bioactive plant polysaccharides in treating disease and gut dysbiosis. Front Microbiol 14:1183130. https://doi.org/10.3389/fmicb.2023.1183130
Noctor G, Foyer CH (2020) Intracellular redox compartmentation and ROS-related communication in regulation and signaling. Plant Physiol 182(4):1695–1710. https://doi.org/10.1104/pp.19.01393
Peng M, Jaeger KE, Lu Y, Fan Z, Zeng W, Sampathkumar A, Wigge PA (2025a) Activation and memory of the heat shock response is mediated by prion-like domains of sensory HSFs in Arabidopsis. Mol Plant 18(3):457–467. https://doi.org/10.1016/j.molp.2025.01.007
Peng Y, Zhu H, Wang Y, Kang J, Hu L, Li L, Zhu K, Yan J, Bu X, Wang X, Zhang Y (2025b) Revisiting the role of light signaling in plant responses to salt stress. Hortic Res 12(1):uhae262. https://doi.org/10.1093/hr/uhae262

Peters S, Keller F (2023) Raffinose family oligosaccharides: functions in plant stress tolerance and signaling. Front Plant Sci 14:1156789. https://doi.org/10.3389/fpls.2023.1156789

Pinski A, Muntazir Mehdi SM, Betekhtin A (2025) Molecular mechanisms underlying abiotic stress responses in buckwheat. Plant Sci 357:112526. https://doi.org/10.1016/j.plantsci.2025.112526

Poonam, Bhardwaj R, Handa N, Kaur H, Rattan A, Bali S, Gautam V, Sharma A, Ohri P, Thukral AK, Sirhindi G (2016) Sugar signalling in plants: a novel mechanism for drought stress management. In: Water stress and crop plants: a sustainable approach. Wiley, pp 287–302. https://doi.org/10.1002/9781119054450.ch19

Pottier D, Roitsch T, Persson S (2023) Cell wall regulation by carbon allocation and sugar signaling. Cell Surface 9:100096. https://doi.org/10.1016/j.tcsw.2023.100096

Price J, Li TC, Kang SG, Na JK, Jang JC (2003) Mechanisms of glucose signaling during germination of Arabidopsis. Plant Physiol 132(3):1424–1438. https://doi.org/10.1104/pp.103.020347

Qi W, Bai J, Yu H, Han G (2025) Physiological adaptations of *Vigna radiata* to heavy metal stress: soluble sugar accumulation and biomass enhancement. Plants 14(8):1191. https://doi.org/10.3390/plants14081191

Ramel F, Sulmon C, Gouesbet G, Couée I (2009) Natural variation reveals relationships between pre-stress carbohydrate nutritional status and subsequent responses to xenobiotic and oxidative stress in Arabidopsis thaliana. Ann Bot 104(7):1323–1337. https://doi.org/10.1093/aob/mcp243

Rathor P, Borza T, Liu Y, Qin Y, Stone S, Zhang J, Hui JPM, Berrue F, Groisillier A, Tonon T, Yurgel S, Potin P, Prithiviraj B (2020) Low mannitol concentrations in *Arabidopsis thaliana* expressing *Ectocarpus* genes improve salt tolerance. Plants 9(11):1508. https://doi.org/10.3390/plants9111508

Rosado A, Li R, van de Ven W, Hsu E, Raikhel NV (2012) Arabidopsis ribosomal proteins control developmental programs through translational regulation of auxin response factors. Proc Natl Acad Sci USA 109(48):19537–19544. https://doi.org/10.1073/pnas.1214774109

Ruan YL (2014) Sucrose metabolism: gateway to diverse carbon use and sugar signaling. Annu Rev Plant Biol 65:33–67. https://doi.org/10.1146/annurev-arplant-050213-040251

Sakr S, Wang M, Dédaldéchamp F, Perez-Garcia MD, Ogé L, Hamama L, Atanassova R (2018) The sugar-signaling hub: overview of regulators and interaction with the hormonal and metabolic network. Int J Mol Sci 19(9):2506. https://doi.org/10.3390/ijms19092506

Salehi M, Rashidinejad A (2025) Multifaceted roles of plant-derived bioactive polysaccharides: a review of their biological functions, delivery, bioavailability, and applications within the food and pharmaceutical sectors. Int J Biol Macromol 290:138855. https://doi.org/10.1016/j.ijbiomac.2024.138855

Salvi P, Agarrwal R, Kajal, Gandass N, Manna M, Kaur H, Deshmukh R (2022) Sugar transporters and their molecular tradeoffs during abiotic stress responses in plants. Physiol Plant 174(2):e13652. https://doi.org/10.1111/ppl.13652

Sengupta S, Mukherjee S, Basak P, Majumder AL (2015) Significance of galactinol and raffinose family oligosaccharide synthesis in plants. Front Plant Sci 6:656. https://doi.org/10.3389/fpls.2015.00656

Siddiqui H, Sami F, Hayat S (2020) Glucose: sweet or bitter effects in plants-a review on current and future perspective. Carbohydr Res 487:107884. https://doi.org/10.1016/j.carres.2019.107884

Supriya L, Dake D, Woch N, Gupta P, Gopinath K, Padmaja G, Muthamilarasan M (2025) Sugar sensors in plants: orchestrators of growth, stress tolerance, and hormonal crosstalk. J Plant Physiol 307:154471. https://doi.org/10.1016/j.jplph.2025.154471

Szilagyi A (2019) Digestion, absorption, metabolism, and physiological effects of lactose. In: Paques M, Lindner C (eds) Lactose. Academic Press, Elsevier, pp 49–111. https://doi.org/10.1016/B978-0-12-811720-0.00002-7

Valluru R (2015) Fructan and hormone connections. Front Plant Sci 6:180. https://doi.org/10.3389/fpls.2015.00180

Valluru R, Van den Ende W (2008) Plant fructans in stress environments: emerging concepts and future prospects. J Exp Bot 59(11):2905–2916. https://doi.org/10.1093/jxb/ern164

Van den Ende W (2013) Multifunctional fructans and raffinose family oligosaccharides. Front Plant Sci 9(4):247. https://doi.org/10.3389/fpls.2013.00247
Wang W, Chen Q, Xu S, Liu WC, Zhu X, Song CP (2020) Trehalose-6-phosphate phosphatase E modulates ABA-controlled root growth and stomatal movement in Arabidopsis. J Integr Plant Biol 62(10):1518–1534. https://doi.org/10.1111/jipb.12925
Wang J, Yu YC, Li Y, Chen LQ (2022) Hexose transporter SWEET5 confers galactose sensitivity to Arabidopsis pollen germination via a galactokinase. Plant Physiol 189(1):388–401. https://doi.org/10.1093/plphys/kiac068
Wayne T (2024) Improved production and benefits of prebiotic oligosaccharides. J Nutr Food Sci 9(3):60
Wingler A, Henriques R (2022) Sugars and the speed of life—metabolic signals that determine plant growth, development and death. Physiol Plant 174(2):e13656. https://doi.org/10.1111/ppl.13656
Xiao F, Zhou H (2023) Plant salt response: perception, signaling, and tolerance. Front Plant Sci 13:1053699. https://doi.org/10.3389/fpls.2022.1053699
Xie JH, Jin ML, Morris GA, Zha XQ, Chen HQ, Yi Y, Li JE, Wang ZJ, Gao J, Nie SP, Shang P, Xie MY (2016) Advances on bioactive polysaccharides from medicinal plants. Crit Rev Food Sci Nutr 56(Suppl 1):S60–S84. https://doi.org/10.1080/10408398.2015.1069255
Xiong Y, Sheen J (2014) The role of target of rapamycin signaling networks in plant growth and metabolism. Plant Physiol 164(2):499–512. https://doi.org/10.1104/pp.113.233234
Yan S, Liu Q, Li W, Yan J, Fernie AR (2022) Raffinose family oligosaccharides: crucial regulators of plant development and stress responses. Crit Rev Plant Sci 41(4):286–303. https://doi.org/10.1080/07352689.2022.2111756
Yuan K, Wysocka-Diller J (2006) Phytohormone signalling pathways interact with sugars during seed germination and seedling development. J Exp Bot 57(12):3359–3367. https://doi.org/10.1093/jxb/erl096
Zeeman SC, Kossmann J, Smith AM (2010) Starch: its metabolism, evolution, and biotechnological modification in plants. Annu Rev Plant Biol 61:209–234. https://doi.org/10.1146/annurev-arplant-042809-112301
Zha L, Wang S, Berglund LA, Zhou Q (2023) Mixed-linkage (1,3;1,4)-β-d-glucans as rehydration media for improved redispersion of dried cellulose nanofibrils. Carbohydr Polym 300:120276. https://doi.org/10.1016/j.carbpol.2022.120276
Zhai Z, Keereetaweep J, Liu H, Xu C, Shanklin J (2021) The role of sugar signaling in regulating plant fatty acid synthesis. Front Plant Sci 12:643843. https://doi.org/10.3389/fpls.2021.643843
Zhang B, Gao Y, Zhang L, Zhou Y (2021) The plant cell wall: biosynthesis, construction, and functions. J Integr Plant Biol 63(1):251–272. https://doi.org/10.1111/jipb
Zhang Y, Xu J, Li R, Ge Y, Li Y, Li R (2023) Plants' response to abiotic stress: mechanisms and strategies. Int J Mol Sci 24(13):10915. https://doi.org/10.3390/ijms241310915
Zhang H, Lang Z, Zhu JK, Wang P (2025a) Tackling abiotic stress in plants: recent insights and trends. Stress Biol 5(1):8. https://doi.org/10.1007/s44154-025-00216-x
Zhang H, Yu C, Zhang Q, Qiu Z, Zhang X, Hou Y, Zang J (2025b) Salinity survival: molecular mechanisms and adaptive strategies in plants. Front Plant Sci 16:1527952. https://doi.org/10.3389/fpls.2025.1527952
Zhao X, Chen G, Wang F, Zhao H, Wei Y, Liu L, Zhang H (2022) Extraction, characterization, antioxidant activity and rheological behavior of a polysaccharide produced by the extremely salt tolerant Bacillus subtilis LR-1. LWT 162:113413. https://doi.org/10.1016/j.lwt.2022.113413
Zheng S, Liu C, Zhou Z, Xu L, Lai Z (2024) Physiological and transcriptome analyses reveal the protective effect of exogenous trehalose in response to heat stress in tea plant (Camellia sinensis). Plants 13(10):1339. https://doi.org/10.3390/plants13101339
Zhou Q, Sun H, Zhang G, Wang J, Tian J (2023) Gene co-expression analysis reveals the transcriptome changes and hub genes of fructan metabolism in garlic under drought stress. Plants 12(19):3357

Zhou J, Zhang Y, Wang X (2024a) Sugar metabolism and signaling in plants: integrating photosynthesis, growth, and stress responses. Front Plant Sci 15:1234567. https://doi.org/10.3389/fpls.2024.1234567

Zhou H, Shi H, Yang Y, Feng X, Chen X, Xiao F, Lin H, Guo Y (2024b) Insights into plant salt stress signaling and tolerance. J Genet Genomics 51(1):16–34. https://doi.org/10.1016/j.jgg.2023.08.007

Chapter 14
Physiological Roles of Cadaverine in Plant Growth and Development

Mahmoud S. Abdelmoneim and Mona F. A. Dawood

Abstract Cadaverine (Cad), a diamine belonging to the polyamine (PA) family, has emerged as a crucial molecule in modulating plant responses to environmental stressors. It is biosynthesized through the activity of lysine-by-lysine decarboxylase (LDC) and is involved in various physiological processes, including cell division, membrane stabilization, and reactive oxygen species scavenging. Recent studies highlight cadaverine's role in enhancing tolerance to both abiotic and biotic factors, often through cross talk with signaling molecules like abscisic acid, salicylic acid, and jasmonic acid. This chapter discusses the biosynthetic pathways and metabolic regulation of cadaverine in plants and explores its involvement in stress signaling networks. We also discuss the transport and accumulation dynamics of cadaverine, its influence on ion homeostasis, and its interaction with antioxidant systems under stress conditions. The potential of cadaverine as a biostimulant in agricultural practices is examined, along with future directions in genetic engineering and breeding strategies aimed at optimizing cadaverine levels to improve crop resilience.

Keywords Abiotic stress · Crop production · Nano-elicitors · Crop resilience · Sustainable agriculture

14.1 Introduction

In both prokaryotic and eukaryotic organisms, the polyamine (PA) cadaverine (Cad) (1,5-diaminopentane) is involved in environmental perception and the regulation of biological processes. Modifications in cadaverine levels cause changes in root architecture (Niemi et al. 2002; Strohm et al. 2015), plant biomass (Cassan et al.

M. S. Abdelmoneim · M. F. A. Dawood (✉)
Botany and Microbiology Department, Faculty of Science, Assiut University, Assiut, Egypt
e-mail: Mo_fa87@aun.edu.eg

H. I. Mohamed (ed.), *Elicitors for Sustainable Crop Production*, Sustainability Sciences in Asia and Africa, https://doi.org/10.1007/978-981-95-3533-0_14

2009; Abdelmoneim et al. 2025), and stress response (Shevyakova et al. 2001; Sziderics et al. 2010; Abdelmoneim et al. 2025) concerning the effects on agricultural productivity and resilience to environmental stresses (Jancewicz et al. 2016).

Cadaverine is synthesized in plastids via the decarboxylation of lysine (Bunsupa et al. 2012), which affects environmental stresses in various plants (Bunsupa et al. 2012; Simon-Sarkadi et al. 2014; Sziderics et al. 2010). Contrarily, *Arabidopsis thaliana* does not show detectable quantities of free cadaverine (Shimizu et al. 2019). Nevertheless, cadaverine can be directly absorbed by plants from their surroundings as molecules released by microorganisms in the phyllosphere, rhizosphere, or endosphere, or even from decaying organic matter (Cassan et al. 2009). It is worth noting that cadaverine, whether produced internally or applied externally, might influence plant development and resilience to stress in response to environmental factors (Jancewicz et al. 2016; Abdelmoneim et al. 2025). The current chapter will discuss the recent advancements of plant growth regulators, with special focus on polyamines and, more specifically, on cadaverine, biosynthesis, the biological role of cadaverine, and the role of cadaverine under abiotic and pathogenic stress.

14.2 Plant Growth Regulators

Meeting the anticipated global food demand for a projected population of nine to ten billion by 2050 requires innovative approaches to strengthen food security, particularly under diverse environmental stresses. A range of strategies has been developed to enhance crop performance under abiotic stress, including both conventional and advanced breeding methods, optimized agronomic practices, seed priming, microbial seed treatments, microorganism inoculation, grafting, and the application of plant growth regulators and osmoprotectants (Rakkammal et al. 2023; Dawood 2025). Recently, low-cost and environmentally sustainable innovations—such as biochar, yeast extract, seaweed extract, and agri-nanotechnology—have gained attention for their potential to boost crop resilience to abiotic stresses (Oyebamiji et al. 2024). However, to maximize productivity, agricultural systems have increasingly relied on the excessive use of synthetic fertilizers and pesticides, which present significant short- and long-term risks to ecosystem and human health (Damalas and Koutroubas 2016). Consequently, there is a growing emphasis on biological alternatives, such as biostimulants, to reduce the dependence on chemical inputs (Van Oosten et al. 2017; Yakhin et al. 2017; Dawood et al. 2025). Moreover, polyamines are emerging as promising plant growth regulators with the potential to enhance tolerance to salt stress in crops (Amiri et al. 2024).

14.3 Polyamines

Polyamines (PAs) are aliphatic amines with phytohormone-like properties that participate in a wide range of essential cellular processes in plants. They are critical regulators of multiple developmental stages, including cell division, germination, morphogenesis, floral bud initiation, embryogenesis, fruit formation, and responses to both biotic and abiotic stresses (Nidhi and Khan 2025). Several types of PAs have been identified in plants, the most common being putrescine (Put), spermidine (Spd), spermine (Spm), and its structural isomer thermospermine (t-Spm), along with cadaverine (Cad) (Wang et al. 2019). Because of their polycationic nature, PAs interact with negatively charged macromolecules, such as nucleic acids, proteins, membrane phospholipids, and cell wall components. These interactions help preserve membrane stability and reduce cellular injury under stressful conditions (Strohm et al. 2015).

Earlier studies highlighted the role of PAs in regulating defense mechanisms against environmental challenges, in addition to supporting fundamental physiological and biochemical processes. They contribute to DNA replication, transcription, translation, enzyme activity, ion homeostasis, cell wall stability, and membrane integrity (Wang et al. 2016). Furthermore, PAs are directly involved in developmental processes such as seed germination, flower bud initiation, embryogenesis, fruit ripening, and tolerance to abiotic stresses (Özmen et al. 2022). The levels of PAs often fluctuate when plants are exposed to unfavorable environmental conditions. These changes depend on the plant genotype, developmental stage, type of stress, stress intensity, and duration (Bouchereau et al. 1999; Liu et al. 2007). In many cases, exposure to stress leads to an accumulation of free or conjugated PAs, suggesting that their biosynthesis and metabolism are tightly linked to stress responses (Nidhi and Khan 2025). For example, tomato plants subjected to salt stress showed a significant increase in PA levels at later developmental stages compared to early stages, indicating developmental regulation of PA-mediated stress adaptation (Botella et al. 2000).

PAs are also recognized as integral components of the antioxidant defense system. Putrescine, spermidine, spermine, and cadaverine function as scavengers of reactive oxygen species (ROS), thereby protecting cells from oxidative damage (Dawood and Abeed 2020). Their ROS-quenching activity is not only due to direct oxidation but also due to their capacity to bind proteins and form conjugates with phenolic compounds, which enhance stability and reduce oxidative damage (Kuznetsov and Shevyakova 2010). Conjugated PAs have been proposed as key players in defense responses, contributing to PA transport and the regulation of free PA pools, thus strengthening plant tolerance mechanisms under environmental stresses (Nidhi and Khan 2025).

14.4 Cadaverine

Cadaverine (1,5-diaminopentane) is a diamine that acts as a growth regulator via activating a distinct biosynthetic pathway. Although Cad has a different structure compared to putrescine, a similar mode of action was recorded for Cad and Put (Rajpal and Tomar 2020a). One way of its biosynthesis found in both prokaryotes and eukaryotes is through lysine decarboxylase (LDC) that directly decarboxylates L-lysine. Cadaverine protects cells that are deprived of inorganic phosphate, Pi, under anaerobic circumstances and is crucial for cell viability at acidic pH levels (Moreau 2007; Samartzidou et al. 2003). Cadaverine has a similar structure to the synthesized petrochemical hexamethylenediamine and, thus, can be used to replace hexamethylenediamine in the production of polyamides or polyurethanes (Becker and Wittmann 2012). Cadaverine accumulates in multiple crops' tissues due to living and nonliving factors (Simon-Sarkadi et al. 2014; Sziderics et al. 2010). Reports suggest that plants may absorb cadaverine from their surroundings and utilize it for subsequent metabolic and signaling processes (Jancewicz et al. 2016). Like other polyamines, cadaverine has been involved in stress response, such as enhancing seed development and growth parameters (Ozmen et al. 2023). Cadaverine induces multifaceted functions by being a precursor for polyamides, facilitating water absorption, promoting cell survival under stress conditions, and serving as a protective agent for cells deprived of orthophosphate under an oxygen-deprived environment (Ma et al. 2017). In plants, it is involved in regulating diverse processes such as plant growth and development, cell signaling, antioxidants, and plant defense responses and genes (Abdelmoneim et al. 2025). Furthermore, cadaverine enhances the ability to tolerate stress conditions, facilitates cellular defense, and increases resistance to insects and other organic and inorganic contaminants (Ozmen et al. 2023). Cadaverine is becoming an important industrial chemical because it exhibits broad prospects for various applications, especially for the synthesis of fully biobased polyamides through polymerization with dicarboxylic acids derived from renewable sources (Ma et al. 2017).

14.4.1 Distribution and Occurrence of Cadaverine

14.4.1.1 Entire Plants and Plant Products

Cadaverine was initially detected in bacteria in minimal quantities and was associated with various roles based on its binding affinity to membranes and other molecules (Wortham et al. 2007). Several plants, their parts, and products have been found to contain this diamine. *Panicum miliaceum*, *Setaria italica*, oats, wheat, barley, sorghum, maize, rye, and *Phleum pratense* are some of the plants that have reported this trait (Bagni et al. 1986; Hamana et al. 1994; Rodrıguez et al. 2000). Barley leaves, pea roots, tomatoes, microcuttings of grapevines, and tea plants have all been found to contain Cad in their tissues (Bailey et al. 2002). Some vegetables that contain Cad are radicchio (Simon-Sarkadi and Holzapfel 1994), iceberg lettuce (Zhu and Jiang 1994), and Chinese cabbage (Zhu and Jiang 1994). According to the research of Broquedis et al. (1989), *Vitis vinifera* either contains trace amounts of Cad (in leaves and bunches) or does not have any at all. *Pisolithus tinctorius* is lacking in this regard as well (Vinci and Antonelli 2002). Unlike soybean nodules and ripening egg plants, Cad is thought to be mostly localized in the radicle and hypocotyl area of chickpea (Caffaro et al. 1993; Torrigiani and Scoccianti 1995). In the root nodules of adzuki bean, a novel polyamine called 4-amino butyric Cad was identified by Fujihara (2009). Table olive, sugarcane juice, and other plant sources are known to contain Cad (de Armas et al. 1999; Rodrıguez et al. 2000). The function of Cad in such items remains unclear. When it comes to its physiochemical relevance, its content variability is still a mystery.

Depending on the species and portions of cereal, the Cad titer can range from less than 1 nmol/g FW to more than 15 mol/g FW. Tobacco's hairy roots may play a role in anabasis synthesis since they accumulate 0.7% Cad on a dry mass basis. In spite of the fact that Cad was detectable in Chinese cabbage, the putative dominating element was Put (Zhu and Jiang 1994). According to Angosto and Matilla (Angosto and Matilla 1993), the presence of Cad was found to be more prevalent in plants grown at higher elevations, suggesting that the diamine titers of three leguminous species might be affected by altitude (Križek et al. 1993). Cad level fluctuations are also observed in certain plants over time. Pea, tomato, millet and corn internodes, corn coleoptiles (Aziz et al. 1999; Shen and Galston 1985), and grapevine leaves all showed a decrease in Cad concentrations as the plants aged (Broquedis et al. 1989). The Cad titer in *Brugmansia candida* may decrease gradually in hairy roots, as shown by Carrizo et al. (2001), who also found that it depends on age.

There is a lot of Cad in the seeds of lupin, cucumbers, chickpeas, and broad beans, and that amount grows as the seeds germinate (Ramakrishna and Adiga 1975; Smith 1975; Villanueva et al. 1978). Pea seedlings similarly have a larger concentration of Cad in the third internode (Goren et al. 1982). This finding supports the work of Villanueva et al. (1978) but runs counter to the findings of Smith and Wilshire (1975), who found no evidence of Cad in any of the tested species (excluding *Trifolium subterraneum* seedlings) of *Phaseolus aureus, P. vulgaris, or Vicia faba*. The existence of Cad in rice phloem sap and shoot exudates was shown by Yokota et al. (1994), who also proposed a larger diffusion of the element from bacteria to plants to mammals.

In addition to crop plants, Cad can be found in the leaves and roots of certain aquatic plants, including *Nymphaea tetragona*, *Nelumbo nucifera* (Smith and Wilshire 1975), *Spirodela polyrhiza,* and *Eichhornia crassipes*. There is a considerable quantity of Cad in the shoots and leaves of *Allium sativum*, but the edible mushrooms *Boletus badius*, *B. chrysontereon*, *B. variegatus*, and *Agaricus bisporus* contain the highest concentrations (165 mg/dry wt.) (Parolo et al. 1997). Nonetheless, the cadaverine concentration is lower than that of putrescine (600 mg/dry wt.) in the plant system (Kalač and Křížek 1997). Killiny and Nehela (2020) observed elevated levels of cadaverine in immature seeds of citrus species and in straw mushrooms

(Yen 1992). The presence of Cad can be observed in the brines of green table olives (Hornero-Mendez and Garrido-Fernandez 1997). The ice plant's young leaves and apex exhibited lower levels of Cad (20–29 nmol/g Fwt) compared to the root and stem (López-Caballero et al. 2002). The ice plant cadaverine in aged leaves, stems, and roots was shown to be heat-labile; exposure to 47 °C for 2 h significantly diminished its concentration (Kuznetsov et al. 2002). Križek et al. (1993) found Cad concentrations ranging from 341 mg/kg to 642 mg/kg in 54 maize silages and 53 grasses, respectively. Lysine, the precursor of LDC enzyme protein, or its expression could be modulated by the physical variables impacting the Cad titer. Rizvi et al. (2025) found differences in Cad when peas were exposed to red and far-red light. Rape leaf discs stressed by osmotic and drought conditions exhibit substantial cadaverine buildup.

Hou and Lin (1997) have also observed varying Cad titers in storage roots, emerged roots, and sprouts of *Ipomoea batatas*, as well as a comparable trend in mung bean, lentil, and radish sprouts. It appears that the differential distribution in germinating/growing tissue may be a result of LDC isoforms, as demonstrated by the application of the inhibitor α-difluoromethyl L-lysine. Charges had an impact on the adoption and distribution of polyamines (Hart et al. 1992). The Cad in cereals has been further verified by the localization of a related gene on the chromosome 5B fragments that are responsible for wheat synthesis (Chen et al. 2021).

14.4.1.2 In Cultured Tissues

A larger quantity of Cad was seen in the cucumber cotyledon culture (Kiełkowska and Dziurka 2021). Cad was found exclusively in converted root cultures of *Brugmansia candida*, according to Carrizo et al. (2001). Additionally, compared to non-differentiating callus, tomato explants bud and root differentiated callus exhibited a higher Cad content (Tomar et al. 2013). The increased production of Cad was observed in the hairy root cultures of *Nicotiana tabacum* by Fecker and group (Fecker et al. 1993). Hydroxy cinnamoyl cadaverine and other secondary metabolites are formed during the metabolism of Cad that was present in the tobacco root culture (Berlin et al. 1998). N, N-bis 3-aminopropyl Cad and aminopropyl (Cad APC) were both found in barley (Tomar et al. 2013). A large concentration of Cad conjugates is also seen in micropropagated carnation plants (Piqueras et al. 2002). A possible function for Cad is growth differentiation/organogenesis, as evidenced by the fact that its titer varies between cell types in culture (Tomar et al. 2013). In similar lines, Hou and Lin (1997) found varying Cad titers in *Ipomoea batatas* storage roots, sprouting roots, and sprouts, as well as in the sprouts of mung bean, lentil, and radish. The use of the inhibitor α-difluoromethyl L-lysine confirmed that LDC isoforms could be the cause of the observed differential distribution in germination and developing tissues (Tomar et al. 2013). Charging influenced polyamine absorption and distribution (Sofian et al. 2021). The Cad in cereals has been further validated by identifying the associated gene on chromosome 5B segments responsible for synthesis in wheat (Galiba et al. 1993).

14.4.2 Detection of Cadaverine

The procedure used to find Cad is the same as that used for the other polyamines. The root nodules of pea, fava bean, and *Sesbania rostrata* were tested for the presence of Cad and other PAs. The results showed that the 15 N values for Cad were + 1 and + 7 per thousand, which were very close to the limits of detection when using an automatic nitrogen and carbon analysis mass spectrometer. Nevertheless, following the preparation of benzoyl amine with benzoyl chloride derivative, a fast high-performance liquid chromatography (HPLC) reverse-phase approach is also recommended. However, at 225 and 254 nM, this has been detected at a limit of 5.60–54.40 pg using a UV-visible spectrometer (Hornero-Mendez and Garrido-Fernandez 1997). Cad was measured in rice (*Oryza sativa* L.) using dansyl chloride derivatives, in conjunction with high-performance liquid chromatography. Alternatively, $^{14}CO_2$ generation from (^{14}C)-lysine can be used to test lysine decarboxylase (LDC) activity (Cassan et al. 2009). The fact that a large variety of plants have not been tested for LDC and Cad is likely due to the high cost of using accurate detection methods.

14.4.3 Chemical Synthesis of Cadaverine

The hydrogenation of glutaronitrile using an amorphous nickel alloy catalyst was initially employed for the production of cadaverine (Hoffer and Moulijn 2009). It was generally acknowledged that glutaronitrile was initially hydrogenated to 5-aminovaleronitrile and, thereafter, further hydrogenated to cadaverine (Fig. 14.1)

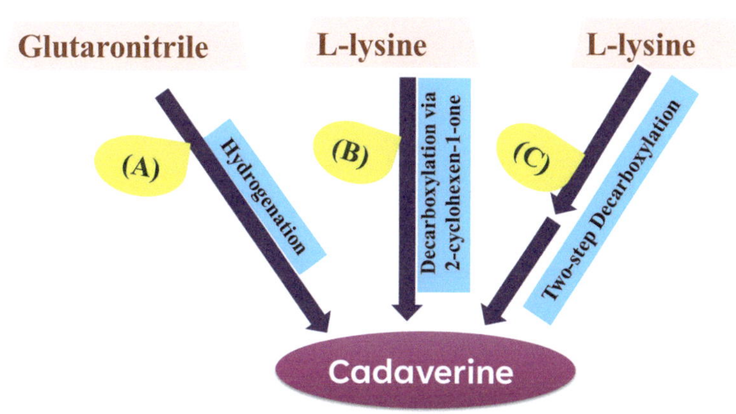

Fig. 14.1 Chemical synthesis of cadaverine using three pathways via the hydrogenation of glutaronitrile (**a**) (De Bellefon and Fouilloux 1994; Hoffer and Moulijn 2009) and from L-lysine decarboxylation using 2-cyclohexen-1-one (**b**) (Hashimoto et al. 1986) or two-step L-lysine decarboxylation in two stages (**c**) (Laval and Golding 2003)

(Chong et al. 2010; De Bellefon and Fouilloux 1994). Nonetheless, the reactive intermediate 5-aminovaleronitrile may yield the unwanted cyclic molecule piperidine. The synthesis of glutaronitrile necessitated the use of highly poisonous cyanide as a precursor, presenting significant security challenges for scale production (De Bellefon and Fouilloux 1994).

14.4.4 Biosynthesis of Cadaverine

In 1942, Gale and Epps initially discovered that *Escherichia coli* may stimulate several amino acid decarboxylases in acidic environments (Gale and Epps 1942), which brought up a new insight into the biosynthesis of cadaverine. Moreover, there is an urgent need for renewable and sustainable resources to produce cadaverine, prompted by the depletion of petroleum resources and the environmental contamination resulting from chemical processes. The quick advancement of biotechnology and synthetic biology over the past decade has created a novel opportunity to investigate cadaverine manufacturing (Huang et al. 2021). Figure 14.2 shows the capability of using different microbes to produce cadaverine from a variety of carbon sources. Cadaverine can be produced from different sources, such as glucose, galactose, starch, xylose, cellobiose, methanol, methane, mannitol, and L-lysine (Fig. 14.2).

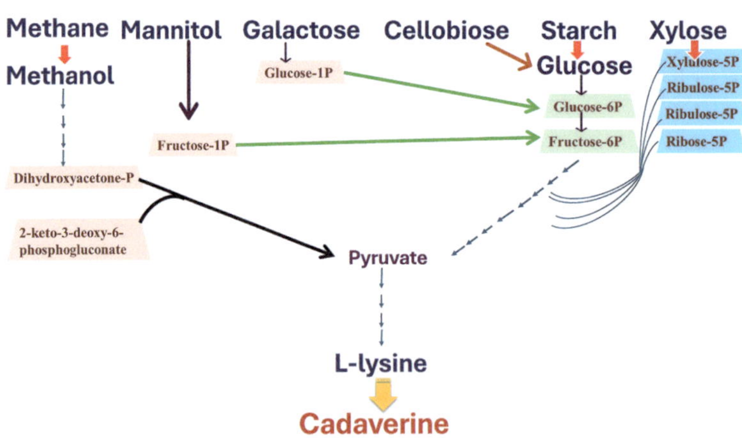

Fig. 14.2 The routes for synthesizing cadaverine from glucose, galactose, starch, cellobiose, xylose, methane, mannitol, and L-lysine. A detailed description of enzymes and genes involved in the biosynthesis of cadaverine can be reviewed in Huang et al. (2021)

14.4.5 *Biological Roles of Cadaverine*

14.4.5.1 Cadaverine's Role in Growth and Development

Even though Cad and Put have different chemical structures, researchers discovered that both of them have a good effect on chickpea seed germination and seedling development (Baraiya et al. 2025). On the other hand, Grzesik (1993) found that electrolyte leaking affected the germination of *Helichrysum bracteatum* seeds due to a substantial increase in Cad during germination. Within the context of stomatal opening and shutting in *Vicia faba*, the inhibitory role of Cad is highlighted (Liu et al. 2000). For soybean, the use of Cad is suggested during rooting (Gamarnik and Frydman 1991). Photosynthetic enzymes, such as pyruvate orthophosphate dikinase, nicotine amide dinuclutide (NADP)-malate dehydrogenase in *Rumex dentatus,* and the lipid metabolizing enzyme lipoxygenase, also exhibit certain negative reactions to Cad (El-Shora 2001). It is noteworthy that there is no effect of Cad on chloroplast outer envelope porins (OEPs), which are important in transporting triosephosphate, dicarboxylic acids, charged amino acids, sugars, adenosine triphosphate (ATP), and Pi. This finding further supports the unique role of Cad in plant growth metabolism. However, it is noted that Cad has an effect on bacterial porins (Onat-Taşdelen et al. 2024).

14.4.5.2 Cadaverine and Plant Development

According to research, the concentration of PAs and their internal ratio are both thought to play a significant role in the development of plants (Shao et al. 2022). Depending on the species of plant, the Cad titer has been found to be varied in different conditions, which suggests that developmental regulation is possible. The cadaverine concentration in *Malus hupehensis* seedlings decreased gradually as the seedlings grew taller (El-Yazal et al. 2018).

Compared to *Glycine max* cotyledons, Gamarnik and Frydman (1991) discovered that embryonic axes have significantly higher Cad concentrations. Applying difluoro methyl ornithine (DFMO) in the first 24 h of soybean germination proved the same. The fact that 1 mM of DFMO had no effect on other PAs but caused anomalous root growth that was counteracted by 1 mM of Cad raises the possibility that the effects of Cad-dependent root growth are plant-specific (Gamarnik and Frydman 1991). While Cadaverine levels increase during pea seed germination but subsequently decline in the shoots of light-grown plants after an extended growth period (6–18 days). The faster metabolic rate and LDC localization in chloroplasts in roots may explain why there are 70 times more Cad in roots than in leaves, according to certain reports (Li et al. 2021a, 2021b). No changes to the titer were seen in *Phaseolus aureus*, *P. vulgaris*, or *Vicia faba* in response to $CaCO_3$, ammonium treatment, or K^+ insufficiency, except for the variable reaction with physical variables (Rajpal and Tomar 2020a). Additionally, the tissue-dependent Cad titer is raised.

The presence of Cad was found to be more pronounced in the flowers of *Momordica charantia*, which could indicate a function in sex differentiation, compared to the vegetative tissue (Kiełkowska and Dziurka 2021). No link was identified between sex reversal and Cad levels in cucumber flowers, according to Xuehao et al. (2001). Additionally, the cultured cucumber cotyledon cells did not exhibit an increase in Cad during growth (Wang ShouFeng et al. 1998). Differentiated cells exhibited a noticeably larger amount than non-differentiated callus; however, tissue-specific Cad content is observed in certain circumstances of cultured cells, such as when Cad grows more during bud differentiation of callus than root differentiation of callus (Rakesh et al. 2021). It has been shown by Caffaro and Vicente (1994) that Cad and other PAs are engaged in the processes of flowering. These findings point to the possibility that other members of the polyamine family influence the physiochemical processes, rather than just Cad alone. When ectomycorrhizal fungi are present in Scotch pine (Niemi et al. 2002), the application of exogenous Cad stimulates roots and changes the bacteroid, which, in turn, affects nodule metabolism (Vassileva and Ignatov 1999).

14.4.5.3 Cadaverine in Senescence and Stress Modulation

Polyamines are commonly recognized for their roles in anti-stress and anti-senescence mechanisms. Three remote doses of Cad (0.2 mM, 1 mM, and 10 mM) showed no impact on leaf senescence in barley under dark conditions (Srivastava 1987). Cadaverine exhibited enhanced or comparable inhibitory feedback relative to that of putrescine (Maccarrone et al. 1998). Vassileva and Ignatov (1999) proposed that putrescine, often regarded as a precursor to spermidine and spermine, has inhibitory feedback on interactions that may be effectively modulated through cadaverine, which surprisingly influences the specifics of cadaverine's role in anti-senescence and anti-stress mechanisms.

Even with high-performance liquid chromatography, no trace of dark-induced senescence in rice was discovered for Cad or 1,3-diaminopropane. However, other PAs were detected (Chen and Kao 1991). There was no evidence of Cad's presence in senescent rice leaves tested in the trials either (Ten Chen and Kao 1991). It would be reasonable to assume that, unlike rice, the endogenous amount of Cad did not significantly influence the regulation of dark-induced senescence (Chen and Kao 1991).

14.4.6 Cadaverine and Stress Response

Very little research has focused on the possibility of Cad in reducing plant stress. Two potential signal systems utilized by the defense system in response to stress are the diamines (putrescine and cadaverine) and polyamines (spermidine and spermine) (Tassoni et al. 1998). To prevent disintegration and maintain stability under stress, PAs with cationic charges typically bind to negatively charged proteins or nucleic acids; at the same time, they inhibit the activity of DNases, RNases, and proteases (Kushad and Dumbroff 1991). Spermine and cadaverine make up for the lower Put levels in stress-tolerant plant species. The function of Cad in stressed plants remains unclear despite its abundant presence in a small number of higher plants. One possible explanation is that, similar to microbial systems, a disruption in Put synthesis could cause lysine to be redirected to Cad synthesis, thus compensating for PAs (Li et al. 2021a, 2021b).

The elevated levels of cadaverine in the germinating seed embryo axis suggest a function in stress management and subsequent development. Cadaverine was identified in elevated amounts in the excised embryonic axis of *Cicer arietinum* during germination (Gallardo et al. 1994). In addition to the presence of cadaverine in many plants, a limited number of higher plant products also contain cadaverine (Fujihara et al. 1995). The necessity of elucidating the role of cadaverine in these products is essential for physiochemical significance, such as in table olives (Garcia-Garcia et al. 2000), sugarcane juices (Rodrıguez et al. 2000), and matured eggplants (del Rodriguez et al. 1999).

The current literature on the impact of cadaverine on the growth and physiochemical responses of crop plants indicates a deficiency of comprehensive research. Research indicated the presence of cadaverine in angiosperms; thus, the physiochemical dimensions of previous research remain to be investigated. The induction of cadaverine under salt stress conditions in maize was found to be much less effective compared to other polyamines (Caldeira and Caldeira 1999). According to the research conducted by Bouchereau et al. (1999), the production of cadaverine in higher plants is increased in response to osmotic and salt stressors. The concentration of cadaverine in the aged leaves, stems, and roots of the ice plant significantly diminished at 47 °C for 2 h, indicating its heat-lability (Kuznetsov et al. 2002). Conversely, cadaverine exhibited a substantial increase in the roots and was translocated to the shoots because of heat shock (Shevyakova et al. 2000). Additionally, polyamines such as Put demonstrated a temperature-sensitive translocation in vegetable products (Kalač et al. 2002). In addition to the conjugation of cadaverine in *Mesembryanthemum crystallinum* roots, oxidative breakage is promoted, simultaneously degrading Cad and its conjugates upon root treatment with exogenous cadaverine (Kuznetsov and Shevyakova 2010).

14.4.7 *The Application of Cadaverine in Mitigating Abiotic Stress*

Both endogenous and exogenously administered cadaverine significantly influence plant development and resilience to environmental stress (Jancewicz et al. 2016). Additionally, it has been noted that cadaverine enhances plant development in response to various stressors, and the presence of stress-triggered proteins in the presence of cadaverine may further aid plant growth in response to various stressors (Rajpal and Tomar 2020a). Similar to other polyamines, Cad tends to build up when exposed to osmotic or salt stress (Bouchereau et al. 1999). Cadaverine builds up slightly more in water-stressed rice (Lefevre et al. 2001) and salt-stressed maize (Caldeira and Caldeira 1999) in comparison with other polyamines. According to Bouchereau et al. (1999), higher plants may produce Cad in response to osmotic and salt stress. The accumulation of Cad in rape and tomato leaf discs when exposed to 100–300 mM NaCl doubled with an increase in KCl concentration (Aziz et al. 1997). As a result of heat shock, root Cad levels rise dramatically, and the translocation of Cad to the shoot is redirected (Shevyakova et al. 2000).

The impact of pollutants on cadaverine levels is also noted in both sensitive and tolerant cultivars of *Phaseolus vulgaris*. For example, *P. vulgaris* cv. 'Slowianka', a resistant variety, demonstrated an increase in Cad levels in the roots and leaves at lower sulfite concentrations but exhibited a decrease as the sulfite concentration increased (Karolewski and Shevyakowa 1990). The herbicide napropamide (10–20 μm) administered over an extended period to pea roots resulted in a twofold rise in Cad (DiTomaso et al. 1988). Nonetheless, there are disputes over the significance of Cad in promoting salt tolerance and the metabolic alterations in plants generated by salt stress due to Cad (Rajpal and Tomar 2020b). On the other hand, Cassan et al. (2009) proposed that *Azospirillum brasilense* Az39 strain, which is applied to rice seedlings, plays a role in root growth and the reversal of osmotic stress in the rhizosphere of young rice plants. However, findings on rice responses are inconsistent. For example, according to Lefevre and group (Lefevre et al. 2001), there is no correlation between salinity and Cad level in *Oryza sativa*. Additionally, it was noted that there was no Cad in the rice cell suspension culture or when K^+ was lacking (Cassan et al. 2009). In conditions of salt stress, a trace quantity of Cad is observed in maize (Simon-Sarkadi et al. 2014). It appears that *Azospirillum* sp. may be involved in the Cad production in rice, which alleviates stress. However, more trials are needed to confirm this possibility, which could depend on the species and physical conditions (Gureeva and Gureev 2023).

Kuznetsov et al. (2007) showed that the incorporation of cadaverine into the stressed tissues of ice plant mitigated DNA damage and enhanced antioxidant enzyme activity. They proposed that PAs' antioxidant effects may involve stimulating superoxide dismutase (SOD) expression, potentially by increasing its mRNA transcript levels. The exogenous Cad's differential response to metabolites and enzymes in Indian mustard subjected to salt stress is also documented (Tomar et al. 2013). *Arabidopsis thaliana* seedlings subjected to exogenous cadaverine exhibit modified root growth patterns, unequivocally demonstrating the compound's involvement in root growth control (Cassan et al. 2009). Cadaverine possesses significant potential to enhance wheat tolerance by mitigating oxidative damage and shielding the biochemical processes of photosynthesis from NO_3^- toxicity (Balci et al. 2023). Furthermore, the implementation of cadaverine led to a substantial enhancement in the total chlorophyll and carotenoid levels in the leaf tissues of tomato seedlings subjected to salinity, lead, and a combination of both, in comparison to the control group (Rajpal and Tomar 2020c). In a previous study, cadaverine facilitated the preservation of cellular redox equilibrium by modulating antioxidant pathways involved in the ascorbate–glutathione (AsA–GSH) cycle. In addition, it was indicated that cadaverine possesses significant potential to enhance wheat tolerance by mitigating oxidative damage and shielding the biochemical processes of photosynthesis from NO_3^- toxicity (Balci et al. 2023).

Tomar and Arora (2023) concluded that the positive efficacy of cadaverine in enhancing the expression of new peptides and alleviating various stressors in *Brassica juncea* indicates its significance in the growth and development of leaf and root tissues of the species. The role of exogenous application of cadaverine in alleviating the toxicity of salinity, bisphenol A, and their combination was elucidated by Abdelmoneim et al. (2025). However, substantial gaps remain to be elucidated regarding plant responses to cadaverine.

14.4.8 Cadaverine's Role in Plant Pathogenicity

Polyamines have been associated with pathogenicity, influenced by their concentration and the specific plant tissues involved (Tomar et al. 2013). Legaz et al. (1998) documented elevated levels of Cad in the leaves, roots, and stems of infected plants compared to noninfected ones, proposing it as a potential indicator of *Ustilago scitaminea*, the causative agent of sugarcane smut. Ghachtouli et al. (1995) have established a positive association between the length of PA chains and fungal development. Foster and Walters (1993) indicated that Cad production was optimized in in vitro cultures of economically significant plant pathogens, including *Erysiphe graminis* and *Puccinia hordei* on barley, *Botrytis fabae* on broad fava beans, *Podosphaera leucotricha* on apples, and *Phytophthora infestans* on potatoes. However, Cad alone showed no effect on powdery mildew and rust infection in barley, although its replacement induced a significant effect.

Cad is involved in the growth and development of microorganisms. The application of DFML, resulting in LDC inhibition, leads to the suppression of bacterial growth by impeding peptidoglycan production (Takatsuka and Kamio 2004). Both diamines, Cad and Put, are important for the peptidoglycan production of certain pathogenic bacteria. This Cad position may hold clinical importance. Cad inhibits the enterotoxins generated by *Shigella* sp. (Day Jr et al. 2001). Therefore, the improved virulence was prevented when LDC was lost in *Shigella* and commercial *E. coli* (Day Jr et al. 2001). The virulence-inducing and potentially pathoadaptive mutation of *Shigella* and enteroinvasive *E. coli* is thought to be caused by the inactivation of the Cad locus gene, which is responsible for LDC (Sokurenko 2017). Tissue cultures with decreased adhesion showed a return of LDC activity following complementation in a negative strain of the Cad operon (Torres et al. 2005). Few details on Cad's pathogenicity in plants have been clarified, even though they are economically significant for several diseases. To create a model for potential loss caused by disease management, it is necessary to address the varied sensitivity of different plant species to different diseases (Tomar et al. 2013).

14.4.9 Interaction of Cadaverine with Other Growth Regulators

The physical properties of molecules, aside from their titer, may also account for variances in physiochemical processes. Chemically, Cad is like Put in charge distribution and, hence, likely to compete for the same sites in several metabolic processes as other plant hormones. Putrescine and cadaverine impede paraquat absorption in the roots of undamaged maize seedlings. Cad is demonstrated to competitively limit the inflow of Put (Hart et al. 1992). Intercommunication between lysine and cadaverine in plants occurs within the absorption and transport mechanisms. Lysine somewhat enhances the mRNA of Cad B (Tanaka et al. 2008), and ornithine has a similar effect. Conversely, the mRNA of the putrescine transporter Pot E is moderately improved by lysine (Tomar et al. 2013). A decrease in putrescine content was observed after applying Cad to *Ricinus communis* cotyledons, and a similar response that was observed when treating with spermidine indicates a molecular connection (Antognoni et al. 1998).

It has been suggested that the suppression of ethylene during tomato ripening is related to the fact that plant Cad production fluctuates in response to ethylene application (Ponnampalam et al. 1993). There has been some speculation that ethylene may play a role in stress-induced plant Cad level control (Shevyakova et al. 2001). Ice plant leaves that were removed showed an increase in Cad production after exposure to ethylene (Kuznetsov et al. 2002). This finding provides more evidence that ethylene regulates Cad production (Kuznetsov et al. 2002). The alterations in *Arabidopsis thaliana* led to a significant rise in Cad in the leaves but no change in the root system (Shevyakova et al. 2000). Kuznetsov et al. (2002) have documented the production of Cad that is dependent on ethylene. A decrease in putrescine is induced by adding more Cad to the cotyledons of *Ricinus communis*, which indicates that there is competition at either the synthesis or the degradation stage. Several substances are recognized as modulators of Cad synthesis. In *Triticum monococcum* L. culture, amino-oxy-3-aminopropane (0.1 mM) elevated Cad levels compared to those of untreated samples or those treated with 1.0 mM. Subsequent research may elucidate the endogenous interactional behavior of Cad in plants, which could have economic implications (Tomar et al. 2013).

14.4.10 Cadaverine: Its Practical Use in Industry

There have been several initiatives to employ Cad for different industrial goals, driven by the emphasis on knowledge application. The incorporation of biotin-labeled cadaverine into N′-N′-dimethyl casein in a Ca^{2+}-dependent manner is an indication of the transglutaminase (protein-glutamyl transferase) activity, and cadaverine is used to detect activities catalyzed in the presence of Ca^{2+} (Lilley et al. 1998). The Cad and other phenolic acids present in sugarcane juices, which bind to phenols and exhibit lower water solubility, are implicated in the coloration of sweet products and influence quality (Legaz et al. 1998). It is beneficial to investigate the application of Cad for the removal of phenols and similar substances from food products to enhance food quality. Moreover, it suggests that the genetic modification of the LDC gene may be utilized in the development of transgenic organisms and molecular farming of alkaloids and/or Cad (Tomar et al. 2013). Pincus et al. (1996) asserted that Cad may be utilized in the development of anti-HIV medications, necessitating immediate attention from scientists for the benefit of society. The treatment of malignant cells may be strategized by modulating Cad titer and/or alkaloid levels, either directly or indirectly, through food quality (Tomar et al. 2013).

14.5 Conclusions

Cadaverine (Cad), although less abundant in plants, has emerged as a molecule with distinct physiological roles, particularly under stress conditions. Its synthesis, primarily from lysine, a vital amino acid, suggests a tightly regulated and potentially resource-sensitive pathway, reinforcing its proposed classification within growth-regulatory compounds. The unique and sometimes limited influence of Cad on phosphorylation events and gene expression points to a context- and species-dependent regulatory capacity. Additionally, recent findings highlight Cad's involvement in microbial siderophore production, and possible modulation via nucleotide signals such as ppGpp open new avenues for understanding its role in plant–microbe interactions. While Cad's presence in plant systems remains underexplored, especially compared to its broader distribution across biological kingdoms, emerging evidence supports its potential in plant growth modulation, stress responses, and biotechnological applications. Future research should aim to delineate its metabolic and signaling roles more precisely, including implications for food quality, storage, and possible pharmacological relevance.

References

Abdelmoneim MS, Dawood MFA, Hafez EE, Hammad SF, Ghazy MA (2025) Unveiling the role of cadaverine in mitigating salinity and/or Bisphenol A toxicity in tomato plants and reduced Bisphenol A accumulation in tomato roots. Plant Physiol Biochem 224:109799

Amiri H, Banakar MH, Hemmati Hassan Gavyar P (2024) Polyamines: new plant growth regulators promoting salt stress tolerance in plants. J Plant Growth Regul 43:4923–4940

Angosto T, Matilla AJ (1993) Variations in seeds of three endemic leguminous species at different altitudes. Physiol Plant 87:329–334

Antognoni F, Fornalè S, Grimmer C, Komor E, Bagni N (1998) Long-distance translocation of polyamines in phloem and xylem of Ricinus communis L. plants. Planta 204:520–527

Aziz A, Martin-Tanguy J, Larher F (1997) Plasticity of polyamine metabolism associated with high osmotic stress in rape leaf discs and with ethylene treatment. J Plant Growth Regul 21:153–163

Aziz A, Martin-Tanguy J, Larher F (1999) Salt stress-induced proline accumulation and changes in tyramine and polyamine levels are linked to ionic adjustment in tomato leaf discs. Plant Sci 145:83–91

Bagni N, Creus J, Pistocchi R (1986) Distribution of cadaverine and lysine decarboxylase activity in Nicotiana glauca plants. J Plant Physiol 125:9–15

Bailey SR, Rycroft A, Elliott J (2002) Production of amines in equine cecal contents in an in vitro model of carbohydrate overload. J Anim Sci 80:2656–2662

Balci M, Alp FN, Arikan B, Ozfidan-Konakci C, Yildiztugay E (2023) Polyamine cadaverine detoxifies nitrate toxicity on the chloroplasts of Triticum aestivum through improved gas exchange, chlorophyll a fluorescence and antioxidant capacity. J Plant Growth Regul 42:4958–4974

Baraiya T, Nathwani SA, Vaishnani B, Panigrahi M, Panigrahi J (2025) Enhancing industrial crop resilience: effects of salinity and exogenous putrescine supplementation on growth, biochemical, and enzymatic attributes of Cicer arietinum L. In: Industrial crops improvement: biotechnological approaches for sustainable agricultural development. Springer, pp 227–237

Becker J, Wittmann C (2012) Bio-based production of chemicals, materials and fuels–Corynebacterium glutamicum as versatile cell factory. Curr Opin Biotechnol 23:631–640

Berlin J, Mollenschott C, Herminghaus S, Fecker LF (1998) Lysine decarboxylase transgenic tobacco root cultures biosynthesize novel hydroxycinnamoylcadaverines. Phytochemistry 48:79–84

Botella MÁ, Del Amor F, Amorós A, Serrano M, Martínez V, Cerdá A (2000) Polyamine, ethylene and other physico-chemical parameters in tomato (Lycopersicon esculentum) fruits as affected by salinity. Physiol Plant 109:428–434

Bouchereau A, Aziz A, Larher F, Martin-Tanguy J (1999) Polyamines and environmental challenges: recent development. Plant Sci 140:103–125

Broquedis M, Dumery B, Bouard J (1989) Determination of polyamines (putrescine, cadaverine, nor-spermidine, spermidine and spermine) in leaves and grape bunches of Vitis vinifera L. Connaissance de la Vigne et du Vin 23:1–6

Bunsupa S, Katayama K, Ikeura E, Oikawa A, Toyooka K, Saito K, Yamazaki M (2012) Lysine decarboxylase catalyzes the first step of quinolizidine alkaloid biosynthesis and coevolved with alkaloid production in leguminosae. Plant Cell 24:1202–1216

Caffaro SV, Vicente C (1994) Polyamine implication during soybean flowering induction and early reproductive transition of vegetative buds. Plant Physiol. Biochem. 32(3):391–397

Caffaro S, Scaramagli S, Antognoni F, Bagni N (1993) Polyamine content and translocation in soybean plants. J Plant Physiol 141:563–568

Caldeira H, Caldeira G (1999) Free polyamine accumulation in unstressed and NaCl-stressed maize plants. Agron Lusitana 47:209–215

Carrizo CN, Pitta-Alvarez SI, Kogan MJ, Giulietti AM, Tomaro Maria L (2001) Occurrence of cadaverine in hairy roots of Brugmansia candida. Phytochemistry 57:759–763

Cassan F, Maiale S, Masciarelli O, Vidal A, Luna V, Ruiz O (2009) Cadaverine production by Azospirillum brasilense and its possible role in plant growth promotion and osmotic stress mitigation. Eur J Soil Biol 45:12–19
Chen CT, Kao CH (1991) Senescence of rice leaves XXX levels of endogenous polyamines and dark-induced senescence of rice leaves. Plant Cell Physiol 32:935–941
Chen C, Chang J, Wang S, Lu J, Liu Y, Si H, Sun G, Ma C (2021) Cloning, expression analysis and molecular marker development of cinnamyl alcohol dehydrogenase gene in common wheat. Protoplasma 258:881–889
Chong LI, Liyu C, Tao LEI (2010) Hydrogenation of glutaronitrile to cadaverine on amorphous Ni alloy catalyst. Pet Tech 39:524–527
Damalas CA, Koutroubas SD (2016) Farmers' exposure to pesticides: toxicity types and ways of prevention, vol 4. MDPI, p 1
Dawood MF (2025) A new frontier for plant-microbe interaction under changing climate conditions. Glob Chang Biol 31(7):e70393
Dawood MFA, Abeed AHA (2020) Spermine-priming restrained water relations and biochemical deteriorations prompted by water deficit on two soybean cultivars. Heliyon 6:e04038
Dawood MF, Mazrou YS, Sheteiwy MS, Nehela Y, Makhlouf AH, Abd El FEZA, Hifney AF (2025) Exploring the role of Arthrospira platensis and earthworms (Lumbricus castaneus) biosystem in mitigating antimony toxicity and enhancing the biochemical properties of cucurbit plants. Plant Physiol Biochem 227:110096
Day WA Jr, Fernández RE, Maurelli AT (2001) Pathoadaptive mutations that enhance virulence: genetic organization of the cadA regions of Shigella spp. Infect Immun 69:7471–7480
de Armas R, Martinez M, Vicente C, Legaz M-E (1999) Free and conjugated polyamines and phenols in raw and alkaline-clarified sugarcane juices. J Agric Food Chem 47:3086–3092
De Bellefon C, Fouilloux P (1994) Homogeneous and heterogeneous hydrogenation of nitriles in a liquid phase: chemical, mechanistic, and catalytic aspects. Catal Rev 36:459–506
del Rodriguez SC, López B, Chaves AR (1999) Changes in polyamines and ethylene during the development and ripening of eggplant fruits (Solanum melongena). J Agric Food Chem 47:1431–1434
DiTomaso JM, Rost TL, Ashton FM (1988) Herbicide-induced diamine accumulation in pea roots. The effect of napropamide on polyamine levels. Plant Cell Physiol 29:1367–1372
El-Shora HM (2001) Activators and inhibitors of photosynthetic enzymes from Rumex dentatus cotyledons. J Plant Physiol 158:1501–1510
El-Yazal MAS, Rady MM, Sa SE-Y (2018) Metabolic changes in polyamines, phenylethylamine, and arginine during bud break in apple flower buds under foliar-applied dormancy-breaking. Int J Empir Educ Res 1:1–18
Fecker LF, Rügenhagen C, Berlin J (1993) Increased production of cadaverine and anabasine in hairy root cultures of Nicotiana tabacum expressing a bacterial lysine decarboxylase gene. Plant Mol Biol 23:11–21
Foster SA, Walters DR (1993) Fungicidal activity of the polyamine analogue, keto-putrescine. Pestic Sci 37:267–272
Fujihara S (2009) Biogenic amines in rhizobia and legume root nodules. Microbes Environ 24:1–13
Fujihara S, Abe H, Yoneyama T (1995) A new polyamine 4-Aminobutylcadaverine: occurrence and its biosynthesis in root nodules of adzuki bean plant Vigna Angularis. J Biol Chem 270:9932–9938
Gale EF, Epps HMR (1942) The effect of the pH of the medium during growth on the enzymic activities of bacteria (Escherichia coli and micrococcus lysodeikticus) and the biological significance of the changes produced. Biochem J 36:600
Galiba G, Kocsy G, Kaur-Sawhney R, Sutka J, Galston AW (1993) Chromosomal localization of osmotic and salt stress-induced differential alterations in polyamine content in wheat. Plant Sci 92:203–211

Gallardo M, Gallardo ME, Matilla AJ, de Rueda PM, Sánchez-Calle IM (1994) Inhibition of polyamine synthesis by cyclohexylamine stimulates the ethylene pathway and accelerates the germination of Cicer arietinum seeds. Physiol Plant 91:9–16

Gamarnik A, Frydman RB (1991) Cadaverine, an essential diamine for the normal root development of germinating soybean (Glycine max) seeds. Plant Physiol 97:778–785

Garcia-Garcia P, Brenes-Balbuena M, Hornero-Mendez D, Garcia-Borrego A, Garrido-Fernández A (2000) Content of biogenic amines in table olives. J Food Prot 63:111–116

Ghachtouli NE, Paynot M, Morandi D, Martin-Tanguy J, Gianinazzi S (1995) The effect of polyamines on endomycorrhizal infection of wild-type Pisum sativum, cv. Frisson (nod+ myc+) and two mutants (nod− myc+ and nod− myc−). Mycorrhiza 5:189–192

Goren R, Palavan N, Flores H, Galston AW (1982) Changes in polyamine titer in etiolated pea seedlings following red light treatment. Plant Cell Physiol 23:19–26

Grzesik M (1993) *Effect of different solutions on the viability of the stored Helichrysum bracteatum seeds*. Paper presented at the Proceedings of the Fourth International Workshop on Seeds: basic and applied aspects of seed biology, Angers.

Gureeva MV, Gureev AP (2023) Molecular mechanisms determining the role of bacteria from the genus Azospirillum in plant adaptation to damaging environmental factors. Int J Mol Sci 24:9122

Hamana K, Matsuzaki S, Niitsu M, Samejima K (1994) Distribution of unusual polyamines in aquatic plants and gramineous seeds. Can J Bot 72:1114–1120

Hart JJ, DiTomaso JM, Linscott DL, Kochian LV (1992) Transport interactions between paraquat and polyamines in roots of intact maize seedlings. Plant Physiol 99:1400–1405

Hashimoto M, Eda Y, Osanai Y, Iwai T, Aoki S (1986) A novel decarboxylation of α-amino acids. A facile method of decarboxylation by the use of 2-cyclohexen-1-one as a catalyst. Chem Lett 15:893–896

Hoffer BW, Moulijn JA (2009) Hydrogenation of dinitriles on Raney-type Ni catalysts: kinetic and mechanistic aspects. Appl Catal A Gen 352:193–201

Hornero-Mendez D, Garrido-Fernandez A (1997) Rapid high-performance liquid chromatography analysis of biogenic amines in fermented vegetable brines. J Food Prot 60:414–419

Hou W-C, Lin Y-H (1997) Polyamine-bound trypsin inhibitors in sweet potato (Ipomoea batatas [L.] lam cv. Tainong 57) storage roots, sprouted roots and sprouts. Plant Sci 126:11–19

Huang Y, Ji X, Ma Z, Łężyk M, Xue Y, Zhao H (2021) Green chemical and biological synthesis of cadaverine: recent development and challenges. RSC Adv 11:23922–23942

Jancewicz AL, Gibbs NM, Masson PH (2016) Cadaverine's functional role in plant development and environmental response. Front Plant Sci 7:870

Kalač P, Křížek M (1997) Formation of biogenic amines in four edible mushroom species stored under different conditions. Food Chem 58:233–236

Kalač P, Švecová S, Pelikánová T (2002) Levels of biogenic amines in typical vegetable products. Food Chem 77:349–351

Karolewski P, Shevyakowa NI (1990) Effect of sulphite ions on the proline and polyamine content in the bean Phaseolus vulgaris. Acta Soc Bot Pol 59:55–64

Kiełkowska A, Dziurka M (2021) Changes in polyamine pattern mediates sex differentiation and unisexual flower development in monoecious cucumber (Cucumis sativus L.). Physiol Plant 171:48–65

Killiny N, Nehela Y (2020) Citrus polyamines: structure, biosynthesis, and physiological functions. Plants 9:426

Križek M, Kalač P, Peterka J (1993) Biogenic amines in silage 3. The occurence of six biogenic amines in farm-scale grass and maize silages. Arch Anim Nutr 45:131–137

Kushad MM, Dumbroff EB (1991) Metabolic and physiological relationships between the polyamine and ethylene biosynthetic pathways. In: Biochemistry and physiology of polyamines in plants, pp 77–92

Kuznetsov VV, Shevyakova NI (2010) Polyamines and plant adaptation to saline environments. In: Ramawat KG (ed) Desert plants: biology and biotechnology. Springer Berlin Heidelberg, Berlin, Heidelberg, pp 261–298

Kuznetsov VV, Rakitin VY, Sadomov NG, Dam DV, Stetsenko LA, Shevyakova NI (2002) Do polyamines participate in the long-distance translocation of stress signals in plants? Russ J Plant Physiol 49:120–130

Kuznetsov V, Shorina M, Aronova E, Stetsenko L, Rakitin V, Shevyakova N (2007) NaCl-and ethylene-dependent cadaverine accumulation and its possible protective role in the adaptation of the common ice plant to salt stress. Plant Sci 172:363–370

Laval G, Golding BT (2003) One-pot sequence for the decarboxylation of α-amino acids. Synlett 4:0542–0546

Lefevre I, Gratia E, Lutts S (2001) Discrimination between the ionic and osmotic components of salt stress in relation to free polyamine level in rice (Oryza sativa). Plant Sci 161:943–952

Legaz ME, Armas R d, Pinon D, Vicente C (1998) Relationships between phenolics-conjugated polyamines and sensitivity of sugarcane to smut (Ustilago scitaminea). J Exp Bot 49:1723–1728

Li B, Liang J, Hanfrey CC, Phillips MA, Michael AJ (2021a) Discovery of ancestral L-ornithine and L-lysine decarboxylases reveals parallel, pseudoconvergent evolution of polyamine biosynthesis. J Biol Chem 297:101219

Li G, Li D, Wang T, He S (2021b) Pyrimidine biosynthetic enzyme CAD: its function, regulation, and diagnostic potential. Int J Mol Sci 22:10253

Lilley GR, Skill J, Griffin M, Bonner PLR (1998) Detection of Ca2+-dependent transglutaminase activity in root and leaf tissue of monocotyledonous and dicotyledonous plants. Plant Physiol 117:1115–1123

Liu K, Fu H, Bei Q, Luan S (2000) Inward potassium channel in guard cells as a target for polyamine regulation of stomatal movements. Plant Physiol 124:1315–1326

Liu J-H, Kitashiba H, Wang J, Ban Y, Moriguchi T (2007) Polyamines and their ability to provide environmental stress tolerance to plants. Plant Biotechnol 24:117–126

López-Caballero M, Álvarez Torres M, Sánchez-Fernández J, Moral A (2002) Photobacterium phosphoreum isolated as a luminescent colony from spoiled fish, cultured in model system under controlled atmospheres. Eur Food Res Technol 215:390–395

Ma W, Chen K, Li Y, Hao N, Wang X, Ouyang P (2017) Advances in cadaverine bacterial production and its applications. Engineering 3:308–317

Maccarrone M, Baroni A, Finazzi-Agrò A (1998) Natural polyamines inhibit soybean (Glycine max) lipoxygenase-1, but not the lipoxygenase-2 isozyme. Arch Biochem Biophys 356:35–40

Moreau PL (2007) The lysine decarboxylase CadA protects Escherichia coli starved of phosphate against fermentation acids. J Bacteriol 189:2249–2261

Nidhi IN, Khan NA (2025) Polyamines interaction with gaseous signaling molecules for resilience against drought and heat stress in plants. Plants 14:273

Niemi K, Häggman H, Sarjala T (2002) Effects of exogenous diamines on the interaction between ectomycorrhizal fungi and adventitious root formation in scots pine in vitro. Tree Physiol 22:373–381

Onat-Taşdelen KA, Öztürkel-Kabakaş H, Yüksektepe E, Çatav ŞS, Güzel G, Çöl B, Kim H, Chae YK, Elgin ES (2024) Functional groups matter: metabolomics analysis of Escherichia coli exposed to trans-cinnamic acid and its derivatives unveils common and unique targets. World J Microbiol Biotechnol 40:47

Oyebamiji YO, Adigun BA, Shamsudin NAA, Ikmal AM, Salisu MA, Malike FA, Lateef AA (2024) Recent advancements in mitigating abiotic stresses in crops. Horticulturae 10:156

Özmen S, Tabur S, Öney-Birol S, Özmen S (2022) The effect of exogenous spermine application on some biochemichal and molecular properties in Hordeum vulgare L. under both normal and drought stress. Biologia 77:193–202

Ozmen S, Tabur S, Oney-Birol S (2023) Alleviation role of exogenous cadaverine on cell cycle, endogenous polyamines amounts and biochemical enzyme changes in barley seedlings under drought stress. Sci Rep 13:17488

Parolo ME, Orioli GA, Croci CA (1997) High-pressure liquid chromatographic determination of free polyamines in garlic (Allium sativum L.). Acta Hortic 433:463–468
Pincus SH, Wehrly K, Cole R, Fang H, Lewis GK, McClure JAN, Conley AJ, Wahren B, Posner MR, Notkins AL (1996) In vitro effects of anti-HIV immunotoxins directed against multiple epitopes on HIV type 1 envelope glycoprotein 160. AIDS Res Hum Retrovir 12:1041–1051
Piqueras A, Cortina M, Serna MD, Casas JL (2002) Polyamines and hyperhydricity in micropropagated carnation plants. Plant Sci 162:671–678
Ponnampalam R, Arul J, Makhlouf J, Castaigne F (1993) Inhibition of 1-aminocyclopropane-1-carboxylic acid-dependent ethylene production in tomato-tissue discs by biochemical treatments. Postharvest Biol Technol 2:291–300
Rajpal C, Tomar PC (2020a) Cadaverine: a diamine presence & role in plants. Plant Arch 20:1754–1763
Rajpal C, Tomar PC (2020b) Cadaverine: a potent modulator of plants against abiotic stresses: Cadaverine: a potent modulator. J Microbiol Biotechnol Food Sci 10:205–210
Rajpal C, Tomar PC (2020c) Effect of cadaverine on cultivars of tomato [lycopersicon esculentum (l.) em. thell] under multiple stress. Plant Arch 20:1706–1709
Rakesh B, Sudheer WN, Nagella P (2021) Role of polyamines in plant tissue culture: an overview. Plant Cell Tissue Organ Cult 145:487–506
Rakkammal K, Maharajan T, Ceasar SA, Ramesh M (2023) Biostimulants and their role in improving plant growth under drought and salinity. Cereal Res Commun 51:61–74
Ramakrishna S, Adiga PR (1975) Amine levels in Lathyrus sativus seedlings during development. Phytochemistry 14:63–68
Rizvi A, Kumar S, Dixit K, Kumar A, Debbarma E, Saxena S (2025) Trailing the prospects of leguminous microgreens as rising functional and future foods. In: Mathur P, Gupta A (eds) Recent trends and applications of leguminous microgreens as functional foods. Springer, Cham, pp 47–65
Rodrıguez CW, De Armas R, Vicente C, Legaz M a-E (2000) Changes in free and conjugated polyamines during starvation of sugarcane juices as analyzed by high-performance liquid chromatography. J Chromatogr A 881:531–541
Samartzidou H, Mehrazin M, Xu Z, Benedik MJ, Delcour AH (2003) Cadaverine inhibition of porin plays a role in cell survival at acidic pH. J Bacteriol 185:13–19
Shao J, Huang K, Batool M, Idrees F, Afzal R, Haroon M, Noushahi HA, Wu W, Hu Q, Lu X (2022) Versatile roles of polyamines in improving abiotic stress tolerance of plants. Front Plant Sci 13:1003155
Shen H-J, Galston AW (1985) Correlations between polyamine ratios and growth patterns in seedling roots. Plant Growth Regul 3:353–363
Shevyakova NI, Rakitin VY, Dam DB, Kuznetsov VV (2000) Cadaverine as a signal of heat shock in plants. Dokl Biol Sci 375:657–659
Shevyakova NI, Rakitin VY, Duong DB, Sadomov NG, Kuznetsov VV (2001) Heat shock-induced cadaverine accumulation and translocation throughout the plant. Plant Sci 161:1125–1133
Shimizu Y, Rai A, Okawa Y, Tomatsu H, Sato M, Kera K, Suzuki H, Saito K, Yamazaki M (2019) Metabolic diversification of nitrogen-containing metabolites by the expression of a heterologous lysine decarboxylase gene in Arabidopsis. The Plant Journal 100:505–521
Simon-Sarkadi L, Holzapfel WH (1994) Determination of biogenic amines in leafy vegetables by amino acid analyser. Zeitschrift fur Lebensmittel-untersuchung und-forschung 198:230–233
Simon-Sarkadi L, Ludidi N, Kocsy G (2014) Modification of cadaverine content by NO in salt-stressed maize. Plant Signal Behav 9:e27598
Smith TA (1975) Recent advances in the biochemistry of plant amines. Phytochemistry 14:865–890
Smith TA, Wilshire G (1975) Distribution of cadaverine and other amines in higher plants. Phytochemistry 14:2341–2346
Sofian ZM, Harun N, Mahat MM, Hashim NAN, Jones SA (2021) Investigating how amine structure influences drug-amine ion-pair formation and uptake via the polyamine transporter in A549 lung cells. Eur J Pharm Biopharm 168:53–61

Sokurenko E (2017) Pathoadaptive mutations in uropathogenic Escherichia coli. In: Mulvey MA, Klumpp DJ, Stapleton AE (eds) Urinary tract infections: molecular pathogenesis and clinical management, American society for microbiology, 2nd edn. Wiley, Washington, DC, pp 331–357

Srivastava BIS (1987) Polyamine changes during senescence and tumorogenesis in plants. Mech Ageing Dev 40:17–30

Strohm AK, Vaughn LM, Masson PH (2015) Natural variation in the expression of Organic Cation Transporter 1 affects root length responses to cadaverine in Arabidopsis. J Exp Bot 66:853–862

Sziderics AH, Oufir M, Trognitz F, Kopecky D, Matušíková I, Hausman J-F, Wilhelm E (2010) Organ-specific defence strategies of pepper (Capsicum annuum L.) during early phase of water deficit. Plant Cell Rep 29:295–305

Takatsuka Y, Kamio Y (2004) Molecular dissection of the Selenomonas ruminantium cell envelope and lysine decarboxylase involved in the biosynthesis of a polyamine covalently linked to the cell wall peptidoglycan layer. Biosci Biotechnol Biochem 68:1–19

Tanaka Y, Kimura B, Takahashi H, Watanabe T, Obata H, Kai A, Morozumi S, Fujii T (2008) Lysine decarboxylase of Vibrio parahaemolyticus: kinetics of transcription and role in acid resistance. J Appl Microbiol 104:1283–1293

Tassoni A, Antognoni F, Luisa Battistini M, Sanvido O, Bagni N (1998) Characterization of spermidine binding to solubilized plasma membrane proteins from zucchini hypocotyls. Plant Physiol 117:971–977

Ten Chen C, Kao CH (1991) Senescence of rice leaves XXIX. Ethylene production, polyamine level and polyamine biosynthetic enzyme activity during senescence. Plant Sci 78:193–198

Tomar PC, Arora K (2023) Effect of cadaverine on Brassica juncea (RH-30) seedlings under multiple stress-A 3 quantitative analysis 4. Songklanakarin. J Sci Technol 45

Tomar PC, Lakra N, Mishra SN (2013) Cadaverine: a lysine catabolite involved in plant growth and development. Plant Signal Behav 8:e25850

Torres AG, Zhou X, Kaper JB (2005) Adherence of diarrheagenic Escherichia coli strains to epithelial cells. Infect Immun 73:18–29

Torrigiani P, Scoccianti V (1995) Regulation of cadaverine and putrescine levels in different organs of chick-pea seed and seedlings during germination. Physiol Plant 93:512–518

Van Oosten MJ, Pepe O, De Pascale S, Silletti S, Maggio A (2017) The role of biostimulants and bioeffectors as alleviators of abiotic stress in crop plants. Chem Biol Technol Agric 4:1–12

Vassileva V, Ignatov G (1999) Effect of polyamines on dicarboxylate and oxygen uptake by symbiosomes and free bacteroids from Galega orientalis nodules. Symbiosis (Rehovot) 27:59–71

Villanueva VR, Adlakha RC, Cantera-Soler AM (1978) Changes in polyamine concentration during seed germination. Phytochemistry 17:1245–1249

Vinci G, Antonelli ML (2002) Biogenic amines: quality index of freshness in red and white meat. Food Control 13:519–524

Wang ShouFeng WS, Liang HaiMan LH, Zhou JuHua ZJ (1998) Changes of some physiological characteristics in detached cucumber cotyledons. J Trop Subtrop Bot 6:253–259

Wang XH, Wang SJ, Chen Z, Gong B, Wang XF, Win M, Shi QH, Li Y, Yang FJ (2016) Effects of exogenous polyamines on nitrate tolerance in cucumber. Russ J Plant Physiol 63:549–557

Wang W, Paschalidis K, Feng J-C, Song J, Liu J-H (2019) Polyamine catabolism in plants: a universal process with diverse functions. Front Plant Sci 10:561

Wortham BW, Oliveira MA, Patel CN (2007) Polyamines in bacteria: pleiotropic effects yet specific mechanisms. In: Perry RD, Fetherston JD (eds) The genus yersinia: from genomics to function. Springer, New York, NY, pp 106–115

Xuehao C, Guangwen Z, Yanping C, Beisheng C (2001) Relationship of plant hormones and polyamine with sex reversion of cucumber (Cucumis sativus L.). J Zhejiang Univ (Agric Life Sci) 27:639–642

Yakhin OI, Lubyanov AA, Yakhin IA, Brown PH (2017) Biostimulants in plant science: a global perspective. Front Plant Sci 7:2049

Yen GC (1992) Effects of heat treatment and storage temperature on the biogenic amine content of straw mushroom (Volvariella volvacea). J Sci Food Agric 58:59–61

Yokota T, Nakayama M, Harasawa I, Sato M, Katsuhara M, Kawabe S (1994) Polyamines, indole-3-acetic acid and abscisic acid in rice phloem sap. Plant Growth Regul 15:125–128

Zhu ZJ, Jiang YT (1994) Effects of different forms of nitrogen fertilizer on growth and nitrate accumulation in nonheading Chinese cabbage. Plant Physiol Commun 3:198–201

Chapter 15
The Role of *Trichoderma* Species in Mitigating Abiotic Stress in Plants

Farah K. Ahmed, Rabab M. El-Sherif, Mohamed G. Farhat, and Ahmed M. Gabr

Abstract Global agricultural productivity is increasingly at risk from abiotic stresses, which include a variety of environmental conditions, like drought, salinity, severe temperatures, and heavy metal (HM) toxicity. Plant growth, development, and production are severely limited by these non-biological stresses, making creative and long-lasting mitigation techniques necessary. Because of their many uses, species within *Trichoderma*, a common genus of filamentous fungi, have become a very promising biological solution. The processes by which *Trichoderma* reduces abiotic stress in plants are thoroughly examined in this research, with particular attention paid to its functions in hormonal regulation, improved nutrient absorption, strong antioxidant defense, and accurate osmotic adjustment. The study describes the precise physiological and biochemical reactions that *Trichoderma* causes under different stressors, highlighting the intricate molecular relationships that support its effectiveness. Additionally, it examines *Trichoderma*'s present uses in agriculture as well as its enormous potential in the future as a vital instrument for promoting crop resilience and guaranteeing food security in a fast-changing climate.

Keywords Abiotic stress · *Trichoderma* · Sustainable agriculture · Plant · Microbe interactions · Drought · Salinity · Growth promotion

---

F. K. Ahmed (✉) · M. G. Farhat
Department of Biotechnology, Faculty of Postgraduate Studies for Nanotechnology, Cairo University, El Sheikh Zayed, Giza, Egypt

R. M. El-Sherif
Postgraduate Studies for Nanotechnology, Cairo University, El Sheikh Zayed, Giza, Egypt

A. M. Gabr
Biotechnology Research Institute, National Research Centre (NRC), Cairo, Egypt

H. I. Mohamed (ed.), *Elicitors for Sustainable Crop Production*, Sustainability Sciences in Asia and Africa, https://doi.org/10.1007/978-981-95-3533-0_15

15.1 Introduction

15.1.1 Global Agricultural Challenges Posed by Abiotic Stress

Abiotic stressors, which are environmental conditions that are not biological and significantly reduce plant yield, provide a significant challenge to agricultural systems around the world. These stresses significantly affect the growth, development, and production of plants, posing a serious risk to the world's food security (Mishra et al. 2015). Extreme temperatures, both hot and cold; drought; flooding; salinity; heavy metal (HM) toxicity; and nutritional deficits are all examples of environmental circumstances that fall under the broad category of abiotic stress. Plants can experience a range of unique physiological and biochemical changes in response to each of these different stresses (Khan et al. 2025). Abiotic stresses' extensive effects on agricultural areas demonstrate how ubiquitous they are. They cause more than 50% of all output losses and impact more than 91% of the world's crops (Younis et al. 2020). For example, on 25% of arable land worldwide, yield reductions due to drought conditions alone are expected to exceed the total losses from all other sources (Deb et al. 2022). These environmental stresses cause harmful changes in a number of plant functions, including reduced germination rates, restricted development, hampered respiration and photosynthesis, and disturbances in vital hormonal and enzymatic functions. Plant cells often experience increased oxidative stress as a result of these disturbances (Wahab et al. 2022). The increasing frequency and intensity of abiotic stress episodes, which are mostly caused by ongoing climate change, are a particularly urgent concern. By 2100, it is anticipated that global air temperatures will have increased by 1.8–4.0°C (Annappa et al. 2023). This rising trend, when combined with altered precipitation patterns, makes preexisting issues worse and increases the risk to agricultural productivity. The increasing and widespread nature of abiotic stress caused by climate change elevates *Trichoderma*'s role as a solely beneficial agricultural amendment to a crucial tool for guaranteeing global food security (Valenzuela-Aragon et al. 2025). Since this is not only a current problem but also one that is only going to get worse in the future, *Trichoderma*'s ability to lessen these pressures becomes crucial for preserving agricultural output. Consequently, it becomes an essential component of climate-resilient agriculture, replacing its role as a supplementary agent (El-Saadony et al. 2024). *Trichoderma* provides a flexible, systemic approach to improving plant resilience, as evidenced by the wide range of abiotic stresses it may mitigate, such as temperature extremes, drought, flooding, salinity, and metal and nutrient imbalances. This all-encompassing ability is more effective and flexible than creating extremely specialized treatments for every stressor (Zaidi et al. 2013).

15.1.2 Overview of Trichoderma Species: Characteristics and Ecological Significance

The filamentous fungal genus *Trichoderma* is well-known for its considerable agricultural potential and ecological adaptability. These fungi are distinguished by their quick growth, copious mycelium, and characteristic green conidia, or spores, which are produced asexually from the fruiting bodies' bare cells (Yao et al. 2023). They can also produce chlamydospores and have phialides and branched conidiophores in their morphology. In addition to their morphological characteristics, *Trichoderma* species are physiologically noteworthy for producing a large number of different enzymes and secondary metabolites, both volatile and nonvolatile. Many of their advantageous interactions with plants and other microbes are based on these metabolic characteristics (Ruano-Rosa et al. 2016). In all climatic zones, *Trichoderma* species are common and often dominant elements of the mycoflora in a variety of habitats, such as soil, litter, organic matter, and rhizospheres. Their primary role is that of saprophytes, which thrive on organic debris that has decomposed (Liu et al. 2024). Recent findings, however, have broadened our knowledge of their functions and demonstrated their potential as avirulent, opportunistic plant symbionts. They can pierce the epidermis and are known to form strong, persistent colonization of root surfaces (Zaidi et al. 2013) (Nassary 2025). Crucially, the use of *Trichoderma* species extends beyond biocontrol. Additionally, they boost overall plant growth and vigor, increase nutrient uptake, confer tolerance to a variety of abiotic stresses, create resistance against plant diseases, and aid in the bioremediation of heavy metals and other environmental contaminants (Sood et al. 2020). *Trichoderma*'s combined characteristics as a strong biocontrol agent, an efficient abiotic stress–mitigating agent, and a plant growth promoter point to an evolutionary adaptation for all-encompassing plant protection. This makes it a very effective and adaptable symbiotic partner for sustainable farming (Manzar et al. 2022). Its wide functionality, which goes beyond a single-function application and includes both biotic and abiotic stress management as well as growth promotion, is a fundamental distinction and a major factor in its growing agricultural promise. Additionally, *Trichoderma*'s natural environmental resilience is demonstrated by its extensive occurrence and adaptability across a variety of climatic zones. The natural hardiness of the fungus is crucial to its effectiveness as a bio-inoculant, as it suggests that the strains are more likely to survive and perform well under the diverse and often harsh conditions of agricultural fields. This resilience enhances the global reliability and consistency of Trichoderma-based solutions in sustainable crop management (Mohd Din et al. 2025) (Fig. 15.1).

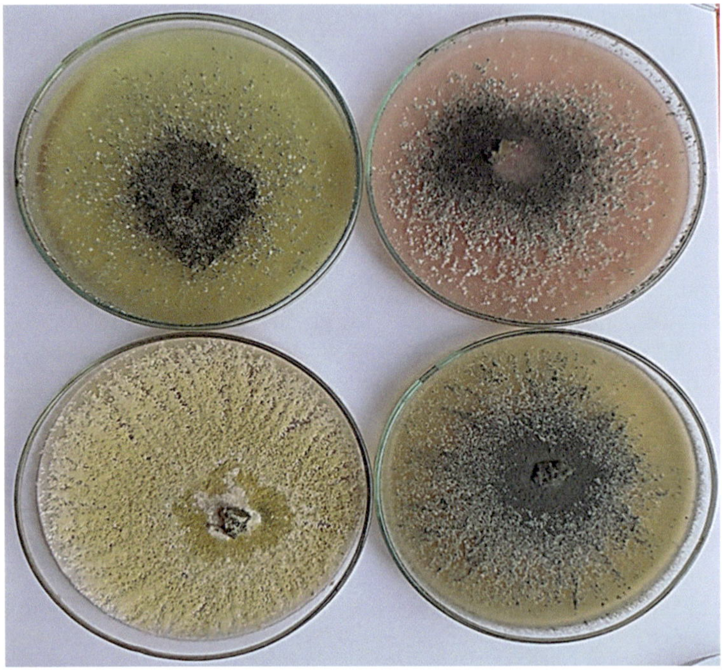

Fig. 15.1 Different types of *Trichoderma*

15.2 General Mechanisms of *Trichoderma*–Plant Interactions Enhancing Stress Tolerance

15.2.1 Root Colonization and Symbiotic Relationships

Trichoderma's capacity to form close bonds with plant roots is an essential component of its advantageous impacts on plants. Numerous *Trichoderma* species are common rhizosphere-dwellers that can colonize plant roots both internally and externally, frequently piercing the outer cortex and epidermis (Halifu et al. 2019). Through this colonization, *Trichoderma* obtains vital nutrients, including sugar, from the plant host, forming an advantageous opportunistic or facultative symbiosis, as shown in Table 15.1. In exchange, the fungus offers the plant several benefits, such as increased resistance to invasive diseases and better photosynthetic capacities (Ahsan et al. 2024). *Trichoderma*'s attraction to plant roots, which is the result of a complex interaction of chemical signals between the two partners, marks the beginning of this symbiotic connection. Despite its importance, this first stage of interaction is still not as well-understood as the others (Afridi et al. 2024). *Trichoderma* adheres to the root surface after being attracted and then enters the root tissues. Certain fungal proteins, particularly expansin-like proteins with cellulose-binding modules and endopolygalacturonase, are secreted to actively aid this penetration (Sood et al. 2020). *Trichoderma* fungus can grow intercellularly

Table 15.1 Mechanism of *Trichoderma* against abiotic stress

Abiotic stress	Mechanism of action by *Trichoderma*	Plant response/benefit
Drought stress	Induction of drought-responsive genes Production of abscisic acid (ABA) Enhanced root growth	Improved water uptake Better stomatal regulation Increased drought tolerance
Salinity stress	Modulation of ion transporters Increased proline accumulation Na^+ exclusion	Maintenance of ionic balance Improved osmotic adjustment
Heavy metal stress	Metal biosorption and chelation Siderophore production Antioxidant enzyme activation	Reduced metal uptake Enhanced detoxification Lower oxidative damage
Heat stress	Stabilization of membranes Production of heat shock proteins (HSPs)	Improved cellular integrity Enhanced heat tolerance
Cold stress (CS)	Upregulation of cold-responsive genes Increased soluble sugar content	Maintenance of membrane fluidity Protection against freezing
Oxidative stress	Activation of antioxidant enzymes (superoxide dismutase (SOD), catalase (CAT), and peroxidase (POD)) Reactive oxygen species (ROS) scavenging	Lower accumulation of ROS Reduced lipid peroxidation
Nutrient deficiency	Phosphate solubilization Siderophore-mediated iron acquisition Improved nitrogen use	Enhanced nutrient uptake Better plant growth under stress

after they are inside the roots, although usually just in the outer cortex and the epidermal layer. It is interesting to note that the fungus seems to be able to invade roots more easily when plant defense mechanisms are first suppressed (Morán-Diez et al. 2021). Improved crop yield, greater nutrient uptake, and improved root growth and development are all results of *Trichoderma* species successfully colonizing roots. Additionally, the plant's hormonal signals can be actively modulated by *Trichoderma*, which speeds up the process of root colonization (Contreras-Cornejo et al. 2024). A coevolutionary strategy that guarantees the stability and permanence of this connection, making it a durable biological solution, is indicated by the reciprocal nature of the *Trichoderma*–plant symbiotic relationship, in which the fungus receives nutrients while also providing growth benefits and stress tolerance (Laevens et al. 2024) (Mukherjee et al. 2012) (Guzmán-Guzmán et al. 2023) (Yang et al. 2025). This complex connection is important because it implies that *Trichoderma* is not only physically present but also actively affecting the cellular functions of the plant, which is necessary to initiate the wide spectrum of positive benefits seen in stress reduction (Contreras-Cornejo et al. 2024).

15.2.2 Direct Plant Growth Promotion

Trichoderma actively supports plant growth and development, in addition to its defenses against diseases and stressors. This increases overall plant vigor and biomass accumulation, which, in turn, improves stress tolerance (Asghar et al. 2024). It is commonly known that certain species of *Trichoderma* can promote plant growth and development, which improves nutrient uptake, crop yield, and plant vigor. More lateral roots, longer roots overall, taller plants, and more leaves and tillers are some of the specific morphological changes that have been noted. For example, research has shown a direct correlation between the presence of *T. virens* and *T. harzianum* with the production of many lateral roots (Tyśkiewicz et al. 2022) (Hasan et al. 2012). Particularly important regulators of root system design, auxins have a direct impact on a plant's capacity to search the soil for supplies. Additionally, by aggressively solubilizing immobile soil minerals and expanding the root surface area, *Trichoderma* improves nutrient uptake (Halifu et al. 2019). *T. asperellum* Q1, for instance, has been shown to facilitate phosphate solubilization, increasing the availability of phosphorus for plant absorption. Additionally, it helps other vital macronutrients and micronutrients like P, K, Ca, Mg, and Mn be absorbed (Timofeeva et al. 2022). Additionally, direct seed treatment with *Trichoderma* species and seed biopriming can cause the production of phytohormones and enzymes that greatly increase seed germination rate and seedling vigor, resulting in a more resilient and consistent field stand (Maitra et al. 2025). *Trichoderma*'s direct promotion of plant growth, especially through improved root development and nutrient mobilization, gives plants a proactive, fundamental resilience that makes them more capable of withstanding and recovering from abiotic stresses rather than just responding to their symptoms (Taha et al. 2020). Even in less-than-ideal circumstances, a more resilient root system enables better water and nutrient absorption, as seen by the recurring emphasis on enhanced root growth, higher nutrient uptake, and phytohormone synthesis. This indicates that *Trichoderma* boosts the plant's health and capacity to acquire resources, which increases its inherent ability to withstand stress, rather than only reducing its symptoms (Sofo et al. 2012). This is an important preventative measure. A complex, multipoint manipulation of plant physiological pathways is indicated by the production of a wide range of phytohormones by *Trichoderma*, including auxins, ethylene (ET), gibberellins, ABA, and salicylic acid (SA). This suggests that *Trichoderma* functions as a "master regulator" of plant development and stress responses rather than a straightforward, single-action stimulant (Singh et al. 2020). A sophisticated, integrated signaling network is suggested by the vast number and diversity of phytohormones listed, as well as their well-established interrelated roles in plant growth and stress responses. This suggests that Trichoderma orchestrates a comprehensive modulation of the plant's internal regulatory system rather than merely supplying a single growth-promoting factor. Such coordination enables the plant to fine-tune its defense and developmental strategies in response to changing environmental stimuli (Martínez-Medina et al. 2014).

15.2.3 Induction of Systemic Resistance (ISR and SAR)

The positive effects of *Trichoderma* also include priming the plant's natural defenses, which makes it more resilient to a variety of stressors. Systemic resistance mechanisms, particularly induced systemic resistance (ISR) and systemic acquired resistance (SAR), are activated in this process (Contreras-Cornejo et al. 2011). Through processes that resemble or involve ISR and SAR pathways, *Trichoderma* species are known to provide abiotic stress tolerance and resistance against plant diseases. *Trichoderma* colonization of the roots causes the plant to develop systemic or localized resistance responses, which significantly alter the metabolism of the plant (Nawrocka and Małolepsza 2013). Interestingly, *Trichoderma* can even initiate a salicylic acid (SA)–mediated SAR response, a broad-spectrum defense mechanism usually triggered by necrotrophic infections, at higher inoculum dosages. Known as "priming," this phenomenon is the increased accumulation of latent cellular proteins involved in intracellular signal amplification (Dutta et al. 2023). When a stressor comes along, this physiological state enables the plant to respond more quickly and robustly, minimizing damage and promoting recovery. Additionally, *Trichoderma* actively regulates the activity of important defensive hormones, including salicylic acid (SA), ethylene (ET), and jasmonic acid (JA), which are essential regulators of downstream plant defense pathways. This complex hormonal cross talk allows an integrated reaction to both biotic and abiotic stresses, emphasizing a unified defense strategy (Ikram et al. 2025). Instead of merely responding to stress events, *Trichoderma*'s induction of systemic resistance pathways (ISR/SAR) is a "preemptive strike" mechanism in which the plant is physiologically primed and ready for future stress events. Long-term protection is provided by this proactive defense (Choudhary et al. 2007). The term "priming" suggests that the plant's defense mechanism is activated to a higher level of alertness. When a stress occurs, this is a quicker and more energy-efficient reaction than creating defenses from the ground up, improving the plant's capacity to lessen damage and sustain output in inconsistent conditions (Liu et al. 2022). Further evidence of a comprehensive interaction that coordinates and integrates biotic and abiotic stress responses is provided by *Trichoderma*'s ability to influence important plant defense hormones (JA, ET, and SA). This shows that the fungus regulates a single plant defensive strategy. Because JA, ET, and SA hormones are known to mediate a variety of plant defensive responses against environmental stresses and pathogens, their regulation is important (Pacheco-Trejo et al. 2022). Instead of having distinct, isolated reactions to various forms of stress, *Trichoderma*'s impact on these central signaling pathways suggests that it functions as a conductor, allowing the plant to effectively prioritize and coordinate its defenses, suggesting a deeper level of biological integration and control (Dutta et al. 2023).

15.3 *Trichoderma*-Mediated Mitigation of Specific Abiotic Stresses

15.3.1 Drought Stress Alleviation

One of the most important abiotic stresses that affects plant water status and overall productivity worldwide is drought. *Trichoderma* species have repeatedly shown a great deal of promise in helping plants adapt to conditions involving water deprivation. These helpful fungi successfully protect plants from dryness and increase their tolerance (Akbari et al. 2024). Enhancing the plant's water status is one of the main ways by which *Trichoderma* reduces drought stress. By encouraging strong root growth, *Trichoderma* inoculation improves water shortage tolerance by increasing the soil's ability to retain water and facilitating nutrient mobilization (Rawal et al. 2022). The relative water content (RWC) of plant leaves, a crucial sign of better hydration, noticeably rises as a result. Physiologically speaking, *Trichoderma* helps increase photosynthetic efficiency (Quazi et al. 2024). Even when there is a water deficit, plants inoculated with *Trichoderma* show higher levels of net photosynthesis (Pn), increased stomatal conductance (Gs), and improved leaf chlorophyll content (including chlorophyll a, chlorophyll b, and total chlorophyll). For instance, research has demonstrated that *T. asperelloides*–NT33 and *T. asperellum*–NT16 markedly enhanced the number of lateral roots, shoot weight, and root biomass in tomato plants exposed to water deficit (Oljira et al. 2020). *Trichoderma* also has a favorable effect on the biochemical reactions to drought. *Trichoderma* causes increased levels of total phenols and the activity of antioxidant enzymes like catalase (CAT) when there is a drought. At the same time, malondialdehyde (MDA) and superoxide radical (O_2-) levels decline, which suggests that oxidative damage in plant cells has significantly decreased (Boorboori and Zhang 2023). A key factor in this mitigation is hormonal modulation; for example, *T. parareesei* has a chorismate mutase, which helps rapeseed tolerate water deprivation. Furthermore, it has been demonstrated that *T. hamatum* affects intricate stress-signaling pathways by modifying the expression of important genes in cacao plants under water deficit, including mitogen-activated protein kinase 3 (MAPK3), sorbitol transporter (SOT), pathogenesis-related 5 (PR5), and lipoxygenase (LOX) (Pérez et al. 2015). In terms of osmotic adjustment, a meta-analysis explicitly observed a decrease in proline production during drought circumstances in the presence of *Trichoderma*, whereas several studies generally claim an increase in proline levels under stress as a coping mechanism (Liang et al. 2013). This insight is a crucial and unexpected discovery. It implies that *Trichoderma* successfully lessens the plant's physiological demand to produce and collect large amounts of osmolytes by lowering the intensity of the drought stress on the plant. This suggests that stress reduction goes beyond symptomatic coping to a deeper level (Yu et al. 2023). Proline is a popular osmolyte that usually builds up in drought-stressed plants to help scavenge reactive oxygen species and maintain cellular turgor. Thus, it stands to reason that *Trichoderma* would raise proline. However, the observed decrease in proline in drought-treated

Trichoderma plants suggests that the fungus's other mechanisms—such as increased photosynthesis and better root water uptake—are so potent that the plant suffers from less severe osmotic stress (Maggio et al. 2002). As a result, proline synthesis requires less energy from the plant, which indicates a real reduction in the stress load rather than merely assisting the plant in managing its symptoms. *Trichoderma*'s effectiveness in reducing drought is frequently "isolate-specific," which means that some strains work better with specific host plants or in particular environments. The different impacts based on environmental conditions and strain and inoculum concentration highlight how crucial it is to choose the right strain and follow the best application procedures in order to get reliable results in real-world agricultural settings (Rawal et al. 2022). This demonstrates that applying *Trichoderma* in a "one-size-fits-all" manner is likely to provide uneven outcomes. Research must keep concentrating on determining which strains are most suited for specific crops and environments, as well as creating customized application techniques, in order for commercial and agricultural adoption to be successful (Wyckhuys et al. 2024).

15.3.2 Salinity Stress Mitigation

Salinity stress is a serious global danger to plant growth and agricultural output, influencing soil physicochemical qualities, restricting seed germination, creating ion imbalance, and slowing photosynthetic rates. *Trichoderma* species have been demonstrated to be beneficial in boosting plant development and increasing salt tolerance, notably by colonizing plant roots in saline soils (Singh et al. 2024). The stimulation of the plant's antioxidant system is one of the unique ways by which *Trichoderma* reduces the effects of salinity stress. For example, *T. asperellum* 22,043 modifies the activity of enzymes that scavenge reactive oxygen species (ROS) by controlling the transcriptional levels of their genes (Hu et al. 2025). *T. asperellum* 22,043 significantly increases the activity of antioxidant enzymes like catalase (CAT), peroxidase (POD), and superoxide dismutase (SOD) in tomato plants under salinity stress. In particular, when compared to seedlings treated with NaCl alone, *T. asperellum* 22,043–treated seedlings exhibited a 10% increase in SOD and POD activities and a 93% increase in CAT activity. This was predominantly due to the higher CAT activity, which improved resistance. A reduction in the amount of malondialdehyde (MDA), a by-product of membrane lipid peroxidation, indicates less oxidative damage as a result (Murshed et al. 2014). In order to preserve cellular homeostasis, *Trichoderma* is also essential for controlling stress-resistant genes, especially those related to ion transport and aquaporins. To keep the Na + equilibrium in tomato cells, *T. asperellum* 22,043 controls the transcriptional level of ion transport genes (Hu et al. 2025). In order to promote Na^+ circulation, this involves downregulating HKT gene expression in roots and upregulating it in seedlings under NaCl stress. Moreover, it lowers the transcriptional level of the SOS1 and LHA4 genes, which are involved in Na + efflux and H + -ATPase activity, respectively, and increases the expression of the NHX1 gene, which stores excess

Na^+ in vesicles (Zhang et al. 2017). *Trichoderma* also controls the expression of aquaporin genes to preserve water balance; for instance, *T. asperellum* 22,043 can boost the expression of the PIP2-9 gene in tomato seedlings treated with salt to improve water absorption while lowering it in roots to stop excessive water entry (El Kelish et al. 2019). Additionally, plants' physiological and biochemical traits are greatly enhanced. It has been demonstrated that *T. asperellum* 22,043 promotes plant height and chlorophyll content while also increasing seed germination and seedling survival rates under salinity stress (Adusumilli and Kolli 2022). *Trichoderma* inoculation significantly increased morphological markers, physiological markers, and activity of enzymes in plants under salinity stress, as determined by a meta-analysis that found that the greatest effects occurred under moderate salinity (4–8 dS m^{-1}) and medium-term (2–4 weeks) stress (Cheng et al. 2023). Additionally, *Trichoderma* can create auxins like auxin Naphthylacetic acid (NAA), and indole-3-acetic acid (IAA), which are plant growth regulators that assist in reducing salinity stress. A mechanism known to improve salt tolerance is the production of 1-Aminocyclopropane-1-carboxylate (ACC) deaminase (ACCD) by some strains, which inhibits excessive ethylene synthesis. Improved salt tolerance is also a result of *Trichoderma*'s capacity to enhance the absorption of vital nutrients like P, K, Ca, Mg, and Mn (Boorboori and Zhang 2023).

15.3.3 Heavy Metal Toxicity Mitigation

Heavy metal (HM) contamination has a major impact on agricultural output, causing stunted growth, chlorosis, decreased nutrient uptake, and oxidative stress–induced plant damage. *Trichoderma* species have emerged as prospective agents for reducing HM toxicity, with some strains exhibiting exceptional metal resistance and bioremediation capacities. *Trichoderma* reduces heavy metal toxicity via a variety of molecular and physiological reactions (Altaf et al. 2024). *Trichoderma* species produce a number of bioactive chemicals that aid plants in dealing with HM stress. These include indole-3-acetic acid (IAA), a phytohormone that fosters root growth and nutrient uptake, and ammonia, which helps with nitrogen uptake (Abdenaceur, 2024). *Trichoderma* strains such as TF-13 can also dissolve insoluble phosphorus, making it more accessible for plants, and create siderophores such as salicylic acid (SA) and 2,3-dihydroxybenzoic acid (2,3-DHBA), which chelate insoluble iron and restrict the mobility of harmful heavy metals, all while reducing oxidative stress. Furthermore, *Trichoderma* produces ACC deaminase (ACCD), an enzyme that lowers harmful ethylene levels in plants under stress, thereby stimulating growth (Altaf et al. 2024). *Trichoderma* inoculation also affects stress indicators in HM-exposed plants. While HM stress normally increases proline, an osmolyte and ROS scavenger, *Trichoderma* inoculation can drastically lower proline concentration in both roots and leaves of metal-stressed plants, indicating efficient oxidative damage prevention (Agarwal et al. 2025). In a similar way, *Trichoderma* significantly lowers hydrogen peroxide (H_2O_2) and malondialdehyde (MDA) concentration, both major reactive oxygen species (ROS) and markers of lipid peroxidation, implying increased stress tolerance and reduced cellular membrane damage (Buono et al. 2021). Although growing HM concentrations normally improve antioxidant enzyme activity as a defense, *Trichoderma* inoculation can limit their buildup, signaling that the fungus's additional advantages reduce the requirement for high antioxidant enzyme activity (Eftekhari et al. 2025). Physiologically, *Trichoderma* inoculation significantly improves plant performance under heavy metal stress. It improves seed germination rates and seedling vitality. Treated plants exhibit enhanced root and shoot length, fresh weight, and dry biomass, which mitigate the negative impacts of Cd and Pb on growth. *Trichoderma* also increases leaf pigment levels (total chlorophyll, chlorophyll a, chlorophyll b, and carotenoids) and enhances the uptake of important nutrients such as nitrogen and phosphorus in both roots and shoots of metal-stressed plants (Adhikari et al. 2024). Importantly, *Trichoderma* species such as TF-13 can greatly reduce Cd and Pb deposition in both root and shoot tissues by reducing metal mobility at the rhizosphere, which is often achieved by capturing them with cell-surface polymeric compounds. The direct reduction in metal absorption is an important detoxifying process (Altaf et al. 2024).

15.3.4 Extreme Temperature Stress Mitigation (Heat and Cold)

High temperatures have a significant impact on plant growth, metabolism, and output all over the world. *Trichoderma* species have demonstrated promising results in increasing plant tolerance to these temperature difficulties (Carillo 2025). Temperatures exceeding 30 °C have a deleterious effect on tomato plants, reducing germination, flowering, and fruit set, resulting in lower yield and quality. Heat stress disrupts photosynthesis, lowers pollen viability, and raises the oxidative damage to plant cells (Pressman et al. 2002). *Trichoderma* enhances plant tolerance to heat stress via a variety of methods. It promotes systemic resistance, releases growth hormones, and increases nutrient intake. For example, *Trichoderma koningii* ITCC 5201 (TK) was demonstrated to be beneficial in conferring resistance to heat stress in tomato plants by regulating antioxidants (Rai and Prasad 2023). While heat stress normally enhances the activity of antioxidant enzymes and the production of reactive oxygen species (ROS) such as H_2O_2, *Trichoderma* therapy reduces these enzymatic activities in stressed seedlings while decreasing H_2O_2 creation and lignin deposition. This suggests that *Trichoderma* helps regulate the generation of ROS, shielding plant cells from oxidative damage (Tripathi et al. 2021). *Trichoderma*-treated plants showed an increase in carbohydrate, protein, and total phenol content while under heat stress. This shows that *Trichoderma* allows the plant to manage stress more efficiently, minimizing the requirement for an excessive antioxidant response and, instead, fostering favorable metabolic changes (Contreras-Cornejo et al. 2024). Cold stress (CS) has a substantial impact on plant growth, development,

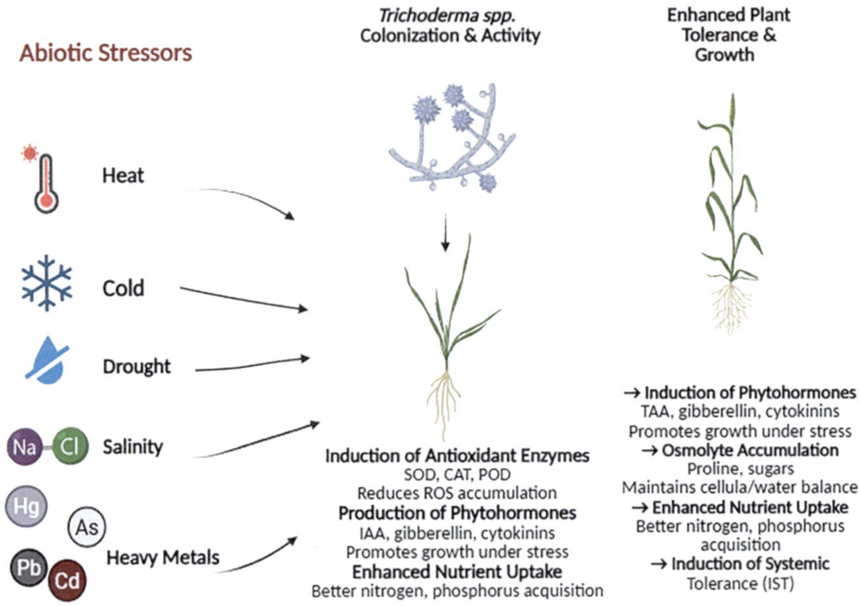

Fig. 15.2 Mechanisms by which *Trichoderma* species mitigate abiotic stress in plants

and production, particularly in temperate and subtropical locations with irregular temperature swings due to climate change. Plants under cold stress produce excessive ROS, causing cell membrane breakdown, fluid leakage, and elevated MDA concentration (Adhikari et al. 2022). Cold also reduces chlorophyll synthesis, breaks down photosynthetic enzymes, and prevents photosynthesis. *Trichoderma* can assist plants in surviving cold stress by strengthening cell osmotic adjustment, improving antioxidant pathways, controlling phytohormones, stimulating photosynthesis, and enriching nutritional status (Feng et al. 2025). Plants react to cold stress by modifying osmoregulatory components such as soluble sugars, soluble proteins, and proline to accomplish osmotic alterations. *Trichoderma* also assists plants in sensing oncoming cold and activating the production of cold stress–related genes via signal cross talks, controlling responses to neutralize the impact and mitigate damage (Ameen et al. 2024) (Fig. 15.2).

15.4 Molecular and Genetic Pathways in *Trichoderma*–Plant Interactions

Complex molecular and genetic processes develop positive interactions between *Trichoderma* and plants, notably those that increase abiotic stress tolerance. Advances in "omics" technologies, including genomics, transcriptomics, proteomics, and metabolomics, have considerably increased the understanding of these complex processes (Doni et al. 2022). Genomic and proteomic investigations have helped find new genes and gene products, such as ABC transporters, enzymes, and other proteins, which serve as elicitors of resistance-producing genes in plants. These findings also uncover *Trichoderma*-induced plant proteins, providing targets for developing strains with improved biocontrol and stress tolerance traits (Huang et al. 2021). *Trichoderma*'s genome has more genes for cell wall–degrading enzymes (CWDEs), such as chitinases and β-1,3-glucanases, than other fungi, indicating their importance in antagonistic capacity and nutrient absorption (Geraldine et al. 2013). Transcriptomics, which studies the entire collection of RNA transcripts, helps researchers uncover the critical regulatory genes involved in stress tolerance by investigating gene expression patterns under stress circumstances (Bimpong et al. 2024). For example, *Trichoderma* infection stimulates local and systemic expression of abiotic stress tolerance–related genes, which are linked with the abscisic acid (ABA) and ethylene (ET) pathways, resulting in enhanced plant responses under stress (Morán-Diez et al. 2021). Proteomics, using techniques such as mass spectrometry, sheds light on how proteins function and interact during stress reactions, finding proteins implicated in signaling networks that activate defensive mechanisms. For example, *T. harzianum* treatment of tomato plants elicits transcriptomic and metabolomic alterations that drive both direct and indirect defense responses, including the overexpression of transcripts encoding for defense-related transcription factors (Coppola et al. 2019). Metabolomics, which identifies important metabolites associated with stress responses, provides insights into the molecular pathways that drive plant resilience. *Trichoderma* regulates the formation of many metabolites, including osmolytes like proline and soluble sugars, as well as antioxidants such as total phenols, ascorbic acid, and glutathione (Ghatak et al. 2018). These substances aid in preserving cellular turgor, protecting cellular components, and scavenging reactive oxygen species, reducing oxidative stress. The combination of multi-omics methods is a transformational paradigm for understanding plant resilience (Hasanuzzaman et al. 2021). This paradigm enables researchers to explore the complex interconnections of diverse biological components and their involvement in stress responses, resulting in more effective crop enhancement tactics. These approaches decode gene regulation and transcriptional networks, revealing the fundamental mechanisms underpinning stress tolerance, providing targets for genetic modification and informing breeding strategies to generate climate-smart crops (Varadharajan et al. 2025).

15.5 Application Methods and Environmental Factors Influencing Efficacy

The successful use of *Trichoderma* in agriculture for abiotic stress mitigation is strongly reliant on adequate application methods and an understanding of the environmental conditions that determine its effectiveness. Several ways have been studied to improve the resilience of *Trichoderma* application in agricultural settings (Tyśkiewicz et al. 2022). In general, *Trichoderma* can be used to promote plant growth and suppress soilborne diseases by applying it to crop growth media, primarily soil, after seedling emergence. This is a potential method, especially for crops directly drilled into soils, because it primes the root system throughout seedling growth, increasing seedling vigor and guarding against soilborne illnesses following transplantation (Bandara and Kang 2024). *Trichoderma* seed treatment has been demonstrated to reduce biotic and abiotic stressors in germinating seeds and seedlings in a variety of field crops, including pea, cotton, wheat, cowpea, common bean, corn, and tomato. For transplanted crops like tomatoes, administering *Trichoderma* at the time of transplantation has been discovered to be particularly favorable for boosting growth (Bandara and Kang 2024). This technique appears to be more efficient than posttransplant treatments, even if the latter increases fungal abundance in the rhizosphere, implying that the timing of initial interaction is critical for maximum plant growth promotion. Direct insertion into the soil is a typical approach, which allows *Trichoderma* to colonize the rhizosphere and interact with plant roots (Ayangbenro et al. 2022). Various environmental variables can have a considerable impact on *Trichoderma* species' effectiveness, as they control their development, sporulation, and overall beneficial activity. *T. harzianum* and *T. viride* grow and sporulate best at 25–40 °C, while *T. asperellum* and *T. hamatum* thrive at 25–35 °C (Zehra et al. 2017) (Gasch et al. 2000). Temperature is an important limiting factor in the natural dispersion and performance of *Trichoderma* strains. Species of *Trichoderma* prefer and thrive in acidic soils rich in organic materials. The optimal pH range for *T. harzianum*, *T. viride*, *T. hamatum*, and *T. asperellum* is between 4.6 and 7.6, with maximum development and sporulation found between a pH of 4.1 and 8.6 (Kumar 2013). Growth and sporulation are greatly reduced outside this range, with growth ending at a pH of 8.6. *Trichoderma* species show variable levels of salt tolerance. For example, *T. harzianum* demonstrates great tolerance, followed by *T. viride* (Zehra et al. 2017). Moderate NaCl concentrations (200 μM and 400 μM) promote growth and sporulation; however, higher concentrations (e.g., 1000 μM) result in poor sporulation. Soil salinity is a known environmental condition that can limit *Trichoderma*'s capacity. However, certain species have developed extrusion systems to regulate intracellular salt levels below dangerous concentrations (Dikilitas and Karakas 2014). The efficiency of *Trichoderma* isolates varies greatly depending on the host plant genotype. This emphasizes the importance of particular strain–plant pairings for the best results. The concentration of *Trichoderma* inoculum influences its efficacy, with greater doses occasionally inducing distinct plant responses (e.g., SA-mediated SAR) (Cournault et al. 2025). *Trichoderma*'s uneven field effectiveness across production systems and environmental conditions remains an issue. Limited understanding of the reasons that cause biocontrol failures, as well as how *Trichoderma* works under multifaceted and intricate biotic and abiotic settings, has impeded efforts to improve its effectiveness and devise dependable deployment techniques (Bandara and Kang 2024). This stresses the need for ongoing research to create highly effective biocontrol products and application tactics that are economically viable, technically simple, and simply adaptable (Chaudhary et al. 2024) (Fig. 15.3)

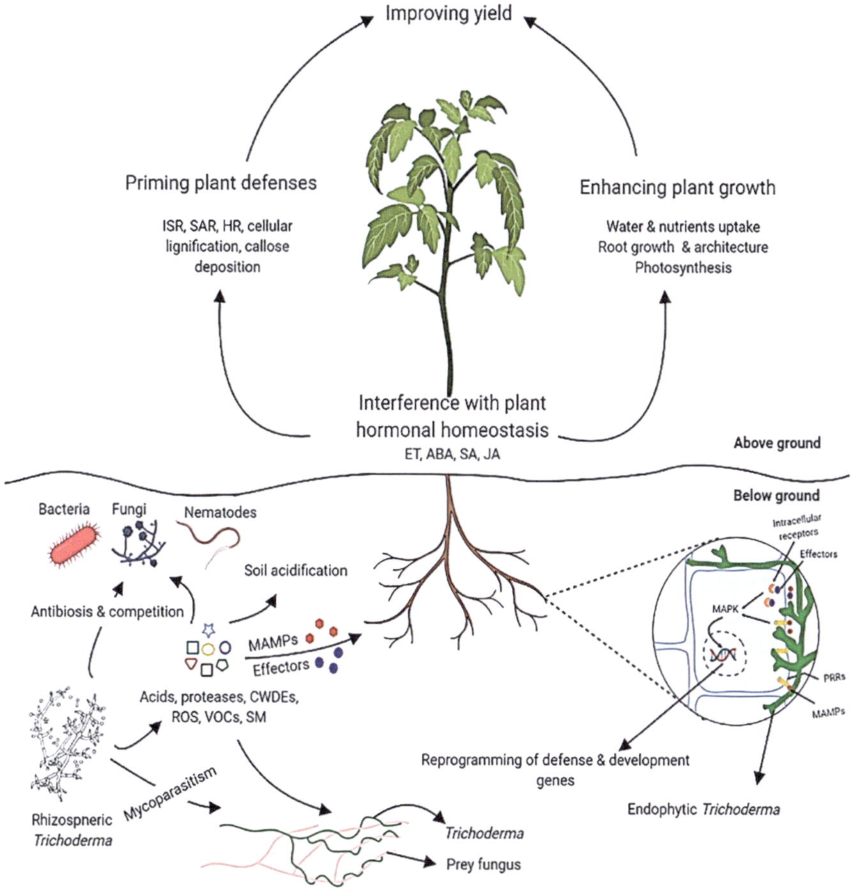

Fig. 15.3 Diagram illustrating the interactions between *Trichoderma*, plant, and pathogen. It shows how *Trichoderma* enhances plant growth and defense mechanisms while inhibiting pathogen activity through competition, mycoparasitism, and the production of antimicrobial compounds. (Alfiky and Weisskopf 2021)

15.6 Challenges and Future Research Directions

Despite tremendous advances, some hurdles remain in fully realizing *Trichoderma*'s promise for abiotic stress reduction. Although numerous investigations show the promising ability of *Trichoderma* to reduce abiotic stresses, enhance quality, and improve yield, the fundamental mechanisms of *Trichoderma*–plant interactions at the morphological, physiological, and molecular levels to overcome stress challenges are unclear in many contexts (Prismantoro et al. 2024). Additional research is needed to understand these complicated pathways. *Trichoderma*'s positive effects tend to be isolate-specific and can differ according to stress level, host type, strain and inoculum concentration, and environmental circumstances. This heterogeneity makes consistent field efficacy challenging (Galletti et al. 2020). Although beneficial interactions between *Trichoderma* and other microorganisms (for example, bacteria) have been shown to produce larger advantages than separate treatments, the particular mechanisms underlying these synergies deserve further exploration. The long-term stimulating effects of *Trichoderma*, which include epigenetic marks for increased resistance and growth promotion, are an intriguing topic that needs to be extensively researched (Shao et al. 2025). Studying the transmission of these epigenetic modifications may lead to new breeding tactics. Genetic intervention to create *Trichoderma* strains that outperform native ones shows promise. This goal could be reached by learning more about the molecular mechanisms that govern interactions with both biotic and abiotic stimuli. Converting laboratory and greenhouse accomplishments into consistent field-scale efficacy remains a challenge (Yanfei et al. 2022). Additional research is needed to understand how *Trichoderma* introduction causes structural and functional changes in the soil microbiome, as well as how these changes affect crop development and pathogen control. Future research should focus on an improved comprehension of the molecular dialogue between *Trichoderma* and plants, using multi-omics data to discover essential genes, proteins, and metabolites that regulate stress tolerance (Kredics et al. 2024). Cultivating targeted *Trichoderma* strains for particular crops and habitats, improving application methods, and studying synergistic microbial consortia will be vital for realizing their potential in sustainable agriculture (Díaz-Rodríguez et al. 2025).

15.7 Conclusion

Trichoderma species are a robust and adaptable biological tool for alleviating the negative impacts of abiotic stress in plants, providing a long-term solution to key agricultural concerns worldwide. Their ubiquitous presence and remarkable flexibility across varied climate zones make them strong candidates for bio-inoculants. *Trichoderma*'s potency is due to a complex network of mechanisms. *Trichoderma* forms a mutually advantageous relationship with roots by colonizing them opportunistically, increasing nutrient intake, and encouraging total plant growth. This immediate development promotion is largely mediated by the fungus's ability to create and modify a diverse range of plant phytohormones, acting as a "master regulator" that coordinates the plant's developmental and stress response pathways. Moreover, *Trichoderma* stimulates the plant's internal defense systems, resulting in systemic resistance (ISR and SAR). This "preemptive strike" mechanism prepares the plant for future stress events, allowing for faster, stronger, and more energy-efficient responses to environmental challenges. *Trichoderma*, in particular, has shown promising results in the treatment of drought, salinity, heavy metal toxicity, and high temperature stress. In drought situations, it enhances plant water status, increases photosynthetic efficiency, and regulates antioxidant defense, with lower proline accumulation indicating effective stress reduction. The introduction of multi-omics methods has transformed our understanding of these intricate interactions at the molecular and genetic levels, revealing important information about gene regulation, transcriptional networks, and metabolic alterations. However, obstacles persist, notably in terms of strain specificity, environmental unpredictability, and the translation of laboratory results into consistent field-scale applications. To summarize, *Trichoderma* species provide a complex, environmentally benign, and commercially viable strategy to improve crop resilience. Ongoing research into targeted strain creation, better application tactics, and a deeper molecular understanding of *Trichoderma*–plant interactions will be critical to fully fulfilling their potential to guarantee food production in an era of increasing environmental constraints.

References

Abdenaceur R, Farida BT, Fatma SH (2024) Volatile organic compounds activities of Trichoderma speciesisolated from olive grove soil against the wilt pathogen, Verticillium dahliae. European J Plant Pathol,170(4):789–803

Adhikari L, Baral R, Paudel D (2022) Cold stress in plants: strategies to improve cold tolerance in forage species. Plant Stress 4:100081. https://doi.org/10.1016/j.stress.2022.100081

Adhikari P, Shrestha S, Manandhar H, Marahatta S (2024) Effect of native Trichoderma as seed treatment on germination and seedling performance of lentil under biotic and abiotic stress conditions. SAARC J Agric 21:145–156. https://doi.org/10.3329/sja.v21i2.68649

Adusumilli N, Kolli S (2022) Management of salinity stress by the application of Trichoderma. pp 303–320

Afridi MS, Kumar A, Javed MA (2024) Harnessing root exudates for plant microbiome engineering and stress resistance in plants. Microbiol Res 279:127564. https://doi.org/10.1016/j.micres.2023.127564

Agarwal H, Chaudhary D, Aggarwal H (2025) Molecular underpinning of heavy metal sequestration through advanced remediation strategies in higher plants. Plant Stress 16:100881. https://doi.org/10.1016/j.stress.2025.100881

Ahsan SM, Injamum-Ul-Hoque M, Das AK (2024) Plant–Entomopathogenic Fungi Interaction: Recent Progress and Future Prospects on Endophytism-Mediated Growth Promotion and Biocontrol. Plants 13:1420. https://doi.org/10.3390/plants13101420

Akbari SI, Prismantoro D, Permadi N (2024) Bioprospecting the roles of *Trichoderma* in alleviating plants' drought tolerance: principles, mechanisms of action, and prospects. Microbiol Res 283:127665. https://doi.org/10.1016/j.micres.2024.127665

Alfiky A, Weisskopf L (2021) Deciphering Trichoderma–plant–pathogen interactions for better development of biocontrol applications. J Fungi 7:61. https://doi.org/10.3390/jof7010061

Altaf M, Ilyas T, Shahid M (2024) Trichoderma inoculation alleviates cd and Pb-induced toxicity and improves growth and physiology of Vigna radiata (L.). ACS Omega 9:8557–8573. https://doi.org/10.1021/acsomega.3c10470

Ameen M, Mahmood A, Sahkoor A (2024) The role of endophytes to combat abiotic stress in plants. Plant Stress 12:100435. https://doi.org/10.1016/j.stress.2024.100435

Annappa NN, Bhavya N, Kasturappa G (2023) Climate change's threat to agriculture: impacts, challenges and strategies for a sustainable future. In: Climate change and agriculture. AkiNik Publications, pp 113–136

Asghar W, Craven KD, Kataoka R (2024) The application of *Trichoderma* spp., an old but new useful fungus, in sustainable soil health intensification: a comprehensive strategy for addressing challenges. Plant Stress 12:100455. https://doi.org/10.1016/j.stress.2024.100455

Ayangbenro AS, Chukwuneme CF, Ayilara MS (2022) Harnessing the rhizosphere soil microbiome of organically amended soil for plant productivity. Agronomy 12:3179. https://doi.org/10.3390/agronomy12123179

Bandara AY, Kang S (2024) Trichoderma application methods differentially affect the tomato growth, rhizomicrobiome, and rhizosphere soil suppressiveness against Fusarium oxysporum. Front Microbiol 15:1366690. https://doi.org/10.3389/fmicb.2024.1366690

Bimpong D, Zhao L, Ran M (2024) Transcriptomic analysis reveals the regulatory mechanisms of messenger RNA (mRNA) and long non-coding RNA (lncRNA) in response to waterlogging stress in rye (Secale cereale L.). BMC Plant Biol 24:534. https://doi.org/10.1186/s12870-024-05234-x

Boorboori M, Zhang H (2023) The mechanisms of Trichoderma species to reduce drought and salinity stress in plants. Phyton 92:2261–2281. https://doi.org/10.32604/phyton.2023.029486

Buono D, Regni L, Del Pino AM (2021) Effects of Megafol on the olive cultivar 'Arbequina' grown under severe saline stress in terms of physiological traits, oxidative stress, antioxidant defenses, and cytosolic ca. Front Plant Sci 11:603576. https://doi.org/10.3389/fpls.2020.603576

Carillo P (2025) Can biostimulants enhance plant resilience to heat and water stress in the Mediterranean hotspot? Plant Stress 16:100802. https://doi.org/10.1016/j.stress.2025.100802

Chaudhary R, Nawaz A, Khattak Z (2024) Microbial bio-control agents: a comprehensive analysis on sustainable pest management in agriculture. J Agric Food Res 18:101421. https://doi.org/10.1016/j.jafr.2024.101421

Cheng L, Xu Z, Zhou X (2023) Application of Trichoderma species increases plant salinity resistance: a bibliometric analysis and a meta-analysis. J Soils Sed 23:1–13. https://doi.org/10.1007/s11368-023-03557-0

Choudhary DK, Prakash A, Johri BN (2007) Induced systemic resistance (ISR) in plants: mechanism of action. Indian J Microbiol 47:289–297. https://doi.org/10.1007/s12088-007-0054-2

Contreras-Cornejo HA, Macías-Rodríguez L, Beltrán-Peña E (2011) Trichoderma-induced plant immunity likely involves both hormonal- and camalexin-dependent mechanisms in Arabidopsis thaliana and confers resistance against necrotrophic fungus Botrytis cinerea. Plant Signal Behav 6:1554–1563. https://doi.org/10.4161/psb.6.10.17443

Contreras-Cornejo HA, Schmoll M, Esquivel-Ayala BA (2024) Mechanisms for plant growth promotion activated by *Trichoderma* in natural and managed terrestrial ecosystems. Microbiol Res 281:127621. https://doi.org/10.1016/j.micres.2024.127621

Coppola M, Diretto G, Digilio MC (2019) Transcriptome and metabolome reprogramming in tomato plants by Trichoderma harzianum strain T22 primes and enhances defense responses against aphids. Front Physiol 10:745. https://doi.org/10.3389/fphys.2019.00745

Cournault Q, Gibot-Leclerc S, Monnier N, Steinberg C (2025) Modes of action and inhibitory activity of *Trichoderma* species on potato and tomato pathogenic *Phytophthora infestans*: a review. Fungal Biol Rev 52:100414. https://doi.org/10.1016/j.fbr.2025.100414

Deb P, Moradkhani H, Han X (2022) Assessing irrigation mitigating drought impacts on crop yields with an integrated modeling framework. J Hydrol 609:127760. https://doi.org/10.1016/j.jhydrol.2022.127760

Díaz-Rodríguez AM, Parra Cota FI, Cira Chávez LA (2025) Microbial inoculants in sustainable agriculture: advancements, challenges, and future directions. Plants 14:191. https://doi.org/10.3390/plants14020191

Dikilitas M, Karakas S (2014) Crop plants under saline-adapted fungal pathogens. In: Emerging Technologies and Management of Crop Stress Tolerance, pp 173–192

Doni F, Miranti M, Mispan MS (2022) Multi-omics approaches for deciphering the microbial modulation of plants' genetic potentials: what's known and what's next? Rhizosphere 24:100613. https://doi.org/10.1016/j.rhisph.2022.100613

Dutta P, Mahanta M, Singh SB (2023) Molecular interaction between plants and Trichoderma species against soil-borne plant pathogens. Front Plant Sci 14:1145715. https://doi.org/10.3389/fpls.2023.1145715

Eftekhari F, Sarcheshmehpour M, Lohrasbi-Nejad A, Boroomand N (2025) Effects of mycorrhizal and Trichoderma treatment on enhancing maize tolerance to salinity and drought stress, through metabolic and enzymatic evaluation. BMC Plant Biol 25:687. https://doi.org/10.1186/s12870-025-06729-x

El Kelish A, Al-Haithloul H, Qari S (2019) Pretreatment with Trichoderma harzianum alleviates waterlogging-induced growth alterations in tomato seedlings by modulating physiological, biochemical, and molecular mechanisms. Environ Exp Bot 171:103946. https://doi.org/10.1016/j.envexpbot.2019.103946

El-Saadony MT, Saad AM, Mohammed DM (2024) Drought-tolerant plant growth-promoting rhizobacteria alleviate drought stress and enhance soil health for sustainable agriculture: a comprehensive review. Plant Stress 14:100632. https://doi.org/10.1016/j.stress.2024.100632

Feng D, Zhang M, Xu J (2025) Revelation of mechanisms associated with strengthening plant cold tolerance through using exogenous substances. Front Plant Sci 16:1478692. https://doi.org/10.3389/fpls.2025.1478692

Galletti S, Paris R, Cianchetta S (2020) Selected isolates of *Trichoderma gamsii* induce different pathways of systemic resistance in maize upon *Fusarium verticillioides* challenge. Microbiol Res 233:126406. https://doi.org/10.1016/j.micres.2019.126406

Gasch AP, Spellman PT, Kao CM (2000) Genomic expression programs in the response of yeast cells to environmental changes. MBoC 11:4241–4257. https://doi.org/10.1091/mbc.11.12.4241

Geraldine A, Lopes F, Costa Carvalho DD (2013) Cell wall-degrading enzymes and parasitism of sclerotia are key factors on field biocontrol of white mold by Trichoderma spp. Biol Control 67:308–316. https://doi.org/10.1016/j.biocontrol.2013.09.013

Ghatak A, Chaturvedi P, Weckwerth W (2018) Metabolomics in plant stress physiology. In: Advances in biochemical engineering/biotechnology

Guzmán-Guzmán P, Kumar A, de los Santos-Villalobos S (2023) Trichoderma species: our best fungal allies in the biocontrol of plant diseases—a review. Plants 12:432. https://doi.org/10.3390/plants12030432

Halifu S, Deng X, Song X, Song R (2019) Effects of two Trichoderma strains on plant growth, rhizosphere soil nutrients, and fungal Community of Pinus sylvestris var. mongolica annual seedlings. Forests 10:758. https://doi.org/10.3390/f10090758

Hasan S, Vibhash D, Manisha M (2012) Trichoderma – a promising plant growth stimulator and biocontrol agent. Mycosphere 3:524–531. https://doi.org/10.5943/mycosphere/3/4/14

Hasanuzzaman M, Raihan MRH, Masud AAC (2021) Regulation of reactive oxygen species and antioxidant defense in plants under salinity. Int J Mol Sci 22:9326. https://doi.org/10.3390/ijms22179326

Hu G, Zhao Z, Wei Y (2025) Trichoderma asperellum 22043: inoculation promotes salt tolerance of tomato seedlings through activating the antioxidant system and regulating stress-resistant genes. J Fungi (Basel) 11:253. https://doi.org/10.3390/jof11040253

Huang J, Li X, Chen X (2021) Genome-wide identification of soybean ABC transporters relate to aluminum toxicity. Int J Mol Sci 22:6556. https://doi.org/10.3390/ijms22126556

Ikram M, Khalid B, Batool M (2025) Secondary metabolites as biostimulants in salt stressed plants: mechanisms of oxidative defense and signal transduction. Plant Stress 16:100891. https://doi.org/10.1016/j.stress.2025.100891

Khan NA, Owens L, Nunez M, Khan AL (2025) Complexity of combined abiotic stresses to crop plants. Plant Stress:100926. https://doi.org/10.1016/j.stress.2025.100926

Kredics L, Büchner R, Balázs D (2024) Recent advances in the use of Trichoderma-containing multicomponent microbial inoculants for pathogen control and plant growth promotion. World J Microbiol Biotechnol 40:162. https://doi.org/10.1007/s11274-024-03965-5

Kumar V (2013) Optimal physical parameters for growth of Trichoderma species at varying pH, temperature and agitation. Virol Mycol 03:127. https://doi.org/10.4172/2161-0517.1000127

Laevens GCS, Dolson WC, Drapeau MM (2024) The good, the bad, and the fungus: insights into the relationship between plants, fungi, and oomycetes in hydroponics. Biology 13:1014. https://doi.org/10.3390/biology13121014

Liang X, Zhang L, Natarajan SK, Becker DF (2013) Proline mechanisms of stress survival. Antioxid Redox Signal 19:998–1011. https://doi.org/10.1089/ars.2012.5074

Liu H, Able AJ, Able JA (2022) Priming crops for the future: rewiring stress memory. Trends Plant Sci 27:699–716. https://doi.org/10.1016/j.tplants.2021.11.015

Liu J, Zeng D, Huang Y (2024) The structure and diversity of bacteria and fungi in the roots and rhizosphere soil of three different species of Geodorum. BMC Genomics 25:222. https://doi.org/10.1186/s12864-024-10143-2

Maggio A, Miyazaki S, Veronese P (2002) Does proline accumulation play an active role in stress-induced growth reduction? Plant J 31:699–712. https://doi.org/10.1046/j.1365-313X.2002.01389.x

Maitra S, Sairam M, Ray S (2025) Chapter 7 – Seed priming with endophytic microbiome enhances crop yield. In: Kumar A, Singh J, Queijeiro López AM, N Kharwar R (eds) Microbial inoculants. Academic, pp. 111–130

Manzar N, Kashyap AS, Goutam RS (2022) Trichoderma: advent of versatile biocontrol agent, its secrets and insights into mechanism of biocontrol potential. Sustainability 14:12786. https://doi.org/10.3390/su141912786

Martínez-Medina A, Del Mar AM, Pascual JA, Van Wees SC (2014) Phytohormone profiles induced by trichoderma isolates correspond with their biocontrol and plant growth-promoting activity on melon plants. J Chem Ecol 40(7):804–815. https://doi.org/10.1007/s10886-014-0478-1

Mishra AK, Mottaleb KA, Khanal AR, Mohanty S (2015) Abiotic stress and its impact on production efficiency: the case of rice farming in Bangladesh. Agric Ecosyst Environ 199:146–153. https://doi.org/10.1016/j.agee.2014.09.006

Mohd Din ARJ, Mohamad Azam Z, Othman NZ, Yong JWH (2025) *Trichoderma*-bacterial network: a balance inter-kingdom interaction for agricultural relevance. The Microbe 7:100360. https://doi.org/10.1016/j.microb.2025.100360

Morán-Diez ME, Martínez de Alba ÁE, Rubio MB (2021) Trichoderma and the plant heritable priming responses. J Fungi 7(3):18. https://doi.org/10.3390/jof7040318

Mukherjee M, Mukherjee PK, Horwitz BA (2012) Trichoderma–plant–pathogen interactions: advances in genetics of biological control. Indian J Microbiol 52:522–529. https://doi.org/10.1007/s12088-012-0308-5

Murshed R, Lopez-Lauri F, Sallanon H (2014) Effect of salt stress on tomato fruit antioxidant systems depends on fruit development stage. Physiol Mol Biol Plants 20:15–29. https://doi.org/10.1007/s12298-013-0209-z

Nassary EK (2025) Fungal biocontrol agents in the management of soil-borne pathogens, insect pests, and nematodes: mechanisms and implications for sustainable agriculture. The Microbe 7:100391. https://doi.org/10.1016/j.microb.2025.100391

Nawrocka J, Małolepsza U (2013) Diversity in plant systemic resistance induced by *Trichoderma*. Biol Control 67:149–156. https://doi.org/10.1016/j.biocontrol.2013.07.005

Oljira AM, Hussain T, Waghmode T (2020) Trichoderma enhances net photosynthesis, water use efficiency, and growth of wheat (Triticum aestivum L.) under salt stress. Microorganisms 8:1565. https://doi.org/10.3390/microorganisms8101565

Pacheco-Trejo J, Aquino-Torres E, Reyes-Santamaría MI (2022) Plant defensive responses triggered by Trichoderma spp. as tools to face stressful conditions. Horticulturae 8:1181. https://doi.org/10.3390/horticulturae8121181

Pérez E, Rubio MB, Cardoza RE (2015) The importance of chorismate mutase in the biocontrol potential of Trichoderma parareesei. Front Microbiol 6:1181. https://doi.org/10.3389/fmicb.2015.01181

Pressman E, Peet M, Pharr D (2002) The effect of heat stress on tomato pollen characteristics is associated with changes in carbohydrate concentration in developing anthers. Ann Bot 90:631–636

Prismantoro D, Akbari SI, Permadi N (2024) The multifaceted roles of *Trichoderma* in managing rice diseases for enhanced productivity and sustainability. J Agric Food Res 18:101324. https://doi.org/10.1016/j.jafr.2024.101324

Quazi S, ferdous J, Shozib H (2024) Role of *Trichoderma asperelloides* and *Trichoderma brevicompactum* in improving drought tolerance in rice. Plant Stress 12:100457. https://doi.org/10.1016/j.stress.2024.100457

Rai S, Prasad R (2023) Trichoderma: multifunctional role in plant defense mechanism. In: New and future developments in microbial biotechnology and bioengineering, pp 271–288

Rawal R, Scheerens JC, Fenstemaker SM (2022) Novel Trichoderma isolates alleviate water deficit stress in susceptible tomato genotypes. Front Plant Sci 13. https://doi.org/10.3389/fpls.2022.869090

Ruano-Rosa D, Prieto P, Rincón A (2016) Fate of Trichoderma harzianum in the olive rhizosphere: time course of the root colonization process and interaction with the fungal pathogen Verticillium dahliae. Biol Control 61:269–282. https://doi.org/10.1007/s10526-015-9706-z

Shao Y, Gu S, Peng H (2025) Synergic interactions between Trichoderma and the soil microbiomes improve plant iron availability and growth. NPJ Biofilms Microbiomes 11:56. https://doi.org/10.1038/s41522-025-00684-z

Singh H, Bhat J, Singh V (2020) Auxin metabolic network regulates the plant response to metalloids stress. J Hazard Mater 405:124250. https://doi.org/10.1016/j.jhazmat.2020.124250

Singh N, Maurya V, Singh H (2024) Salinity stress in crop plants: Effects and eco-friendly management. Adv. Food Secur. Sustain 9:103–143

Sofo A, Tataranni G, Xiloyannis C (2012) Direct effects of Trichoderma harzianum strain T-22 on micropropagated shoots of GiSeLa6 (R) (Prunus cerasus x Prunus canescens) rootstock. Environ Exp Bot 76:33–38. https://doi.org/10.1016/j.envexpbot.2011.10.006

Sood M, Kapoor D, Kumar V (2020) Trichoderma: the "secrets" of a multitalented biocontrol agent. Plants 9:762. https://doi.org/10.3390/plants9060762

Taha N, Kamel S, Elsakhawy T (2020) Sustainable approaches of Trichoderma under changing environments for vegetable production. Env Biodivers Soil Secur 4:291–311. https://doi.org/10.21608/jenvbs.2020.45046.1109

Timofeeva A, Galyamova M, Sedykh S (2022) Prospects for using phosphate-solubilizing microorganisms as natural fertilizers in agriculture. Plants (Basel) 11:2119. https://doi.org/10.3390/plants11162119

Tripathi R, Keswani C, Tewari R (2021) Trichoderma Koningii enhances tolerance against thermal stress by regulating ROS metabolism in tomato (Solanum lycopersicum L.) plants. J Plant Interact 16:116–126. https://doi.org/10.1080/17429145.2021.1908634

Tyśkiewicz R, Nowak A, Ozimek E, Jaroszuk-Ściseł J (2022) Trichoderma: the current status of its application in agriculture for the biocontrol of fungal Phytopathogens and stimulation of plant growth. Inter J Mol Sci 23:2329. https://doi.org/10.3390/ijms23042329

Valenzuela-Aragon B, Cardinale M, Rolli E (2025) The role of arbuscular mycorrhizal fungi in abiotic stress management in viticulture under climatic shifts. Plant Stress 16:100863. https://doi.org/10.1016/j.stress.2025.100863

Varadharajan V, Rajendran R, Muthuramalingam P (2025) Multi-omics approaches against abiotic and biotic stress—a review. Plants (Basel) 14:865. https://doi.org/10.3390/plants14060865

Wahab A, Abdi G, Saleem MH (2022) Plants' physio-biochemical and Phyto-hormonal responses to alleviate the adverse effects of drought stress: a comprehensive review. Plants (Basel) 11:1620. https://doi.org/10.3390/plants11131620

Wyckhuys KAG, Gu B, Ben Fekih I (2024) Restoring functional integrity of the global production ecosystem through biological control. J Environ Manage 370:122446. https://doi.org/10.1016/j.jenvman.2024.122446

Yanfei W, Xie L, Jiang Y, Li T (2022) Prediction of effector proteins and their implications in pathogenicity of phytopathogenic filamentous fungi: a review. Inter J Biol Macromol 206:188–202. https://doi.org/10.1016/j.ijbiomac.2022.02.133

Yang Y, Zhao M, Li G (2025) Molecular tactics of biocontrol fungi to hack plant immunity for successful host colonization—a focus on Trichoderma Fungi. Microorganisms 13:1251. https://doi.org/10.3390/microorganisms13061251

Yao X, Guo H, Zhang K (2023) Trichoderma and its role in biological control of plant fungal and nematode disease. Front Microbiol 14:1160551. https://doi.org/10.3389/fmicb.2023.1160551

Younis A, Ramzan F, Ramzan Y (2020) Molecular markers improve abiotic stress tolerance in crops: a review. Plants 9:1374. https://doi.org/10.3390/plants9101374

Yu C, Jiang X, Xu H, Ding G (2023) Trichoderma longibrachiatum inoculation improves drought resistance and growth of Pinus massoniana seedlings through regulating physiological responses and soil microbial community. J Fungi (Basel) 9:694. https://doi.org/10.3390/jof9070694

Zaidi N, Dar M, Singh S, Singh U (2013) Trichoderma species as abiotic stress relievers in plants. In: Biotechnology and biology of trichoderma

Zehra A, Dubey M, Meena M, Upadhyay R (2017) Effect of different environmental conditions on growth and sporulation of some Trichoderma species. J Environ Biol 38:197–203. https://doi.org/10.22438/jeb/38/2/MS-251

Zhang W-D, Wang P, Bao Z (2017) SOS1, HKT1;5, and NHX1 synergistically modulate Na+ homeostasis in the halophytic grass Puccinellia tenuiflora. Front Plant Sci 8:576. https://doi.org/10.3389/fpls.2017.00576

MIX
Papier aus verantwortungsvollen Quellen
Paper from responsible sources
FSC® C105338

If you have any concerns about our products,
you can contact us on
ProductSafety@springernature.com

In case Publisher is established outside the EU,
the EU authorized representative is:
Springer Nature Customer Service Center GmbH
Europaplatz 3, 69115 Heidelberg, Germany

Printed by Libri Plureos GmbH
in Hamburg, Germany